Nanoparticle Technologies

From Lab to Market

Interface Science and Technology

Series Editor: Arthur Hubbard

In this Series:

Vol. 1: Clay Surfaces: Fundamentals and Applications
Edited by F. Wypych and K.G. Satyanarayana

Vol. 2: Electrokinetics in Microfluids
By Dongqing Li

Vol. 3: Radiotracer Studies of Interfaces
Edited by G. Horányi

Vol. 4: Emulsions: Structure Stability and Interactions
Edited by D.N. Petsev

Vol. 5: Inhaled Particles
By Chiu-sen Wang

Vol. 6: Heavy Metals in the Environment: Origin, Interaction and Remediation
Edited by H.B. Bradl

Vol. 7: Activated Carbon Surfaces in Environmental Remediation
Edited by Teresa J. Bandosz

Vol. 8: Tissue Engineering: Fundamentals and Applications
By Yoshito Ikada

Vol. 9: Particles at Interfaces: Interactions, Deposition, Structure
By Zbigniew Adamczyk

Vol. 10: Interface Science in Drinking Water Treatment: Theory and Applications
Edited by G. Newcombe and D. Dixon

Vol. 11: Surface Complexation Modelling
Edited by Johannes Lützenkirchen

Vol. 12: Theory of Colloid and Interfacial Electric Phenomena
By Hiroyuki Ohshima

Vol. 13: Sorbent Deformation
By A.V. Tvardovskiy

Vol. 14: Advanced Chemistry of Monolayers at Interfaces: Trends in Methodology and Technology
Edited by Toyoko Imae

Vol. 15: Macromolecules in Solution and Brownian Relativity
By Stefano A. Mezzasalma

Vol. 16: The Properties of Water and their Role in Colloidal and Biological Systems
By Carel Jan van Oss

Vol. 17: Chemistry on Modified Oxide and Phosphate Surfaces: Fundamentals and Applications
By Robson Fernandes de Farias

Vol. 18: Interface Science and Composites
By Soo-Jin Park and Min-Kang Seo

Vol. 19: Nanoparticle Technologies: From Lab to Market
By Farid Bensebaa

Interface Science and Technology

Volume 19

Nanoparticle Technologies

From Lab to Market

Farid Bensebaa

National Research Council of Canada, Ottawa

York University, Toronto

ELSEVIER

Amsterdam • Boston • Heidelberg • London • New York • Oxford

Paris • San Diego • San Francisco • Singapore • Sydney • Tokyo

Academic Press is an imprint of Elsevier

Academic Press is an imprint of Elsevier
Radarweg 29, PO Box 211, 1000 AE Amsterdam, The Netherlands
The Boulevard, Langford Lane, Kidlington, Oxford, OX51GB, UK
32, Jamestown Road, London, NWI 7BY, UK
225 Wyman Street, Waltham, MA 02451, USA
525 B Street, Suite 1900, San Diego, CA 92101-4495, USA

First edition 2013

ISBN: 978-0-12-369550-5
ISSN: 1573-4285

For information on all Academic Press publications
visit our website at store.elsevier.com

Printed and bound by CPI Group (UK) Ltd, Croydon, CR0 4YY

Transferred to digital print 2013

Dedication

To my Mother, to my Queen and Princesses, and to my late Father

Contents

Preface

My interest in nanoparticle science and technology started when I was investigating self-assembled monolayers of ultrasmall silver alkanethiolate particles in the mid-1990s [1,2]. In my efforts to understand and control the synthetic process, I was carried so far away that I found myself reading the controversial book of Eric Drexler, *Engines of Creation*. I was reading so much about nanotechnology that I quickly realized that the line between nanotechnology and nanomaterials is quite blurred. These days, these two words become somehow synonymous particularly to the general public, decision makers, and even to some scientists.

This book is neither about nanotechnology nor about nanomaterials, although it may cover some aspects in both fields. Nanotechnology, originally coined in 1974 by Norio Tanugushi, has been receiving a lot of interest from both academia and industry. Hundreds of reports and books have described the potential of this new field. However, as it often happens with new technologies, the line between hype and reality is sometimes fuzzy. Furthermore, an unattended consequence of this hype is the raise of voices opposing or at least trying to slow down nanotechnology development. Public fears are often based on perception and misunderstanding, which could be addressed through open discussion on the advantages and risks of wider use of nanomaterials.

Another source of confusion in this field is related to the difference between nanomaterials and nanoparticles. Nanomaterials are materials that have at least one physical dimension less than 100 nm. This may include carbon nanotubes, nanofibers, nanoclays, nanocomposites, nanoporous materials, nanowires, and nanoparticles. The definition of a nanoparticle is more restrictive. Nanoparticles are spherical (or quasispherical) particles having a diameter of less than 100 nm. Thus, the size in all three geometrical dimensions should be less than 100 nm without any significant aspect ratio. This restriction is fundamental and practical. With the intent of using nanoparticles as building blocks, nanoparticles are considered as artificial atoms, referring to their small size and spherical shape. Furthermore, limiting this book to nanoparticles will allow us to focus on a relatively smaller field. Otherwise, other nanomaterials with a wider scope such as carbon nanotubes will obscure some specific characteristics of nanoparticles. In some cases nanomaterials do not add any value when compared to conventional materials.

Nanoparticle science and technology is a wide and far-reaching field. It encompasses all the conventional scientific disciplines. Their applications include most industrial sectors. Around 40 words have been found in the literature referring to spherical or quasispherical materials and with submicron size features: aerogel, aerosol, cluster, colloid, crystallite, giant crystal, nanoaggregate, nanobead, nanocapsule, nanocolloid, nanocomposite, nanocluster, nanocompound, nanocrystal, nanodisperse, nanodot, nanogel, nanograin, nanogranule, nanomaterial, nano-object, nanoparticle, nanoparticulate, nanophase, nanophosphor, nanopowder, nanoscale particle, nanosized particle, nanospecies, nanosphere, nanostructured, nanosuspension, quantum dot, submicron, super-cluster, ultrafine, ultrasmall, zero-dimensional. This could be quite confusing even for an expert in this field.

Nanoparticle science and technology predated the age of nanotechnology. Practical utilization of nanoparticles could be traced back to the Roman Empire days. Nanoparticle-based catalysts are also used since 1940s in petrochemical industry and later as catalytic converters in cars. Numerous nanoparticle technologies are under development. Today nanoparticles are faced with similar challenges that any new technology needs to address. Most current nanoparticle applications are based on technology push. Curiosity-driven research in academia dominates this field. New discoveries provide opportunities for new products or improvements of existing products. As the field matured, different nanoparticle value chains are developed. In general, a value chain includes production, formulation, component fabrication, and system integration. Integration of nanoparticles into products and systems brings new and unforeseen scientific and engineering challenges. These challenges appear when using conventional processes in nanoparticle fabrication and integration. These challenges often lead to delays and frustrations of the proponents and/or skepticism from consumers and investors.

When I started working on practical applications of nanoparticles, I realized that integration is the most critical challenge. Only recently this challenge and potential solutions are recognized and described. Indeed, there is no overview of the different assembling techniques to form cost-effective functional film and device. In a few publications, this field was mentioned and/or presented often in a narrow view. A new reader in the field will have difficulty in grasping the actual state of the art and finding the most appropriate solution to current challenges. Integration technology should be adapted to the nanoparticle characteristics and/or applications. This book intends to fill the gap. We have dedicated one chapter to integration of nanoparticles into 1D, 2D and 3D structures. Three application sectors are described in some detail in three different chapters: (i) energy, (ii) optoelectronics, and (iii) biomedical. The last chapter of this book is dedicated to health, safety, and environmental effects of nanoparticles. A general overview of scalable nanoparticle production techniques is also provided.

This book could be used as a general introduction to nanoparticle technology. It also provides brief description of recipes on how to produce and integrate nanoparticles. The reader is implicitly invited to consult the reference for more details. Throughout this book, we will limit ourselves to illustrative examples instead of aiming at a broad coverage of all publications. Although I have not spared any effort in acknowledging every relevant work, one has to recognize the limitations particularly with the modern age of multiple sources of information, with some significant amount of background noise. Besides higher cost and reliability, nanoparticle-based productions and applications are also facing numerous health and safety issues. Indeed, the lack of data and understanding of nanoparticle properties has raised legitimate concerns from the general public. This challenge stems from the particularity of nanoparticles. Their small size makes them difficult to fully characterize. For example, toxicity of nanoparticles is found to depend on the size and surface chemistry. Unfortunately, measuring the size and chemical composition of these nanoparticles is challenging, given the limitation of the current instrumentation.

Simplicity is another striking difference between nanoparticle technology and other emerging technologies. Indeed, numerous methods used at the different stages of nanoparticle value chain do not require high capital equipment and installations. Furthermore, these methods are not technology intensive. Thus, adoption of these technologies is within reach of small and underfunded startups around the world, researchers in nondeveloped countries, and small-budget research institutions in developed countries. However, this does not mean success will be very easy. Defining clear objectives that could address local issues and understanding current nanoparticle technologies are very important for a successful endeavor in this field.

This book took over 6 years to complete. This does not mean that I have been working all this period on this book. Numerous times I have been carried away by social and professional obligations. Revisiting numerous times the content of this book has helped me improve the overall scope, organization, and presentation of the different chapters. We hope this final version will be helpful for newcomers as an introductory text, and to established researchers and managers as a more complete overview, of this challenging and fast moving field. We certainly welcome any feedback and suggestions by sending your inquiries to: nanoparticletechnologies@gmail.com.

Farid Bensebaa
November 1st 2012

REFERENCES

[1] F. Bensebaa, T.H. Ellis, E. Kruss, R. Voicu, Y. Zhou, Langmuir 14 (1998) 6579–6587.

[2] F. Bensebaa, Y. Zhou, Y. Deslandes, E. Kruus, T.H. Ellis, Surf. Sci. including Surf. Sci. Lett. 405 (1998) L472–L476.

Chapter 1

Nanoparticle Fundamentals

1 INTRODUCTION

For centuries, scientists and nonscientists have been searching for new approaches to improve functionalities, performance, and cost-effectiveness of engineering materials. In the last few decades, numerous groups of advanced materials have been developed for different applications. They include biomaterials, ceramics, composites, carbon nanotubes, fibers, and coatings. The end of the last century has seen the emergence of nanomaterials. These novel materials are likely to overtake plastic and composites as the base materials for future components and products (Fig. 1.1). In a few cases, nanomaterials have actually been used for several centuries, although often inadvertently. Nanomaterials will likely become the dominant building blocks in several industries in the future [2,3]. These nanomaterials have at least one geometrical dimension less than 100 nm. They include carbon nanotubes, nanofibers, nanoclays, nanocomposites, nanoporous materials, nanowires, and nanoparticles (NPs). NPs are the focus of this book; sometimes referred to as artificial atoms, NPs are spherical (or quasi-spherical) particles having a diameter less than 100 nm.

NPs are often embedded in a solid, liquid, or gas matrix. NPs are omnipresent around us in various forms. Whether they occur naturally (e.g., lipoprotein and volcanic particles), incidentally (carbon nanoparticle (NP) from diesel combustion), or are engineered (catalysts), they can have beneficial and/or harmful effects on humans and the environment. NPs have been used for centuries, mostly in colloidal form [4–6]. Colloidal particles consist of solid dispersion in a liquid, with at least one dimension less than 1 μm (=1000 nm or 10,000 Å). Note that the geometrical size of colloids has been first defined in the 5–200 nm range, which is quite arbitrary [4]. These values are within the currently accepted range of NP dimensions.

The Romans used submicron metal particles for glass staining, which allowed a colored mosaic [7]. The intense light scattering of metallic NPs in a stained-glass window provided beautiful features in numerous historic

Interface Science and Technology, Vol. 19. http://dx.doi.org/10.1016/B978-0-12-369550-5.00001-X

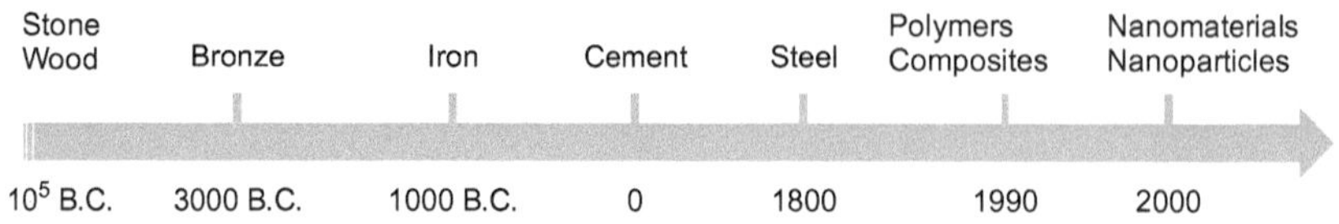

FIGURE 1.1 Evolution of material development and applications. The timeline axis is not to scale. *Adapted from Ref. [1].*

landmarks including the Notre Dame Cathedral in Paris (France) [8]. The Lycurgus glass cup, containing metal NPs, dates back to the fourth century A.D. It appears green in daylight (reflected light) and red when light is transmitted from the inside of the vessel [7]. This special effect has recently been ascribed to the presence of mono and alloyed gold and silver NPs within the bulk glass.

Faraday's report published more than 150 years ago [9] is considered as the first scientific publication detailing the synthesis of NPs. Indeed, colloidal gold is obtained by reducing gold chloride in aqueous solution. A century later, Turkevich's group [10] and others [6] laid the scientific ground for today's NP-based research by shedding light on the so-called world of neglected dimension [5]. Colloidal particles had tremendous commercial success in the food, painting, pharmacy, soap, photography, and plastic industries during the second half of the twentieth century. In 1949, Universal Oil Products started using platinum NPs as catalysts for the first time in a commercial reforming operation [11]. The catalyst system contains platinum NPs with an average size of about 1 nm. These NPs are dispersed in alumina supports containing a pore structure with an average pore size of about 10 nm. This new design was so efficient that numerous other petroleum companies integrated it in their petrochemical processing plants [12,13].

The role of NPs in catalysis is important from the technological and scientific points of view. A fundamental understanding of the catalysis processes and the relationship between surface composition and structure on the one hand and the catalytic properties on the other has received a lot of interest. To help shed light and quantify the importance of parameters at the nanometer level, model catalyst systems have been devised. This fundamental study will, for example, help understand the so-called three-way catalyst used in automotive exhausts (Fig. 1.2). Fundamental studies on well-controlled planar model catalysts allow the understanding of the structure–reactivity relationships and also improve catalytic efficiency. Unfortunately, there is still a significant gap between the understanding of the model catalyst system and the operational conditions of commercial NP-based catalysts. This gap stems mostly from the difficulty in characterizing these nanocatalysts *in situ*. The majority of advanced characterizations require specific sample preparation and operation conditions.

Metal and oxide NP preparations attracted more attention than semiconducting and insulating NPs until the early 1980s. Scientists and technologists took advantage of the geometrical, optical, and catalytic characteristics of

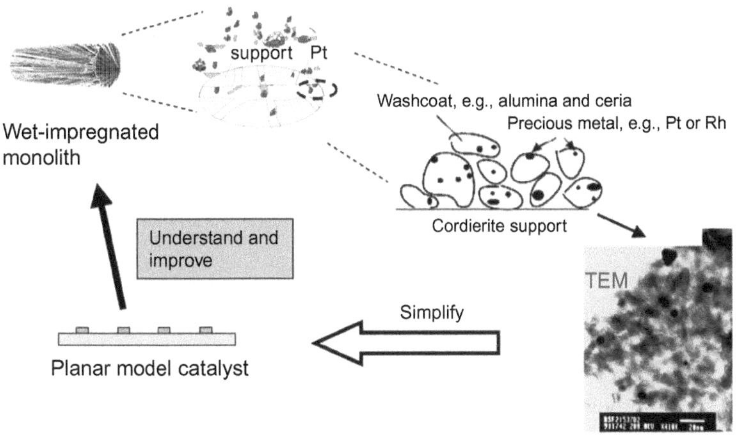

FIGURE 1.2 Macroscopic and microscopic structures of a three-way heterogeneous catalytic converter used in an automotive exhaust system [14]. (For color version of this figure, the reader is referred to the online version of this chapter.)

some metal NPs. In recent days, semiconductor quantum dot (QD) has come into the spotlight. QD properties were first reported by research groups from Russia in 1981 [15] and the USA [16]. However, this field remained marginal until the publication of the seminal work by Alivisatos's [17] and Bawendi's [18] research groups. The possibility of controlling the optical emission wavelength by changing the particle size using simple wet synthesis methods has led to an exponential growth of activity in this field. This renewed interest has been amplified by significant financial support from private and public institutions. These investments have been targeted toward potential applications in energy, microelectronics, and life science. Although synthesis of monodispersed NPs and/or large-scale production continues to receive a lot of interest, other downstream processes are attracting more attention. Commercialization of NPs is hindered by significant shortcomings and challenges particularly in the area of integration, durability, cost, and toxicity. NPs based on II–VI and III–V semiconductors have been the most widely studied systems in the recent decades.

Several books on NPs have been published during the last few decades. The authors of these books have attempted to summarize this rapidly evolving subject focusing mostly on synthesis and characterization [4,19–31,6,32]. In these books, attempts have also been made to cover specific short- and long-term potential applications. Not surprisingly, QDs have received significant coverage because of their potential applications in the health and optoelectronic sectors. These application areas have large and high value-added market potential. Very important R&D activities related to scalability, durability, and integration processes have been overlooked. In this book, an attempt is made to describe technologies at bench, pilot, and commercial scales. For example, there is a strong interest in alternative approaches to integrate NPs to achieve successful commercial products and systems. Until recently, this field had also been overlooked.

There is no restriction in this book to fit the strict definition of nanotechnology as defined by the US National Nanotechnology Initiative [33]. The NP field has been closely linked to nanotechnology, to the point that these two words have somehow become synonymous, particularly among the general public. The majority of the published books on nanotechnology deal mostly with nanomaterials that include NPs. However, it is important to stress that not every activity involving NPs can be categorized as nanotechnology and vice versa. Nanotechnology based on top-down approaches is currently being used in the microelectronic industry. Such approaches are used to control functionalities at the nanoscale without NPs as building blocks. Several applications of NPs do not fulfill the definition of nanotechnology [33]. This is the case when NPs are used simply to take advantage of their high surface area, without any attempt to control their properties by specifically changing the size at the nanometer scale.

Different particle shapes occur naturally or are engineered for specific applications (Fig. 1.3). In this book, the focus is on spherical particles with dimensions less than 100 nm. However, cubic, angular, polygonal, porous, and aggregate shapes [35] are also considered. Particles based on cylindrical, acicular, flakey, fibrous, and dendritic structures are not considered. For

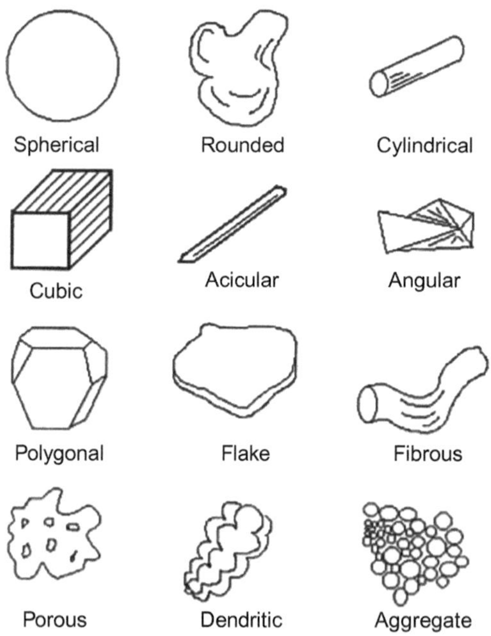

FIGURE 1.3 Shapes of different particulates [34].

example, carbon nanotubes, whiskers, and flakes with at least one dimension much longer than 100 nm are not considered here. In these structures, one or two geometrical dimensions dominate the remaining dimensions. It is worth noting that these various particulates have been widely used. This restriction will allow focusing on the NP characteristics.

NPs in powder or colloidal states are often either agglomerated or aggregated depending on the strength of the inter-NP interaction. Free or loosely interconnected NPs (or agglomerates) are often desired. Strongly bonded NPs (or aggregates) are considered mostly when integration leads to higher density in specific applications. For example, aggregated NPs are needed to allow efficient inter-NP charge transfer. In the first case, the overall surface area is equal to (or slightly less than) the sum of the individual NP's surface area. In the second case, the overall surface area is significantly less than the sum of the individual NP's surface area.

The word "nanoparticle" remains ill-defined. QD, colloid, nanomaterial, and nanocrystal are often used in the literature to define particles with sub-1000 nm (more often smaller than 100 nm) size in at least one dimension. The first group (QD) refers to sub-100 nm particles that show size dependence quantization. The second one refers to particles in liquid solution. The third refers to particles with at least one dimension in the sub-100 nm regime. The fourth and last one refers to single-phase crystalline particles dispersed in a gas, liquid, or solid matrix. These definitions are either restrictive (QD) or too broad (nanomaterials) for the scope of this book.

Dozens of words are found in the literature describing sub-100 nm particles. This rich vocabulary is adding a lot of confusion to an already rich and fast-paced area. In this book, the following broad definition of NP is adopted: any ensemble of atom forming an independent material with some similarities to the bulk materials. This definition parallels the one used for molecular materials [36–38]. The only difference is with relation to the size and shape of the materials' entity. Indeed, the dimensions of an NP in all geometrical directions should be around a nominal value of 100 nm or less. In some cases, the upper limit is not imposed religiously. A few reports involving NPs in the 100–200 nm size range are discussed. There is no size restriction in the case of molecular electronics. Furthermore, the aspect ratio should not be more than 1:2 in the case of NPs. Thus, sub-100 nm particles with high aspect ratio such as nanorods and carbon nanotubes do not fit within this definition, and so are not discussed in this book.

Although not referred to specifically as NPs, submicron particles have been studied for centuries in colloidal science [4,5,39]. About 40 words have been found in the literature referring to materials with features of submicron size: aerogel, aerosol, cluster, colloid, crystallite, giant crystal, nanoaggregate, nanobead, nanocapsule, nanocolloid, nanocomposite, nanocluster, nanocompound, nanocrystal, nanodisperse, nanodot, nanogel, nanograin, nanogranule, nanomaterial, nano-object, NP, nanoparticulate, nanophase, nanophosphor, nanopowder, nanoscale particle, nano-sized particle, nanospecies, nanosphere, nanostructured, nanosuspension, QD, submicron, super-cluster, superstructure, ultrafine, ultrasmall, and zero-dimensional. When possible, the word nanoparticle or NP is used here to refer to spherical particles with a mean diameter of 100 nm or less.

A few attempts have been recently made to clarify the semantic interpretation of the words used to identify nanoscale materials [40–42]. In the case of particle dispersion, a well-established nomenclature is considered [40]. Indeed, NPs could be suspended in a gas (e.g., aerosol), liquid (e.g., colloid), or solid (e.g., sol–gel). NPs are rarely used alone. Often they are integrated into a continuous matrix. NPs are often prepared in three different forms: core-only NP, core–shell NP, hollow (or shell-only) NP. Given its high surface energy and area, it is difficult to prepare and isolate core-only NPs without surface oxidation and/or advantageous absorption of surrounding molecules. NPs with surface carbon contamination can be considered as having a core–shell structure. Thus, it is quite difficult to clearly distinguish the first two categories. In this book, the focus is mostly on the first two categories, although a few examples of studies and applications of hollow NPs are mentioned.

Given the excessively large amount of peer-reviewed publications and non-peer-reviewed reports, it is impossible to do justice to everyone and provide a complete set of definitions for each word. It is quite evident that NPs share a lot of similarities with colloids. A colloid consists of a dispersion medium (gas, liquid, or solid) and a disperse phase. The disperse phase is usually ultrafine, with dimensions from 1 to 1000 nm. Up to nine different colloidal categories can be distinguished. Table 1.1 provides a few examples of colloidal systems. Excluding duplications, five different groups of dispersion are distinguished: smoke (dispersion of solid particles in a gas), cloud (dispersion of liquid droplets in a gas), foam (dispersion of gas bubbles in a liquid),

TABLE 1.1 Examples of Different Types of Colloidal Systems

		Nanoparticle (dispersion phase)		
		Solid	*Liquid*	*Gas*
Matrix (dispersion phase)	Solid	Ruby glass, alloy	Sol–gel	Aerogel, solid foam
	Liquid	Sol, ink	Emulsion, milk	Foam
	Gas	Aerosol, smoke	Aerosol, mist	Gas

emulsion (dispersion of liquid drops in another liquid), and suspension (dispersion of solid particles in a liquid). Most NPs are thus dispersed in a gas, liquid, or solid matrix. Compared to colloidal systems, NPs have a more restrictive size range. Indeed, NPs are particles with diameters less than 100 nm. All sub-100 nm particles dispersed in either gas, liquid, or solid are considered as NPs in this book.

NPs of different forms and structures are produced. Often these NPs are synthesized in a well-dispersed medium. This medium is either a gas, a liquid, or a solid. It is possible to transfer NPs from one medium to another. This flexibility may help integrate these NPs into bulk materials, films, devices, and/or systems. Formulations are often required after NP synthesis. NPs in solid state may include one or more phases with either ordered or random phase distribution. NPs are often monolithic and randomly dispersed, forming a macroscopically uniform phase. Bulk materials consisting of ordered and disordered multiphase systems have also been reported [43].

Arresting growth and agglomeration of NPs is a key factor allowing size distribution control. There are usually three approaches to control NP growth. The first approach is based on lowering the critical concentration of the precursor source. The second approach uses electrical charge on the surface of NPs to increase coulomb repulsions. The last approach is by adding a passive shell layer to cover the surface of the NPs. These approaches are carried out via wet or dry processes. Wet processes, often involving solvent, provide more flexibility in terms of functionalization and formulation. On the other hand, dry processing methods are often carried out under vacuum or under controlled atmospheric gas to prevent surface degradation. This approach is often more scalable.

An obvious advantage of NPs is the dramatic increase in the ratio of surface atom to bulk atom. For example, a 1-cm-diameter particle divided into 10 nm particles will give rise to a six orders of magnitude increase in a relative surface area. Other physical and chemical properties of the NPs may also change with size. Using the bottom-up approach, one could take advantage of NP attributes as building blocks. NP properties can be tailor-made and assembled on nearly any surface. Indeed, particles with sizes from 1 nm or below to 100 nm or above can be made with almost all elements and their combinations. Shape and other functionalities can also be adjusted. Higher surface areas of interest in catalysis and the color effect for architectural glass and biosensing have attracted a lot of interest. Numerous properties of these sub-100 nm particles are currently used at commercial stages [44]:

a. High surface area leading to increased reactivity, important in catalysis and batteries;
b. Reduced percolation threshold, critical for low-temperature processing of patterned and nonpatterned conductive coating;
c. Improved hardness, an important property for tools and other devices.

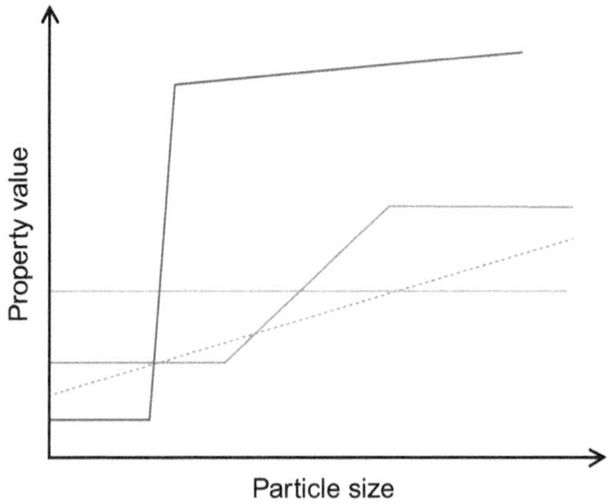

FIGURE 1.4 Possible patterns of property–size relationship in particle-based materials. (For color version of this figure, the reader is referred to the online version of this chapter.)

Numerous other NP properties are currently being considered for different applications:

a. Bandgap tunability could improve biosensing characteristics;
b. Magnetic properties could be important for therapeutics and drug delivery.

There is some debate about the scale factor when particle size is reduced to below 100 nm. As illustrated in Fig. 1.4, there are several types of relationships between the properties and particle size. For example, a given property (melting temperature) will be constant when the size is changed. In another case, the melting temperature will be reduced continuously as a function of the particle size. Another possible relationship consists of continuous reduction around a critical size before leveling off. The fourth option is related to a significant reduction in the melting temperature around a critical size. This last option is the one most widely sought after. For example, significant reduction in the melting temperature will help in reducing the thermal budget during material sintering processes. Other types of property-size variations are also of interest in some applications.

In this chapter, different NP classifications are discussed. Classification according to the properties (applications) of NPs receive particular interest. Current and future developments in the NP market are also briefly assessed. Issues related to NP functionalization, characterization, metrology, optimization, scale-up, and quality control are discussed in this chapter. These aspects are critical to the successful commercialization of NP-based products.

2 CLASSIFICATION

The field of NP science and technology is rich and diverse. It is beyond the scope of this book to provide a comprehensive classification. Instead, broad classifications of NPs are provided with a few examples to highlight their depth and potential. In this section, an attempt is made to classify NPs according to their composition, properties, fabrication process, and applications. The choice of the best classification may depend on the specific field of NP applications.

2.1 Composition

Klabunde has classified early work on clusters and nano-sized particles according to their composition following the periodical table division [19]. Recent research is this field has rendered this classification quite obsolete, particularly in the case of alloyed and composite NPs. A simpler classification could be obtained by dividing NPs into four groups: (i) inorganic metals, (ii) inorganic semiconductors, (iii) inorganic insulators including oxides and sulfides, and finally (iv) organics and polymers. Other categories such as hybrid NPs should also be considered. It is also worth noting that a variety of combinations of these four groups are reported owing to the advantage of their different compositions. Good examples are the various core–shells, compounds, and alloyed structures. For example, a core–shell composition may consist of a metallic core and an organic shell. The core composition is used for the classification. This composition is practical and is widely reported.

NPs could be produced and used in gas, liquid, or solid phase, depending on whether the surrounding medium is gas, liquid, or solid. The most difficult group of NPs to classify and also the most widely used in different industrial applications is related to solid NPs embedded in a matrix consisting of a second solid phase. These NPs are often referred to as nanocrystalline materials [45]. These nanocrystalline materials consist mostly of two subgroups depending on the structure and composition of the solid matrix. The first subgroup is related to crystalline NPs surrounded by a noncrystalline solid matrix phase [46]. The second subgroup consists of nanocomposites of distinctly different phases and compositions. In the first group, the two phases have similar compositions, albeit different structures, whereas in the second group, they are multiphase. Another difference is the interface between the two phases at the macroscopic level. In one case, the interface is blurred, and in the second case, the interface is well defined.

2.1.1 Inorganic Metals and Alloys

Noble metal NPs have been studied since the early twentieth century [50]. Noble metals often refer to chemically resistant and precious metals. They include gold, platinum, silver, rhodium, and tantalum. One of the first scientific reports on metallic NPs with specific applications dates to the early twentieth century [47], although Faraday's paper appeared about half a century earlier [9]. More recently, Turkevich et al. have also made tremendous contributions to the preparation and understanding of colloidal gold NPs [48].

The antimicrobial and electrical conductivity properties of silver NPs have opened up numerous opportunities for industrial applications. Silver NPs are already integrated into washing appliances, allowing efficient cleaning and sterilization, even under low-temperature operation conditions. Silver-based colloidal inks have also been developed to produce patterned films for sensing and radio frequency identification systems on flexible substrates. Colloidal inks are easily deposited using low-cost printing techniques and sintered by low-temperature postprocessing.

Reducing transition metal salts in aqueous solution is the most widely used approach for colloidal NP synthesis. Shape control is also possible using suitable surfactant and surfactant-to-precursor ratio. The polyol method has also been widely used to produce monometallic and alloyed NPs [49]. Water-in-oil microemulsions have also been used to prepare metal NPs with sizes ranging from 5 to 50 nm [50]. A plethora of other techniques have also been developed (see Chapters 2 and 3).

For various reasons, gold NPs have attracted attention the most [51]. With its inertness and stability, gold offers numerous possibilities for applications in biology and catalysis [52]. Even when ultrasmall NPs are produced, gold remains chemically stable. Furthermore, gold NPs are easily prepared and their properties are reproducible. In their seminal work, Brust et al. devised a new method for preparing monodispersed metal NPs using a two-phase (water–toluene) reduction of tetrachloroauric acid $HAuCl_4$ in the presence of alkanethiol [53]. This method has been used to precisely control the colloidal size from 1.5 to 9.7 nm [54]. Functionalized and tunable gold NPs are suitable for building 2D and 3D structures for different applications.

Numerous bimetallic NPs have been developed [55]. Bimetallic composition improves the properties of single-phase NPs. In particular, Pt-based bimetallics have received increasing attention because of their utilization as bifunctional catalysts in fuel cells and other applications. Indeed, alloyed Pt–Ru NPs have been shown to be more resistant to deactivation than Pt NPs. Fe–Pt NPs have also been reported to provide tunable magnetic properties [56]. Alloyed gold–silver NPs with tunable optical properties have also been investigated [57]. Silver halide NPs (AgI, AgBr) have also received some interest because of their potential applications in photography [58].

Because of their propensity toward oxidation, zero-valent transition metal NPs are not easy to synthesize and use. Zero-valent metallic iron NPs have been reported [59]. Cobalt NPs of tunable size have also been obtained [60]. Copper and cobalt NPs with different sizes and shapes have been prepared using the micellar approach [61]. Magnetic alloy NPs such as $CoPt_3$ have also been reported [62].

2.1.2 Inorganic Semiconductors

The early work of Brus' [16] and Ekimov's [63] groups provided theoretical and experimental foundations for today's semiconductor NP science and technology. The seminal work of Murray et al. [18] described a relatively simple process to produce monodisperse semiconductor NPs with tunable particle size and optical properties. This process, which is easily reproducible in any lab with minimal capital investment, has been widely adopted in academia. As the size of the semiconductor NP is reduced, the bandgap increases and thus the optoelectronic properties are tunable. This property has been observed on monoatomic silicon [64], and on diatomic [18] and multiatomic [65] semiconductor NPs.

The quantum confinement in NPs has been widely studied [66,67]. These semiconductor NPs are often referred to as zero-dimensional (0D) or QD materials. As illustrated in Fig. 1.5, quantum confinement obtained with NPs should be distinguished from quantum confinement obtained with wire (1D or quantum wire), surface (2D or quantum well), and bulk (3D) structures. The variation of density states as a function of energy levels is different for these different low-dimensional structures. Going from 3D to 0D structures, a continuing discretization of the energy states has been predicted theoretically and observed experimentally. Discreet energy states have been observed in the case of 0D NPs. This phenomenon has attracted significant interest from the scientific community, although no commercial application has been successfully achieved so far. Furthermore, it is possible to build 1D, 2D, and 3D structures using only 0D (or QD) structures (Fig. 1.5).

Quantum confinement effects, including electronic excitation shifts to higher energy, are observed in semiconductor NPs. This effect occurs when the particle size ($d = 2R$) is reduced to below the Bohr radius a_B of the bulk material defined by

$$a_B = \hbar^2 \varepsilon [1/m_e^* + 1/m_h^*]/e^2,$$

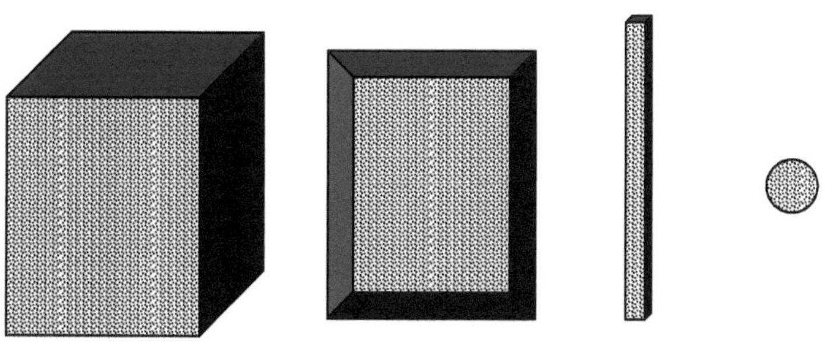

FIGURE 1.5 Schematic representation of bulk and low-dimensional structures (not to scale).

where m_e* and m_h* are the effectives masses of the electron and hole, respectively. ε is the high-frequency dielectric constant of the material. The bandgap shift with respect to the bulk value (E_g) and NP radius (R) is derived from Ref. [16]:

$$\Delta E = E_R - E_g = h^2\pi^2[1/m_e^* + 1/m_h^*]/2R^2 - 1.8e^2/\varepsilon R.$$

The first term ($1/R^2$) is related to quantum localization energy, whereas the second term ($1/R$) is related to coulomb repulsion energy. Bandgap shifting as a function of size has attracted a lot of interest. Depending on the value of the exciton Bohr radius, bandgap shift can amount to up to 5 eV in the case of Ge and 3.2 eV for Si. Lower bandgap shifts are reported for II–VI systems. In some cases, no bandgap shift has been reported as a function of particle size.

Using different semiconductor compositions, a wide span of energy bandgap values can be covered, for example, starting with low bandgap materials such as PbS (0.41 eV) and Ge (0.67 eV), to medium bandgap materials consisting of Si (1.1 eV), GaAs (1.4 eV), CdTe (1.5 eV), and CdSe (1.7 eV), and then wide bandgap materials such as CdS (2.5 eV). Quaternary and pentanary semiconductor NPs have also been reported [65,68]. Chalcopyrite NPs based on $Cu(In,Ga)(S,Se)_2$ compositions give rise to a wide bandgap range from 1 to 3 eV or more. $Cu(In,Ga)(S,Se)_2$ NPs have been synthesized using aqueous solvents [68] instead of organic solvents such as dichlorobenzene [64] and methanol [69].

Measured and predicted absorption cross-sections of CdSe QD in hexane have been found to follow a quasi-linear variation as a function of size [70]. Impurity doping of II–VI NPs has been reported [71]. Doping not only improves luminescence but may also shift the luminescence wavelength. When compared to visible fluorophores (e.g., organic dyes), inorganic QD are highly fluorescent with very high absorption cross-sections and tunable color properties. The size dependence of the optical and electronic properties of semiconductor NPs provides promising applications such as light-emitting diodes (LEDs), solar cells, lasers, and tunable sources of infrared radiation. These size-dependent properties have attracted the interest of several hundreds of research groups around the globe. This has provided the possibility for every chemist to produce his own "electronic chip" using the bottom-up approach starting from tunable semiconductor NPs. For a detailed review and account of this field, see, for example, Refs. [17,66,67,72–75,64,76].

Synthesis conditions seem to affect not only the physical properties but also their toxicity level. This behavior has been illustrated in the case of CdTe [77]. Higher growth rate of particles has been obtained in air when compared to those obtained under nitrogen atmosphere. However, both types of NPs exhibit similar high PL signals. Cytotoxicity studies of the CdTe NPs obtained in air showed relatively higher toxicity than those made in a nitrogen environment. This difference has been assigned to the excess of free cadmium released from the surface following air oxidation.

2.1.2.1 Monoatomic

Silicon nanocrystals have been used to build single-electron transistors [78], vertical transistors [79], floating-gate memory devices [80,81], QD lasers [82], and solid-state LEDs [83] prototypes. Submicron silica particles occur naturally or could be obtained as a by-product of metallic silicon production. In the case of silica, it could be obtained using precipitation of sodium silicate or high-temperature silicon tetrachloride hydrolysis processes. While the former synthesis technique is a batch process, the latter is a continuous process.

Silicon NPs have been synthesized using both wet and dry methodologies. Among wet methods, anodizing silicon wafers in hydrofluoric acid solution [84,85], inverse micelles [86], high-temperature supercritical fluid [87], oxidation of metal silicide [88], and the reduction of alkylsilicon halides [89] have been reported. Although they offer advantages in their ability to control the size distribution (and thus electronic properties), wet techniques are limited in their yield and overall output. For example, surface oxidation and other contaminations are limiting factors.

Sub-100 nm single-crystalline silicon NPs have been obtained using nonthermal plasma approaches [90]. Several gas phase techniques have been developed. Pyrolysis of silane has been achieved in furnace flow reactors using high-power infrared lasers. Laser photolysis of silane or disilane has been obtained. Tetragonal Sn(0) particles with diameters of 3.2 and 40.0 nm were prepared by heating the tetrahydrofuran (THF) solution containing Sn precursor at 65 °C for 2 and 4.5 h, respectively [91]. Note that Sn NPs could be potentially used in high-energy density lithium batteries.

Group IV NPs give rise to efficient luminescence with the possibility of new materials for optoelectronic applications. Si NPs with a quantum yield (QY) exceeding 60% have been reported [92]. With a relatively high exciton Bohr radius (11.5 nm), Ge QD provides the largest bandgap shift. Indeed, bandgaps starting from 0.66 to 3.5 eV are obtained with 20 and 5 nm NPs, respectively [93]. These tunable Ge NPs have been obtained using a photon-enhanced etching process in an aqueous solution. Relatively large Ge particles obtained by bead milling are used as starting materials. Visible photoluminescence obtained with Ge nanocrystals could be used in numerous optoelectronic applications.

2.1.2.2 II–VI NP Series

Binary semiconductors (II–VI, III–V, and IV–VI) have received the most interest. This is mainly due to the ease in preparing tunable optical emissivity. Thiol-stabilized and hot injection methods are widely used [94]. Numerous aqueous and organic media-based approaches have been developed to synthesize CdS, CdSe, and CdTe NPs. Reducing the particle size to below 10 nm allowed bandgap tuning of CdS QD from 2.5 to 4 eV [95]. At the same time, the melting temperature can be reduced from 1600 down to 400 °C [96].

Recent developments have been focusing on long-term stability of the NP in various media, reducing production cost through scalable processes, and using environment-friendly precursors and solvents.

Given the high cost of organic solvents and potential applications of aqueous QD, CdSe synthesis using water as solvent has been sought [97]. Environmental issues related to toxic precursor utilization have also been addressed. Peng's group has proposed several cadmium precursors to replace toxic Cd $(CH_3)_2$ precursors. They have showed that CdO gives rise to high-quality CdS, CdSe, and CdTe NPs [98]. Other precursors such as cadmium carbonate and cadmium acetate have been successfully used to prepare CdSe NPs [99].

The wide range of CdTe QD absorption and emission wavelengths is of particular interest [100]. CdO and Te powders are used to prepare monodisperse NPs with sizes ranging from 3.2 to 9.1 nm, covering a spectral range of more than 200 nm (Fig. 1.6). However, CdTe are more prone to degradation than CdSe and CdS. Appropriate coatings are often added to prevent degradation of optical properties.

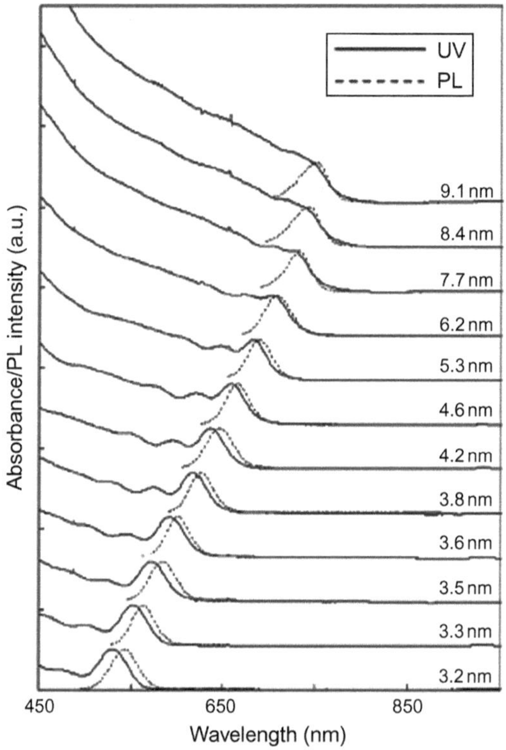

FIGURE 1.6 UV–visible absorption and PL emission spectra of CdTe nanoparticles of various sizes. *Reprinted with permission from Yu et al. [100]. Copyright 2003, American Chemical Society.*

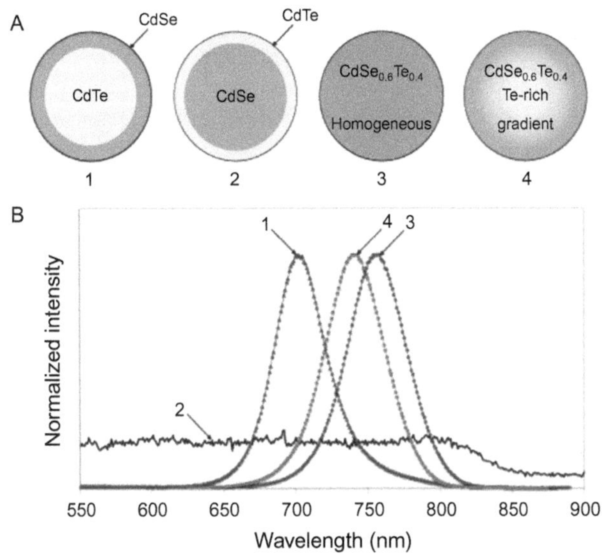

FIGURE 1.7 Relationship between structures and optical properties of core–shell and alloyed $CdSe_{1-x}Te_x$ nanoparticles (NP). Schematic representations of (A) four different structures of NPs and (B) their corresponding fluorescence emission spectra are shown. *Reprinted with permission from Bailey and Nie [101]. Copyright 2003, American Chemical Society.* (For color version of this figure, the reader is referred to the online version of this chapter.)

Differences in optical properties between core–shell and alloyed structures are expected. As shown in Fig. 1.7, differences in fluorescence features are observed between the conventional core–shell CdTe–CdSe (1), reversed core–shell (2), homogeneous alloyed (3), and gradient alloyed (4) NPs. All these different NPs have an average size of 5.9 nm (core and shell) and an overall composition of $CdSe_{0.6}Te_{0.4}$ [101]. Alloyed semiconductor NP $CdSe_xTe_{1-x}$ with tunable bandgap depending on the alloying value x has been reported [101]. Note that in the case of NPs consisting of CdSe core, no fluorescence is observed (Fig. 1.7).

Similar to CdX (where X = S, Se, and Te), ZnX (where X = O, S, Se, and Te), and HgX (where X = S, Se, and Te) NPs have also been synthesized with tunable bandgaps. These compositions are discussed later under oxide and sulfides. Usually they give rise to a higher bandgap, approaching values obtained with insulating materials. ZnS and ZnSe systems, and particularly Mn^{2+}-doped NPs, have been extensively studied because of their magnetic and optoelectronic properties. Indeed, new physical properties are obtained by doping QD including spin-polarizable excitonic photoluminescence, magnetic circular dichroism, exciton storage, and excitonic magnetic polaron formation [102]. In the case of colloidal and epitaxial grown CdSe QD films, Mn^{2+} doping gives rise to new spin-electronic and spin-photonic properties [102]. For example, Mn^{2+}-doped CdSe showed long-lived Mn^{2+} PL signal that is efficiently sensitized by semiconductor excitation.

2.1.2.3 III–V NP Series

The majority of reports on II–V semiconductors are based on InP and GaAs NPs. Sub-5 nm InP NPs have been reported with promising optical properties [94]. GaX (where X = N, P, As, Sb) NP series are obtained using different methods. GaN nanocrystals have been synthesized using a solid-state displacement reaction of Ga_2O_3 with Mg_3N_2 [103]. Transmission-electron microscopy (TEM) characterizations showed a wide particle size distribution ranging from 4 to 20 nm. However, an average size of around 5 nm has been assigned. Optical characterization indicates a bandgap of 3.43 eV consistent with the bulk bandgap value. GaN NPs with an optical bandgap consistent with that of bulk materials have also been synthesized using mechanical methods [103].

2.1.2.4 II–III–VI NP Series

$CuIn_{1-x}Ga_xS_{2-y}Se_y$ materials (with x and y varying from 0 to 1 and 0 to 2, respectively), also called CIGS or chalcopyrite, are currently used for photovoltaic solar cells and other optoelectronic-related applications [104]. Several low-temperature and nonvacuum approaches have been recently reported to produce chalcopyrite NPs and microparticles [105–109]. These materials are particularly of interest for low-cost integration into solar cell modules [110,111].

There are very few reports of monodispersed chalcopyrite NPs with sub-10 nm diameter. Gurin [112] showed that using polyvinyl alcohol as solvent, $CuInS_2$ NPs can be obtained. Based on the diffraction peak width, a crystallite size of 7 nm could be inferred. Utilization of toxic gases (H_2S and H_2Se) and the presence of difficult-to-remove polymer coatings are major weaknesses in microelectronic applications. Using a modified TOPO method, O'Brien's group showed that NPs of $CuInSe_2$ are obtained with a diameter size in the range of 4.5 nm [107]. A band edge of 420 nm (2.95 eV) is also measured that is outside the optimum solar absorption range. Using a single-source precursor, Castro et al. [113] prepared colloidal $CuInS_2$ NPs between 2.7 and 4 nm, although the fabrication process involves numerous steps that are difficult and costly to scale up. Taking advantage of high microwave absorption by water, a green chemistry process has been developed for sub-10 nm NP production [104]. This aqueous process gives rise to well-dispersed and composition-tunable CIGS NPs using nontoxic materials [104]. This microwave technique has been also used in nonaqueous synthesis of multimetallic alloys using nontoxic precursors [110].

2.1.3 Inorganic Insulators

Oxide and sulfide NPs are often considered as insulators or large bandgap semiconductors [114–118]. Oxide NPs such as TiO_2, SiO_2, Al_2O_3, and ZnO have already been obtained in pilot, demonstration, and/or commercial-scale productions. A method developed by Strober et al. [114] is widely used today to synthesize monodisperse and size-tunable SiO_2 NPs. Often these monodispersed silica NPs are used as model systems for nanotemplating. Submicron titania and alumina particles are also readily obtained and commercially available. Commercial applications include coating paints, solar cells, catalysts support, reinforcing agents of soft materials, and polishing.

Several metal sulfide QDs have been developed with potential application in photooxidation [115]. However, oxides remain the most widely studied inorganic NPs. Various oxide NPs are currently used in several applications because of the advantages provided by their specific properties. Metal oxide NPs have been used in packaging (Al_2O_3), capacitors ($BaTiO_3$, rare earths), optoelectronic components ($LiNbO_3$, $PbTiO_3$), transparent electrodes (SnO_2), gas sensing (ZrO_2, SnO_2), display devises (WO_3), and pyroelectric detectors ($PbTiO_3$) [118]. They are often used for their dielectric and optical properties. This includes mechanical (Al_2O_3, SiO_2, CeO), UV stabilizer (ZnO, TiO_2, CeO, and Fe_2O_3), antimicrobial (ZnO, TiO_2, CuO), magnetism (Fe_2O_3), and solar radiation control (indium or antimony-doped SnO).

Scarcity combined with the strategic and economic importance of rare earth (RE) has led to little interest in RE NPs. Very few investigations of RE NP structures have been reported so far, although RE NPs may optimize raw material utilization. Solution-based techniques have been used to synthesize RE compound NPs [116]. Sub-50 nm EuO_x NPs have been reported [117].

Currently, the TiO_2 NP market is estimated at 50,400 tons, representing 0.7% of the overall TiO_2 market [119]. TiO_2 is used as a white pigment because of its advantages, such as brightness, very high refractive index, and resistance to discoloration under UV radiation. TiO_2 powder has been produced commercially using mostly the sulfate and chloride processes [120,121]. The Knoll process has been used to produce titanate NPs [122]. Surface modification of TiO_2 NPs is not only important to prevent agglomeration but may also improve the sintering properties [123]. Using diethylene glycol monomethyl ether as the surface modifier, anatase NPs remain dispersed (nonagglomerated) up to 760 °C.

Zeolite NPs, often referred to as nanozeolites, have also been obtained using size reduction techniques, diluted clear solutions containing a large amount of tetramethylammonium cations, and fluid gel systems containing tetraethylammonium hydroxide in the absence of alkali metal cations [124]. Synthesis and crystallization processes and applications of zeolite nanocrystals have been reviewed [124]. Nonaggregated zeolite NPs are often obtained at low crystallization temperatures under supersaturation and steric stabilization conditions. Two different approaches have often been used to obtain NPs. In the first approach, nanocrystals are grown from clear solution and gels. The second approach uses a confined space to grow these nanocrystals. Numerous factors contributing toward stabilization of colloidal zeolite nanocrystals have been identified [125].

Transparent conductive oxides (TCO) are omnipresent in several application areas, particularly those related to microelectronic and photonic devices.

Fluorine-doped tin oxide, antimony-doped tin oxide, zinc-doped tin oxide, and indium-doped tin oxide NPs have been developed recently to reduce the cost of fabrication of TCO films. Indeed, TCO NPs will allow not only nonvacuum deposition integration but also potential low-temperature processing. Low fabrication temperatures will facilitate the incorporation of a plastic-based flexible substrate in the new display devices.

Some oxide NPs are known for their magnetic and dielectric properties. Oxide-based magnetic NPs provide a wealth of physical properties [126–128]. Oxide ferrofluids with large specific loss power are also reported [129]. Metal oxide NPs are obtained via forced hydrolysis following thermal heating of metal salt dissolved in water and diethylene glycol [130]. Polycrystalline perovskite $Pb(Zr_xTi_{1-x})O_3$ NPs in the range of 2–10 nm are readily obtained by sol–gel. This is often achieved by hydrolysis of mixed metal alkoxide precursors [131]. Other bottom-up (co-precipitation, hydrothermal) and top-down (combustion) techniques are also used. Various polyol methods have been widely used to produce different metal oxide NPs [132,133]. For example, cobalt–ferrite NPs have been obtained by co-precipitation in 1,2-propanediol using microwave heating [133]. Microwave heating time has been used to control the magnetic properties of these cobalt–ferrite NPs [134].

2.1.4 Organics and Polymers

Organics including polymers could provide a sustainable solution for the development and application of NPs. For example, renewable biomass stock provides an unlimited source of precursors for NP production. Furthermore, organic materials are potentially biocompatible and relatively easier to recycle. Organic NPs are used as active substances (pharmaceuticals) and pigments.

2.1.4.1 Polymer

Microemulsion and miniemulsion polymerizations are used to prepare polymeric NPs starting from a different ratio of monomers and surfactants. Miniemulsion polymerization has received a lot of interest because polymeric formulation without using organic solvents provides some advantages [135]. Indeed, polymeric formulation in water provides safety, health, and environmental advantages. Polymeric NPs have been developed for biological applications including drug-delivery systemsc [136].

Besides surfactants, the presence of ionic groups also contributes toward stabilizing polymeric NPs. These ionic groups could be added during initiation, copolymerization, and surface modification [137]. The presence of hydrophilic polymer chains at the particle surface also contributes to microemulsion stabilization. The resulting polymeric NPs may consist of neutral and/or polyelectrolytes.

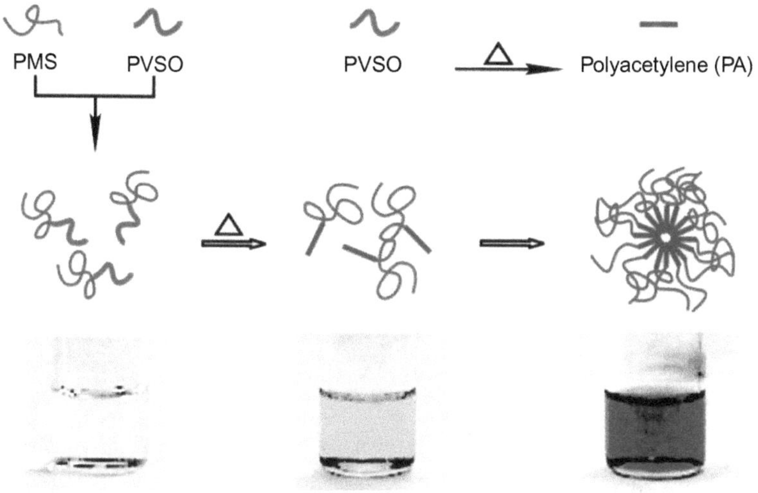

FIGURE 1.8 Schematic representation of poly(4-methyl styrene-block-phenyl vinylsulfoxide) (PMS-*b*-PVSO) diblock copolymer nanoparticle formation based on a chemical reaction-induced self-assembly. The color change reflects the conversion of PVSO to polyacetylene upon heating. *Reprinted with permission from Zhang et al. [137]. Copyright 2001, American Chemical Society.* (For color version of this figure, the reader is referred to the online version of this chapter.)

Block copolymers such as poly(ethylene oxide) and poly(propylene oxide) self-assemble into core–shell NPs. Core–shell with the polyacetylene (PA) as the core and polymethyl styrene (PMS) as the shell is obtained by heating diblock copolymer poly(4-methyl styrene-*block*-phenyl vinylsulfoxide) in THF [137]. A color change is observed upon PVSO (yellow) conversion to a PA (red) by heating (Fig. 1.8). The size of polymeric NPs depends on the chain length, solvent, and other parameters [138]. Spherical polystyrene particles with diameters varying from about 20 nm to 5 μm are commercially available. Emulsion polymerization is used to synthesize polystyrene and other polymer latex NPs. These polymeric NPs can be recycled at the end of their life cycle. Surfactant-free polystyrene NPs have also been prepared in water using potassium persulfate as the reaction initiator [139]. In this case, microwave seems to provide uniform heating, allowing monodispersed NPs.

Conducting polymer NPs have also been reported [140]. Polypyrole and graphite NPs with tunable size have been obtained using low-temperature microemulsion polymerization [141]. NPs composed of an elastomeric core and a rigid shell are also receiving some interest. Core–shell polymer NPs can be used as reinforcing agent polymers. Core–shell NPs of 40–50 nm have been obtained using copolymerization of butyl acrylate with dimethyl*meta*-isopropenylbenzyl isocyanate at 40 °C in a semicontinuous microemulsion batch process [142].

Biodegradable and biocompatible polymer-based NPs are of interest with regard to their applications in drug delivery and health-related applications [143–145]. Monodispersed β-carotene NPs with an average size of 25 nm

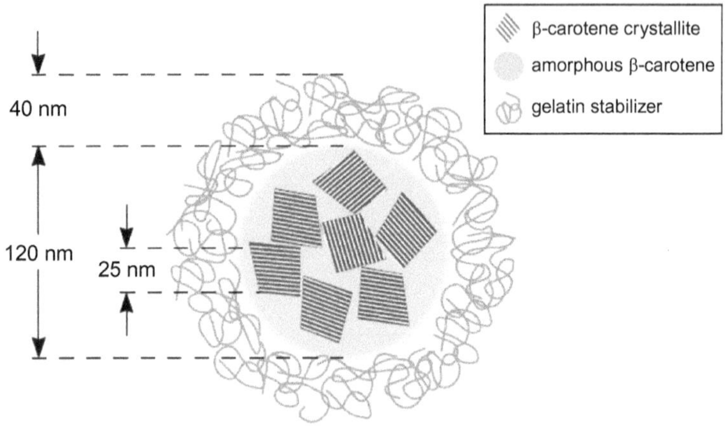

FIGURE 1.9 Schematic representation of the core–shell structure of a β-carotene nanocomposite [146]. (For color version of this figure, the reader is referred to the online version of this chapter.)

dispersed in an amorphous matrix have been obtained (Fig. 1.9). This crystalline-amorphous β-carotene nanocomposite is capped with a gelatine stabilizer forming a 160-nm particle easily detected by TEM [146].

2.1.4.2 Organic and Monomer

Organic NPs are obtained by laser ablation [147], sol–gel [148], and reprecipitation [149] techniques. Organic NP synthesis has been recently reviewed [146,150]. C_{60} NPs have been formed in water with tunable size and shape [151]. C_{60} has no liquid phase, with solid C_{60} formed by nucleation from the vapor phase. C_{60} is expected to be amorphous rather than crystalline [152,153]. Aqueous colloidal suspension of C_{60} NPs with a bimodal distribution around 19 and 150 nm has been reported [154]. C_{60} particles with sizes varying from 25 to 400 nm are readily obtained [151]. These particles can be dispersed and remain stable in water. Other solvents could also be used to form stable C_{60} aggregates. Clusters of C_{60} showed bandgap in the range of 2.02–2.24 eV, slightly different from the gas phase C_{60} value [155]. There are numerous potential applications of fullerene buckyballs, including supraconductor, photovoltaic, and energy storage. As of 2007, C_{60} materials production capacity is estimated at 1500 tons/year [156]. Gas phase processes are believed to be scalable with potentially monodisperse distribution. An aerosol route has been used to generate C_{60} NPs of about 20 nm in diameter [157].

Carbon black with particle size ranging from 10 nm to 1 μm is widely used for numerous applications including pneumatic, electrodes, printing, and painting dyes. Its high conductivity and high surface area are key factors contributing to their commercial success. These carbon black materials are often obtained using the furnace heating process. Typically, a carbon source is heated up to 2000 °C. Alternatively, plasma heating processes have been used, although to a lesser extent. Modified carbon-based QDs have been developed to probe cancer cells with high specificity [158]. These QDs do not show any apparent cytotoxicity. Nanofluid consisting of polar organic dye and a surfactant has been reported [159]. Fluorescent organic NPs have been reported [160]. Liquid crystal particles with a cubic shape and an average particle size of about 100 nm have been reported [161].

2.2 Structure

There are very few reports on NP structure changes as a function of size. Structural differences between bulk and NP materials have been reported in the case of oxides [162]. In some cases, below a certain critical size, the structure of powder materials changes quite significantly. In the case of Y_2O_3, ZrO_2, and TiO_2, a transition is observed below 100 nm [162]. A transition from rutile to anatase has been reported in the case of TiO_2 around particle diameter of 50 nm. Therefore, it is critical to keep the size of the NP below a critical value if we need to retain a specific NP structure during the postprocessing steps.

X-ray diffraction and computer simulations have been used to determine the 3D structure of gold NPs with different sizes [163]. This is the case of aqueous colloidal gold. Although the bulk fcc-type structure is dominant, a considerable local structural disorder is observed when the particle size is increased. When increasing the particle size from 3 to 15 nm, this local disorder is reduced [163].

Controlling the structure of metal NPs in an oxide-supported catalyst is very critical. Using density functional theory calculations, lowest-energy structures have been simulated for different Ni clusters deposited on the MgO(001) substrate [164]. Different structures have been predicted for different cluster sizes and substrate compositions [164]. Exposed facets may reveal different adsorption sites that may lead to different catalytic properties. Another dramatic change is observed in the case of Fe, Fe–Ni, Fe–Mn, Fe–Ni–Mn, Fe–Pt, and Fe–Co alloy NPs obtained by evaporation in Ar atmosphere [165]. While the small (5–8 nm) alloyed NPs containing 60–65% Fe have a bcc structure, larger NPs (15 nm) show an fcc structure. Furthermore, the change in structure is also followed by significant variation in magnetic properties.

2.3 Properties

Numerous properties are found to depend on the particle size [166–169]. A peak maximum is predicted for most bulk properties as a function of the particle size [169]. This behavior is related to the skin depth (or proximity length) of the properties, which is defined by the surface layer thickness defining the overall NP attributes. The nominal value of the skin depth depends on the specific property ranging from about 0.5 to 10 nm or more. Even in the early

1960s, NP-based materials had been predicted to exhibit superior physical properties when compared to their bulk counterparts [167]. However, according to a recent review [168], a direct relationship between the melting temperature of NPs on the one hand and their size and surface energy on the other has been established since the early twentieth century. On the other hand, due to high surface energy, NPs are highly unstable with high surface stress that is inversely proportional to the radius of curvature. A sintering stress of 300 and 3 MPa has been estimated in the case of 10 and 1000 nm particles, respectively [162]. Binding energy and forces increase linearly with diameter and could become much larger than the thermal energy (kT) for NPs in the range of 1–2 nm.

Size dependence of the optical properties of QDs is well documented. Other size-dependent properties, including magnetic, electronic catalytic, and mechanical, have also been reported. The size range where these properties vary is quite large [6]. Magnetic properties vary over a much larger size range spanning up to 10 μm. In comparison, the optical properties vary up to the size of 100 nm [6]. In some cases, the changes are quite dramatic. For example, gold NPs are endowed with catalytic [170] and bioactivity [171] properties that are absent or significantly reduced in their bulk counterpart. Applying the scaling law developed for the macro-world to NPs can provide some insight [172]. It is quite interesting that the magnitude of different energies (thermal, chemical, mechanical, magnetic, and electrostatic) converges as the critical size reaches the nanometer regime [173]. This convergence provides opportunities for coupling different types of excitation modes.

Improvements in material properties are key for further market penetration of NP-based products. More precisely, to justify significant changes in the fabrication process, the improvement in material properties should be significant. A survey of enterprises working with nanomaterials showed that material properties are the key success factor in 78% of the cases [174]. Other important factors include quality improvement, intellectual properties, and cooperation with other organizations. Increasing the surface energy and surface-to-bulk atomic ratio and decreasing the particle size remain the most obvious and important material characteristic of NPs.

Number of surface atoms (N_s) is approximately given by

$$N_s = 400(N_b)^{-1/3},$$

where N_b corresponds to the total number of atoms [6]. A typical metal particle of 150 nm diameter will have around 10^7 atoms, of which less than 1% consists of surface atoms [6]. As a consequence, surface area is significantly increased when the particle size is lowered. Indeed, lowering particle size from 1 μm to 1 nm increases the theoretical surface area from 6 to 6000 m^2 [175]. This property has been used for decades to improve catalyst efficiency.

In the case of NPs embedded in bulk materials, other material features may change as a function of NP size. Indeed, the particle diameter (d) will differently affect NP concentration, interparticle distance, and surface area

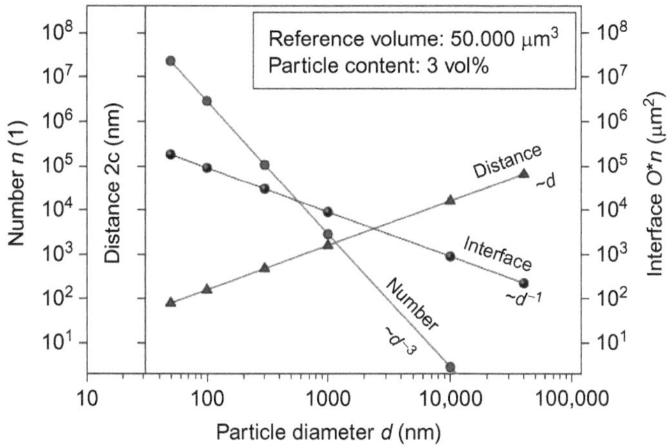

FIGURE 1.10 Variation of the absolute number n of particles, the interparticle distance ($2c$), and the resulting interface area $O*n$ (O: particle surface) as a function of the particle diameter. Sample volume is set at 50,000 μm^3, and the filler volume content is set at 3 vol% [176]. (For color version of this figure, the reader is referred to the online version of this chapter.)

(Fig. 1.10). For a given sample volume and filler volume content, the interface and number of particles vary as d^{-1} and d^{-3}, respectively [176]. Note that the distance between the particles will be proportional to the particle diameter for a given concentration (Fig. 1.10).

Different concepts have been used to quantify size dependence of NP properties. Indeed, different size dependency relations have been reported (Fig. 1.4). For any given property I, the maximum value is reached at a critical skin depth (λ), also called proximity length [177,169]:

$$I(x) = I_\infty - (I_\infty - I_r)e^{-(x-r)/\lambda}.$$

Hodes recently summarized the properties directly affected by the size of the various nanomaterials [178]. The typical NP length scale giving rise to a significant enhancement factor has also been provided for each property. Usually, this length scale is around 10 nm in quantization properties. In the case of light and ballasting electron scattering, the length scale increases to several 100 nm [178]. Other properties could be affected by size reduction. Blocking and Curie temperatures of magnetic nanomaterials are known to be size dependent. It has also been observed and understood that the formability of ceramic materials improves when grain sizes are reduced to below 1 μm. Creep properties of metal materials decrease at the nanoscale grain size. Smaller size does not always provide advantages. In catalysis, below a critical size, metallic properties are significantly reduced. Reducing particle size to below 100 nm has also been shown to affect the toxicity level. Indeed, polytetrafluoroethylene (PTEE) NPs with an average size of 20 nm are shown to be lethal upon inhaling. No ill effects have been reported in the case of 130-nm-sized PTFE particles [179]. However, controversy still surrounds the potential improvement in thermal conductance when NPs are used in a nanofluid [180].

2.3.1 Electric and Electronic

Significant changes in the electronic states are expected with size reduction. Metals and semiconductors respond differently to size reduction. The continuum energy level above the Fermi level (E_F) in bulk metal is disrupted below a critical NP size. In the case of semiconductors, discrete energy levels are created at the band edge below a critical NP size. In both cases, discrete energy levels are created with similar interlevel spacing [6]. In the case of Pd NPs, the energy shift is lowered from around 1.1 eV with NP size of 1 nm to around 0.2 eV for NP size of 6 nm [6]. Decreasing grain size of metallic materials may lead to higher resistivity. This property could be detrimental in some cases as it may lead to the loss of catalytic properties. However, this property could also be used in some applications related to sensing and passive components.

Ionization potential (IP) or electron affinity (E_a) refers to the energy difference between the vacuum level and the conduction band in bulk metals. IP and E_a are slightly different in the case of ultrasmall particles [181]. The variation of IP and E_a values with metal NP radius R is given by [6]

$$S(d)/S_0 = S_0 \exp[\gamma V_m / RTd],$$

$$E_a(Z) = W + (\beta)e^2/R,$$

where α and β are two constants with $\alpha + \beta = 1$, and W is the work function of the corresponding bulk metal. Thus, the charging energy

$$U = \text{IP} - E_a = (\alpha - \beta)e^2/R,$$

is inversely proportional to the NP radius.

The difference between the highest occupied and the lowest unoccupied state, also referred as the Kubo gap (δ), is an important parameter. Electric properties are directly related to the relative Kubo gap δ value versus the thermal energy (kT) level [166]. Its value may change with particle size: $\delta \gg kT$ for atoms and molecules [166]. In the case of insulating ultrasmall clusters, $\delta > kT$ [166]. For metallic clusters, $\delta < kT$ [166]. When reducing the particle size, a value below a critical threshold is reached. This occurs when the Kubo gap equals the thermal energy (kT). Electrical properties are affected by the particle size. Depending on the value of δ, a metal-insulator or metal-semiconductor transition has been reported.

Electric current flow through semiconducting NPs occurs when applying an external voltage (V) greater than the critical voltage $V_c = e/2C$, where C is the capacitance. Below this critical voltage, Coulomb blockade occurs. It has been found that the energy for charging one NP with a single electron scales with d^{-1}, where d is the particle diameter. For CdSe NPs with 5.5 nm diameter,

E_c is estimated to be around 0.1 eV. To carry single-charge effect experiments, two conditions are required to be met. First, experimentation should be operated at low temperature to minimize thermal noise effects. Small semiconductor NPs with diameter sizes of 20 nm or less should be used. In this case, the Kubo gap (δ) value is comparable to thermal energy at room temperature. Single QD devices have been pursued to build single-electron transistors [182] and single-electron memories [183].

NPs and the matrix in nanocomposites are often used for their functional and mechanical properties, respectively. It has also been reported that addition of NPs to superconducting materials enhances their flux-pinning properties, thus increasing the critical current density [184]. For some applications such as membranes and supercapacitors, it is important that the base material is proton- and/or electron-conductive. For example, Machida et al. showed that hydrous RuO_2 NPs embedded in polyaniline matrix provide high capacitance and excellent rate capability [185]. In this case, interparticle conductivity is facilitated by the conductive polymer matrix.

Besides surface area effects, catalysis has also recently been shown to be affected by particle size through quantum size effects [166]. Indeed, the CO oxidation peak is proportional to the Pt diameter. Quantum chemical calculation showed that catalysis turnover frequency varies with substrate-to-NP binding energy, which is itself dependent on NP size [166].

2.3.2 Magnetic

Size characteristics of magnetic NPs have a direct effect on the surface, magnetization, and magnetic relaxation properties. Magnetic NPs could be relevant in high-density magnetic recording, magnetic bioseparation, radio frequency and magnetic components, radar-absorbing composites, permanent magnets, magneto-rheological fluids, and magnetic fluid sensor applications.

Coercive forces, sometimes also referred to as magnetic memory, that are required to reverse the internal magnetic field are often dependent on particle size in the nanometer regime. Particle size dependence of coercivity (H_c) of NP-based magnetic materials has been well documented [186–191]. A peak in coercivity is often observed below 100 nm. Depending on the relative size of the particle and magnetic domains, two types of magnetic NPs can be distinguished. Indeed, single-domain ferromagnetic (FM) NPs and multiples of superparamagnetic (SPM) NPs have been identified [191]. Size and structure directly affect the coercivity value [187]. The variation of the H_c value as a function of the size is different when using multidomain and single-domain materials. In the case of iron particles, the SPM region of zero coercivity occurs with particle sizes up to 10 nm. Above this critical size, the FM state starts to dominate. Maximum coercivity coincides with the development of multiple magnetic domains with particle sizes in the range of 100 nm. When particle size reaches 1000 nm or more, the coercivity value is essentially the same as that of bulk iron [187]. Variation of coercivity with temperature of nanocrystalline magnetic materials has confirmed this description. Below a critical size of 100 nm, H_c is proportional to d^6 [189]. Above the nominal size of 100 nm, coercivity scales with d^{-1}.

Contrasting values of the magnetic properties have been reported for bulk nickel (2 μm) and nickel NPs (10 nm) [190]. Saturation magnetization (M_s) values of 22.5 and 55.0 Oe have been measured at room temperature for NP and bulk states, respectively. Coercivity (H_c) values of 7 and 100 Oe have been reported at room temperature for NP and bulk states, respectively.

The difference between FM and SPM NPs consists in the presence or absence of remnant magnetization (M_r) and coercivity (H_c). Indeed, a hysteresis loop is observed in an array of single-domain FM NPs, but no hysteresis loop is observed in an SPM material [191]. The single-domain limit defining the critical size of magnetic NPs is provided by [191]

$$d_c = 36\sqrt{(AK)}/\mu_0 M_s^2,$$

where A, K, and M_s are the exchange constant, the effective anisotropy constant, and the saturation magnetization, respectively. Thus, the critical domain size plays a critical role in defining the properties of magnetic NPs. Depending on the material, d_c is typically around 10–20 nm.

A linear relation between blocking temperature (T_B) and particle size is observed [192]. Blocking temperature (T_B) is defined as the temperature at which a maximum magnetic susceptibility is reached. In the case of $CoFe_2O_4$, T_B increases from 100 to 350 K when particle size is increased from 4 to 9.5 nm [192]. The variation of the blocking temperature with particle size is of interest in biomedicine. Variation in defect concentration with size, and the relation between magnetic properties and the number of defects have been observed [193]. These defects will likely be critical to some applications. In the case of metals and oxide-based magnetic NPs, the variations of magnetic properties versus size and concentration versus defects have similar trends [193].

Magnetic viscosity coefficient, normalized magnetization decay, thermal stability factor, and anisotropy field distribution of Fe–Co tape films were also found to depend on the particle size [194]. Numerous ferrofluids, defined as colloids with a permanently magnetized single domain, have been developed and commercially used. Magnetic NP stability is obtained via functionalization using organic surfactants such as oleic acids. This stability is achieved via steric hindrance and electrostatic repulsion. Furthermore, stability toward gravity settling of ferrofluid is achieved if the thermal energy is larger than the attractive magnetic energy [195]:

$$kT > \mu_0 M_d H\left(\pi d^3/6\right).$$

For a magnetic field of $M_d = 4.46 \times 105$ A/m, $H = 104$ A/m, $T = 298$ K, the maximum particle diameter for which NPs are stable is $d = 11.2$ nm [196].

A linear or quasi-linear relationship has been established between NP size and magnetic properties. There are few reports showing a drastic change in properties for ultrasmall NPs. This is the case of the magnetic moment per atom of rhenium NPs [196]. A sharp increase in the magnetic moment per atom is observed for particles containing less than 20 atoms. Indeed, about a 10-fold increase in the atomic magnetic moment is measured when the number of atoms is reduced by a factor of 2 below 20 atoms. Note that there is no change in the atomic magnetic moment for NPs with 20–80 atoms.

2.3.3 Optical

Optical property changes as a function of semiconducting NP size have spurred a large interest in the development of NP technology. In the case of CdSe, a wide optical range is covered by varying the particle size in the 2.5–6.3 range [175]. Indeed, emission wavelengths of 480, 520, 560, 590, 610, and 640 nm are observed with NP sizes of 2.5, 3.1, 3.4, 4.0, 5.2, and 6.3 nm, respectively [175]. These lumiphor properties can be used in luminescent (optical) tags, optoelectronic components, and magneto-optical materials (diluted magnetic semiconductors). However, current production costs, reproducibility, and stability of these NPs are hindering their commercial success.

Besides sensing, numerous optoelectronic applications are currently being considered at a pre-commercial scale. For example, semiconductor NPs are potentially ideal candidates for replacing existing bulk RE as luminescent materials. High financial and environmental costs of RE have prompted this interest. There are three types of NPs with improved luminescence properties [175]:

a. Semiconductor QDs, such as II–VI semiconductor CdSe that cover the entire visible range;
b. Doped materials, e.g., $LaPO_4$:Ce,Tb (green), ZnS:Mn (yellow);
c. Organic–inorganic hybrids, consisting of fluorescent dye embedded in NPs.

QY, corresponding to the ratio of the number of emitted photons to the absorbed photons, is often used as the figure of merit of different lumiphors. Existing bulk lumiphors showed QY up to 80%. Up to 100% and even more (with multiple-photon luminophors) has been obtained with NP lumiphors. Other important factors including stability and reproducibility are considered when choosing lumiphor materials. In the case of medical applications, biocompatibility and noninterference with other optical emissions within the tissue and cells are important. Inorganic QDs may address these requirements. Furthermore, extinction coefficients of CdTe, CdSe, and CdS nanocrystals were found to be strongly dependent on the size of the nanocrystals, between a square and a cubic dependence [197].

2.3.4 Electromagnetic

Dielectric, piezoelectric, pyroelectric, and electro-optic properties of ferroelectric materials are used in a wide range of applications including nonvolatile memories, multilayered ceramic capacitors, modulators, switches, thermal sensors, and transducers. The size-dependent dielectric properties of ceramic material have been known for more than half a century [198]. The dielectric properties of bulk ferroelectric materials are quite different from their corresponding NP films. For example, the transition temperature from the ferroelectric phase to the paraelectric phase of lead titanate ($PbTiO_3$) decreases when the temperature is reduced. A significant change in the ferroelectric–paraelectric transition temperature is measured with NPs of sizes below 20 nm [199]. The critical temperature of barrium titanate ($BaTiO_3$) NP is at least an order of magnitude higher [200]. Dielectric constant is also shown to depend on particle size [201]. The fabrication process was found to impact the dielectric constant [201].

2.3.5 Mechanical

Engineers have been researching high-strength and ductile materials for decades. These two properties often behave differently when composition and structure are varied. Indeed, high-strength materials often have low ductility properties. Alloyed and nonalloyed NPs embedded in matrices to form nanocomposites may enhance several mechanical properties of the final product [202]. The following properties could be affected by the presence of NPs: yield stress, ductility, strain rate sensitivity, creep, and fatigue. Mechanical properties of green body based on NPs depend on the density. In the case of metal NPs, green body density increases with pressure of compaction before reaching saturation around 1 GPa [202]. It has been reported in the case of Ni, Si_3N_4, and Al_2O_3 that the green density obtained by 1 GPa compaction is inversely proportional to the NP size [202]. Copper materials with bimodal grain size distribution showed high tensile ductility [203]. Deformation and heat treatment are two basic processes used to improve the mechanical properties of base materials. Deformation under high stress leads to accumulation of dislocation concentration and increased interaction between dislocations. As a consequence, hardening properties are improved. Annealing induces dislocation rearrangements, allowing stress relief. This could lead to improved ductility properties. In some cases, heat treatment and deformation could lead to opposite consequences. This is particularly the case with aluminum-based materials [204].

Often, details such as size distribution of the grain and contamination in the grain boundaries can play an important role. It is well established that reducing the grain size (d) is beneficial to the mechanical properties of polycrystalline bulk materials [205]. As illustrated in Fig. 1.11, strength, toughness, and ductility increase with decreasing grain size. Welding properties remain relatively unchanged as a function of grain size.

Numerous mechanical properties such as yield stress, flow stress, and hardness of metals and alloys increase when the critical dimensions for the

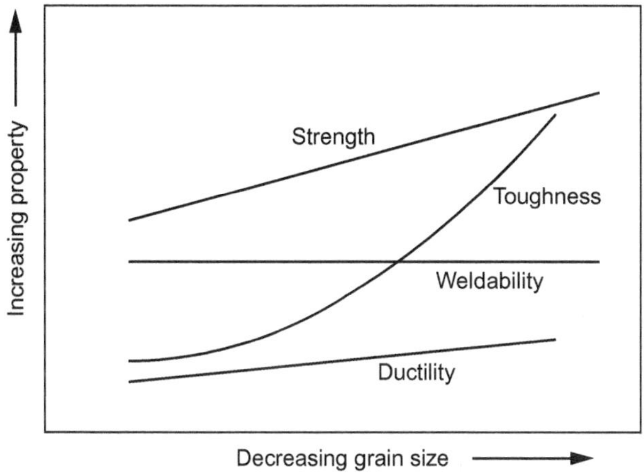

FIGURE 1.11 Variation of different mechanical properties as a function of the grain size [205].

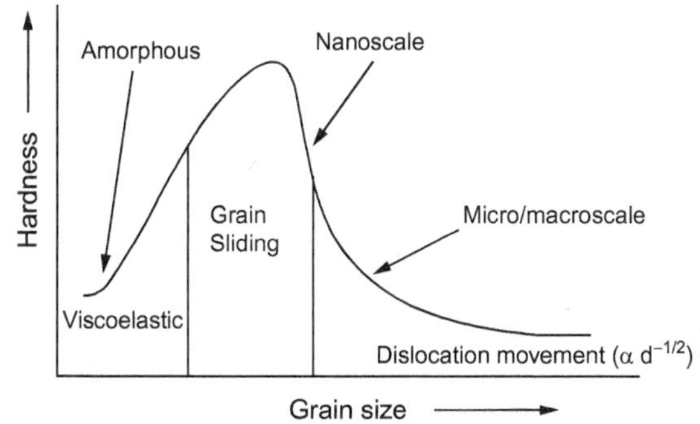

FIGURE 1.12 Evolution of the hardness as a function of the grain size [209].

grain size are reduced. Very small grain size improves metallic material strength and toughness at low temperature. Optimal grain size is usually found in the range of 5–10 nm. Contrarily to micro-sized FeAl materials, their NP counterparts are magnetic with a relatively improved combination of strength and ductility [206]. In the case of silicon, NP-based films are four times harder than bulk Si films [207]. It has also been shown that NP addition improves the lubricating properties of oils [208]. Hardness and tribology properties of aluminum-based alloys are improved by incorporating NP grain [209]. However, reduction of the grain size of the dispersed particle below a critical value will degrade the mechanical properties (Fig. 1.12). The dispersion strengthening process has been used to enhance creep properties of metals and alloys at high temperatures [210,211]. Yttrium oxide and carbonitride NPs have been successfully used to improve creep properties of steel-based components.

NP alloys dispersed in a glassy matrix provide superior mechanical properties [212]. Zr-based alloy nanocomposites show improved mechanical properties when NP size and the relative volume of the intergranular amorphous phase are reduced. A wide range of microstructural features and mechanical strength is obtained by adding Zr in Al–Ni–Cu–M (where M = Ad, Pd, Au, Pt, or Nb) to bulk amorphous alloys (Fig. 1.13). Improved mechanical properties have been achieved over bulk amorphous alloys. Controlling starting composition and thermal treatment allowed obtaining nanostructured quasi-crystalline materials with different geometrical features. In particular, high-strength amorphous alloy materials containing 5 nm NPs are easily obtained [212].

Metal–matrix composites are believed to provide high strength and high ductility. This is achieved by reinforcing a ductile metal (aluminum, titanium, copper, nickel, and iron) matrix using rigid ceramic materials (particles, whiskers, or fibers). In situ formations of NP-based reinforcements have been

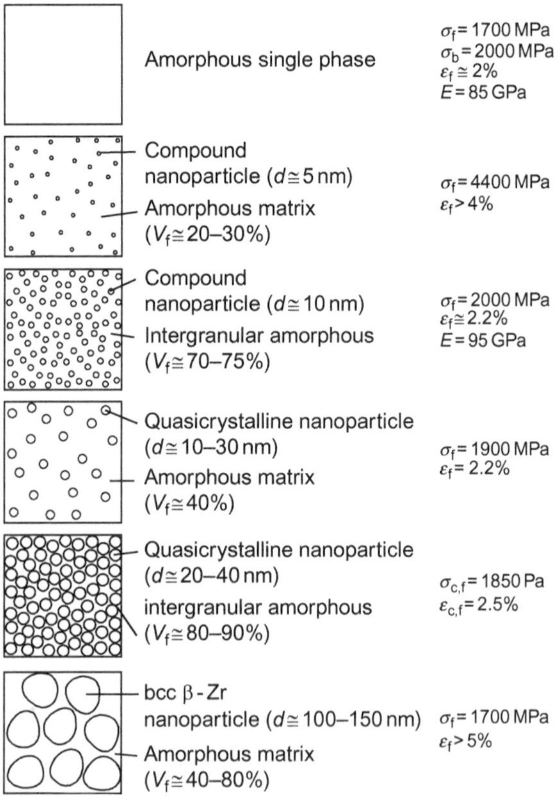

FIGURE 1.13 Relationship between microstructure and mechanical strength of nonequilibrium Zr-based bulk alloys. Changes in the size of the nanoparticles (d) and the nanoparticle-to-matrix volume fraction (V_f) allowed tunability of the tensile strength (σ), strain (ε), and Young modulus (E) [212].

shown to improve mechanical properties significantly. The homogeneity of particle-reinforced metals is critical for obtaining improved mechanical properties. Nonuniform distribution of the NP reinforcements can lead to lower ductility, strength, and toughness of the nanocomposites [213]. Uniform dispersion of ceramic NPs in a metal matrix gives rise to significant improvements in the yield strength, stiffness, and resistance to creep and wear of the materials. TiC reinforcement of aluminum-based matrices improves the tensile properties of the final product. Furthermore, strength (σ) increases with increasing volume fractions or decreasing diameters of TiC particles. Strength values in the range of 120–270 and 200–350 MPa have been obtained using volume fractions of dispersed particles of about 15 and 30 vol%, respectively [213].

2.3.6 Thermal

The potential effect of the particle size on the melting point has been discussed since the nineteenth century [214]. However, experimental demonstration was reported only in 1954 [215]. Today it is well established that both melting and heat capacity are directly affected by the NP size. Size effect on the melting temperature of gold NPs has been reported since the 1970s [216]. A variety of metallic and semiconducting NPs obtained using different synthesis methods have been shown to have a size-dependent melting temperature [217]. Abnormal thermal properties have been observed for ultrasmall NPs with a mean diameter of less than 2 nm [217].

A lower melting temperature process using metallic NPs is critical for developing low-temperature processing for ohmic contact on plastic substrates [218]. For gold NPs around 2 nm, T_m is lower than 400 °C [219]. Gold NPs have been shown to reach the bulk material's melting point ($T_m(\infty) \sim$ 1132 °C) with particle sizes above 20 nm. A similar trend is also observed with semiconductor NPs [219]. The effect of size on the melting temperature of indium NPs dispersed in two different types of porous silica glass (CPG and Vycor) has been observed [220]. In the case of aluminum NPs smaller than 9 nm, the melting temperature is size dependent. Above 9 nm, the melting temperature is size independent. Nonetheless, defects such as voids have an impact on the thermal properties [221]. It is not yet clear, however, whether increasing defects in small NPs have a significant effect on the melting temperature.

A direct correlation has been established between the evolution of the relative concentration of bulk (inner) atoms as a function of NP diameter on the one hand and the variation of the melting temperature (T_m) as a function of NP particle diameter on the other [216]. This correlation is probably related to the role of the loosely bonded surface atoms. Molecular-dynamics simulations of small gold particles containing 100–900 atoms showed the formation of a liquid skin on the surface of the NPs [222]. This liquid skin, obtained with NPs of 350 atoms or more, is considered the precursor for the melting stage. These simulations also confirmed that the melting mechanism is induced by increased atomic diffusivity. Low T_m value is mostly explained by the size dependence of the cohesive energy of NPs [223]. The shape of NPs has also been found to affect the melting temperature [224].

Two regimes have been established for the melting process [225]. In the first regime, the melting process is continuous and reversible. In this regime, the melting point $T_m(R)$ variation is described approximately by the following law:

$$T_B - T_m(R) \sim 1/R,$$

where R is the radius of the NP and T_B is the melting temperature of the corresponding bulk material. In the second regime, below a critical radius, the melting proceeds irreversibly [225].

Effects of the top layer coating composition and thickness on these two regimes are not yet well understood. The effect of oxide coating and size distribution on heat of fusion and melting temperature of aluminum NPs has been examined using differential scanning calorimetry [226]. The critical melting temperature and the heat of fusion decrease when particle size is reduced in the range of 8–50 nm. It has been suggested that the oxide shell layer applies a compressive pressure over the metallic aluminum layer when temperature is increased [226].

A melting temperature (T_m) difference of 30 °C between bulk materials and NPs with an average size of 8 nm has been measured [138]. This difference is lower than that reported on other NP compositions. For example, a difference of about 30 °C has been measured when the polyisobutylene particle size is varied from 6 to 9 nm (Fig. 1.14). Indeed, a continuous decrease in T_m and T_g (glass transition temperature) values as the particle is reduced has been observed. There is no discontinuity in the T_m diameter curve.

Specific heat and melting point are directly related to the particle size [219]. Compared to their microcrystalline counterparts, nanocrystalline materials have higher specific heat [219]. Wang et al. modelized the heat capacity increases as the size of NPs decreased [227]. This behavior has been recently explained by softening of vibrational state energies due to higher contributions from a larger number of free surface atoms [228]. Besides particle size, shape and structure have also been found to affect the heat capacity [227]. Specific heat of $Zr_{90}Al_{10}$ samples decreases when grain sizes are lowered in the range of 4–21 nm [229].

2.3.7 Superconducting

Characteristic parameters of superconducting NPs are found to change with their size. Particle size differently affects the superconducting transition temperature T_C of In and Pb NPs [230]. A maximum T_C is obtained using indium

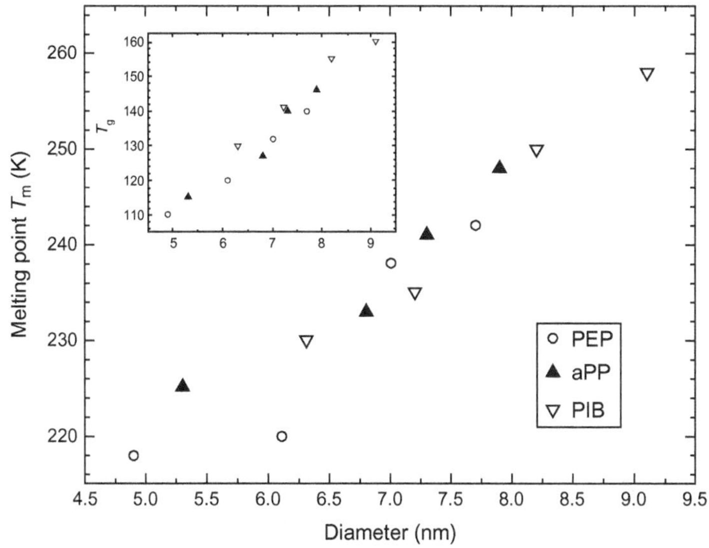

FIGURE 1.14 Variation of T_m and T_g (inset) as a function of particle size in poly(ethylene)-propylene (PEP), atactic polypropylene (aPP), and polyisobutylene [138].

NPs with an average size of around 40 nm although the variation is quite small. In the case of Pb, there is a sharp increase in T_C value for sub-10 nm NPs before leveling and this remains constant above 10 nm [231].

Critical magnetic field Nb-based NPs with a mean particle size of 10–70 nm obtained by microwave plasma discharge is about 2.5 times higher when compared to bulk samples of the same composition [232]. Furthermore, this critical magnetic field is inversely proportional to the particle size in the 10–90 nm range [232]. Adding NPs improves superconducting properties of YBaCuO materials [233]. Pinning forces are shown to increase when the grain size is reduced [234].

2.3.8 Surface and Chemical Properties

Particle surface energy changes drastically when the diameter of the particle is reduced below the 1 μm nominal value [235]. Wettability, measured by the surface contact angle, is significantly reduced when particle size is reduced below 10 nm. Specific surface area and other surface chemical properties of NPs change significantly when compared to microparticles and macroparticles. Nitridation, hydrogenation, and Grignard reactions on Mg NPs and Mg powders have been reported [236]. Consistently, Mg NPs with an average size of 76 nm showed larger chemical reactivity when compared to 266 nm and 155 μm particles [236]. In the case of 155 μm particles, there are no nitridation, hydrogenation, or Grignard reactions.

Higher surface area accounts mostly for the differences in catalytic properties. However, improved catalysis properties with peculiar properties have been reported with transition metal NPs [237,238]. Thermal stability is a significant issue with small metal NPs. Indeed, with high surface energy, these small NPs tend to agglomerate, forming aggregation/agglomeration NPs or even microparticles and macroparticles.

The most interesting size-dependent properties are related to the chemical activity as a function of catalyst particle size. In the case of expensive noble metals, mass activity (amperes per gram, A/g) should be used instead of specific activity (amperes per unit surface, A/cm^2) as a figure of merit. Indeed, this will allow reduced rates of utilization of rare and expensive elements. In the case of oxygen reduction reaction (ORR), mass activity of Pt/C catalysts is optimum with Pt NP size around 2.5 nm [239]. Note that the specific activity increases when the particle size is increased up to 20 nm, where it starts to level off. More detailed investigations of particle size effect on oxygen reduction of highly dispersed platinum in acid electrolytes showed that the relative fraction of Pt surface atoms on the (111) and (100) faces are critical [240]. A similar behavior is expected for anodic catalysts, although Ru alloying renders the measurements and evaluation more difficult.

2.3.9 Solubility

Particle solubility is critical in the formulation industry. This is particularly the case in the ink and pharmaceutical industries. For example, reducing the size of drugs and/or drug carriers will increase their efficiency while reducing their side effects. Saturation solubility increases dramatically below 100 nm [241]. Some materials give rise to a lower critical particle size. For example, Gypsum has a critical solution size of less than 10 nm. In the case of pyrite, the critical size is found to be around 100 nm [241]. The variation has been shown to follow the Ostwald–Freundich relation:

$$S(d)/S_0 = S_0 \exp[\gamma V_m / RTd],$$

where $S(d)$, S_0, V_m, and γ are the solubility (mol/kg H_2O) of the mineral with a diameter size d, solubility of the bulk material, molar volume (m^3/mol), surface free energy or surface tension (mJ/m^2), and gas constant (8314.5 mJ/(mol K)), respectively.

2.3.10 Tribology

Tribology, including friction, wear, and lubrication of NPs, has not been studied much or understood in spite of its industrial relevance in cases where moving bodies are in contact [242]. Friction properties depend on both tip and surface materials. The energy dissipation at the contact area of antimony NPs on two different substrates (HOPG and MoS_2) has been studied. Power dissipation in moving Sb NPs on HOPG (filled triangles) and MoS (empty circles) as a function of the particle surface area has been measured [243]. A linear increase in the dissipated power as a function of particle size is reported for both substrates.

2.4 Fabrication Process

Hundreds of NP fabrication processes have been reported. Some of these techniques have promising attributes for industrial scale-up. However, very few have been used in commercial settings so far. There are numerous ranking methodologies based on process strategy, process nature, energy sources, and medium [244]. Bottom-up (e.g., physical method), top-down (e.g., chemical process), or a combination thereof (e.g., biological process) strategies have been developed. The bottom-up approach is attracting a lot of interest in academia given its simplicity, low capital cost requirement, and flexibility. The top-down approach is already in commercial use due to its ease in scalability and reproducibility. There are some techniques that combine both approaches. Other process strategies include the nature of the process (chemical, physical, or biological), media (gas, liquid, solid), and energy source (sputtering, laser, microwave, convection, and sonochemical). Each classification has its own merits and drawbacks.

Monodispersity is a critical factor when producing and using NPs. As produced, NPs are often in a polydisperse state. As properties of these NPs are size-dependent, posttreatment is often required. Multiple nucleation sites and Ostwald ripening are two mechanisms leading to polydispersity of these NPs [245]. However, when carefully planned, the nucleation step may lead to monodispersed NPs [245].

Current NP synthesis processes used mostly in laboratories are energy- and resource-intensive. The energy input is often dependent on the composition. In the case of carbon nanomaterials, energy requirements are up to 2 orders of magnitude higher than for aluminum [246]. An overall LCA of these synthesis processes should be considered when comparing the cost–benefit advantage of NPs. In the case of wet chemical synthesis, large amounts of solvent are used during the synthesis, cleaning, and separation steps. Furthermore, the overall yield should be estimated for all processing steps. The environmental impact including chemical waste should be considered when comparing the different synthesis approaches.

Starting from precursors (ion, atom, and/or molecule), it may take up to five process steps to produce NPs. Following the clustering and complexation steps, a nucleation process leads to the formation of nuclei with uniform or nonuniform size distribution. These nuclei give rise to NPs following a diffusional growth process. The resulting NPs could be monodisperse, polydisperse, or aggregates.

In this book, the synthesis media phase is used as the main ranking parameter to classify the different fabrication processes. Indeed, three different synthesis media are often used for NP production:

a. Gas phase: vapor condensation, flame reaction, and aerosol;
b. Solid phase: mechanical milling;
c. Liquid phase: sol–gel, chemical precipitation, electrochemical.

Solid and gas phases are grouped in one category, often referred to as dry processes, whereas the last one is referred to as a wet process. However, a third category is distinguished that combines liquid and nonliquid phases. This is the case, for example, when a solution is used as precursor in gas phase processing.

Various factors are considered in the choice of one process over another. Convenience is the main factor when a small amount is needed. Cost, technical characteristics of the product, and quantity are important factors. Scale-up, quality control along with health, and environmental impact are also important.

Several physical and chemical processes have been developed for NP fabrication. For the same NP characteristics, two important parameters should be considered. Cost and environmental impact of the industrial fabrication processes should be quantified. Classification of fabrication processes according to their environmental and financial costs has been reported [247]. Physical methods often require large capital cost with reduced environmental impact. On the other hand, the chemical methods are relatively cheaper due to reduced energy input. Indeed, chemical reaction often relies on spontaneous reaction. This difference is related to the nature of starting materials. In the chemical processes, metal salts are often used in combination with a reducing agent, whereas in the case of physical processes, metallic powders are commonly used. Waste materials are a big issue with chemical processes, particularly when organometallic compounds are used instead of salts. Low conversion efficiency to NPs gives rise to a significant amount of waste.

Successful marketing of NPs requires the understanding of the concept of a value chain. Using NPs as building blocks instead of conventional bulk materials can significantly affect the value chain leading to the final commercial product. If the changes in the postprocessing and assembling increase the cost and/or the quality of the product, commercialization will be challenging, if not impossible.

2.5 Application Sectors

There are very few successful commercial applications of NPs. Utilization of catalytic properties of metallic NPs in fuel cells and petrochemistry provides rare examples of commercial success. However, there are hundreds of other potential applications at different stages of market development. Prototype and pilot plant demonstrations have been reported in energy, biomedical, optoelectronics, and environment sectors.

Optoelectronics including consumer electronics, pharmaceutics, cosmetics, catalysis, batteries, and chemicals constitute the backbone of the NP market. Some specific applications on the market or under development include [248]:

a. Energy: Pd NPs for automotive catalyst; TiO_2-based dye-sensitized solar cell;
b. Environment: TiO_2-based photocatalyst; magnetic NPs for remediation; catalytic converters;
c. Microelectronics: Al_2O_3-based chemical-mechanical polishing; magnetic NPs for electromagnetic interference (EMI) shielding; metal NPs for conductive patterns;
d. Biomedical: NP-based drug delivery; magnetic NP-based magnetic resonance imaging;
e. Others: light materials; sensing; ceramic membrane.

At this stage, it is important to distinguish hype (long term) from reality (short term). NP fabrication and integration into commercial products is part of a new industry [249]. Based on laboratory and pilot plant demonstrations, a large number of NP-enabled products will be commercialized in the future. Three large industries in which NPs already have or will have a significant impact have been identified:

a. Manufacturing, including NP-based material production and integration for transportation and packaging. Some manufactured products are already in the market although with relatively slow market penetration;
b. Optoelectronics, with some commercial products to be developed in the medium term. Once successful, a very fast market penetration is expected in this sector;
c. Life sciences, where a more long-term development is expected with rapid market penetration afterwards.

Low risk and low rewards, medium risk and medium rewards, and high risk and high rewards are expected for those investing in the manufacturing, electronics, and life science sectors, respectively.

3 FUNCTIONALIZATION, FORMULATION, CHARACTERIZATION, AND NANOMETROLOGY

NPs are seldom used as prepared substances. Often they require functionalization and formulation before their integration into films and objects. Both aqueous and nonaqueous solvents are used to disperse NPs. Often, the type of coating dictates the choice of solvents (Fig. 1.15). In some cases, solvent is not required. Lipophilic NPs are capped with alkyl chain (fatty acids, amines, phosphine oxides, and phosphonic acids) and are soluble in nonpolar solvents (toluene and chloroform). Hydrophilic NPs are capped with ligands (citrate, mercapto) in polar solvent (water, alcohol).

In most cases, NP formulation consists of dispersed particles, dispersion media, and other additives. Formulation strategy will depend on the application.

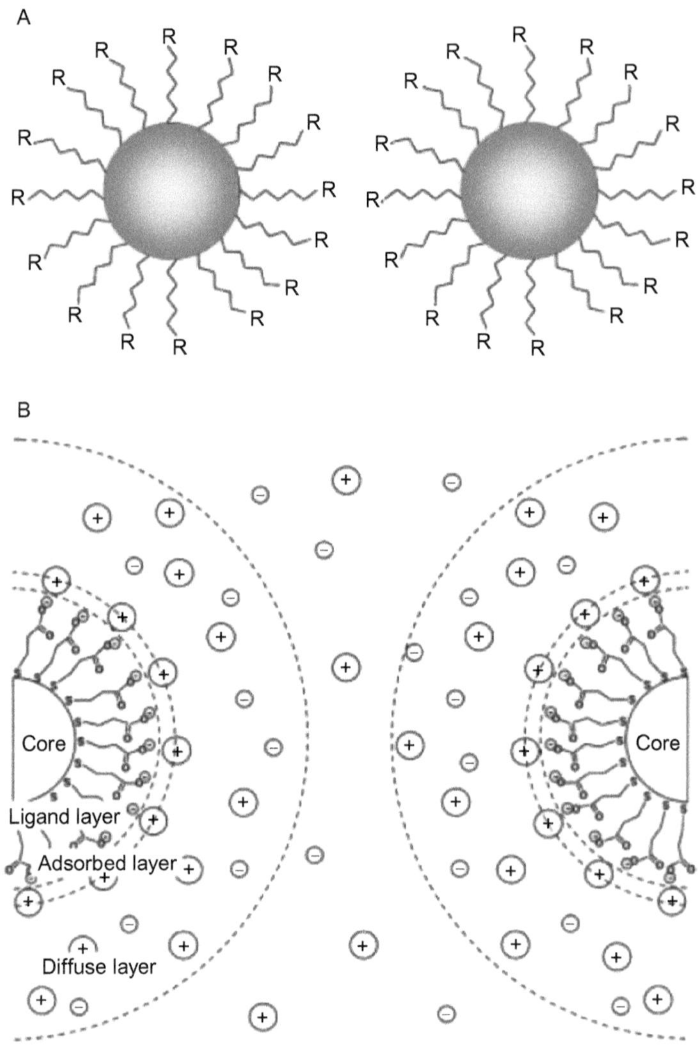

FIGURE 1.15 Schematic representation of (A) oil-soluble and (B) water-soluble nanoparticles [250].

Surface functionalization and dispersion medium are critical for medical applications. Utilization of a solid matrix is required when improved mechanical and rheological properties are sought. Thermodynamic instability of NPs is the biggest challenge. For example, aggregation may lead to a degradation of the functional properties.

Colloidal dispersions should be distinguished from suspension with critical size of 1 μm or more. Colloids consist of submicron particles as the dispersed phase, distributed in a second phase (dispersion medium). These colloidal dispersions have been used for centuries. Controlling aggregation and agglomeration of unstable and metastable NPs in a media is challenging due to their high surface area and high energy. Interfacial forces are adjusted using surface modification approaches. Tensides such as anionic (e.g., fatty acid), cationic (e.g., alkyltrimethyl-ammonium chloride), zwitterionic (e.g., *N*-alkylbetaine), nonionic (e.g., polyethylene oxide), and H-bond-active (e.g., fatty acyl derivatives) are used [251]. These tensides are already being used in the industry as wetting agents, detergents, foam formers, dispersants, and emulsifiers [251].

Detailed characterization of NP compositions and properties is very important. Without rigorous characterization, it will be difficult to evaluate and quantify the toxicity of NPs. Although conventional characterization techniques provide valuable information, new methods of analysis have to be developed to provide a more accurate picture at both nanometric (spatial) and molecular (chemistry) levels. Indeed, X-ray, optical, and electronic-based techniques often provide average information that could be misleading in the case of NPs. For example, if NPs with different sizes and compositions are analyzed, specific NP characteristics are not obtained using these conventional techniques. Very few techniques provide molecular content information with high spatial resolution. Efforts have been made recently to improve the spatial resolution of molecular spectroscopy techniques. For example, Raman characterization of nanomaterials not only allows phase identification but also provides information related to the disordered state, particle size, and even mechanical properties [252]. Recent developments in laser and detection technologies allow high spatial resolution Raman spectroscopy. However, when probed volume and/or surface area are reduced, the signal-to-noise ratio is also reduced. Thus, improvement in sensitivity is required to allow meaningful characterizations. Single-molecule detection and characterization is possible using surface-enhanced Raman spectroscopy (SERS) based on NP substrates.

Simulation could complement the information obtained by different characterization techniques. This is particularly the case of surface composition and size distribution characterizations when NPs are in a relative state of aggregation. Agglomeration (weakly bonded) and aggregation (strongly bonded) are two important phenomena with industrial implication during the fabrication, integration, and utilization of NPs. Understanding and controlling aggregation phenomena could provide practical solutions for the formulation of NPs before deposition into films. Atomic force microscopy (AFM) can be used to probe NPs at nanoscale and even at atomic-scale levels. Light scattering is used to probe microscale, mesoscale, and macroscale features. Different simulation approaches are used to understand the phenomena occurring at different scales [253]. The NP surface forces (<1 nm), Brownian aggregation (1–100 nm), and shear-induced restructuring (0.1–10 μm) are probed using quantum calculations, molecular dynamics, and Brownian dynamics, respectively [253]. In the case of NP surface forces and Brownian aggregation, interparticle potentials are the dominant factor. These simulations could be used to bridge our understanding of NP composition and bulk functional properties.

A complete NP characterization requires chemical composition, molecular structure, crystal structure, physical form at room temperature, pressure at room temperature, surface area, particle size distribution and average diameter, solubility in water and biologically active fluids, bulk density, agglomeration state, porosity, surface charge sites, particle density, and dispersibility. However, a complete characterization of NPs is complicated by the dependence on the probing conditions. It has been shown that probed properties are affected by the probe and/or the environment [173]. For example, impinging TEM's electron beam can induce sputtering, melting, and oxidation. TEM and to a lesser extent AFM have been used extensively to characterize NPs. AFM provides mostly geometrical information under ambient conditions, even without extensive sample preparation. TEM may also provide structural and composition information, although under high vacuum on relatively well-dispersed NPs.

3.1 Coating and Functionalization

Surface modification prevents NP agglomeration, but it allows compatibility with a second phase and enables self-organization [254]. Noble metal NPs have been modified by thiols, disulfides, amines, nitriles, carboxylic acids, and phosphines [254]. The choice of the modifying molecules depends on the solvent, applications, and other factors. In the case of metal oxides, carboxylates, silanes, and phosphonates have been used [254].

Oxide or hydroxide layers are naturally formed on the surface of numerous powder particles. This activity is often controlled by the isoelectric point (IEP) corresponding to the pH value at which there is no charge at the surface. The IEP value varies over a wide range [255]. The hydroxyl group formed on the surface of NPs will accept or donate protons in the presence of water molecules. Surface charging will directly affect dispersion and further functionalization. Coating of NPs has also been used to passivate the surface to prevent material degradation. A two steps process was used in the case of InGaP NPs to address the stability issues in deep-tissue fluorescent imaging (Fig. 1.16).

Carbon coating of metallic and oxide NPs is of paramount importance in numerous applications. For example, carbon coating is used not only as a shell but also as a means for high electrical conductivity. Carbon coating could be easily obtained through a facile two-step process [250]. First, a ligand exchange is performed using a new ligand with stronger affinity to NPs, with a possibility of further functionalization. For example, thiol-based ligands could be used in the case of anionic citrate-modified gold NPs. The

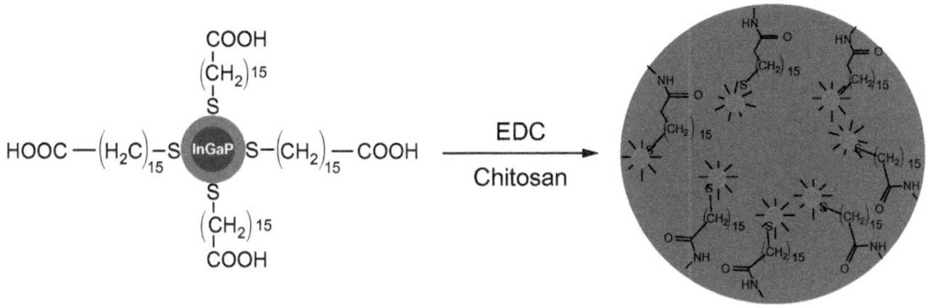

FIGURE 1.16 A two-step functionalization process leading to the chitosan-encapsulated MHA-capped InGaP–ZnS [256]. (For color version of this figure, the reader is referred to the online version of this chapter.)

second step consists in surface conjugation by covalent linkages or weak supramolecular interactions [250].

To prevent fluorescence instabilities and enhance their QY, different passivation approaches have been developed [257]. As shown in Table 1.2, the

TABLE 1.2 Quantum Yields (QY) of Different CdSe Nanoparticles and Organic Dyes in Different Solvents [257]

Quantum dot coating	QY (%)
Octadecylamine (in hexane)	79.3
Mercaptopropionic acid[a]	63.4
Polyethylenimine[a]	43.0
Alkylated polycarboxylic acid[a]	64.1
Lipid–PEG[a] [1,2-dipalmitoyl-sn-glycero-3-phosphoethanolamine-*N* [methoxy (polyethylene glycol)-2000]]	52.8
Diblock copolymer[a] Poly(methyl methacrylate)–polyethylene oxide (PMMA–PEO; 20,300 Da; 43% PMMA, 57% PEO)	20.9
CdSe–ZnS ($CdMe_2$)	30–50
CdSe–CdS ($CdMe_2$)	60
CdSe–CdS ($CdAc_2$)	85
Core CdSe ($CdMe_2$)	5–15
Core CdSe (CdO)	80

[a]*QY measured in aqueous borate buffer.*

fluorescence QY of CdSe compares quite well with that of currently used dye fluorescents. Indeed, QY of up to 85% has been reported in the case of CdSe NPs, surpassing that of most organic dyes. Two groups of shell coating are used to optimize the fluorescence properties of CdSe NPs. One is based on molecular (e.g., mercaptopropionic acid) and polymeric (e.g., polyethylenimine) materials. The second group is based on inorganic materials such as CdS. It is worth noting that QY also depends on the solvent type.

Key to the printing electronics industry is the formulation of NP-based inks. Polar and nonpolar surfactants are used to stabilize powder particles in aqueous and nonaqueous solvents, respectively. Typical solvents used in nonaqueous suspensions are organic acids, thiols, and polymers with reactive polar functional groups. Binders are often added to allow highly cross-linked structures. Direct-write technology using NP-based inks allows cost-effective patterning without using mask or resist. To compete with conventional patterning, direct-write technology needs to lower the high melting point of current metal particles, coagulation, impurities, and cost [258].

3.1.1 Core–Shell

Numerous core–shell structures have been reported in the literature. They include semiconductor on semiconductor, metal on metal, metal on semiconductor or vice versa, oxide on metal or semiconductor, and organic on metal or semiconductor. Core–shell structures can serve several purposes including protection, dual properties, further chemical modification, and/or compatibility with the surrounding environment.

Semiconducting materials are used for both core and shell. Depending on the bandgap and the relative conduction band/valence band (CB/VB) positions, two types (type I and type II) of core–shell structures have been designed (Fig. 1.17). Both type I and type II core–shell QDs have been prepared and characterized [259]. The radial wave function distributions of CdTe/CdSe and CdSe/ZnTe core/shell QDs are modeled using a 20 Å core radius and a 4 Å shell thickness, respectively. The relative CB/VB energy levels seem to have a direct effect on the spatial distribution of the wave function (Fig. 1.17). To satisfy energy level requirements and long-term stability, the choice of shell composition for any given semiconductor is quite limited [260]. In the case of InAs cores, QD, III–V semiconductor shells (InP and GaAs), and II–VI semiconductor shells (CdSe, ZnSe, and ZnS) with various radii are obtained using a two-step synthesis approach [260].

Because of their potential applications in microelectronics and photonics, core–shell QDs have been widely studied. Deposition of semiconducting shells on gold core NPs has been achieved by wet chemistry [261] and atomic layer deposition (ALD) [262] techniques. NPs with two different shell layers have also been developed [263]. The intermediary layer in the InAs/CdSe/ZnSe (core/shell1/shell2) structure is used to reduce the strain between the

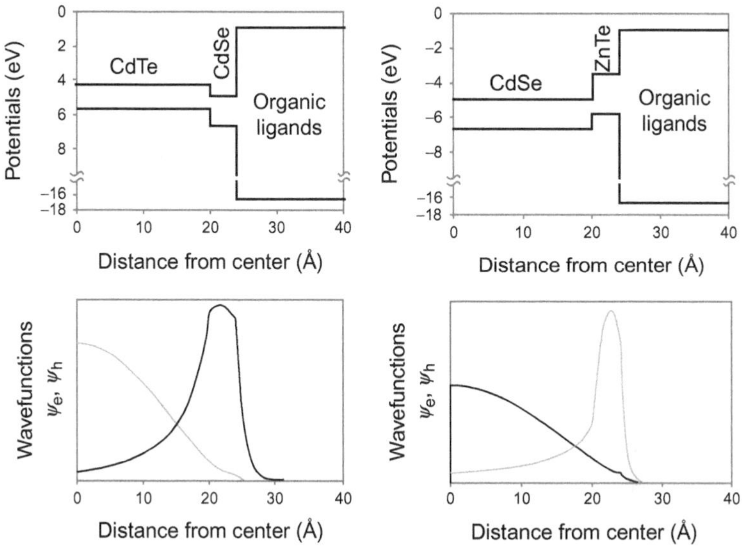

FIGURE 1.17 Energy diagrams (top) and their corresponding modeled electron (gray) and hole (black) radial wave functions (bottom) for CdTe/CdSe(core/shell) QD (left) and CdSe/ZnTe (core/shell) QD (right). *Reprinted with permission from Kim et al. [259]. Copyright 2003, American Chemical Society.*

InAs core and the top ZnSe shell layer [263]. Lattice mismatches of 0.003% and 6.6% have been measured in the case of InAs and CdSe, respectively. Furthermore, band offsets of 1.38 and 2.33 eV have been predicted for InAs and CdSe, respectively [263]. WS_2 and MoS_2 core NPs with 10 or more shell layers have also been reported [264].

A scalable process has been reported in the case of carbon-coated copper NPs [265], magnetic NPs coated with semiconducting, metallic, and insulating shells [266], and monodispersed core/shell Fe_3O_4/Au NPs [267]. As shown in Fig. 1.18, it is possible to screen large and small core/shell structures by varying the centrifugation speed. Indeed, small and large monodispersed Fe_3O_4/Au NPs have been obtained following redispersion of Fe_3O_4 NPs and addition of a reducing agent and a capping agent [267].

3.1.2 Functionalization

Functionalization is critical for the integration stage and also for the particles to satisfy the application requirements. Biomedical applications present the biggest challenge in terms of functionalization. Besides stability in biological media, modified NPs should also provide specific physicochemical properties. Four different components of the NPs are critical to providing all the required properties. The core and shell compositions provide the main properties of the QDs [268]. The core determines the absorption wavelength, whereas the shell

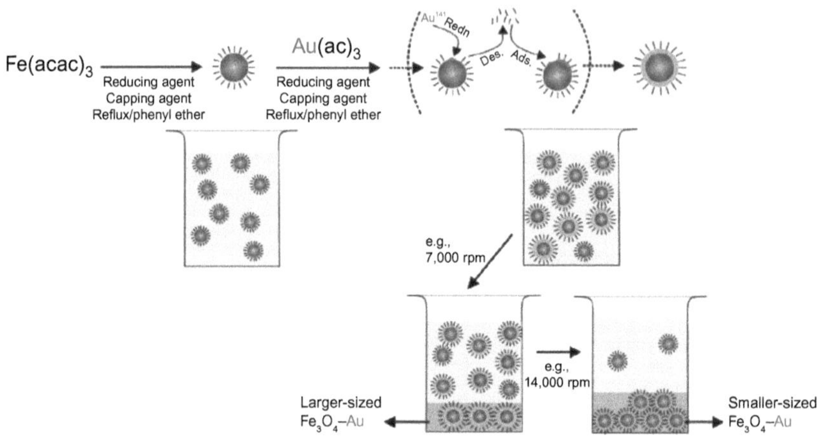

FIGURE 1.18 Schematic representation of the process used to prepare Fe_3O_4 and Fe_3O_4–Au nanoparticles. *Reprinted with permission from Wang et al. [267]. Copyright 2005, American Chemical Society.* (For color version of this figure, the reader is referred to the online version of this chapter.)

determines the emission brightness. The top coating empowers the NPs with the required characteristics that include biocompatibility, solubility, and specificity. In some cases (e.g., drug targeting), additional biomolecules are attached to the top layer to provide specific binding.

Targeted applications require specific functionalization approach, structure, and composition. In some optoelectronic applications, ligand groups keep the nanocrystals separated from adjacent NPs. Heat treatment and exchange with shorter ligands have been used in the past with limited success. Indeed, this often leads to surface contamination or degradation of the device. In the case of PbSe/CdSe core/shell NPs, octyldithiocarbamate is stripped from the surface by briefly heating to 140–205 °C, instead of 330 °C when oleate ligands are used [269]. In the case of latex NPs, a variety of functionalization compositions have been developed (Table 1.3). Specific functional groups are used to provide some properties. For example, charged functional groups (e.g., –COO–) provide stability against agglomeration and aggregation. Oligosaccharides and nucleic acids provide recognition attributes for antigen, DNA, and cells.

Functionalization is particularly critical for the application of NPs in biomedicine [271]. For example, cross-linked iron oxide (CLIO) SPM particles that are conjugated to a peptide (Tat) allow imaging of cell tracking using a fluorescent tag [271]. These NPs are attached to the peptide, and feature a fluorescent tag, fluorescein isothiocyanate. Blinking, defined as PL signal turning on and off stochastically, has been observed on several QD materials. This phenomenon is hindering their applications in areas requiring nonstop or on-demand single-photon emission. In particular, devices based on single-molecule imaging (SERS) and single-photon logic devices will be thus

TABLE 1.3 Characteristics of Different Functional Groups Available for QD Fictionalization

Functionality	Property
Charged groups ($-OSO_3^-$, $-SO_3^-$, $-COO^-$, $-PO_3^-$, $-N^+(CH_3)3Cl^-$, etc.)	Colloidal stability, electrostatic interactions (adsorption of polyelectrolytes, biomolecules)
Reactive groups (–COOH, –CHO, $-CH_2Cl$, –OH, $-NH_2$, –SH, epoxide, acetal, activated ester, tosylate, etc.)	Covalent immobilization on plain supports (silica wafers); grafting of biomolecules; chemical modification of preformed particles
Hydrophilicity (PEO, PAA, PAM, PMAA, PSSNa, hydroxyethylcellulose, etc.)	Steric stabilization, depletion of biomolecules, stealth effect
Sensitivity to stimulus (*T*, pH, ionic strength, light or UV, electric or magnetic fields, stress, etc.)	Shape, swelling behavior (microgels); reaction rates; ionic charge, recognition
Dye label (color, fluorescence)	Detection of a molecular interaction
Conductive polymers (pyrrole, aniline)	Conductivity, optical absorbance
Complexation (PEO, PMAA, metal chelates, etc.)	Protein purification, oriented immobilization (proteins etc.)
Ligand (oligosaccharide, lipid, peptide, nucleic acid, antibody, protein)	Recognition of antigen, specific cells, DNA, RNA, protein, lectin, etc.

PEO, polyethylene oxide; PAA, polyacrylic acid; PAM, polyacrylamide; PMAA, polymethacrylic acid; PSSNa, polysodium styrene sulfonate [270].

unreliable. Several solutions have been developed with limited practical success. Recently, CdSe/ZnS QDs blinking was suppressed by the addition of TiO_2 NPs, without changing the rate of radiative relaxation [272]. Arrest of blinking is achieved by suppression of the nonradiative relaxation in charged QDs. Electron transfer to TiO_2 NPs and back-electron transfer to QD allows charge neutralization.

Metallic NPs are modified with bifunctional ligands to allow monodispersity and biofunctionalization [273,274]. Versatility of gold NPs allowed functionalization with various molecules targeting different applications [275]. Indeed, DNA, peptide, antibody, protein, and organic dye are used to functionalize gold NPs [275]. Depending on the composition of the functionalizing group, modified gold NPs can be used as optical probes, targeting ligands and drug delivery. It is possible that multifunctionalization of a single NP-based system could be used for targeting, imaging, and tumor treatment simultaneously. Gold NPs modified with different monofunctional groups provide enhanced yield and high purity [274].

Silica NPs have also been modified with different functional groups, including carboxylate, amine, amine/phosphonate, poly(ethylene glycol), octadecyl, and carboxylate/octadecyl groups to prevent aggregation and nonspecific binding [276]. Ni NPs are successfully coated with a carbon shell following sucrose modification [277]. QD biofunctionalization is used to provide water solubility and a linking functionality for sensing [278]. Water-soluble monodisperse gold NPs have been modified by a silica layer followed by bioconjugation [279]. ALD is potentially a well-suited NP surface coating when a thin and pinhole-free layer is required [280]. A silica coating provides good chemical and thermal stability for NPs over a wide range of polar and nonpolar solar cells and also over a large pH range because of its low IEP. Furthermore, this coating provides the possibility for further derivatization.

Functionalization of dry NPs is often very challenging [281–283]. Mixing ceramic fillers with polymeric materials requires appropriate surface modification. Several functionalization methods are reported in the case of metal hydroxide surfaces obtained via the flame-aerosol method. These NPs do not readily mix with relevant polymers, often nonpolar, such as polyethylene, polypropylene, and polystyrene. Several alkyl/organo (methyl, vinyl, 3-acryloxy-propyl) functional groups can improve the compatibility of hydroxide-terminated NPs with several polymers. Functionalization of metal oxides (M) with chloro- or alkoxysilanes via hydrolysis of silicone functional groups gives rise to hydrochloric acid, methanol, or ethanol release [283]. Photo-induced coating of NPs in a gas phase has been reported [281]. After introducing the mixture of aerosolized NPs with coating precursors into a reactor, the mixture was exposed to a 172-nm excimer lamp [281]. Growth species from reactant decomposition allow NP growth. Depending on the flow rate, coating thickness can be varied from less than 1 nm to 20 nm using this photo-assisted chemical vapor deposition technique. NP size variation is measured online using tandem differential mobility analysis.

3.2 Characterizations

It is very easy to draw some similarities between molecules and NP-based materials. NPs are used as building blocks for structured systems such as molecular systems. It is often desirable to directly follow the fabrication process of NPs and their modification. This allows optimization of properties. While this is quite obvious for molecule synthesis using nuclear magnetic resonance (NMR), it is very difficult to duplicate this approach in the case of NPs. Often, characterization of NPs may appear quite difficult and some contradictions are easily identified between the results of different techniques [284,285]. This could have some negative impact. For example, the lack of reliable measurement methods is the main reason why there are no air quality regulations in place to control the concentrations of airborne atmospheric NPs [286].

Four parameters are important when analyzing the chemical composition. They include speciation, concentration, chemical sensitivity, and spatial resolution. A variety of micro- and nano-characterization techniques have been developed. Today electronic and vibration spectroscopies provide valuable information about these parameters. Techniques with lowest detection limits and highest spatial resolution are specifically sought. Basic techniques for elemental analysis include secondary ion mass spectroscopy (SIMS), time of flight (ToF) SIMS, Auger electron spectroscopy, electron probe microanalysis (EPMA), energy-dispersive X-ray analysis, and X-ray fluorescence (XRF). These techniques should be divided between bulk (XRF) and surface (SIMS, AES, XPS (X-ray photoelectron spectroscopy)) specific procedures. XRF using radiation of 1–100 keV or more is typically used. In the case of SIMS, bulk and surface analyses are possible. For example, ToF–SIMS is surface-specific, whereas dynamic SIMS provides a depth profile of up to several microns. Advanced TEM, SIMS, and XRF techniques provide high spatial resolution and low detection limits. In the case of alloys, a quantitative chemical analysis is critical to verify that it is within the specified composition range. Recent advances in synchrotron radiation have allowed improvement in sensitivity and spatial resolution of several techniques including XRF, XPS, and EPMA.

NP characterizations often involve information about size, composition, and structure. Detection limits related to the size and concentration are two critical parameters (Fig. 1.19). Several surface and bulk characterization

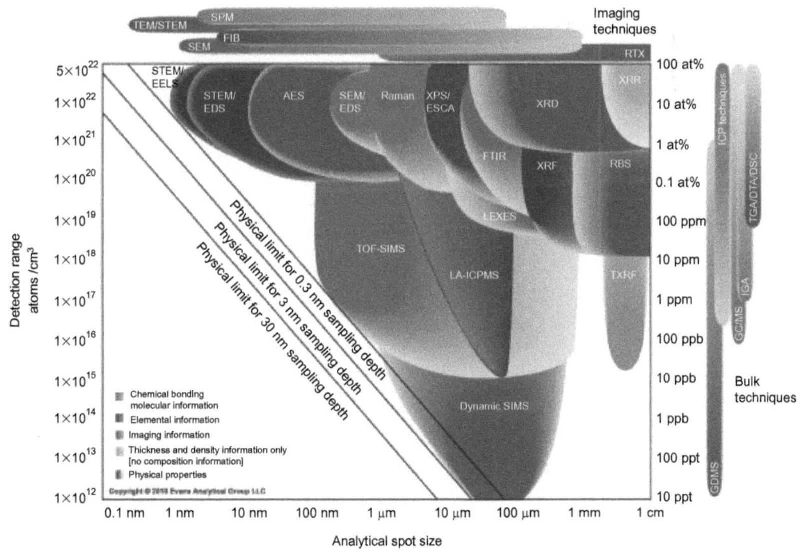

FIGURE 1.19 Detection limit and analytical spot ranges of different imaging and bulk techniques [287]. (For color version of this figure, the reader is referred to the online version of this chapter.)

techniques are amenable to NPs. Available techniques allow either high spatial resolution (AFM and TEM) or low detection limit (SIMS or XRF). Current techniques do not allow sub-100 nm with a very good detection limit. However, recent developments in ToF–SIMS and SERS may address this gap, particularly with the newly developed excitation sources [288].

Measuring the size of NPs is not straightforward. Given the vagueness with which size parameters are measured and reported, it is difficult to compare data from the literature using different methods. Photon correlation spectroscopy (PCS) and TEM have been used to characterize the core and the shell dimensions of polymer-coated platinum NPs [289]. PCS is used to measure the hydrodynamic diameter directly in solution under ambient conditions. Surface charging and aggregation will affect the average size. Furthermore, mathematical models are required to extract size distribution. Caution should be applied when comparing PCS to dry and high vacuum-based techniques such as TEM. Indeed, unless an extremely large number of particles are analyzed from representative sample areas, the outcome is often subjective with regard to interpretation and operator. It is worth noting that TEM requires more careful sample preparation compared to PCS.

3.2.1 Geometrical and Structural

There are a myriad of approaches and techniques to measure particle size [290–293]. Each technique has its merits and limitations. Depending on the NP type, various size ranges are conservatively estimated for each technique [294]:

a. Laser diffraction ($\geq$40 nm);
b. Dynamic light scattering ($\geq$4 nm);
c. Photon migration ($\geq$100 nm);
d. AFM and electron microscopy ($<$0.5 nm);
e. Centrifugal sedimentation ($\geq$30 nm);
f. Scanning mobility particle sizer ($\geq$10 nm);
g. Fast field fractionation ($\geq$10 nm);
h. Acoustic techniques ($\geq$30 nm);
i. Capillary hydrodynamic fractionation ($\geq$15 nm);
j. ToF techniques ($\geq$100 nm).

For ultrasmall NPs ($<$5 nm), AFM and TEM are widely used techniques. However, these two techniques do not provide a good sampling as they probe only a very small part of the sample. Mass spectrometry provides m/z (with z being the electric charge of the detected ion) value. Although theoretically there is no upper mass limit, mass spectrometry is often limited to particle sizes of 10 nm depending on the element. In some cases, PCS and XRD can also provide geometrical size distribution.

Geometrical characterizations that include SEM, TEM, and AFM have been reviewed recently [295]. Light-scattering techniques provide several advantages in terms of flexibility and versatility (Table 1.4). Scattering techniques that include dynamic light scattering (DLS or PCS), small-angle X-ray scattering (SAXS), wide-angle X-ray diffraction, and small-angle neutron scattering (SANS) can be used to probe the structural and dynamic properties of NPs [297].

TABLE 1.4 Characteristics of the Main Particle Size Analysis [296]

Method	Principle	Measured size[a]	Limitations
LS	Light interaction	50 nm to 1 μm	Nonappropriated for very polydisperse populations
			Indirect method
			Great influence of aggregates or larger particles
			Many influencing parameters
LLD	Light interaction	1–1000 μm	Large amount of sample required
			Indirect method
SEM, TEM	Microscopy	50 nm to 100 μm	Time-consuming
			Influence of the preparation sample
AFM	Microscopy	10 nm to 1 μm	Sampling
			Nonautomated
			Complexity of the setup
			Image treatment
			Subjective
ANUC	Centrifugation	–	Complex data processing
FFF	Elution	20 nm to 1 μm	Difficult to handle
			Optimization needed for each kind of particle
CE	Electrophoresis	20–500 nm	Complexity of the setup
PCH, SEC	Chromatography	<100 nm	Long steps of optimization
			Time-consuming

LS, light scattering; LLD, laser light diffraction; SEM, scanning electron microscopy; TEM, transmission electron microscopy; AFM, atomic force microscopy; ANUC, analytical ultracentrifugation; FFF, field flow fractionation; CE, capillary electrophoresis; PCH, packed column hydrodynamic; SEC, size exclusion chromatography.
[a]*Size range for which the method can be applied with accuracy.*

There are three categories of characterization techniques depending on the NPs surrounding the medium [298]. Often NP size distribution is carried out either in gas phase, liquid phase, or a solid surface. In the case of gas phase NPs, a dozen techniques have been reported [298]. Generating and collecting NP schemes are critical and often specific for each type of NP. Measurement of NP size is not straightforward, even impossible in some cases. This is why these techniques are not so widespread, as they require specific technical skills to measure and extract meaningful information. Differential mobility analysis (DMA) and, to a lesser extent, ToF–SIMS remain the two most widely used techniques. Liquid-based techniques are easier to operate, although they are more prone to pitfalls. Higher signal-to-noise ratio and stability are often a positive attribute of these wet techniques. DLS is the most used wet technique. In the third category, NPs are first collected onto a substrate for analysis using TEM and, to a lesser extent, SEM or AFM. Specific techniques have been developed for liquid-phase NPs and surface-embedded NPs [298].

Another important geometrical parameter that has received less attention is the shape of NPs. Shape has a strong correlation with the exposed face of NPs [299,300]. For example, {100} and {210} facets of Pt NPs are found to be highly active and give rise to selective H_2 and CO catalysis reactions [301]. The shape is often dictated by the seed crystallinity and growth rate.

Although TEM has been widely used for NP characterization, it has one obvious shortcoming. Besides the high capital cost and the sophistication of sample handling, only a very small part of the samples is analyzed. Typically, less than one part per million of the materials is sampled. Image analysis software has been developed to accurately estimate size distribution based on automated method analysis of images containing up to thousands of particles [292]. Particle size distribution analysis is sometimes more difficult with supported NPs [302].

TEM and XRD remain the most popular characterization techniques of NPs. TEM can be used to provide size distribution, although often not representative of the sample. TEM also provides composition and structural information. State-of-the-art TEM instruments are also equipped with electron energy loss spectroscopy (EELS), allowing the characterization of the chemical bonding and electronic structure at the surface and interface. XRD provides an average size of the crystallites instead of the particle. SAXS is used to measure the overall particle size, although this value could be overestimated for smaller NPs (<2 nm). However, when these techniques are correctly applied, good agreement between the different techniques is obtained [303]. Correlation between TEM and XRD information has been reported in the case of $CoPt_3$ NPs (Fig. 1.20). This is particularly the case with monocrystalline NPs. Similar observations have been reported in the case of Pt-based catalyst NPs [289].

For NPs integrated in film, cross-sectional studies using cross-sectional scanning tunneling microscopy are often required [304]. SANS could be used

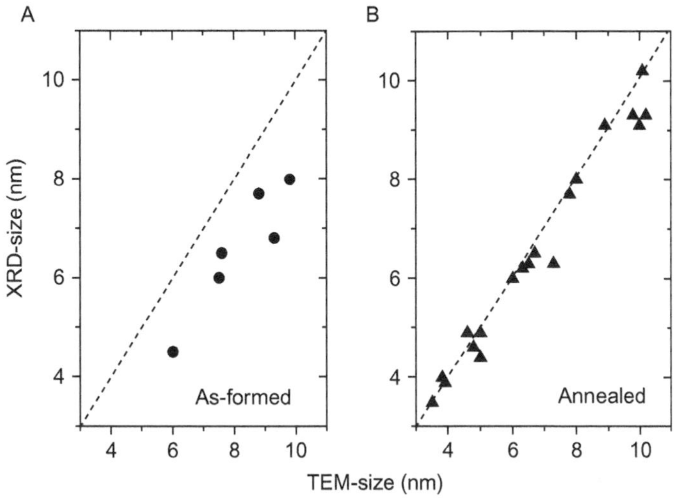

FIGURE 1.20 Variation of $CoPt_3$ nanoparticle crystallite size obtained by XRD as a function of the diameter measured by TEM (A) before and (B) after annealing. *Reprinted with permission from Shevchenko et al. [62]. Copyright 2003, American Chemical Society.*

to probe average particle size, molecular weight, surface area, volume fraction, and particle shape [297]. Mesoscopic structures of NPs in colloidal systems have also been studied using ultrasmall-angle X-ray scattering and ultrasmall-angle neutron scattering [305].

XRD is used not only for measuring crystallite size, but also for phase determination. XRD is based on elastic scattering of impinging X-ray photons by the materials periodic lattice. Bragg relation is used to estimate the lattice spacing (d_{hkl}):

$$2d_{hkl} = n\lambda / \sin\theta.$$

It is well established that small NPs having a diameter of $2R$ give rise to peak broadening that could be quantified by the oft-used Scherer's equation:

$$2R = K\lambda / \beta \cos\theta,$$

where $0.9 \leq K \leq 1.0$ is a correction factor to account for the specimen's shape, and β is the full width at half maximum in radians. As suggested by Pernicone [306], XRD provides average size distribution that could be misleading in some applications. It is important to consider instrumental broadening and strain broadening when extracting the crystallite size from XRD data.

XRD is often used for phase determination and lattice parameter determination of NPs [307]. Indeed, with the large number of atoms at the surface, interface, and grain boundaries, it is difficult to consider these materials as single-phase materials. These factors contribute to peak broadening. A detailed Rietveld analysis is often required to reconcile the different characterization techniques [285].

A schematic representation of the different sizes provided by XRD, DLS, and TEM is shown in Fig. 1.21. Often, the following relation will apply when analyzing geometrical size (GD) using these techniques:

$$\mathrm{GD(DLS)} > \mathrm{GD(TEM)} > \mathrm{GD(XRD)}.$$

SAXS was used to characterize gold NPs embedded in glass a long time ago [308]. Furthermore, with smaller wavelength (about 0.003 nm) in comparison to conventional X-ray radiation (0.15 nm), the Ewald sphere is much larger. This will allow direct observation of the reciprocal lattice. Monolayer (2D) and rod (3D) ordering of cobalt NPs has been probed using grazing incidence small-angle X-ray scattering (GISAXS) [309]. Structural properties of NPs are also characterized using diffractive electron imaging within a TEM [310].

Elastic and inelastic laser scattering are used to obtain information on particle size, composition, and morphology directly from single-aerosol droplets [311]. Measurement of particle size distribution in colloidal solution is very important and challenging. Two techniques are often used: correlation spectroscopy and UV-visible absorption. PCS, often referred to as DLS, belongs to a group of scattering techniques [297] used to measure the hydrodynamic diameter (H_d) of submicrometer particles in diluted solution. The Stokes–Einstein equation links H_d to diffusion coefficient (D) and viscosity (η):

$$H_\mathrm{d} = kT/3\pi\eta D,$$

where k and T are the Boltzmann constant and absolute temperature, respectively. In this technique, diluted suspensions are required to prevent multiple

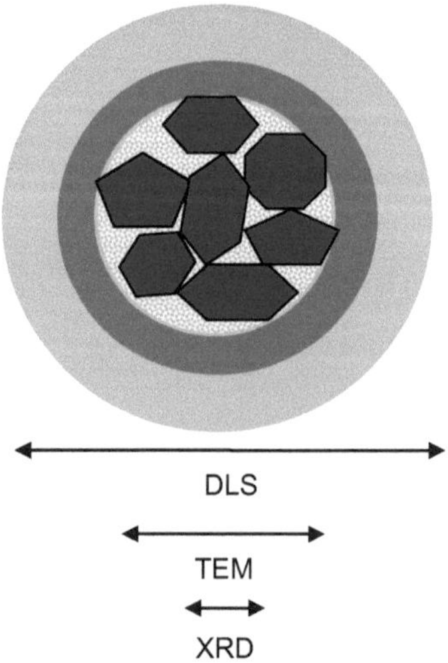

FIGURE 1.21 Size measurement of nanoparticles using various techniques. (For color version of this figure, the reader is referred to the online version of this chapter.)

scattering. With higher NP concentrations, mathematical models have been developed to account for multiple scattering and agglomeration effects [312].

DLS offers some advantages over TEM; for example, it can sample a much larger number of particles. This provides better precision even when smaller numbers of larger particles are present in a bimodal distribution [313]. Small and weakly scattering particles can be detected using high-powered excitation lasers and new-generation avalanche photodiodes. Unfortunately, there are very few data comparing microscopic (TEM) and macroscopic (DLS) techniques. The so-called Turkevich method [10] has been widely used and improved to synthesize gold NPs. It has been used to precisely measure size distribution using gold-to-reductant ratio as a key parameter [314]. As shown in Fig. 1.22, relatively good agreement between microscopic and macroscopic measurements has been achieved over a wide size range. Consistent NP size measurements obtained by XRD, TEM, and DLS have been reported in the case of Pt–Ru NPs [110].

Most synthesis methods do not provide highly monodisperse NPs. Post-preparative size fractionation is performed taking advantage of size-dependent particle properties. Typically, narrowing the size distribution takes care of the difference in solubility of large and small particles in different solvents [18,231,315]. Metallic NP sizes are measured directly in solution using UV–visible absorption in combination with theoretical peak fitting [316]. Scattered light (I) from incident light (I_0) is reduced by a factor A given by [316]:

$$A = \log(I_0/Id) = N^* C_{\mathrm{ext}}(l/2.303),$$

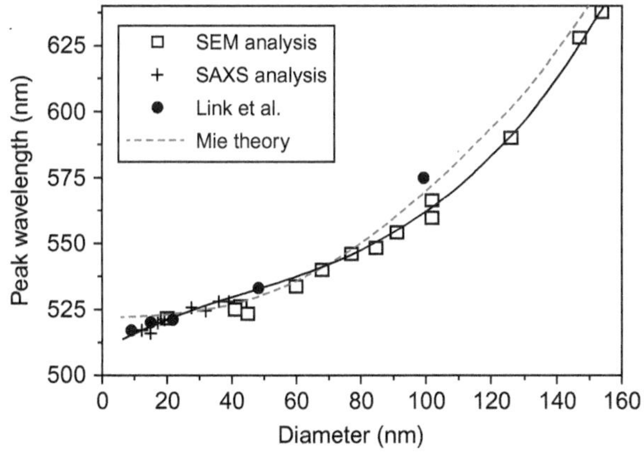

FIGURE 1.22 Variation of the surface plasmon resonance (SPR) peak position as a function of the diameter of citrate-prepared gold nanoparticles. Solid dots are obtained from the literature. The dashed line is obtained using Mie calculation. Open squares and crosses are derived from SEM and SAXS measurements, respectively. The solid line corresponds to a polynomial least-square fit. *Reprinted with permission from Kimling et al. [314]. Copyright 2006, American Chemical Society.*

where N is the number of diluted colloidal NPs per unit volume, C_{ext} is the particle extinction cross-section, and l is the path length. Considering the fact that the NP size ($2R$) is at least five times smaller than the excitation wavelength (λ):

$$C_{ext} = 24\pi^2 R^3 {\varepsilon_m}^{3/2} \varepsilon'' / \left[\lambda(\varepsilon' + 2\varepsilon_m)^2 + {\varepsilon''}^2\right]$$

where ε_m and $\varepsilon = \varepsilon' + \varepsilon''$ are the medium and NP dielectric constants, respectively.

UV–visible absorption is used to control the formation of gold NPs directly from the mother solution in the presence of tetraoctylammonium bromide [54]. UV–visible spectra of solutions (a) containing $[AuCl_4]^-$ and dodecanethiol (DT) before reduction and (b) immediately after reduction and DT–Au NPs after heat treatment at (c) 150, (d) 190, and (e) 230 °C are shown in Fig. 1.23. UV–visible spectroscopy is limited by the spectral resolution. Static light-scattering spectroscopy in the UV range (from 250 to 390 nm) has been used for making very accurate measurements [317]. Indeed, sizes of standard polystyrene nanospheres of 33, 63, and 81 nm have been confirmed using this approach.

Absorption and photoluminescence excitation spectra have been used to probe various QD properties [318]. CdTe nanocrystals obtained via colloidal solution gives rise to photoluminescence properties which depend on the growth conditions [319]. A direct relationship between UV–visible reflectance and TiO_2 QD size distribution has also been established (Fig. 1.24). Position of the absorption maximum can be used to measure the size distribution of some metallic and semiconductor NP suspensions. This method is often limited to a few compositions. Using modified Mie's equations, UV–visible absorption has been used to measure precisely the diameter of

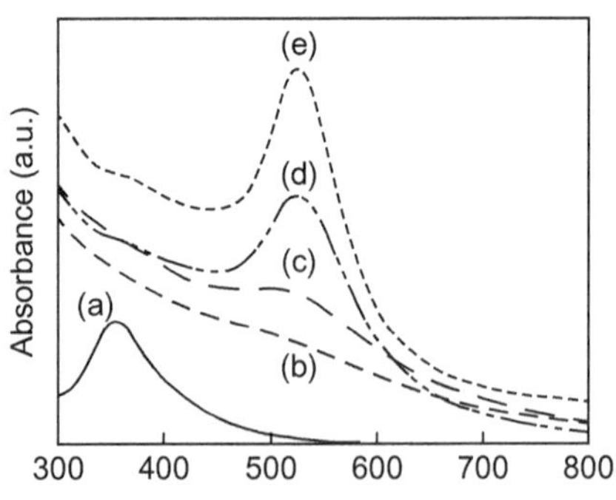

FIGURE 1.23 UV–visible absorption spectra of gold nanoparticles corresponding to (a) 1.5 nm, (b) 3.4 nm, (c) 5.4 nm, (d) 6.8 nm, and (e) 8.7 nm particle size. Spectrum (A) corresponds to non-reduced gold precursor. *Reprinted with permission from Shimizu et al. [54]. Copyright 2003, American Chemical Society.*

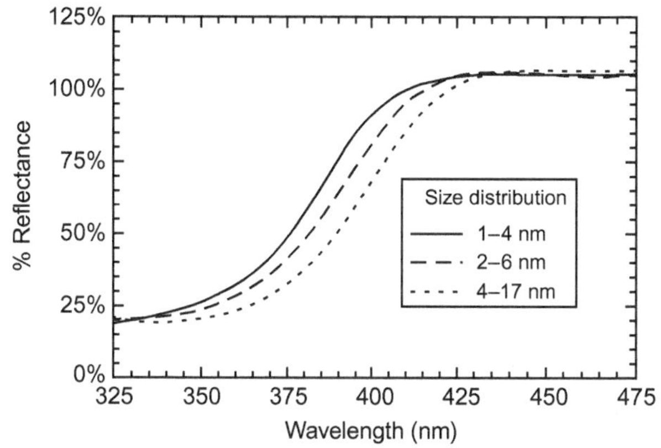

FIGURE 1.24 UV–visible reflectance spectra of TiO_2 QD with different size distribution ranges. *Reprinted with permission from Hoffmann et al. [320]. Copyright 2005, American Chemical Society.*

nonabsorption NPs in the UV–visible region [316]. A good correspondence has been established between size distribution obtained by TEM and optical data [321].

Matrix-assisted laser desorption ionization (MALDI) analysis of aqueous gold NPs in the mass range of 50–250 kDa provides reliable data [322]. Sizes inferred from MALDI are comparable to those obtained by XRD and TEM. MALDI could also provide information about the state of aggregation of these NPs.

The majority of size characterization techniques are based on off-line measurements. However, there are a few exceptions. Online size distribution measurement during gas phase NP generation and growth is critical. Three techniques are often used including light scattering, sampling probes, pressure impactor, and scanning mobility particle sizer (SMPS) [323]. SMPS can detect NP sizes from 3 to 1000 nm in a concentration range up to 108 particles per cubic centimeter within 1 min.

DMA provides aerosol size measurements in real time [324]. DMA could be coupled to other techniques for complementary analysis. DMA measures the electrical mobility of charge particles. This electrical mobility is inversely proportional to the particle size. To extend size measurement to a lower size range, a more efficient aerosol charging is required [325]. Unipolar charging with high ion concentration and long charging time has been obtained by directing ion and aerosol flow in the same direction using sheath airflow [325]. This charging approach allowed NP size measurement in the range of 3–50 nm using DMA.

There are many other indirect measurement techniques allowing size distribution measurements. Laser-induced incandescence) is also used to probe NP size directly inside a shock wave reactor [326]. Thermogravimetric analysis has also been used to measure size distribution of nanometer- and micro-sized aluminum powders [327]. The Brunauer–Emmett–Tell (BET) method allows indirect measurement of NP size. Plotting pressure versus gas uptake at liquid N_2 temperature allows measuring the surface area of powder using N_2 gas adsorption. BET also allows morphologies and fractal dimensions. This provides some understanding of the degree of agglomeration of the NPs.

Comparison between online SMPS and *ex situ* TEM measurements have been undertaken on silica NPs obtained from H_2/O_2/tetraethylorthosilicate diffusion flame [328]. Two sampling procedures have been considered. One uses a local thermophoretic method inside the flame. The second one uses an electrostatic collector sampling method after the dilution sampling probe. A relatively close correlation between the two measurements has been found in the 30–50 nm size range [328].

3.2.2 *Separation and Postprocessing*

Variation in solubility due to difference in particle diameter has been used in size selection [18]. Drop-wise addition of a nonsolvent to the freshly synthesized NPs will induce precipitation of relatively larger particles. After separating the precipitate consisting of larger particles by centrifugation or filtering, another series of drop-wise nonsolvent addition to the supernatant will allow separation of the next largest NPs. At the end, several fractions of the initial solution are obtained with different average particle size. Although this technique provides high-quality monodispersed NPs, the yield and the large amount of organic solvent make it very expensive. Furthermore, a significant environmental impact is expected due to the large amount of solvent utilization and low yield.

Selective centrifugation could be used as an alternative approach to reduce solvent utilization. Initially developed for polymeric systems, analytical ultracentrifugation (AUC) is a powerful tool allowing precise distribution following fractionation [329]. AUC is also used to characterize the surface chemistry of NPs.

High-pressure liquid chromatography could also be used to separate particles with different sizes [330]. In the case of CdS NPs, a calibration curve has been developed showing a linear relationship between the retention time and the particle size [330]. For example, a retention time of 4–6 min is measured for NP sizes up to 10 nm.

Other postprocessing techniques have also been developed. Capillary electrophoresis has been used to separate gold and polystyrene NPs into different sizes and compositions [331]. Sieving is also used to separate particles according to their size and shape, although this has been done mostly for microparticles [332]. Surface contamination is often undesirable and requires postprocessing [333].

3.2.3 Composition

Surface composition of NPs is critical. For example, the presence of an oxide layer affects interparticle interactions. This delays the onset of sintering and retaining defects at the grain boundary in the fully dense product [334]. Detailed knowledge of the surface composition and structure is critical to enable the oxide layer characterizations. A combination of XRD, XPS, and TEM are needed to provide a detailed composition picture of technologically relevant Al and Ni NPs [334]. In a 100-nm-diameter Al NP, a 3-nm-thick compact amorphous oxide was measured around a crystallite size of about 85 nm. This is consistent with the low-temperature oxidation mechanism of Al NPs [335]. In the case of Ni particles, a nonuniform and crystalline surface oxide layer was measured with relatively small metal crystallite.

Besides determination of NP surface composition, XPS is also used to follow electronic properties as a function of size. However, XPS of core–shell NPs is not quite as straightforward as one might expect [336]. Indeed, one should include a geometrical factor when using angle-resolved XPS data. Al NPs that can be used as propellants have been found to oxidize, forming a 2- to 4-nm Al_2O_3 shell layer [335]. XPS has been used extensively to distinguish between the chemical structure of surface and bulk layers [337]. The surface composition of NPs is often different from that of the bulk material. This is particularly the case with NPs larger than 5 nm. CdTe NPs capped with thioglycolic acid have been analyzed using variable wavelength XPS. This study allowed explaining the dependence of PL properties on the sample preparations [319]. Attempts to probe the sulfur oxidation state in gold [338] and CdS [339] have been reported. Transition metals and their oxides are challenging to analyze because of the multiplicity of oxidation states and the presence of satellite peaks. Synthesis methods may have significant consequences on NP surface structure. A nonlocal screening process from the Ni7O36 cluster due to the importance of the local environment and/or defect distribution gives rise to characteristic XPS features not observed in bulk nickel [340].

X-ray absorption spectroscopy including XANES and near-edge X-ray absorption fine structure (NEXAFS) have been used to probe Pt and non-Pr-based NP catalysts used in fuel cells [341]. NEXAFS has mostly been used to probe 2D nanostructures [342], although it has been recently applied to NPs as well [343].

Besides size, TEM provides valuable structural and composition information using EDXA, diffraction, and EELS options [344]. Modern TEM instruments are equipped with EELS. This provides a unique tool to probe composition and structure at the sub-10 nm resolution [345,346]. This elemental mapping using transmitted electrons with specific energy loss could also provide valuable chemical information at the nanoscale. The spatial resolution of energy-dispersive X-ray spectroscopy (EDS) is limited to about 1 μm.

To evaluate the characterization ability of field emission scanning electron microscope (FE-SEM), AFM, TEM, EELS, and EDS, various mixtures of known NPs have been studied [347]. FE-SEM is particularly suitable for powder mixtures with different particle shapes and/or atomic numbers. The EELS attachment provides chemical information of nanopowder mixtures of SiO_2 and TiO_2 at the nanoscale [347].

Raman spectroscopy has been used to probe nanomaterials including NPs [252,348]. SERS is used to characterize and compare aggregated and non-aggregated silver NPs [349]. A combination of near-field scanning optical microscopy and SERs has been used to probe the composition of nanomaterials at the nanoscale level [350].

3.2.4 Others

Electron paramagnetic resonance (EPR) [351–353] and NMR [354] are used to characterize various types of NPs to complement structural characterizations obtained by TEM and/or XRD. The nature of trapped electrons and holes in TiO_2 NPs has been studied using EPR [351]. Nanoindentation is often used to probe the mechanical properties of films and solids [354,355]. A real-time detection scheme of low-index NPs in water has been developed [356].

Essential to separation is the relative weight of NPs. The core weight of iron oxide core in the amino-CLIO NPs has been estimated using a combination of three measurements [357]. First, viscosity over a wide concentration has been measured to estimate the partial specific volume. Second, particle diameter is obtained by light scattering, allowing estimation of the volume of an individual particle. Third, the concentration of NPs (w/v) is obtained.

3.3 Nanometrology

Only recently, nanometrology has been recognized as a key to the future of nanotechnology in general and the NP market development in particular. Nanometrology deals with the measurement of dimensions within nanomaterials and devices. Nanometrology is critical for production quality control and toxicology studies. Size reporting is complicated by the numerous parameters influencing the measurement readings. For example, the state of NP agglomeration due to change in the surrounding media and aging will have a direct effect on size distribution. Existing microscale metrology techniques and methodologies are hardly applicable at the nanoscale level. The complication goes beyond the size measurement.

A specific metrology approach is required for most NP-based products. Often, size, shape, and roughness are key physical features in microscale devices [332]. The relative importance of the different physical features changes at different scales. At the nanoscale, roughness plays a limited role. Recent developments in nanomaterials call for improved measurement methodologies.

As the particle size decreases to below 100 nm, it becomes increasingly challenging to provide accurate measurements [358]. There is a need for appropriate NP size standards for instrument calibration and measurement and reporting protocols. Besides the scientific and technological values of a global standard, there are numerous other business-related values. Indeed, standards help build confidence among end users and help overcome trade barriers.

The importance of nanometrology has been growing since it was first coined [359,360]. Accuracy of length measurement, the most important parameter in nanometrology, is limited by the Abbe-sine and Abbe-cosine errors [360]. Numerous techniques have been used to provide measurements at the nanoscale [295].

A suite of techniques is available to carry out metrology investigations (Table 1.5). AFM has also been used to provide more accurate measurements of microscope calibration standards [362]. It is very important to realize the limitations of some metrology techniques when they are used in the nanoscale regime. For example, in the case of SEM, the dimensions of some small details are overestimated [363]. Besides the spatial resolution limits of some techniques, there are other challenges to nanometrology. For example, high-resolution techniques such as TEM are available only for off-line measurements. Thus, there is a need to establish protocols for linking these results with online data. A complete metrological system is described by Danzebrink et al. [364].

Different ISO standards have been developed to measure particle size. Some of these standards can be used for NPs. This is the case of DMA and PCS. Except for polystyrene NPs covering a limited NP size range, very few reliable material standards have been developed so far. These materials are important for proper calibration. Different techniques require different particle feeding, analysis conditions, and software.

Besides size, NP surface composition and other NP property measurements are important. There are very few reliable methodologies to monitor chemical compositions. Fluorescence NP metrology has been proposed to provide high resolution of the kinetic reaction pathway [365]. Gold and silver NPs have been recently used as internal probes in fluorescence resonance energy transfer instead of organic fluorophore [366].

4 SCALE-UP AND QUALITY CONTROL

Recently, the importance of precision in reporting NP size parameters has been stressed, particularly in the area of drug delivery [296]. Indeed, although it is well established that smaller particles have an enhanced drug-targeting ability, it is still difficult to provide accurate particle size distribution. For example, the required size limit and other related parameters for drug delivery to a specific organ are poorly defined. It is also very important to compare the functional properties of NPs obtained using different preparation methods. This comparison should also be fair and standardized. This is particularly

TABLE 1.5 Characteristics of Selected Nanometrology Techniques [361]

Technique	Range	Measured properties	Nanomaterial examples
Atomic force microscopy (AFM). AFM involves scanning a probe tip across the surface under investigation. The force originating between the surface's atoms and the probe tip results in deflection of the tip, which is recordable by a laser beam, cantilever, and photodetector	Depth: 0.5–5 nm; lateral resolution: 0.2–130 nm	Topology, surface roughness and elasticity, grain size, surface molecular interactions, surface magnetic features, surface friction, density of electron states	Researchers have used AFM to characterize CNTs, diamond-like carbon nanostructures, and various nanocrystals, nanoparticles, dendrimers, and nanowires
Scanning-tunneling microscopy (STM). STM scans a metal tip across the surface under investigation. Unlike AFM, application of an electric voltage between the tip and the surface results in a measurable tunneling current	Depth: 1–5 nm; lateral resolution: 2–10 nm	3D topology: size, shape, roughness, defects, electronic structure, density of electron states	Researchers have used STM to characterize various nanomaterials, including CNTs
Scanning near-field optical microscopy (SNOM). SNOM is a combination of scanning-probe microscopy and optical microscopy	Resolution: 50–100 nm	Chemical structure, refractive index, and local stress	Researchers have used SNOM to characterize quantum dots, nanospheres, and nanowires
Scanning-electron microscopy (SEM). SEM bombards the surface of a sample with a high-energy beam of electrons. Interactions of these electrons and the sample's surface atoms are detectable	Depth: 1–5000 nm; lateral resolution: 1–20 nm	Topography, morphology (particle size and shape), composition (sample elements and compounds), and crystallography (atom arrangement)	Researchers have used SEM to characterize various nanowires, nanoparticles, and nanofibers

TABLE 1.5 Characteristics of Selected Nanometrology Techniques [361]—Cont'd

Technique	Range	Measured properties	Nanomaterial examples
Transmission-electron microscopy (TEM). TEM transmits a monochromatic beam of electrons through an ultrathin sample, interacting with the sample as it does so	Depth: 200 nm; lateral resolution: 2–20 nm	Morphology (particle size and shape), crystallography (atomic-scale defects), composition (elements and compounds), and phases (lattice spacing)	Researchers have used TEM to characterize various nanowires, quantum dots, nanoparticles, CNTs, dendrimers, core–shell nanocrystals, and nanorods
Small-angle X-ray scattering (SAXS). SAXS focuses a monochromatic beam of X-rays on a sample, in which some of the X-rays scatter, providing a detectable scattering pattern	Between 1 nm and several hundred nanometers	Particle sizes, shapes, distribution, and surface-to-volume ratio	Researchers have used SAXS to characterize various nanoclusters, nanoparticles, nanocomposites, and nanorods
X-ray diffraction (XRD). XRD relies on the elastic scattering of X-rays from structures that have long-range order	Resolution (vertical direction): sub-nm	Crystallographic structure, crystallite size (grain size), preferred orientation in polycrystalline or powdered solid samples, film thickness, interface roughness, and surface topology	Researchers have used XRD to characterize various nanowires, nanoparticles, quantum dots, CNTs, nanocrystalline materials, nanocomposites, and nanorods
Auger electron spectroscopy (AES). AES sends electrons to a sample, resulting in the ejection of excited electrons from the sample, and the measured energy enables surface analysis	Depth: 0.3–3 nm; resolution: about 30 nm	Composition of surface layers	Researchers have used AES to characterize CNTs and nanowires
Fourier transform infrared spectroscopy	–	Structural and chemical	Researchers have used FTIR to

Continued

TABLE 1.5 Characteristics of Selected Nanometrology Techniques [361]—Cont'd

Technique	Range	Measured properties	Nanomaterial examples
(FTIR). FTIR uses infrared light to obtain the infrared spectrum of a sample based on its resonant frequencies		information, chemical bond types, vibrational energies of molecules	characterize numerous nanomaterials, such as nanocrystalline materials and nanoparticles
Raman spectroscopy. Raman relies on inelastic scattering, or Raman scattering of monochromatic light, usually from a laser in the visible, near-infrared, or near-ultraviolet range	–	Chemical information: impurity concentration, vibrational and rotational information	Researchers have used Raman spectroscopy to characterize numerous nanomaterials, such as nanofibers, nanocrystalline materials, nanowires, CNTs, quantum wells, quantum dots, and nanoparticles
Secondary ion mass spectrometry (SIMS). SIMS involves sputtering the surface of a sample with a focused primary ion beam and collecting and analyzing ejected secondary ions using a mass spectrometer	A few atomic layers of 1–3 nm	Elemental, isotopic, or molecular composition of the surface (can detect elements present in the parts-per-billion range)	Researchers have used SIMS to characterize various nanomaterials, such as CNTs

the case when these NPs are commercially available. A recent study showed how photocatalytic properties of TiO_2 NPs obtained by wet chemical reaction compare with those already commercially available [367]. This study also reported the effect of nitrogen doping/alloying on the photocatalyic properties of both sources of TiO_2.

There are very few published studies related to the reproducibility of NP synthesis and their impact on the device performance. A recent report on titania NP-based solar cells showed that different colloidal TiO_2 from different batches resulted in a deviation from the average power efficiency of at most 3.2% [368]. Similar studies should be conducted and published to allow a comparison between various recipes.

Current synthesis routes of NPs are hardly scalable [369]. Postsynthesis processing steps are required to remove by-products and other contaminations. Purification steps are costly, particularly those involving organic solvents. The precipitation–dissolution method for NP purification is often used in nonaqueous media. This approach is also difficult to scale. Indeed, significant amounts of time and materials are required. Dialysis [370], ultrafiltration [371], difiltration [372], and ternary solvent system [373] methods are also reported although with significant scale-up challenges. Recently, a low-cost purification technique has been reported using a reversible electrophoretic deposition (Fig. 1.25). A variety of NPs in nonaqueous media are purified using this technique within a few minutes with a small amount of solvent. This technique could be used in batch and continuous processes.

Quality control is even more important in the case of bioscience and medicine. Indeed, the interaction between biological systems and NPs will depend on purity, size, and shape. Standards certification of NP properties will certainly help in this regard [374]. Due to the proximity effect, the density of the NPs during handling and storage could affect the overall properties. A so-called auto-reduction has been observed in some core–shell structures [375]. It has been shown that contaminated gold NPs have a negative impact on the maturation and functionality of dendritic cells [376].

Contrary to the general belief within the scientific community, it may take more than 10 years to successfully mature a laboratory NP technology fabrication into successful commercial-scale manufacturing. In the case of a single-wall nanotube (SWNT), nearly two decades of development were needed to successfully start a commercial-scale plant [377]. In the late 1980s, the lab-scale production capacity of SWNT was less than 100 g/year. A start-up pilot plant capacity of 5 kg/year was achieved in 2000. In 2003, a production capacity of 5 tons/year was achieved with a demo plant [377].

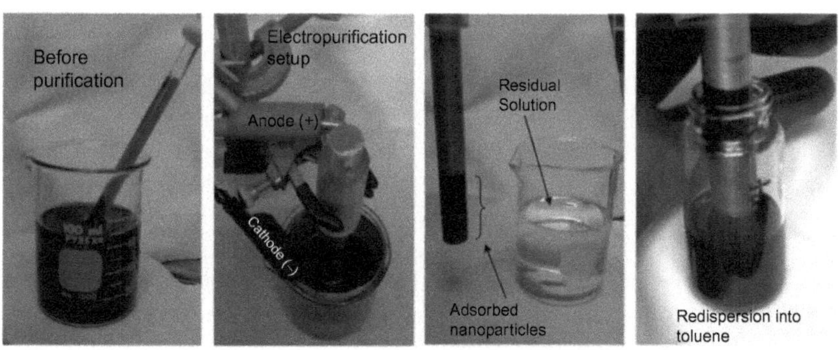

FIGURE 1.25 Pictures showing the main steps used in the electro-purification process. As-synthesized nanoparticle solution (CdTe) before, during, and after reversible deposition on the electrode is shown (from left to right). The last picture on the far right shows the redispersion of the nanoparticles into clean solvent [373]. (For color version of this figure, the reader is referred to the online version of this chapter.)

Today's commercial production capacity of SWNT is about an order of magnitude higher. Within this time span, unit production cost has been reduced by more than two orders of magnitude. Similar trends are expected for NP manufacturing.

NP handling and storage are critical, and require some specific procedures [378,379]. Time and history are important variables that could be important, particularly for relatively unstable NPs. Their high surface area and highly reactive surface are challenging. NPs are very sensitive to the environment [380]. Structural changes have been observed on ZnS NPs dispersed in methanol [381]. These changes are observed following methanol desorption.

5 VALUE CHAIN—FROM RAW MATERIALS TO MARKET

NPs have been used for decades and even centuries, although with very limited scope. Today, NPs with very wide compositions, shapes, and structures have been synthesized. In the recent past, several overhyped technologies failed to deliver their promises. Cold fusion and, to some extent, room temperature superconductivity are two examples of technologies that did not deliver their promises. In a few cases, NP technology proponents have also promised too much. However, it should be recognized that there are some successful commercial products based on NPs. Depending on the application, the states of commercial applications of NP-based products are quite different. Replacement of current top-down microelectronic manufacturing processes with bottom-up NP-based processes is still far from the commercialization stage. Indeed, it is still not possible to manufacture defect-free transistors using these bottom-up solutions. Successful products have already been commercialized in other sectors. Indeed, there are numerous NP-based catalysts in use in petrochemistry, fuel cell, and converter products. With the maturity of the chemical industry based on established commodity products, NPs offer numerous opportunities through specialty chemicals and alternative products and processes.

Understanding the value chain of any NP technology and product is critical. Scientists within academia and industry working in this field have a limited understanding of the breadth of this field and what it takes to bring laboratory findings to the commercial stage. Although patenting is an important step, it is by far not the most important one. Unless the goal is to file a patent and try to block others from entering the field, several years if not decades of development are required to take a protected invention to the market. Protecting one's own intellectual property (IP) is more important in order to have the freedom of use, particularly when one is challenged in courts.

In spite of the numerous breakthroughs, very few new NP-based technologies have found market success so far. There are numerous obstacles preventing these technologies from reaching their full potential. For example, financial support available for start-ups is often limited to venture capital

(VC) investors, who often prefer to invest in short-term and less risky technologies. Potential hazards combined with public skepticism, inadequate regulation, and ineffective patenting are other important obstacles.

5.1 Concept of Value Chain for NPs

Different concepts of the value chain have been developed over time. The value may depend on the commercial product. For some start-ups, NPs are the final product; for others, the final product is a device containing NPs. Figure 1.26 shows a high-level value chain of commercial products involving NPs. Often, NPs are integrated into intermediate products before being assembled into finished goods. These intermediate products include 3D materials, 2D coatings, wires, chips, and fabrics. For each specific product, a more detailed value chain can be constructed.

NP marketing and commercialization face two major hurdles. The first hurdle is related to the potential hazard of NPs that is often difficult to quantify (see Chapter 8). The second hurdle is related to the IP. Indeed, IP protection and enforcement of NP manufacturing, integration, and applications are quite challenging [382,383].

Whether they are sold as powder or embedded in a system, the commercialization process of NPs is similar to that of other advanced materials. Commercialization consists of two interrelated activities: (i) product and process development and (ii) marketing and distribution strategy. In large corporations, usually lawyers, scientists, and engineers work together to protect the IP, develop the product, and improve its development, respectively. This process could also involve external players such as original equipment manufacturers (OEMs), who could incorporate these NPs into a car component. For example, an OEM producing a component using a standard material could add these NPs to reduce weight and improve performance. An OEM may also work directly with a small start-up to incorporate these NPs into the standard component. The most glaring difficulty facing start-ups is the fact that usually it takes up to 10 years after the concept demonstration for them to start generating revenues. This is usually a NoGo for the majority of venture capitalists. Two alternatives are pursued. For example, the new product could be used first in a niche market (e.g., defense applications), which is the case when the performance of the new product is more important than the cost. The other avenue is to structure the product development into several phases

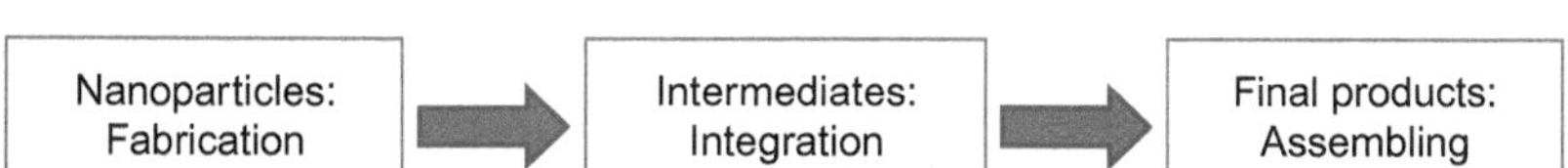

FIGURE 1.26 High level of the value chain of nanoparticle-based products. (For color version of this figure, the reader is referred to the online version of this chapter.)

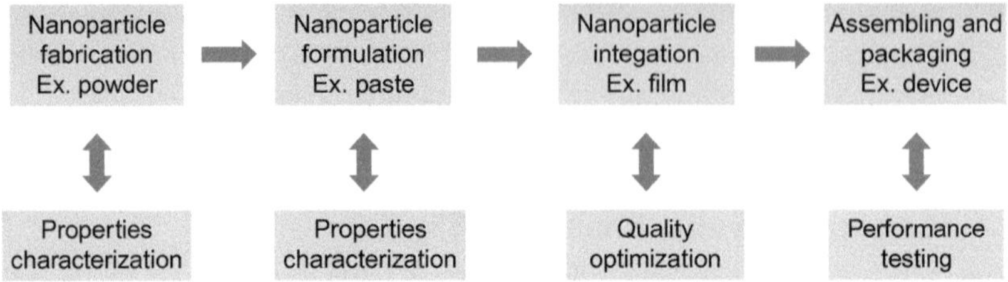

FIGURE 1.27 The three main steps of the nanoparticle value chain and the optimization process of their properties and the performance of the manufactured product. (For color version of this figure, the reader is referred to the online version of this chapter.)

with exit strategies at each phase. In this case, venture capitalists will invest in a specific development phase.

Market valuation of a new product is always challenging and difficult to quantify. It is important to differentiate between direct economic impact and global economic impact [384]. The market value of inorganic NPs has been valued at nearly US$1 billion for 2005 [248]. The overall market value of NPs as primary products is even higher. The global production of NPs for sunscreen products was estimated to be approximately 1000 tons in 2004 [385]. ZnO and TiO_2 NPs dominate this market. The sunscreen market valuation is obviously much higher than the market value of the NPs used in their fabrication.

As shown in Fig. 1.27, the value chain indicates that NPs are the building blocks in existing and new products [386]. The road toward commercialization of new materials and systems based on NPs is not an easy or straightforward one. There are three main issues that need to be addressed [387]:

a. Conditioning and compounding;
b. Forming and shaping;
c. De-bindering and sintering.

In terms of market value, NPs could be used as raw material (e.g., powder and solution), intermediate fabrication (e.g., film), products (e.g., fuel cell), and tools and equipments (e.g., AFM and ink-jet printer). In most cases, this value chain may include a means to synthesize the NPs, another means to fabricate film or consolidate material, and, in the end, product fabrication and packaging. For new markets such as application of gold NPs, the value chain could depend on the specific niche [388].

5.2 Market and Commercialization

The number of products on the market based on nanotechnologies has increased relatively linearly from 54 in 2005 to more than 1300 in 2010 [389]. NPs are the key components of the current nanotechnology products, and are expected to play a major role in this fast-growing sector. Present in

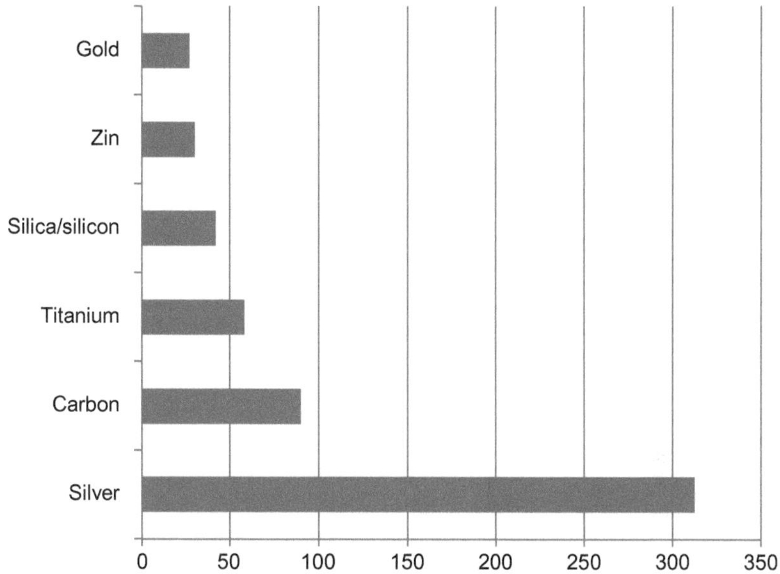

FIGURE 1.28 Number of products for each material. (For color version of this figure, the reader is referred to the online version of this chapter.) *Based on data from Ref. [389].*

more than 300 products, silver NPs are the most widely used materials. Other NP materials include carbon, titanium, silica, silver, and gold (Fig. 1.28).

Two marketing approaches are considered depending on the type of NP technology. In the case of existing markets and customers, a validation study is required to confirm the value proposition for using NPs instead of conventional materials. This is the case when using high surface area NPs in catalysis. In the case of a new market, an exploratory marketing study is undertaken. In this case, the value proposition is relatively fuzzy, as the market and the customers are not well identified. This is the case, for example, of NP-based photovoltaic and cancer treatment solutions. Often exploratory market studies start by interviewing few potential customers.

Commercialization of new NP-based products is quite challenging. Most of these new technologies arise from curiosity research within academia. As such, they are often considered as technology-push products. The production process is not optimized to provide completive cost of pure and high-quality NPs. The lack of quality control, public resistance and approval delays, lack of long-term VC funding, and reluctance of large corporations are a few of the other challenges facing new NP-based products.

As discussed earlier, the biggest challenge for new NP technologies is the lengthy lab-to-market timeline. With expenditure outweighing income, developers and investors cannot afford several years of development before reaching the market. Thus, reducing time-to-market should be the main focus of new start-ups in this field. However, in some cases, it is difficult to reduce the development stage beyond a certain limit. Thus, some start-ups explore alternative income. This is possible, for example, by providing products and services that are not related directly to the final product.

Several coating and material solutions based on NP technologies have already seen some real-world commercial applications. Most biomedical and energy solutions using NPs are still in the conceptual lab development stages. The commercialization of any product requires several steps. Developed by the NASA, the technology readiness level (TRL) concept applies to the NP technologies [390]. The nine TRL stages could be grouped into five major development phases [390] in the case of NPs:

a. Basic technology research;
b. Feasibility;
c. Technology development;
d. Technology demonstration;
e. System development, testing, and launch.

Typically, the first two phases are carried out in academia and the last three within a start-up or established corporation. Thus, one should expect a lab-to-market timeline of 5–10 years. Additional delay could be incurred in the case of health-related products.

The price of NPs may depend on several factors. Often particles of smaller size distribution are more expensive. Furthermore, the manufacturer of these NPs must provide assurance to the customers that it can respond to large orders with the same quality. This last issue could be addressed by first demonstrating the technology at a pilot plant production scale. In the case of aluminum, about 200,000 tons/year are sold for metallurgical, chemical, and paint and pigment industries, of which only 100 tons consist of submicron powder [391]. A differential in market price of more than an order of magnitude has been reported between micron and submicron Al powder (Fig. 1.29). A reduced demand for submicron powder will keep the price relatively high. However, there are some niche markets such as high covering pigments for uniform coatings. Currently, gas-atomizing technology allows higher yields of of submicron materials. However, this technique is not very efficient.

The market for nanomaterials has been predicted to reach US$14 billion in 2008 in the USA [392]. The market share of NPs is relatively small. However, it is predicted to increase at a relatively faster pace with the commercialization of technologies currently evaluated at the pilot plant stage. The market size of the silver NP is expected to increase from about 0.2 tons/year in 2005 to more about 6 tons/year in 2010 [393]. The price will be reduced from $100/kg down to less than $70/kg. Besides their applications for lustrous and conductive patterns, silver NPs have strong antimicrobial activity. Silver compounds are also used in dental amalgam, photography, catalysts, window coatings, and platings. Similar trends are expected in the case of metal and metal oxide NPs.

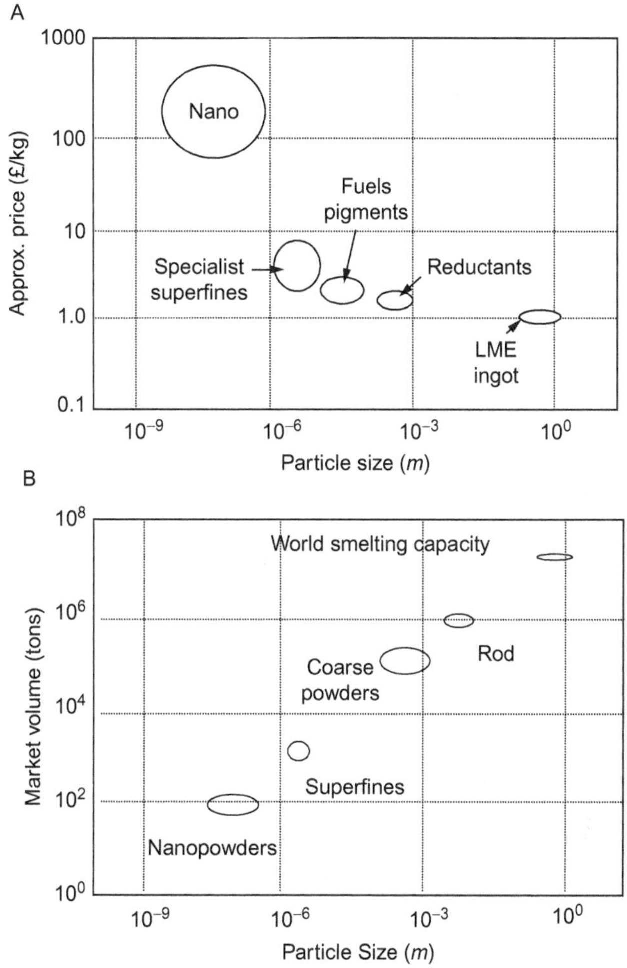

FIGURE 1.29 (A) Price and (B) market size of aluminum powder with different size range [391].

Silica, platinum, and carbon black submicron particles have been used as additives in commercial products for decades [394]. High performance grades of TiO_2 are expected to find a growing market in the cosmetics, sunscreen, toothpaste, and photocatalyst industries. As shown with the introduction of TiO_2 NPs into sunscreen, small size should not be the driver for marketing of new NPs. Indeed, improved performance-to-cost ratio when compared to conventional materials should be the key factor. Currently, the TiO_2 NP market is estimated at 50,400 tons, representing 0.7% of the overall TiO_2 market [119,395]. In the case of ceria and nickel NPs, market potentials are estimated to be €30 billion and €11 billion, respectively [396].

It is sometimes important to showcase the real economic advantage of NPs. In the case of zero-valent iron technology for remediation, the cost of NPs versus granular particles should be based on the cost relative to the surface area [397]. Indeed, NPs are expected to cost 50 times cheaper on the basis of surface area.

5.3 Intellectual Property

IP is part of the intellectual asset of a company. The intellectual asset is part of the intellectual capital (knowledge) that has been identified, documented, and is available for use for the benefit of the organization. IP could be in the form of a patent, trademark, copyright, industrial design, know-how, or confidential information.

Patents are the most recognized IP for the general public. They have received attention the most. Patents are not only a measure of R&D investment and output, but also an indirect measure of the current and potential market. The patent activity of any company in a specific domain could be related to future technological and commercial interest, and also provide some information about the company's strategy. To obtain a patent, the invention must be useful, new, and nonobvious. A patent is often granted by a governmental organization following an application. Patent rights are violated when products and services embodying the claims of the patents are used or sold by persons or organizations other than the applicant. Typically, a patent contains a written description and the claims. In the written description, field, background, summary, and a detailed description of the invention are provided. The detailed description should allow any skilled person in the pertaining art to make and use the invention. The scope and limitations of any patent are defined in the claims.

Patent claims, which are comprised of elements (or limitations), determine the scope of the patents. In the specific case of NPs, the claims could be based on the chemical compositions (e.g., CdSe NPs), product (solar cell), and process (e.g., microwave synthesis). Recently, a new category of claims, referred as product-by-process, has been used. In this case, the product features are limited by the process. This is particularly the case of NPs whose characteristics are affected by the fabrication and integration processes.

As described in the introduction of this chapter, there are several dozens of words that are often used to designate NPs with some nuances of difference. If the applicant does not use the appropriate word in the claim, he may reduce or even lose his rights.

5.4 Life Cycle Analysis

Base and coating materials play a critical role in the overall technology–economy–environment scheme. About 15 billion tons of raw materials are extracted from the earth every year. There is an increasing demand with increasing cost for raw materials. The environmental impact for extraction, refining, manufacturing, and disposal of the final products are important. A greater awareness of more efficient processes can be gained by carefully studying the LCA of a product from extraction and processing to usage and disposal (Fig. 1.30). LCA is a quantitative tool for balancing the economic prosperity and environmental impacts from the development of a new technology. In particular, careful analysis of inputs (energy and materials) and outputs (products, by-products, and waste) is important. This will help, for example, to identify the best material and the process steps. However, there are two major challenges. First, a system boundary definition is required. Second, data inputs (life cycle inventory) for each process step are needed. Unfortunately, there are very few data available for energy requirements in NP-based product processes. This is particularly the case with NP manufacturing and integration.

Besides the economic impact, there are also long-term consequences related to the ecosystem to be considered. Each step of the process gives rise to wastes that pollute the air we breathe and the water we drink. With finite resources, effective utilization of raw materials and fossil energy is very important. The fabrication of materials consumes about 50% of the energy used in the manufacturing industries. Comparing current and alternative requirements for each cycle stage is thus critically important. LCA data of NP-based products should not be analyzed on absolute value. Comparison with conventional processes with a well-defined system boundary is the best approach.

Product disposal including recycling is also important to reduce raw material extraction. Furthermore, proper disposal reduces the potential environmental impact when it is land-filled. Materials have different degrees of recyclability and biodegradability. Event NPs are expected to have differences depending on their composition and structure. For example, some metals such

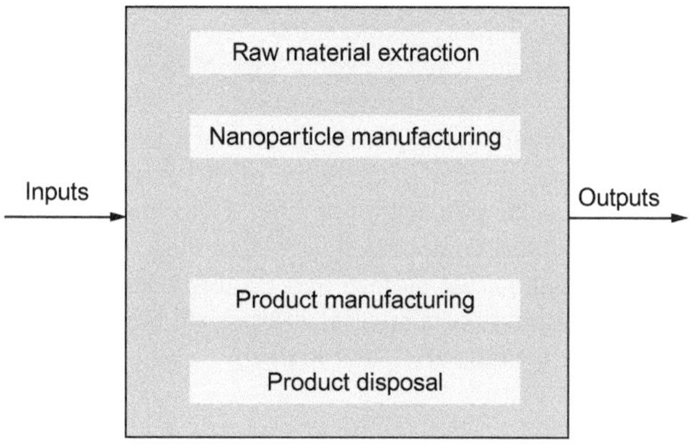

FIGURE 1.30 Simplified view of system boundary and main process steps of nanoparticle-based product life cycle analysis (LCA). (For color version of this figure, the reader is referred to the online version of this chapter.)

as iron are easier to biodegrade because of their higher corrosion rate. These metal NPs could be easier to recycle given the higher surface-to-bulk ratio. However, other metals such as Pb are toxic and thus should be recycled at the end of their shelf life. There are some additional complications associated with alloyed metals and composite materials.

6 IMPLICATIONS OF WIDER USE OF NPs

There are a wide variety of NP-based products and technologies. Some provide just a continuation in the improvement of existing technologies. Other applications involve new products and markets. As a new general-purpose technology [398], NPs will not only generate new growth but will also give rise to structural changes. On average, past technology revolution cycles have lasted for around 60 years. This has been the case in the information and communications technology, oil and automobile, steel/electricity and engineering, steam and railways, and the industrial revolution. It is expected that NP technology will require another 5–30 years for its devolvement and another 5–30 years for its deployment.

Like any other new technology, NP-based technologies will inevitably present risks. This could result from unintentional consequences of beneficial applications. Although NPs have been synthesized and used for more than a century, only recently a more systematic evaluation of the health risks has been undertaken. However, the discussion has been carried out in public space involving little scientific data. Thus, the biggest challenge is to develop regulatory policies in the context of the limited data including information and knowledge of risks.

Understanding NP characteristics and their relationship to their functional properties is critical. Several parameters may contribute to the toxicity of NPs [399]. They include specific surface area, number of particles, size and granulometric distribution, concentration, chemical composition (purities and impurities), surface properties, Zeta charge/potential, reactivity, functional groupings, presence of metal redox potential, potential to generate free radicals, and surface coverage. Other potential factors include solubility, shape, porosity, degree of agglomeration/aggregation, biopersistence, crystalline structure, hydrophilicity/hydrophobicity, pulmonary deposition site, age of particles, and producer, process, and source of the material used. The same factors also drive the value proposition of most NP technologies and products. Thus, besides the chemical composition, size, and shape of NPs, their state of agglomeration/aggregation, the surface and coating composition, the dispersion media, and other physicochemical properties are the main characteristics that contribute to the overall safety and toxicity.

REFERENCES

[1] W.A. Herrmann, Chem. Eng. Technol. 21 (1998) 549–558.
[2] M.A. Kastner, Phys. Today 46 (1993) 24–31.

[3] Chemical Industry R&D Roadmap for Nanomaterials by Design: From Fundamentals to Function; Chemical Industry Vision 2020 Technology Partnership, 2003.
[4] H.N. Holmes, Laboratory Manual of Colloid Chemistry, third ed, John Wiley & Sons, London, 1934.
[5] W. Ostwald, An introduction to theoretical and applied colloid chemistry: the world of neglected dimensions; authorized translation from the 8th German edition by M.H. Fischer, Wiley, New York, 1922.
[6] C.N.R. Rao, P.J. Thomas, G.U. Kulkarni, Nanocrystals: Synthesis, Properties and Application, Springer, Berlin, 2007, pp. 1–23.
[7] F.E. Wagner, S. Haslbeck, L. Stievano, S. Calogero, Q.A. Pankhurst, K.-P. Martinek, Nature 407 (2000) 691–692.
[8] A.J. Haes, D.A. Stuart, Sh. Nie, R.P. Van Duyne, J. Fluoresc. 14 (2004) 355–367.
[9] M. Faraday, Philos. Trans. R. Soc. Lond. 147 (1857) 145–181.
[10] B.V. Enustun, J. Turkevich, J. Am. Chem. Soc. 85 (1963) 3317–3328.
[11] V. Haensel, B.T. Brooks, C.E. Boord, S.S. Kurtz, L. Schmerling (Eds.), The Chemistry of Petroleum Hydrocarbons, vol. II, Reinhold, New York, 1955, pp. 189–219.
[12] F.G. Ciapetta, R.M. Dobres, R.W. Baker, P.H. Emmett (Ed.), Catalysis, vol. 6, Reinhold, New York, 1958, pp. 495–692.
[13] J.H. Sinfelt, Surf. Sci. 5000 (2002) 923–946.
[14] M. Zach, C. Hagglund, D. Chakarov, B. Kasemo, Curr. Opin. Solid State Mater. Sci. 10 (2006) 132–143.
[15] A.D. Yoffe, Adv. Phys. 42 (1993) 173–262 and reference therein.
[16] L.E. Brus, J. Chem. Phys. 80 (1984) 4403–4409 and reference therein.
[17] A.P. Alivisatos, J. Phys. Chem. 100 (1996) 13226–13239 and references therein.
[18] C.B. Murray, D.B. Norris, M.G. Bawendi, J. Am. Chem. Soc. 115 (1993) 8706–8715.
[19] K.J. Klabunde, Free Atoms, Clusters, and Nanoscale Particles, Academic Press, San Diego, 1994.
[20] C. Hayashi, R. Uyeda, A. Tasaki (Eds.), Ultra-Fine Particles: Exploratory Science and Technology, Noyes Publications, Westwood, NJ, 1997.
[21] T. Sugimoto (Ed.), Fine Particles: Synthesis, Characterization, and Mechanisms of Growth, Marcel Dekker, New York, 2000.
[22] D.L. Feldheim, C.A. Foss Jr. (Eds.), Metal Nanoparticles: Synthesis, Characterization and Applications, Marcel Dekker, New York, 2002.
[23] M.I. Baraton, Synthesis, Functionalization and Surface Treatment of Nanoparticles, American Scientific Publishers, Stevenson Ranch, CA, 2003.
[24] L.M. Liz-Marzán, P.V. Kamat (Eds.), Nanoscale Materials, Kluwer Academic Publishers, Boston, 2003.
[25] Y. Waseda, A. Muramatsu (Eds.), Morphology Control of Materials and Nanoparticles: Advanced Materials Processing and Characterization, Springer, Berlin, 2004.
[26] G. Cao, Nanostructures & Nanomaterials: Synthesis, Properties & Applications, Imperial College Press, London, 2004.
[27] H.S. Nalwa (Ed.), Encyclopedia of Nanoscience and Nanotechnology, American Scientific Publishers, Stevenson Ranch, CA, 2004.
[28] C.N.R. Rao, A. Muller, A.K. Cheetham (Eds.), The Chemistry of Nanomaterials: Synthesis, Properties and Applications, Wiley-VCH, Weinheim, 2004.
[29] G. Schmid (Ed.), Nanoparticles: From Theory to Application, Wiley-VCH, Weinheim, 2004.
[30] F. Caruso (Ed.), Colloids and Colloid Assemblies, Wiley-VCH, Weinheim, 2004.
[31] M.-P. Pileni (Ed.), Nanocrystals Forming Mesoscopic Structures, Wiley-VCH, Weinheim, 2005.

[32] K. Nogi, M. Naito, T. Yokoyama (Eds.), Nanoparticle Technology Handbook, Elsevier, Amsterdam, 2007.
[33] US National Nanotechnology Initiative, The National Nanotechnology Strategic Plan. http://www.nano.gov/html/facts/whatIsNano.html, 2006.
[34] I. Gheorghe, H.J. Rack, A. Kelly, C. Zweben (Eds.), Comprehensive Composite Materials, Elsevier, Amsterdam, 2000 (Chapter 3.25).
[35] B. Lim, M. Jiang, J. Tao, P.H.C. Camargo, Y. Zhu, Y. Xia, Adv. Funct. Mater. 19 (2009) 189–200.
[36] D. Jerome, H.J. Schutz, Adv. Phys. 31 (1982) 299–490.
[37] J. Simon, J.-J. André, Molecular Semiconductors, Springer, Berlin, 1985.
[38] J.S. Brooks, Adv. Mater. Opt. Electron. 8 (1998) 269–276.
[39] S.C. Lock, Aerosols in Theory and Practice, Aerosols Ltd., London, 1947.
[40] V.A. Hackley, C.F. Ferraris, The Use of Nomenclature in Dispersion Science and Technology, NIST Special, Publication 960-3, 2001.
[41] K.J. Klabunde, Nanoscale Materials in Chemistry, Wiley & Sons, New York, 2001.
[42] W.H. Hunt, JOM 56 (2004) 13–18.
[43] H. Fukumori, H. Ichikawa, Adv. Powder Technol. 17 (2006) 1–28.
[44] W. Luther, A. Hoffknecht, D. Holtmanspötter, V. Wagner, N. Malanowski, T. Heimer, W. Luther, M. Werner (Eds.), Growth Market Nanotechnology: An Analysis of Technology and Innovation, Wiley-VCH, Weinheim, 2006, p. 32.
[45] C. Syryanarayana, Int. Mater. Rev. 40 (1995) 41–64.
[46] R. Birringer, Mater. Sci. Eng. 7 (1989) 33–43.
[47] J.C. Maxwell-Garnet, Philos. Trans. R. Soc. 203 (1904) 385–420.
[48] J. Turkevich, P.C. Stevenson, J.A. Hillier, Discuss. Faraday Soc. 11 (1951) 55–75.
[49] F. Bensebaa, N. Patrito, Y. LePage, P. L'Ecuyer, D. Wang, J. Mater. Chem. 14 (2004) 3378–3384.
[50] I. Capek, Adv. Colloid Interface Sci. 110 (2004) 49–74.
[51] M.C. Daniel, D. Astruc, Chem. Rev. 104 (2004) 293–346.
[52] R.A. Sperling, P.R. Gil, F. Zhang, M. Zanella, W.J. Parak, Chem. Soc. Rev. 37 (2008) 1896–1908.
[53] M. Brust, M. Walker, D. Bethell, D.J. Schiffrin, R. Whyman, J. Chem. Soc. Chem. Commun. (1994) 801–802.
[54] T. Shimizu, T. Teranishi, S. Hasegawa, M. Miyake, J. Phys. Chem. B 107 (2003) 2719–2724.
[55] N. Toshima, T. Yonezawa, New J. Chem. 11 (1998) 1179–1201.
[56] S. Sun, Adv. Mater. (2006) 1–11.
[57] N.N. Kariuki, J. Luo, M.M. Maye, S.A. Hassan, T. Menard, H.R. Naslund, Y. Lin, C. Wang, M.H. Engelhard, C.-J. Zhong, Langmuir 20 (2004) 11240–11246.
[58] H. Gong, M. Liu, Chem. Mater. 14 (2002) 4933–4938.
[59] Y.P. Sun, X. Li, J. Cao, W. Zhang, H.P. Wang, Adv. Colloid Interface Sci. 120 (2006) 47–56.
[60] C.B. Murray, S. Sun, W. Gaschler, H. Doyle, T. Betley, C.R. Kagan, IBM J. Res. Dev. 45 (2001) 47–56.
[61] I. Lisiecki, J. Phys. Chem. B 109 (2005) 12231–12244.
[62] E.V. Shevchenko, D.V. Talapin, H. Schnablegger, A. Kornowski, O. Festin, P. Svedlindh, M. Haase, H. Weller, J. Am. Chem. Soc. 125 (2003) 9090–9101.
[63] D.I. Chepic, A.L. Efros, Al.I. Ekimov, M.G. Ivanov, V.A. Kharchenko, I.A. Kudriavtsev, T.V. Yazeva, J. Lumin. 47 (1990) 113–127.

[64] S.P. Gubin, N.A. Kataeva, G.B. Khomutov, Russ. Chem. Bull. Int. Ed. 54 (2005) 827–852.
[65] M.G. Panthani, V. Akhavan, B. Goodfellow, J.P. Schmidtke, L. Dunn, A. Dodabalapur, P.F. Barbara, B.A. Korgel, J. Am. Chem. Soc. 130 (2008) 16770–16777.
[66] A.L. Stroyuk, A.I. Kryukov, S.Y. Kuchmii, V.D. Pokhodenko, Theor. Exp. Chem. 41 (2005) 67–91 and reference therein.
[67] J. Bukowski, H. Simmons, Crit. Rev. Solid State Mater. Sci. 27 (2002) 119–142.
[68] F. Bensebaa, C. Durand, A. Aouadou, L. Scoles, X. Du, D. Wang, Y. Le Page, J. Nanopart. Res. 12 (2010) 1897–1903.
[69] S.J. Ahn, K.H. Kim, K.H. Yoon, Colloids Surf. A Physicochem. Eng. Asp. 313–314 (2008) 171–174.
[70] C.A. Leatherdale, W.-K. Woo, F.V. Mikulec, M.G. Bawendi, J. Phys. Chem. B 106 (2002) 7619–7622.
[71] Y. Kanemitsua, A. Ishizumi, J. Lumin. 119–120 (2006) 161–166.
[72] A. Henglein, Chem. Rev. 89 (1989) 1861–1873.
[73] T. Trindade, P. O'Brien, N.L. Pickett, Chem. Mater. 13 (2001) 3843–3858.
[74] T.J. Bulowski, J.H. Simmons, Crit. Rev. Solid State Mater. Sci. 27 (2002) 119–142.
[75] H. Chandler, Mater. Sci. Eng. R 49 (2005) 113–155.
[76] A.L. Rogach, D.V. Talapin, H. Weller, F. Caruso (Ed.), Colloids and Colloid Assemblies: Synthesis, Modification, Organization and Utilization of Colloid Particles, Wiley-VCH, Weinheim, 2004, pp. 52–95.
[77] Y. Liu, W. Chen, A.G. Joly, Y. Wang, C. Pope, Y. Zhang, J.-O. Bovin, P. Sherwood, J. Phys. Chem. B 110 (2006) 16992–17000.
[78] Y. Fu, M. Willander, A. Dutta, S. Oda, Superlatt. Microstruc. 28 (2000) 177–187.
[79] K. Nishiguchi, S. Oda, J. Appl. Phys. 88 (2000) 4186–9410.
[80] S. Banerjee, S. Huang, T. Yamanaka, S. Oda, J. Vac. Sci. Technol. B 20 (2002) 1135–1138.
[81] M.L. Ostraat, J.W. De Blauwe, M.L. Green, L.D. Bell, M.L. Brongersma, J. Casperson, R.C. Flagan, H.A. Atwater, Appl. Phys. Lett. 79 (2001) 433–435.
[82] N.N. Ledentsov, Semicond. Sci. Technol. 26 (2011) 1–8.
[83] N.M. Park, T.S. Kim, S.J. Park, Appl. Phys. Lett. 78 (2001) 2575–2577.
[84] L. Canham, Nature 408 (2000) 411–412.
[85] M. Nayfeh, O. Akcakir, J. Therrien, Z. Yamani, N. Barry, W. Yu, E. Gratton, Appl. Phys. Lett. 75 (1999) 4112–4114.
[86] J.P. Wilcoxon, G.A. Samara, P.N. Provencio, Phys. Rev. B 60 (1999) 2704–2714.
[87] Z. Ding, B.M. Quinn, S.K. Haram, L.E. Pell, B.A. Korgel, A.J. Bard, Science 296 (2002) 1293–1297.
[88] K.A. Pettigrew, Q. Liu, P.P. Power, S.M. Kauzlarich, Chem. Mater. 15 (2003) 4005–4011.
[89] R.K. Baldwin, K.A. Pettigrew, J.C. Garno, P.P. Power, G.-Y. Liu, S.M. Kauzlarich, J. Am. Chem. Soc. 124 (2002) 1150–1151.
[90] A. Bapat, C. Anderson, Ch.R. Perrey, C.B. Carter, S.A. Campbell, U. Kortshagen, Plasma Phys. Control. Fusion 46 (2004) B97–B109.
[91] L. Balan, R. Schneider, D. Billaud, J. Ghanbaja, Nanotechnology 16 (2005) 1153–1158.
[92] D. Jurbergs, E. Rogojina, L. Mangolini, U. Kortshagen, Appl. Phys. Lett. 88 (2006) 233116–233118.
[93] S. Sato, T. Ikeda, K. Hamada, K. Kimura, Solid State Commun. 149 (2009) 862–865.
[94] N. Gaponik, S.G. Hickey, D. Dorfs, A.L. Rogach, A. Eychmuller, Small 6 (2010) 1364–1378.
[95] V.L. Colvin, A.P. Alivisatos, J.G. Tobin, Phys. Rev. Lett. 66 (1991) 2786–2789.
[96] A.N. Goldstein, C.M. Echer, A.P. Alivisatos, Science 256 (1992) 1425–1427.

[97] D.W. Deng, J.-S. Yu, Y. Pan, J. Colloid Interface Sci. 299 (2006) 225–232.
[98] Z.A. Peng, X. Peng, J. Am. Chem. Soc. 123 (2001) 183–184.
[99] L. Qu, Z.A. Peng, X. Peng, Nano Lett. 1 (2001) 333–337.
[100] W.W. Yu, Y.A. Wang, X. Peng, Chem. Mater. 15 (2003) 4300–4308.
[101] R.E. Bailey, S. Nie, J. Am. Chem. Soc. 125 (2003) 7100–7106.
[102] R. Beaulac, P.I. Archer, S.T. Ochsenbein, D.R. Gamelin, Adv. Funct. Mater. 18 (2008) 3873–3891.
[103] S. Cai, T. Tsuzuki, T.A. Fisher, B.D. Nener, J.M. Dell, P.G. McCormick, J. Nanopart. Res. 4 (2002) 367–371.
[104] L. Al Juhaiman, L. Scoles, D. Kingston, B. Patarachao, D. Wang, F. Bensebaa, Green Chem. 12 (2010) 1248–1252.
[105] R.P. Raffaelle, S.L. Castro, A.F. Hepp, S.G. Bailey, Prog. Photovolt. Res. Appl. 10 (2002) 433–439.
[106] S. Bereznev, I. Konovalov, A. Opik, J. Kois, Synth. Metals 152 (2005) 81–84.
[107] M. Azad-Malik, P. O'Brien, N. Revaprasadu, Adv. Mater. 11 (1999) 1441–1444.
[108] H. Nakamura, W. Kato, M. Uehara, K. Nose, T. Omata, S. Otsuka-Yao-Matsuo, M. Miyazaki, H. Maeda, Chem. Mater. 18 (2006) 3330–3335.
[109] D.P. Dutta, G. Sharma, Mater. Lett. 60 (2006) 2395–2398.
[110] F. Bensebaa, N. Patrito, Y. Le Page, P. L'Ecuyer, Da. Wang, J. Mater. Chem. 14 (2005) 3378–3384.
[111] S.J. Roh, R.S. Mane, H.M. Pathan, O.-S. Joo, S.-H. Han, Appl. Surf. Sci. 252 (2005) 1981–1987.
[112] V.S. Gurin, Colloids Sci. 142 (1998) 35–40.
[113] S.L. Castro, S.G. Bailey, R.P. Raffaelle, K.K. Banger, A.F. Hepp, J. Phys. Chem. B 108 (2004) 12429–12435.
[114] W. Stober, A. Fink, E. Bohm, J. Colloid Interface Sci. 26 (1968) 62–69.
[115] B.L. Arams, J.P. Wilcoxon, Crit. Rev. Solid State Mater. Sci. 30 (2005) 153–182.
[116] Z.-G. Yan, C.-H. Yan, J. Mater. Chem. 18 (2008) 5046–5059.
[117] G. Wakefield, H.A. Keron, P.J. Dobson, J.L. Hutchison, J. Colloid Interface Sci. 215 (1999) 179–182.
[118] T. Athar, W. Ahmed, M.J. Jackson (Eds.), Emerging Nanotechnologies for Manufacturing, Elsevier, United Kingdom, 2009 (Chapter 14).
[119] www.salisonline.org/market-research/the-world-market-for-nanoparticle-titanium-dioxide/ (accessed 09.08.11).
[120] K.K. Sahu, T.C. Alex, D. Mishra, A. Agrawal, Waste Manag. Res. 24 (2006) 74–79.
[121] D.J. Fray, Int. Mater. Rev. 53 (2008) 317–325.
[122] Q. Chen, L.M. Peng, Int. J. Nanotechnol. 4 (2007) 44–65.
[123] S.A. Simakov, Y. Tsur, J. Nanopart. Res. 9 (2007) 403–417.
[124] L. Tosheva, V.P. Valtchev, Chem. Mater. 17 (2005) 2494–2513.
[125] V. Nikolakis, Curr. Opin. Colloid Interface Sci. 10 (2005) 203–210.
[126] U. Jeong, X. Teng, Y. Wang, H. Yang, Y. Xia, Adv. Mater. 19 (2007) 33–60.
[127] A.K. Gupta, M. Gupta, Biomaterials 26 (2005) 3995–4021.
[128] S.D. Tiwari, K.P. Rajeev, Thin Solid Films 505 (2006) 113–117.
[129] R. Muller, R. Hergt, M. Zeisberger, W. Gawalek, J. Magn. Magn. Mater. 289 (2005) 13–16.
[130] S. Yu, C.-J. Sun, G.-M. Chow, C.C. Koch (Ed.), Nanostructured Materials: Processing, Properties, and Applications, William Andrew Publishing, Norwich, USA, 2006, pp. 3–46.
[131] C.-Y. Lee, N.-H. Tai, S.-H. Hsieh, J. Nanopart. Res. 8 (2006) 287–292.
[132] C. Feldmann, Adv. Funct. Mater. 13 (2003) 101–107.

[133] F. Bensebaa, F. Zavaliche, P. L'Ecuyer, R.W. Cochrane, T. Veres, J. Colloid Interface Sci. 277 (2004) 104–110.
[134] F. Bensebaa, unpublished data.
[135] K. Landfester, Top. Curr. Chem. 227 (2003) 75–123.
[136] E. Allemanna, J.-C. Leroux, R. Gurny, Adv. Drug Deliv. Rev. 34 (1998) 171–189.
[137] G. Zhang, A. Niu, S. Peng, M. Jiang, Y. Tu, M. Li, C. Wu, Acc. Chem. Res. 34 (2001) 249–256.
[138] M.D. Barnes, A. Mehta, P. Kumar, B.G. Sumpter, D.W. Noid, J. Polym. Sci. B Polym. Phys. 43 (2005) 1571–1590.
[139] W. Zhang, J. Gao, C. Wu, Macromolecules 30 (1997) 6388–6390.
[140] C. Szymanski, C. Wu, J. Hooper, M.A. Salazar, A. Perdomo, A. Dukes, J. McNeill, J. Phys. Chem. 109 (2005) 8543–8546.
[141] J. Jang, J.H. Oh, G.D. Stucky, Angew. Chem. 114 (2002) 4188–4191.
[142] M.E. Trevino-Martınez, Y. del Angel-Vargas, L.F. Ramos-de Valle, R.G. Lopez-Campos, J. Vinyl Additive Technol. 11 (2005) 132–134.
[143] M.J. Meziani, P. Pathak, T. Desai, Y.-P. Sun, Ind. Eng. Chem. Res. 45 (2006) 3420–3424.
[144] J.P. Salvage, S.F. Rosea, G.J. Phillipsa, G.W. Hanlon, A.W. Lloyd, I.Y. Ma, S.P. Armes, N.C. Billingham, A.L. Lewis, J. Controll. Release 104 (2005) 259–270.
[145] S.A. Agnihotri, N.N. Mallikarjuna, T.M. Aminabhavi, J. Controll. Release 100 (2004) 5–28.
[146] D. Horn, J. Rieger, Angew. Chem. Int. Ed. Engl. 40 (2001) 4330–4361.
[147] T. Sugiyama, T. Asahi, H. Masuhara, Chem. Lett. 33 (2004) 724–725.
[148] A. Ibanez, S. Maximov, A. Guin, C. Chaillout, P.L. Baldeck, Adv. Mater. 10 (1998) 1540–1543.
[149] H. Kasai, H.S. Nalwa, H. Oikawa, S. Okada, H. Matsuda, N. Minami, A. Kakuta, K. Ono, A. Mukoh, H. Nakanishi, Jpn. J. Appl. Phys. Part 2 31 (1992) L1132–L1134.
[150] R. Jagannathan, G. Irvin, T. Blanton, S. Jagannathan, Adv. Funct. Mater. 16 (2006) 747–753.
[151] J.D. Fortner, D.Y. Lyon, C.M. Sayes, A.M. Boyd, J.C. Falkner, E.M. Hotze, L.B. Alemany, Y.J. Tao, W. Guo, K.D. Ausman, V.L. Colvin, A.B. Hughes, Environ. Sci. Technol. 39 (2005) 4307–4316.
[152] M.H.J. Hagen, E.J. Meijer, G. Mooij, D. Frenkel, H.N.W. Lekkerkerker, Nature 365 (1993) 425–426.
[153] C. Rey, L.J. Gallego, Phys. Rev. E 53 (1996) 2480–2487.
[154] J. Brant, H. Lecoanet, M.R. Wiesner, J. Nanopart. Res. 7 (2005) 545–553.
[155] H. Zhang, C. Wu, L. Liang, Y. Chen, Y. He, Y. Zhu, N. Ke, J.B. Xu, S.P. Wong, A. Wei, S. Peng, J. Phys. Condens. Matter 13 (2001) 2883–2889.
[156] P.J.A. Borm, D. Robbins, S. Haubold, T. Kuhlbusch, H. Fissan, K. Donaldson, R. Schins, V. Stone, W. Kreyling, J. Lademann, J. Krutmann, D. Warheit, E. Oberdorster, Part. Fibre Toxicol. 3 (2006) 1–35.
[157] J. Joutsensaari, P. Ahonen, U. Tapper, E.I. Kauppinen, J. Laurila, V.-T. Kuokkala, Synth. Met. 77 (1996) 85–88.
[158] Q. Li, T.Y. Ohulchanskyy, R. Liu, K. Koynov, D. Wu, A. Best, R. Kumar, A. Bonoiu, P.N. Prasad, J. Phys. Chem. C 114 (2010) 12062–12068.
[159] R. Jagannathan, G.C. Irbin Jr., Adv. Funct. Mater. 15 (2005) 1501–1510.
[160] L. Wang, L. Dong, G. Bian, T. Xia, H. Chen, L. Wang, Q. Cao, L. Li, Anal. Lett. 37 (2004) 1811–1822.
[161] D. Yang, B. Armitage, S.R. Marder, Angew. Chem. Int. Ed. Engl. 43 (2004) 4402–4409.

[162] J.R. Groza, Carl C. Koch (Ed.), Nanostructured Materials, Processing, Properties and Applications, second ed, William Andrew Publishing, Norwich, USA, 2007, pp. 173–234.
[163] V. Petkov, Y. Peng, G. Williams, B. Huang, D. Tomalia, Y. Ren, Phys. Rev. B 72 (2005) 195402–195409.
[164] R. Ferrando, G. Rossi, F. Nita, G. Barcaro, A. Fortunelli, ACS Nano 2 (2008) 1849–1856.
[165] Y.I. Petrov, E.A. Shafranovsky, J. Nanopart. Res. 3 (2001) 419–432.
[166] E. Roduner, Chem. Soc. Rev. 35 (2006) 583–592.
[167] R. Kubo, J. Phys. Soc. Jpn. 17 (1962) 975–986.
[168] F.D. Fischer, T. Waitz, D. Vollath, N.K. Simha, Prog. Mater. Sci. 53 (2008) 481–527.
[169] Y. Min, M. Akbulut, K. Kristiansen, Y. Golan, J. Israelachvili, Nat. Mater. 7 (2008) 527–538.
[170] M. Haruta, CATECH 6 (2002) 102–115.
[171] J. Wilsdon, IEEE Technol. Soc. Mag. (2004) 16–21.
[172] M. Wautelet, Eur. J. Phys. 22 (2001) 601–611.
[173] D.R. Baer, J.E. Amonette, M.H. Engelhard, D.J. Gaspar, A.S. Karakoti, S. Kuchibhatla, P. Nachimuthu, J.T. Nurmi, Y. Qiang, V. Sarathy, S. Seal, A. Sharma, P.G. Tratnyek, C.-M. Wang, Surf. Interface Anal. 40 (2008) 529–537.
[174] Nanomaterial Roadmap 2015, European Survey on Success Factors, Barriers and Needs for the Industrial Uptake of Nanomaterials in SMEs, Steinbeis, Europa, Zentrum, Karlsruhe (Germany), 2005.
[175] H. Goesmann, C. Feldmann, Angew. Chem. Int. Ed. Engl. 49 (2010) 1362–1395.
[176] B. Wetzel, P. Rosso, F. Haupert, K. Friedrich, Eng. Fract. Mech. 73 (2006) 2375–2398.
[177] I.V. Antonets, L.N. Kotov, S.V. Nekipelov, E.N. Karpushov, Tech. Phys. 49 (2004) 1496–1500.
[178] G. Hodes, Adv. Mater. 19 (2006) 639–655.
[179] R.F. Service, Science 300 (2003) 242–243.
[180] P. Keblinski, R. Prasher, J. Eapen, J. Nanopart. Res. 10 (2008) 1089–1097.
[181] M. Seidl, J.P. Perdew, M. Brajczewska, C. Fiolhais, J. Chem. Phys. 108 (1998) 8182–8188.
[182] C.S. Wu, C.D. Chen, S.M. Shih, W.F. Su, Appl. Phys. Lett. 81 (2002) 4595–4597.
[183] W. Wu, J. Gu, H.X. Ge, C. Keimel, S.Y. Chou, Appl. Phys. Lett. 83 (2003) 2268–2270.
[184] T.J. Haugan, P.N. Barnes, T.A. Campbell, J.M. Evans, J.W. Kell, L.B. Brunke, J.P. Murphy, Ch. Varanasi, I. Maartense, W. Wong-Ng, L.P. Cook, IEEE Trans. Appl. Supercond. 15 (2005) 3770–3773.
[185] K. Machida, K. Furuuchi, M. Min, K. Naoi, Electrochemistry 72 (2004) 402–404.
[186] N. Lee, T. Hyeon, Chem. Soc. Rev. 41 (2012) 2575–2589.
[187] D. Huber, Small (2005) 483–502.
[188] D. Peddis, C. Cannas, A. Musinu, G. Piccaluga, Chem. Eur. J. 15 (2009) 7822–7829.
[189] G. HerZer, K.H.J. Buschow (Ed.), Handbook of Magnetic Materials, vol. 10, Elsevier, 2007, pp. 415–463.
[190] X.C. Sun, X.L. Dong, Mater. Res. Bull. 37 (2003) 991–1004.
[191] N.A. Frey, S. Peng, K. Cheng, S. Sun, Chem. Soc. Rev. 38 (2009) 2532–2542.
[192] C. Liu, A.J. Rondinone, Z.J. Zhang, Pure Appl. Chem. 72 (2000) 37–45.
[193] P. Suzdalev, V.N. Buravtsev, Yu.V. Maksimov, A.A. Zharov, V.K. Imshennik, S.V. Novichikhin, V.V. Matveev, J. Nanopart. Res. 5 (2003) 485–495.
[194] H. Nishio, Y. Ota, M. Maekawa, H. Yamamoto, J. Magn. Magn. Mater. 287 (2005) 234–238.
[195] C. Rinaldi, T. Franklin, M. Zahn, T. Cader, In: Dekker Encyclopedia of Nanoscience and Nanotechnology, C.I. Contescu, K. Putyera (Eds.), Taylor & Francis Group, LLC (New York), 2009, pp. 1922–1938.

[196] A.J. Cox, S.E. Apsel, L.A. Bloomfield, Phys. Rev. B 49 (1994) 12295–12298.
[197] W.W. Yu, L. Qu, W. Guo, X. Peng, Chem. Mater. 15 (2003) 2854–2860.
[198] R. Grigalaitis, J. Banys, S. Lapinskas, E. Erdem, R. Bottcher, H.J. Glasel, E. Hartmann, IEEE Trans. Ultrason. Ferroelectr. Freq. Control 53 (2006) 2270–2274.
[199] K. Ishikawa, K. Yoshikawa, N. Okada, Phys. Rev. B 37 (1988) 5852–5855.
[200] K. Ishikawa, T. Uemori, Phys. Rev. B 60 (1999) 11841–11845.
[201] S. Wada, M. Ohishi, K. Takizawa, T. Hoshina, H. Kakemoto, T. Tsurumi, Key Eng. Mater. 320 (2006) 139–142.
[202] J.R. Groza, C.I. Contescu, K. Putyera (Eds.), Dekker Encyclopedia of Nanoscience and Nanotechnology, Taylor & Francis Group, LLC, 2009, pp. 173–234.
[203] Y. Wang, M. Chen, F. Zhou, E. Ma, Nature 419 (2002) 912–915.
[204] X. Huang, N. Hansen, N. Tsuji, Science 312 (2006) 249–251.
[205] E. Ma, Scr. Mater. 49 (2003) 663–668.
[206] Y.B. Pithawalla, M.S. El-Shall, S.C. Deevi, V. Strom, K.V. Rao, J. Phys. Chem. B 105 (2001) 2085–2090.
[207] W.W. Gerberich, W.M. Mook, C.R. Perrey, C.B. Carter, M.I. Baskes, R. Mukherjee, A. Gidwani, J. Heberlein, P.H. McMurry, S.L. Girshick, J. Mech. Phys. Solids 51 (2003) 979–992.
[208] Y.Y. Wu, W.C. Tsui, T.C. Liu, Wear 262 (2007) 819–825.
[209] N.B. Dahotre, S. Nayak, Surf. Coat.Technol. 194 (2005) 58–67.
[210] S. Ukai, T. Nishida, T. Okuda, T. Yoshitake, J. Nucl. Mater. 258 (263) (1998) 1745–1749.
[211] M. Taneike, F. Abe, K. Sawada, Nature 424 (2003) 294–296.
[212] A. Inoue, C. Fanb, J. Saidab, T. Zhang, Sci. Technol. Adv. Mater. 1 (2000) 73–86.
[213] X.C. Tong, H.S. Fang, Metall. Mater. Trans. A 29 (1998) 893–902.
[214] W. Thomson, Philos. Mag. 42 (1871) 448–452.
[215] M. Takagi, J. Phys. Soc. Jpn. 9 (1954) 359–363.
[216] Ph. Buffat, J.P. Borel, Phys. Rev. A 13 (1976) 2287–2298.
[217] Q.S. Mei, K. Lu, Prog. Mater. Sci. 52 (2007) 1175–1262.
[218] J.R. Greer, R.A. Street, Acta Mater. 55 (2007) 6345–6349.
[219] O. Koper, S. Winecki, K.J. Klabunde (Ed.), Nanoscale Materials in Chemistry, John Wiley & Sons Inc, New York, 2001, pp. 263–277.
[220] K.M. Unruh, T.E. Huber, C.A. Huber, Phys. Rev. B 48 (1993) 9021–9027.
[221] P. Puri, V. Yang, J. Nanopart. Res. 11 (2009) 1117–1127.
[222] F. Ercolessi, W. Andreoni, E. Tosatti, Phys. Rev. Lett. 66 (1991) 911–914.
[223] W.H. Qi, M.P. Wang, J. Mater. Sci. Lett. 21 (2002) 1743–1745.
[224] W.H. Qi, M.P. Wang, Mater. Chem. Phys. 88 (2004) 280–284.
[225] R. Kofman, P. Cheyssac, Y. Lereah, A. Stella, Eur. Phys. J. D 9 (1999) 441–444.
[226] J. Sun, S.L. Simon, Thermochim. Acta 463 (2007) 32–40.
[227] B.-X. Wang, L.-P. Zhou, X.-F. Peng, Int. J. Thermophys. 27 (2006) 139–151.
[228] V.N. Likhachev, T.Yu. Astakhova, G.A. Vinogradov, M.I. Alymov, Russ. J. Phys. Chem. B 1 (2007) 74–78.
[229] U. Herr, G.K. Samwer, Philos. Mag. A 77 (1998) 641–652.
[230] W.H. Li, C.C. Yang, F.C. Tsao, S.Y. Wu, P.J. Huang, M.K. Chung, Y.D. Yao, Phys. Rev. Lett. 72 (2005) 214516–214521.
[231] A.L. Rogach, D.V. Talapin, E.V. Shevchenko, A. Kornowski, M. Haase, H. Weller, Adv. Funct. Mater. 12 (2002) 653–664.
[232] V.N. Troitskiy, I.A. Domashnev, E.N. Kurkin, O.M. Grebtsova, V.I. Berestenko, I.L. Balikhin, S.V. Gurov, J. Nanopart. Res. 5 (2003) 521–528.

[233] T. Haugan, P.N. Barnes, R. Wheeler, F. Meisenkothen, M. Sumption, Nature 430 (2004) 867–870.
[234] R.M. Scanlan, W.A. Fietz, E.F. Koch, J. Appl. Phys. 46 (1975) 2244–2249.
[235] A. Navrostsky, J.F. Banfield, A. Navrostsky (Eds.), Reviews in Mineralogy and Geochemistry, Mineralogical Society of America, 2001, pp. 73–103.
[236] Y. Zhang, S. Liao, Y. Fan, J. Xu, F. Wang, J. Nanopart. Res. 3 (2001) 23–26.
[237] T. Ishida, M. Haruta, Angew. Chem. Int. Ed. Engl. 46 (2007) 7154–7156.
[238] J.M. Campelo, D. Luna, R. Luque, J.M. Marinas, A.A. Romero, ChemSusChem 2 (2009) 18–45.
[239] A.S. Aricò, V. Baglio, V. Antonucci, J. Garcia-Martinez (Ed.), Nanotechnology for the Energy Challenge, Wiley-VCH, Weinheim, 2009, pp. 79–109.
[240] K. Kinoshita, J. Electrochem. Soc. 137 (1990) 845–848.
[241] Y. Sasson, G. Levy-Ruso, O. Toledano, I. Ishaaaya, I. Ishaaya, R. Nauen, A.R. Horowitz (Eds.), Insecticides Design Using Advanced Technologies, Springer-Verlag, Berlin and Heidelberg, 2007, pp. 1–5.
[242] A. Schirmeisen, U.D. Schwarz, Chemphyschem 10 (2009) 2373–2382.
[243] C. Ritter, M. Heyde, B. Stegemann, K. Rademann, U.D. Schwarz, Phys. Rev. B 71 (2005) 085405–085411.
[244] X. Feng, M.Z. Hu, H.S. Nalwa (Ed.), Encyclopedia of Nanoscience and Nanotechnology, vol. 1, American Scientific Publishers, Stevenson Ranch, 2004, pp. 687–726.
[245] I. Robinson, L.D. Tung, N.T.K. Thanh, C.I. Contescu, K. Putyera (Eds.), Dekker Encyclopedia of Nanoscience and Nanotechnology, Taylor & Francis Group, LLC, 2009, pp. 354–363.
[246] D. Kushnir, B.A. Sanden, J. Ind. Ecol. 12 (2008) 360–375.
[247] Y. Hayashi, M. Inoue, H. Takizawa, K. Suganuma, J.E. Morris (Ed.), Nanopackaging: Nanotechnologies and Electronics Packaging, Springer, New York, 2008 Chap. 6.
[248] N. Malanowski, T. Heimer, W. Luther, M. Werner (Eds.), Growth Market Nanotechnology: An Analysis of Technology and Innovation, Wiley-VCH, Weinheim, 2006, p. 39.
[249] C.C. Doumanidia, Proc. IMechE 218 (2005) 51–70.
[250] H. Zhang, J. Han, B. Yang, Adv. Funct. Mater. 20 (2010) 1533–1550.
[251] H. Mollet, A. Grubenmann, Formulation Technology, Wiley-VCH, Weinheim, 2001, p. 29.
[252] G. Gouadec, Ph. Colomban, Prog. Cryst. Growth Charact. Mater. 53 (2007) 1–56.
[253] R. Fox, S. Garrick, M. Gordon, M. Lamm, C. Sorensen, NSF Nanoscale Science and Engineering Grantees Conference, 2005.
[254] M.-A. Neouze, U. Schubert, Monatsh Chem. 139 (2008) 183–195.
[255] R.H. Yoon, T. Salman, G. Donnay, J. Colloid Interface Sci. 70 (1979) 483–493.
[256] M.G. Sandros, M. Behrendt, D. Maysinger, M. Tabrizian, Adv. Funct. Mater. 17 (2007) 3724–3730.
[257] S.F. Lee, M.A. Osborne, Chem. Phys. Chem. 10 (2009) 2174–2191.
[258] K.N. Han, N.S. Kim, KONA Powder Part. J. 27 (2009) 73–84.
[259] S. Kim, B. Fisher, H.J. Eisler, M. Bawendi, J. Am. Chem. Soc. 125 (2003) 11466–11467.
[260] Y. Cao, U. Banin, J. Am. Chem. Soc. 122 (2000) 9692–9702.
[261] Y. Lu, Y. Yin, Z.Y. Li, Y. Xia, Nano Lett. 2 (2002) 785–788.
[262] C. Gua, H. Xu, M. Park, C. Shannon, ECS Trans. 16 (2008) 181–190.
[263] A. Aharoni, T. Mokari, I. Popov, U. Banin, J. Phys. Chem. 128 (2006) 257–264.
[264] L. Rapoport, N. Fleischer, R. Tenne, J. Mater. Chem. 15 (2005) 1782–1788.
[265] E.K. Athanassiou, R.N. Grass, W.J. Stark, Nanotechnology 17 (2006) 1668–1673.
[266] H. Zeng, S. Sun, Adv. Funct. Mater. 18 (2008) 391–400.

[267] L. Wang, J. Luo, Q. Fan, M. Suzuki, I.S. Suzuki, M.H. Engelhard, Y. Lin, N. Kim, J.Q. Wang, C.-J. Zhong, J. Phys. Chem. B 109 (2005) 21593–21601.
[268] K.J. Morrow, R. Bawa, Ch. Wei, Med. Clin. North Am. 91 (2007) 805–843.
[269] A.W. Wills, M.S. Kang, A. Khare, W.L. Gladfelter, D.J. Norris, ACS Nano. 4 (2010) 4523–4530.
[270] C. Pichot, Curr. Opin. Colloid Interface Sci. 9 (2004) 213–221.
[271] C.C. Berry, A.S.G. Curtis, J. Phys. D Appl. Phys. 36 (2003) R198–R206.
[272] M. Hamada, S. Nakanishi, T. Itoh, M. Ishikawa, V. Biju, ACS Nano. 4 (2010) 4445–4454.
[273] W. Shia, Y. Sahoob, M.T. Swihart, Colloids Surf. A Physicochem. Eng. Asp. 246 (2004) 109–113.
[274] J.G. Worden, Q. Dai, A.W. Shaffer, Q. Huo, Chem. Mater. 16 (2004) 3746–3755.
[275] W. Cai, T. Gao, H. Hong, J. Sun, Nanotechnology, Science and Applications, Dove Medical Press Ltd, 2008, pp. 17–32.
[276] R.P. Bagwe, L.R. Hilliard, W. Tan, Langmuir 22 (2006) 4357–4362.
[277] G.-X. Zhu, X.-W. Wei, S. Jiang, J. Mater. Chem. 17 (2007) 2301–2306.
[278] I.L. Medintz, H. Tetsuo Uyeda, E.R. Goldman, H. Mattoussi, Nat. Mater. 4 (2005) 435–446.
[279] S. Liu, M. Han, Adv. Funct. Mater. 15 (2005) 961–966.
[280] L.F. Hakim, S.M. George, A.W. Weimer, Nanotechnology 16 (2005) S375–S381.
[281] B. Zhang, Y.-C. Liao, S.L. Girshick, J.T. Roberts, J. Nanopart. Res. 10 (2008) 173–178.
[282] S.L. Girshick, J. Nanopart. Res. 10 (2008) 935–945.
[283] A. Camenzinda, W.R. Caserib, S.E. Pratsinis, Nano Today 5 (2010) 48–65.
[284] S. Beguin-colin, G. LeCaer, T. Girot, J. Phys. IV France 12 (2002) 441–455.
[285] G. Fagherazzi, P. Canton, P. Riello, N. Pernicone, F. Pinna, M. Battagliarin, Langmuir 16 (2000) 4539–4546.
[286] P. Kumar, P. Fennell, A. Robins, J. Nanopart. Res. 12 (2010) 1523–1530.
[287] Available from World Wide Web: http://www.eaglabs.com/techniques/analytical_techniques.
[288] J. Grams, New Trends and Potentialities of ToF-SIMS in Surface Studies, Nova Science Publishers, New York, 2007.
[289] F. Bensebaa, N. Patrito, Y. Le Page, P. L'Ecuyer, D. Wang, J. Mater. Chem. 14 (2004) 3378–3384.
[290] R. Davies, Microscopy 62 (1970) 87–93.
[291] M.A. Elsayad, Acc. Chem. Res. 37 (2004) 326–333.
[292] R. Fisker, J.M. Carstensen, M.F. Hansen, F. Bødker, S. Mørup, J. Nanopart. Res. 2 (2000) 267–277.
[293] A. Jillavenkatesa, S.J. Dapkunas, L.-S.H. Lum, Particle size Characterisation, NIST special, publication 960-1.
[294] A. Jillavenkatesa, J.F. Kelly, J. Nanopart. Res. 4 (2002) 463–468.
[295] R.J. Colton, J. Vac, J. Vac. Sci. Technol. B 22 (2004) 1609–1635.
[296] M. Gaumet, A. Vargas, R. Gurny, F. Delie, Eur. J. Pharm. Biopharm. 69 (2008) 1–9.
[297] B. Chu, T. Liu, J. Nanopart. Res. 2 (2000) 29–41.
[298] D.Y.H. Pui, J.R. Brock, D.-R. Chen, H. Fissan, C.D. Frisbie, C.E. Lyman, J.C. Miller, G.W. Mulholland, R. Pecora, O. Preining, T. Vo-Dinh, J. Nanopart. Res. 2 (2000) 103–112.
[299] Y. Xiong, Y. Xia, Adv. Mater. 19 (2007) 3385–3391.
[300] Y. Xia, Y. Xiong, B. Lim, S.E. Skrabalak, Angew. Chem. Int. Ed. Engl. 48 (2009) 60–103.
[301] A.-C. Shi, R.I. Masel, J. Catal. 120 (1989) 421–431.
[302] R.J. Matyi, L.H. Schwartz, J.B. Butt, Catal. Rev. Sci. Eng. 29 (1987) 41–99.

[303] H. Borchert, E.V. Shevchenko, A. Robert, I. Mekis, A. Kornowski, G. Grubel, H. Weller, Langmuir 21 (2005) 1931–1936.
[304] R.S. Goldman, J. Phys. D Appl. Phys. 37 (2004) R163–R178.
[305] T. Harada, H. Matsuoka, Curr. Opin. Colloid Interface Sci. 8 (2004) 501–506.
[306] N. Pernicone, CatTech 7 (2003) 196–2004.
[307] S. Begin-Colin, G. Le Caer, T. Girot, J. Phys. IV France 12 (2002) 441–454.
[308] J.A. Williams, G.E. Rindone, H.A. McKinsky, J. Am. Ceram. Soc. 64 (1981) 709–713.
[309] Y. Chushkin, L. Chitu, Y. Halahovets, S. Luby, E. Majkova, A. Satka, G. Leo, M. Giersig, M. Hilgendorff, V. Holy, O. Konovalov, Mater. Sci. Eng. C 26 (2006) 1136–1140.
[310] J. Wu, U. Weitstall, J.C.H. Spence, Nat. Mater. 4 (2005) 912–916.
[311] J.P. Reid, L. Mitchem, Annu. Rev. Phys. Chem. 57 (2006) 9.1–9.27.
[312] R. Pecora, J. Nanopart. Res. 2 (2000) 123–131.
[313] M. Kaszuba, J. Nanopart. Res. 1 (1999) 405–409.
[314] J. Kimling, M. Maier, B. Okenve, V. Kotaidis, H. Ballot, A. Plech, J. Phys. Chem. B110 (2006) 15700–15707.
[315] A. Chemseddine, H. Weller, Ber. Bunsen-Ges. Phys. Chem. 97 (1993) 636–637.
[316] A.V. Gaikwad, P. Verschuren, E. Eiser, G. Rothenberg, J. Phys. Chem. B 110 (2006) 17437–17443.
[317] K. Chen, A. Kromin, M.P. Ulmer, B.W. Wessels, V. Backman, Opt. Commun. 228 (2003) 1–7.
[318] D. Tonti, F. van Mourik, M. Chergui, Nano Lett. 4 (2004) 2483–2487.
[319] H. Borchert, D.V. Talapin, N. Gaponik, C. McGinley, S. Adam, A. Lobo, T. Moller, H. Weller, J. Phys. Chem. B107 (2003) 9662–9668.
[320] M.R. Hoffmann, S.T. Martin, W. Choi, D.W. Bahnemannt, Chem. Rev. 95 (1995) 69–96.
[321] R. Lake, Superlatt. Microstruct. 22 (1996) 559.
[322] L. Maya, C.H. Chen, K.A. Stevenson, E.A. Kenik, S.L. Allman, T.G. Thundat, J. Nanopart. Res. 4 (2002) 417–422.
[323] M. Choi, J. Nanopart. Res. 3 (2001) 201–211.
[324] V. Abdelsayed, M.S. El-Shall, T. Seto, J. Nanopart. Res. 8 (2006) 361–369.
[325] D.-R. Chen, D.Y.H. Pui, J. Nanopart. Res. 1 (1999) 115–126.
[326] R. Starke, B. Kock, P. Roth, Shock Waves 12 (2003) 351–360.
[327] Curtis E. Johnson, Stephen Fallis, Andrew P. Chafin, Thomas J. Groshens, Kelvin T. Higa, Ismail M.K. Ismail, Tom W. Hawkins, J. Propulsion Power 23 (2007) 669–682.
[328] K.H. Ahn, C.H. Jung, M. Choi, J.S. Lee, J. Nanopart. Res. 3 (2001) 161–170.
[329] Cölfen, H., Analytical ultracentrifugation of NPs In: Encyclopedia of Nanoscience and Nanotechnology, vol. 1, H.S. Nalwa (Ed.), American Scientific Publishers, Los Angeles, 2004, pp. 67–88.
[330] C.H. Fischer, M. Giersig, Langmuir 8 (1992) 1475–1478.
[331] W. Hwang, C. Lee, D.W. Boo, J.-G. Choi, Bull. Korean Chem. Soc. 24 (2003) 684–686.
[332] M. Konert, J. Vandenberghe, Sedimentology 44 (1997) 523–535.
[333] J.L. Barr, R.L. Axelbaum, D.M.E. Macias, J. Nanopart. Res. 8 (2006) 11–22.
[334] X. Phung, J. Groza, E.A. Stach, L.N. Williams, S.B. Ritchey, Mater. Eng. A359 (2003) 261–268.
[335] N. Eisenreich, H. Fietzek, M.D.M. Juez-Lorenzo, V. Kolarik, A. Koleczko, V. Weiser, Propellants Explosives Pyrotech. 29 (2004) 137–145.
[336] J.-N. Gillet, M. Meunier, J. Phys. Chem. B 109 (2005) 8733–8737.
[337] J. Nanda, S. Sapra, D.D. Sarma, H.S. Nalwa (Ed.), Encyclopedia of Nanoscience and Nanotechnology, vol. 10, American Scientific Publishers, Stevenson Ranch, 2004, pp. 711–724.

[338] M.-C. Bourg, A. Badia, R.B. Lennox, J. Phys. Chem. B 104 (2000) 6562–6567.
[339] U. Winkler, D. Eich, Z.H. Chen, R. Fink, S.K. Kulkarni, E. Umbach, Chem. Phys. Lett. 306 (1999) 95–102.
[340] V. Biju, M. Abdul Khadar, J. Nanopart. Res. 4 (2002) 247–253.
[341] A.E. Russell, A. Rose, Chem. Rev. 104 (2004) 4613–4635.
[342] T. Hemraj-Benny, S. Banerjee, S. Sambasivan, M. Balasubramanian, D.A. Fischer, G. Eres, A.A. Puretzky, D.B. Geohegan, D.H. Lowndes, W. Han, J.A. Misewich, S.S. Wong, Small 2 (2006) 26–35.
[343] J. Rockenberger, L. Troger, A. Kornowski, T. Vossmeyer, A. Eychmuller, J. Feldhaus, H. Weller, J. Phys. Chem. B 101 (101) (1997) 2691–2701.
[344] W. Sigle, Annu. Rev. Mater. Res. 35 (2005) 239–314.
[345] C. Colliex, Electron Microsc. 45 (1996) 44–50.
[346] J. Liu, J. Electron Microsc. 54 (2005) 251–278.
[347] D. Wei, R. Dave, R. Pfeffer, J. Nanopart. Res. 4 (2002) 21–41.
[348] A. Wei, B. Kim, B. Sadtler, S.L. Tripp, Chem. Phys. Chem. 2 (2001) 743–745.
[349] K. Faulds, R.E. Littleford, D. Graham, G. Dent, W.E. Smith, Anal. Chem. 76 (2004) 592–598.
[350] N. Anderson, A. Hartschuh, L. Novotny, Mater. Today 50 (2005) 50–54.
[351] C.P. Kumar, N.O. Gopal, T.C. Wang, M.S. Wong, S.C. Ke, J. Phys. Chem. B 110 (110) (2006) 5223–5229.
[352] I. Bertholon, H. Hommel, D. Labarre, C. Vauthier, Langmuir 22 (2006) 5485–5490.
[353] Y. Koseoglu, J. Magn. Magn. Mater. 300 (2006) 327–330.
[354] J.W.M. Bulte, R.A. Brooks, B.M. Moskowitz, AUR 5 (1998) S137–S140.
[355] S.J. Bull, J. Phys. D Appl. Phys. 38 (2005) R393–R413.
[356] F.V. Ignatovich, L. Novotny, Phys. Rev. Lett. 96 (2006) 13901–13904.
[357] F. Reynolds, T. O'Loughlin, R. Weissleder, L. Josephson, Anal. Chem. 77 (2005) 814–817.
[358] George W. Mulholland, Barry J. Bauer, J. Nanopart. Res. 2 (2000) 5–15.
[359] H. Kunzmann, Metrologia 28 (1992) 443–455.
[360] K. Nakayama, M. Tanaka, F. Shiota, K. Kuroda, Metrologia 28 (1992) 483–502.
[361] Adeogun, Nanometrology: An Overview, September 2008, SRI Reports; Nanoforum.org—2006, Nanometrology Report.
[362] Jong-Ahn Kim, Jae Wan Kim, Byong Chon Park, Tae Bong Eom, Meas. Sci. Technol. 17 (2006) 1792–1800.
[363] J. Cazaux, Nanotechnology 15 (2004) 1195–1199.
[364] H.-U. Danzebrink, F. Pohlenz, G. Dai, C. Dal Savio, G. Wilkening, L. Koenders (Eds.), Nanoscale Calibration Standards and Methods: Dimensional and Related Measurements in the Micro- and Nanometer Range, Wiley-VCH, Weinheim, 2005, pp. 2–21.
[365] D.J.S. Birch, C.d. Geddes, Proc. Indian Acad. Sci. (Chem. Sci.), 112 (2000) 311–322.
[366] C. Sonnichsen, Nat. Biotechnol. 23 (2005) 741–745.
[367] X. Chen, Y. Lou, A.C.S. Samia, C. Burda, J.L. Cole, Adv. Funct. Mater. 15 (2005) 41–49.
[368] M. Spath, P.M. Sommeling, J.A.M. van Roosmalen, H.J.P. Smit, N.P.G. van der Burg, D.R. Mahieu, N.J. Bakker, J.M. Kroon, Prog. Photovolt. Res. Appl. 11 (2003) 207–220.
[369] O. Masala, R. Sehadri, Annu. Rev. Mater. Res. 34 (2004) 41–81.
[370] H. Xie, Y. Gu, H.J. Ploehn, Nanotechnology 16 (2005) S492.
[371] J.F. Weng, X.T. Song, L. Li, H.F. Qian, K.Y. Chen, X.M. Xu, C.X. Cao, J.C. Ren, Chin. Chem. Lett. 17 (2006) 675–678.
[372] S.F. Sweeney, G.H. Woehrle, J.E. Hutchison, J. Am. Chem. Soc. 128 (2006) 3190–3197.
[373] M.J. Hollamby, J. Eastoe, A. Chemelli, O. Glatter, S. Rogers, R.K. Heenan, I. Grillo, Langmuir 26 (2010) 6989–6994.

[374] P. Gould, Nano Today 1 (2006) 34–39.
[375] K. Rittner, J. Contam. Hydrol. 554 (2002) 87–111.
[376] H. Vallhov, J. Qin, S.M. Johansson, N. Ahlborg, M.A. Muhammed, A. Scheynius, S. Gabrielsson, Nano Lett. 6 (2006) 1682–1686.
[377] J.H. Koo, Polymer Nanocomposites: Processing, Characterization, and applications, McGraw-Hill, New York, 2006 (Chapter 2).
[378] J.T. Numi, Environ. Sci. Technol. 39 (2005) 1221–1230.
[379] R. Mieher, Environ. Sci. Technol. 38 (2005) 139–147.
[380] T.L. Hill, Nano Lett. 1 (2001) 273–275.
[381] H. Zhang, Nature 424 (2003) 1025–1029.
[382] J.C. Miller, R.M. Serrato, J. Represas-Cardenas, G.A. Kundahl, The Handbook of Nanotechnology: Business, Policy, and Intellectual Property Law, John Wiley & Sons, Inc, New York, 2005.
[383] J.A. Lindeman, C.I. Contescu, K. Putyera (Eds.), Dekker Encyclopedia of Nanoscience and Nanotechnology, Taylor & Francis Group, LLC, 2009, pp. 3318–3327.
[384] S.L. Moskowitz, Kirk-Othmer Encyclopedia of Chemical Technology, vol. 1, Copyright John Wiley & Sons, Inc, 2003, pp. 691–730.
[385] P. Borm, D. Robbins, S. Haubold, T. Kuhlbusch, H. Fissan, K. Donaldson, R. Schins, V. Stone, W. Kreyling, J. Lademann, J. Krutmann, D. Warheit, G. Oberdoerster, Part. Fibre Toxicol. 3 (2006) 11.
[386] M.J. Pitkethly, Nanotoday December (2003) 36–42.
[387] R. Gadow, F. Kern, A. Killinger, Mater. Sci. Eng. B 148 (2008) 58–64.
[388] C.W. Corti, R.J. Holliday, R.D.T. Thompson, Gold Bull. 35 (2002) 111–117.
[389] http://www.nanotechproject.org (accessed 02.07.12).
[390] www.en.wikipedia.org/wiki/Technology_readiness_level Wikipedia Aug. 2011.
[391] M. Kearns, Mater. Sci. Eng. A 375–377 (2004) 120–126M. Kearns, Materials Science and Engineering A 375–377 (2004) 120–126
[392] G. Wakefield, Nano Today 3 (2008) 48–48.
[393] Roadmap Report on Nanoparticles, Willems & van den Wildenberg (W&W), November 2005.
[394] M.J. Pitkethly, Nano Today December (2004) 20–29.
[395] The World Market for Nanoparticle Titanium Dioxide (TiO_2), http://www.researchandmarkets.com/reports/1651709/the_world_market_for_nanoparticle_titanium.pdf.
[396] U. Sutter, J. Loeffler, Roadmap Report Concerning the Use of Nanomaterials in the Energy Sector. http://www.nanoroad.net/download/roadmap_e.pdf, 2006, (accessed 13.07.08).
[397] D.W. Elliott, H.L. Lien, W. Zhang, G.E. Fryxell, G. Cao (Eds.), Environmental Applications of Nanomaterials, Imperial College Press, England, 2008, pp. 25–48.
[398] M. Knell, E. Cozzens, J.M. Wetmore (Eds.), Nanotechnology and the Challenges of Equity, Yearbook of Nanotechnology in Society, vol. 2, Springer Science+Business Media B.V, 2001, p. 127.
[399] C. Ostiguy, B. Roberge, L. Ménard, C.-A. Endo, Best Practices Guide to Synthetic Nanoparticle Risk Management, Chemical Substances and Biological Agents, Report R-599, IRSST 2009.

Chapter 2

Wet Production Methods

1 INTRODUCTION

Nanoparticle (NP) synthesis in the liquid phase can be traced back to more than a century. However, solvent-based NP synthesis research remained very limited in scope until the end of the twentieth century. The ability to synthesize monodisperse and highly luminescent quantum dots (QDs) with discrete optical transitions, using nonionic organometallic precursors (Cd and Se or S or Te) in a high-boiling apolar coordinating solvent (trioctylphosphine oxide or TOPO) [1] has revived interest in NP science and technology [2–4]. Recent breakthroughs using this so-called TOPO method and its various adaptations showed high quality and tunable QD [5]. Cost and scale-up remain major issues of this TOPO approach. Alternative TOPO methods using single pot synthesis [6] and continuous processes [7] have been recently demonstrated.

NP synthesis approaches are often divided into two main categories depending on the starting material (Fig. 2.1). The first category consists of bottom-up synthesis, generally using chemical-based processes. Although atmospheric conditions are often used, controlled environmental pressure is also used. In this first category, we could include precipitation techniques, usually starting from salt solutions, and condensation approaches, from the vapor phase. The second category consists of top-down methods using mostly physical processes, often under vacuum conditions. Size-reducing methods such as ball milling are typical top-down techniques used in NP production.

Kinetic (transport and reaction rates) and thermodynamic factors often compete during the NP growth process. Fortunately, these factors can be controlled by varying reactant concentration, pressure, temperature, energy input, pH, and/or order of precursor introduction. Following initial reaction, the degree of supersaturation will dictate whether monodispersal (low supersaturation) or agglomeration (high supersaturation) is achieved [9]. Small size nanoparticles (NPs) give rise to high surface energy value. Particle surface tension is controlled using surfactant and charging to prevent agglomeration

Interface Science and Technology, Vol. 19. http://dx.doi.org/10.1016/B978-0-12-369550-5.00002-1

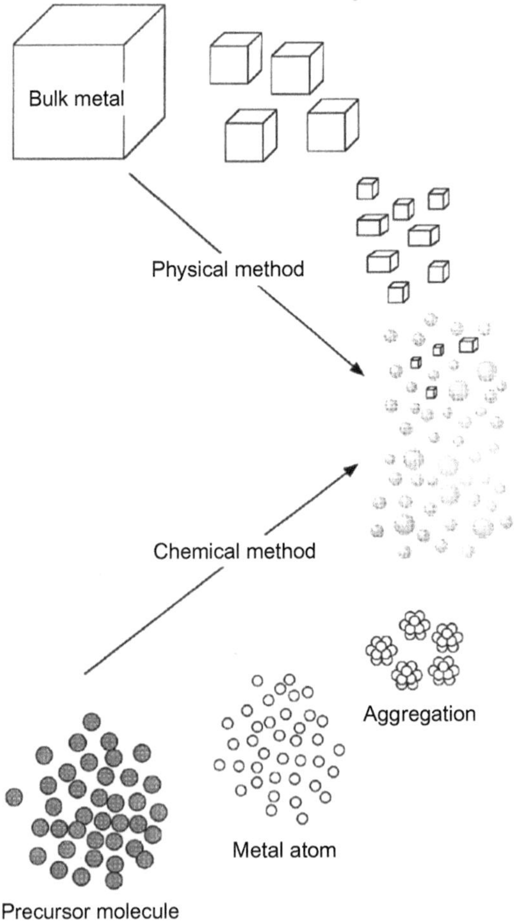

FIGURE 2.1 Schematic representation of the top-down (physical method) and bottom-up (chemical method) fabrication processes of nanoparticles. *Reproduced from Ref. [8] with permission from the Centre National de la Recherche Scientifique (CNRS) and The Royal Society of Chemistry.*

and Ostwald ripening. Controlling surface tension allows achieving polydispersed size distribution.

Bottom-up synthesis processes provide flexibility to produce a wide variety of tunable monodisperse NPs. According to LaMer's diagram, four distinct growth stages are distinguished [10]. Formation of nucleate (solute) following a stimulus is considered the first step toward forming NP. The next steps consist of nucleation (seeds or monomers) and growth (NP) stages. Once NPs are formed, it is important to prevent growth beyond a critical size, which may lead to nonuniform distribution. Uncontrollable growth through nucleation and Ostwald ripening remain the main challenges of monodisperse NP synthesis. Controlling synthesis conditions is critical to obtain the desired product characteristics. Size and functionalization of NPs can be controlled directly in the solution.

In some cases, it is quite difficult to classify some fabrication techniques along the bottom-up and top-down dichotomy. For example, laser evaporation uses bulk materials to produce atomic and/or molecule vapors inside a flow reactor. Then NPs are formed following contact with an inert gas or cold surface inside the flow reactor. In this example, both bottom-up and top-down processes are involved. Laser technique could be used to produce NPs by bombarding a bulk material in a solution containing surfactants. Numerous other synthesis techniques are also difficult to classify as either top-down or bottom-up. Instead of the top-down versus bottom-up dichotomy, we will divide the fabrication processes according to the synthesis medium. As discussed in Chapter 1, there are other possible classification strategies. Wet and dry processing methods are distinguished depending on whether the synthesis process involves a solvent or not. The majority of dry methods use top-down processes, and the majority of wet methods are bottom-up processes. However this is not always the case, as shown in the laser technique example.

The criteria for comparing wet and dry methods are not well defined. In view of their commercialization, these methods should be scalable. Scalability of these synthesis techniques depend on the synthesized NP characteristics. Synthesis using the TOPO method is difficult to scale up, although it provides high-quality NPs with a relatively wide range of surface functionalizations. On the other hand, dry techniques allow large-scale NP production, although size distribution is quite broad. Wet methods offer more flexibility in terms of functionalization than dry methods. However, dry-based NPs have also been successfully functionalized [11].

Very few reports comparing these two approaches are provided in the literature. ZnO_2 NPs prepared by gas phase and liquid phase showed similar geometrical features, although their compaction sintering properties are different [12]. This is likely due to different surface compositions. This example may illustrate the fact that the choice of synthesis methods should be adapted to the targeted application. Indeed, compact and porous nanoparticulate materials are used for different applications. Thus, comparing these two synthesis approaches without specifying the application is not of interest.

Numerous wet-based approaches have been used to synthesize NPs [13]. These methods are often classified on the basis of either the source of energy or size selection. The following methods have been used in the past:

a. salt reduction and co-reduction;
b. ligand reduction and displacement from organometallics;

c. microemulsion;
d. sol–gel;
e. thermal decomposition of organometallics compounds;
f. photochemical reduction;
g. metal vapor synthesis;
h. electrochemical synthesis;
i. hydrothermal;
j. sonochemistry.

Other techniques have been developed for NP synthesis. It is beyond the scope of this book to cover all the production methods. NPs obtained by reduction (or co-reduction) of salts is widely used and studied in academia. Easy preparation and versatility are two attributes of salt reduction. The scalability of these techniques remains a big challenge, particularly when conventional bath heating processes are used. Often, reduction is performed by the solvent. This is the case of the alcohol reduction process. When nonreducing solvents are used, as in the case of an aqueous solution, reducing agents are added. For example, metal salt reduction is achieved by adding reducing agents such as hydrazine hydrate. The so-called Turkevich method has been widely used to synthesize gold NPs with relatively precise size control [14,15]. Tunable and monodisperse gold particles from 9 to 120 nm are obtained by varying the gold-to-reductant ratio, although various shapes are obtained. This technique, based on reduction by citrate and ascorbic acid, remains relatively difficult to scale up. Co-precipitation methods obtained by reduction and decomposition routes have also been reported.

Wet-based techniques provide the simplest approach to produce small and monodispersed particles. Nucleus formation induced through energy input or supersaturation should be arrested to prevent uncontrollable nucleation. Two standard colloidal chemistry approaches have been developed to control aggregation and/or agglomeration [16]. Electrostatic stabilization by surface incorporation or addition of anions (OH^-, SO_4^{2-}, NO_3, $RCOO^-$, RSO_3^-) and cations (H^+, R_4N^+) has been used. Steric stabilization obtained using alkyl chain molecules (oleyl amine, oleic acid, trioctyl phosphine, trioctyl phosphine oxide) and polymers (EVAP, PVP) has received more interest as it allows further functionalization and integration. In some cases, both stabilization techniques are use simultaneously. However, in the absence of these stabilization techniques, controlled growth of the seed particles in a homogenous solution is difficult. Furthermore, Ostwald ripening will lead to precipitation of large particles.

The ability to control growth rate and orientation provides some advantages for wet-based synthesis [17,18]. NP shape is directly correlated to the exposed facets. The ability to control exposed facets and NP shapes is useful in catalysis. For example {100} and {210} facets of Pt NPs are found to be highly active and selective for H_2- and CO-catalyzed reactions [19]. It is worth noting that Pt and other noble metals have been widely studied because of their catalytic properties and commercialization possibilities. Controlling the reduction and growth rates has been used to tune the shape of these NPs [17,18]. As shown in Fig. 2.2, Pd particles with rod, plates, bar, cube, or octahedron shapes can be obtained by taking advantage of preferential/selective and controlled growth on different facets. Controlling the facet geometry of the different intermediate particles allows the preparation of various anisotropic nanostructures. Single crystal, twinned, and plate-like seeds have been obtained at the early growth stages. Thermodynamic and kinetic parameters are used to control the shape of intermediate nanocrystals and final nanostructured materials [17]. Besides nanocrystals, 2D and 3D bulk structures are achievable from these intermediates.

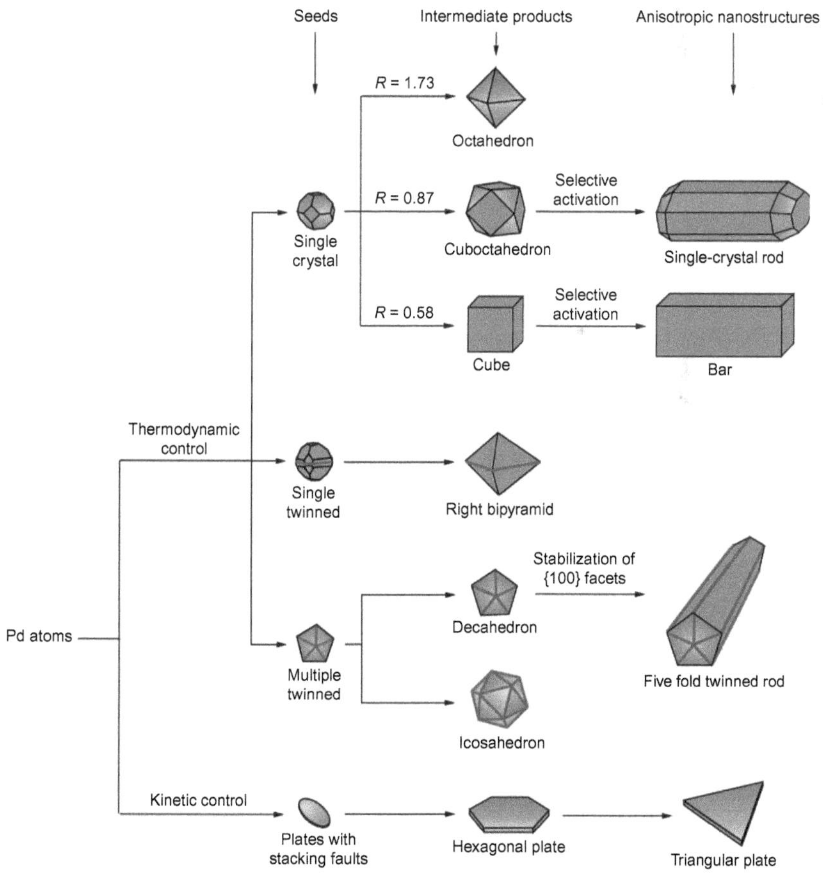

FIGURE 2.2 Possible Pd anisotropic nanostructures and the intermediate synthesis steps [17]. (For color version of this figure, the reader is referred to the online version of this chapter.)

Two parameters are often used to control the properties of NP synthesized in liquid solution: temperature and pressure. These two parameters are key to optimizing NP properties for a given application. Available excitation sources cover wide temperature and pressure ranges [20]. Synthesis temperatures range from around room temperature using solution processes up to 2000 °C or more using plasma processes. Low pressure (10–100 Pa) and extreme pressure (10^{11} Pa) are obtained using evaporation and shockwave techniques, respectively. Chemical vapor deposition or CVD (10^4 Pa) and solvothermal (10^6 Pa) techniques are intermediary pressure-based synthesis techniques with operation temperature reaching up to 1000 °C. The choice of appropriate temperature and pressure values is often dictated by the desired NP characteristics, scale, cost, and nature of precursors. A third key manufacturing parameter is the presence of surfactant and/or other anti-agglomeration techniques. Although surfactants are key to obtaining highly monodispersed NPs, they contribute to increased surface impurities in solution-based NPs. Often, postprocessing steps are required to provide higher quality NPs.

Synthesized NPs should be monodisperse, monophase and crystalline, easy to functionalize, and defect free. Often, when synthesized, many of these criteria are not fulfilled. A significant issue with several wet-based NP preparation methods is the additional cleaning and purification step. Most cleaning and purification methods are resource wasteful. This process step is probably one reason why wet chemistry synthesis methods are not yet widely adopted commercially. For example, a significant amount of organic solvents are used to separate and purify the final product. Recent novel purification processes have been proposed to address these challenges [21–23]. Using water as solvent and/or for solvent recycling remains the best option [7]. Other green synthesis techniques have also been developed.

In this chapter, wet methods that have (or will have) a significant potential for large scale manufacturing are discussed more extensively. We will not try to discuss in detail all the wet-based approaches for NP synthesis. Instead, we focus on methods with potential for commercialization in the future. We will also provide possible avenues to address the bottlenecks hindering the scale-up of current synthesis methodologies. For different reasons, several synthesis techniques are not discussed in this book. First, it is quite impossible to cover a field that has exploded in the last decade. Second, some techniques do not appear to be scalable or offer any apparent technological advantage.

2 SONOCHEMISTRY

Ultrasonic wave is an efficient energy source for heating solutions to induce chemical reactions. The field of ultrasonic wave and its applications have been reviewed [24,25]. It gives rise to high-temperature and high-pressure conditions and allows covering a wide range of operational conditions not possible with most techniques. Besides its scalability and simplicity, high-temperature and high-pressure conditions could provide advantages for NP synthesis. Several reports describing the attributes of this technique have been published [26–28].

Sonochemistry involves application of ultrasound radiation in the frequency (f) range of 0.020–10 MHz [29]. Typically, an ultrasonic system consists of a power source, a piezoelectric system including electrodes, a horn with a stainless steel collar, a reaction vessel, and a bath container [29]. Application of ultrasonic radiation to a solution containing chemical precursors initiated the formation and growth of bubbles, following the diffusion of the solute vapor. The critical size D of the bubble is inversely proportional to the ultrasound frequency (f):

$$D \sim 1/f.$$

The bubble collapses and forms shockwaves once it reaches a critical size. This collapse releases enough energy to allow chemical bond scission and reaction.

Metal, metal oxide, sulfide, and alloy NPs have been prepared by sonochemistry, using aqueous and nonaqueous solutions [30]. This sonochemical approach has been used to prepare NPs embedded in a nanocomposite in a single step. CdS NPs with a sharp emission band onset have been prepared using this technique [31]. Semiconductor HgTe NP dispersed in p-type poly (N-vinylcarbazole), or PVK, semiconducting matrix has been reported [32]. This HgTe-PVK nanocomposite showed enhanced photocurrent levels. A one-step sonochemical approach with trioctylphosphineoxide/trioctylphosphine (TOPO/TOP) solvent has been used to synthesize Fe_3O_4–FeP core–shell and FeP hollow NPs [33].

MnO_2 NPs have been prepared by application of an ultrasonic wave under atmospheric conditions. Particle size in the range of 3–4 nm is readily obtained after incorporating $KMnO_4$ precursor inside the pore channels of the mesoporous carbon CMK-3 [34]. The three-step process comprising surface modification, $KMnO_4$ precursor embedding within the pore channel, and finally ultrasonic application is summarized in Fig. 2.3. Larger particle size could be obtained through increased precursor concentration.

Higher ultrasound frequency has been shown to increase the rate of precursor Au(III) reduction in an aqueous solution [35]. This work showed also that cavitations are responsible for the gold reduction, instead of the direct physical effects from ultrasound-induced cavitations. Ultrasonic frequency of about 200 kHz gives rise to the lowest average (15 nm) particle size [35]. Lower and higher frequencies slightly increase the average particle size [35].

Core/shell Pt/Ru NPs have also been successfully prepared by sonochemistry [36]. Incubation time and the type of stabilizer are two parameters used to control size and morphology of the core–shell NPs. Poly(methyl methacrylate)

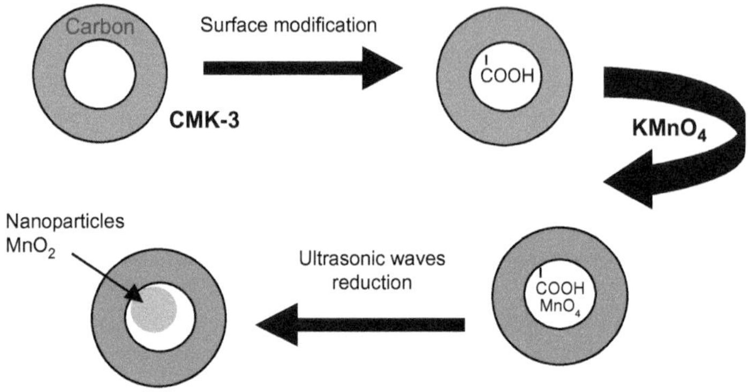

FIGURE 2.3 The three steps of sonochemical mechanism for MnO_2 nanoparticle synthesis within the CMK-3 mesoporous structure [34]. (For color version of this figure, the reader is referred to the online version of this chapter.)

and poly(butyl acrylate) latex NPs (miniemulsion) with variable size distribution range of 40–150 nm have been obtained using ultrasound involving emulsion polymerization. As described in Fig. 2.4, impinging acoustic waves induce cavitations. Formed bubbles grow until they reach critical size and then they collapse, generating primary radical formation (•H, •OH). Implosion also induces the formation of monomeric radical species in an emulsion (M•). Following emulsification of the monomer (M), secondary radicals M• react with monomer droplets to produce latex particles.

As discussed later, combined ultrasonic power and microemulsion conditions provide a powerful process to synthesize high-quality organic and inorganic NPs. For example, conductive polymer NPs have been prepared using this approach [38]. In the presence of additives, ultrasound also allows doping of the conductive polymer NPs.

3 SUPERCRITICAL FLUID

Supercritical fluid (SCF) synthesis gives rise to large scale, tunable, and dispersed NPs. Depending on the ambient temperature (T) and pressure (P), synthesis is carried out at the gas or liquid phase. Under certain conditions, two phases could coexist. Beyond the critical point ($T > T_c$ and $P > P_c$), only a single-phase fluid exists with density, viscosity, and diffusivity intermediate between those of liquid and gas phases [39]. This is referred to as the SCF phase. Under supercritical conditions, solvent strength is tunable by varying the temperature and pressure of the fluid. Indeed, varying temperature and pressure give rise to variable solvent density. In the case of water, the supercritical point occurs at $T_c = 374$ °C and $P_c = 22.1$ MPa. The dielectric constant of water decreases from around 78 at room temperature to less than 10 at high

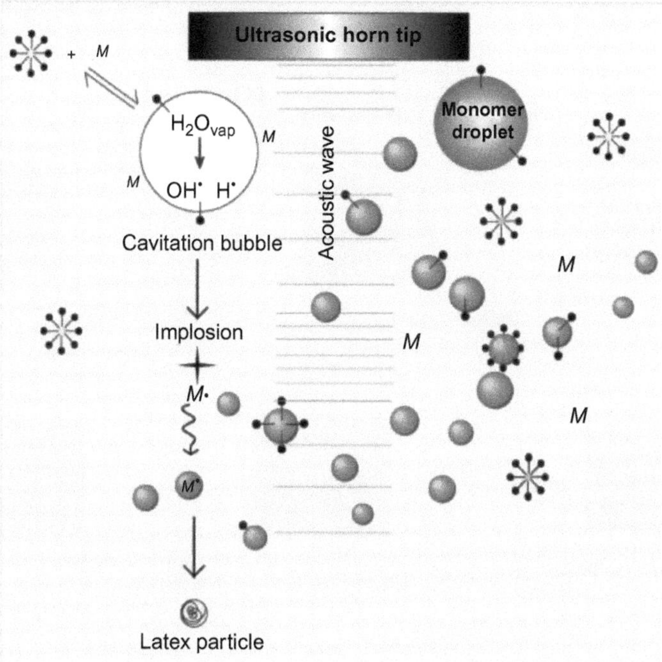

FIGURE 2.4 Schematic mechanism of miniemulsion polymerization induced by ultrasound [37].

temperature [40]. Tunability of the solvent quality (density) is a practical route to control NP characteristics [41].

Preparation methods for metallic NPs using SCF have been recently reviewed [42]. CO_2 and H_2O remain the most widely used fluids for NP synthesis [43,44,42]. SFC has been used in the past to synthesize supported NPs. In a conventional SFC process, metal precursors are first dissolved in a supercritical (sc) fluid such as CO_2 (e.g., scCO_2). The solution is then deposited onto a support followed by a chemical reduction to a metal state. Availability and dielectric constant range are two parameters that could dictate the fluid choice.

Three different SCF synthesis approaches have been developed. They include chemical reduction in the presence of a reducing agent, thermal reduction, and thermal decomposition or chemical conversion following the depressurization step (Fig. 2.5). The first two approaches consist of chemical and thermal reduction following OM (metal precursor) impregnation within the substrate. In the third approach, depressurization is first applied and is followed by thermal or chemical conversion. In all three cases, metal NPs dispersed in the supporting matrix are obtained. Furthermore, all these approaches provide the possibility of direct NP integration into the final product.

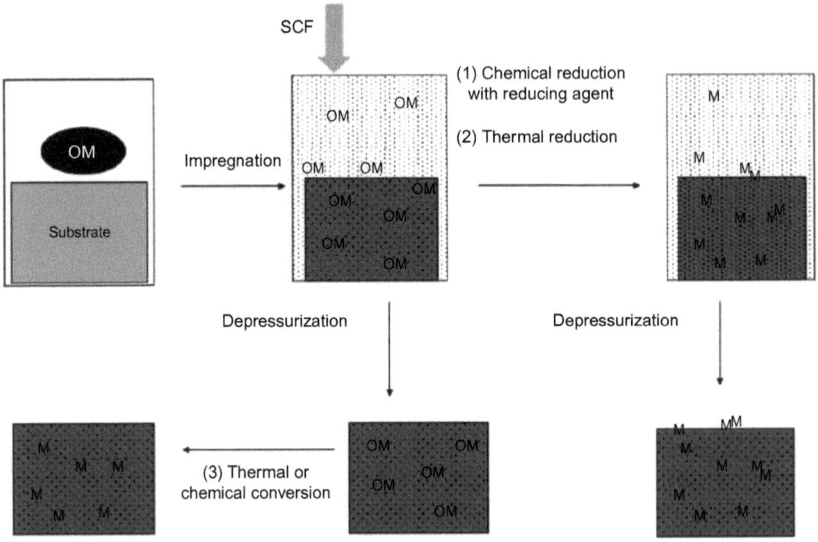

FIGURE 2.5 Supercritical fluid synthesis of supported nanoparticles, consisting of the metal precursor (OM) impregnation step followed by the conversion step [42].

In practice, there are several SCF setups used to synthesize NPs. Indeed, at least 15 design setups have been reported in the literature [43,45,46]:

- **a.** static supercritical fluid (SSF) process;
- **b.** rapid expansion of supercritical solutions (RESS);
- **c.** particles from gas-saturated solutions (PGSS);
- **d.** gas antisolvent process (GAS);
- **e.** precipitation from compressed antisolvent (PCA);
- **f.** aerosol solvent extraction system (ASES);
- **g.** supercritical antisolvent process (SAS);
- **h.** solution enhanced dispersion by supercritical fluids (SEDS);
- **i.** supercritical antisolvent process with enhanced mass transfer (SAS-EM);
- **j.** depressurization of an expanded liquid organic solution (DELOS);
- **k.** supercritical-assisted atomization (SAA);
- **l.** hydrothermal synthesis under supercritical conditions via flow reactor (HTSSF);
- **m.** hydrothermal synthesis under supercritical conditions via batch reactor (HTSSB);
- **n.** supercritical fluids drying (SCFD);
- **o.** supercritical fluid extraction emulsions (SFEE).

Some of these techniques involve physical processes. This is the case of RESS and PGSS. Others are chemical based. This is the case of GAS, SAS, and SEDS, where a chemical reaction occurs. We will not discuss all these SCF techniques in detail. A few examples are provided, without attempting

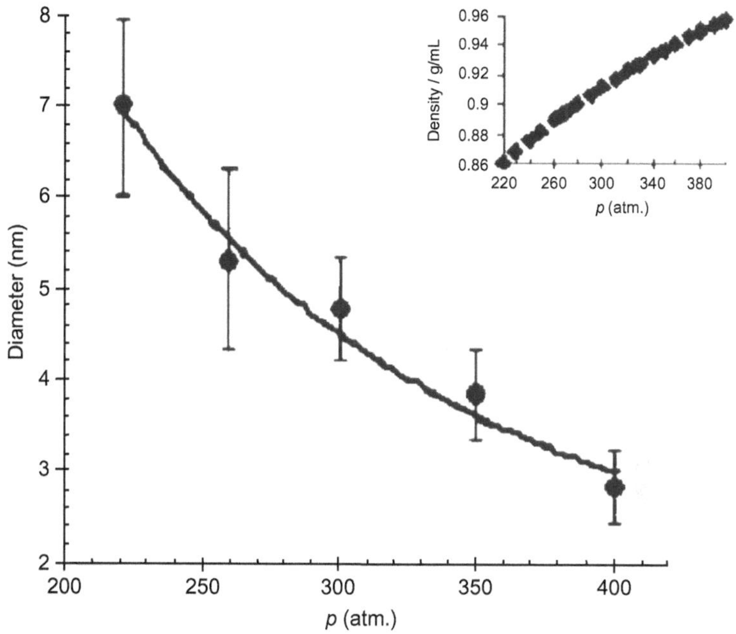

FIGURE 2.6 Variation of average size and size distribution of silver nanoparticles as a function of pressure at 40 °C. The inset shows the variation of the CO_2 density as a function of pressure [49].

to discuss in detail their advantages and disadvantages [43]. The supercritical hydrothermal and solvothermal synthesis processes are discussed further in the next paragraph.

Numerous oxide NPs are obtained using SCF techniques. In the case of TiO_2 NPs, there are numerous parameters that could affect the final product characteristics. Solvent, pressure, temperature, and residence time are used to control the size, crystallinity, and yield of the product [47]. Ge NPs that are relatively difficult to synthesize have been obtained by the SCF method [48].

A continuous tuning of silver NP size over a large range was achieved by the supercritical CO_2 microemulsion-templating method [49]. Using a surfactant concentration of 20 mM, a precursor to stabilizer concentration ratio $M = 6$, and temperature of 40 °C, synthesized silver NPs were continuously tuned from 3 to 8 nm with reducing pressure (Fig. 2.6).

RESS technique (Fig. 2.7) has been successfully used to prepare explosive microparticles and NPs such as RDX, TNT, and HMX [51]. A supercritical solution rapid expansion technique has been used to synthesize sub-100 nm polymeric NPs [50]. Using supercritical CO_2 in combination with ethanol as a solvent, and water as the receiving medium of the rapid expansion, poly (L-lactic acid) and poly(methyl methacrylate) NPs have been synthesized.

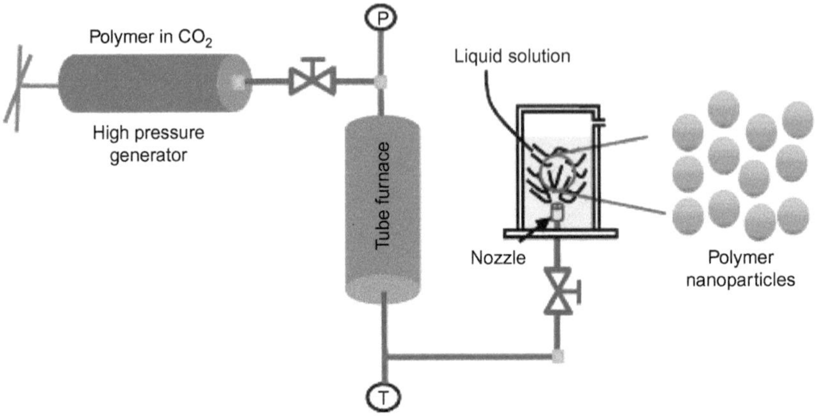

FIGURE 2.7 Schematic setup of a rapid expansion of supercritical solutions (RESS) used to synthesize polymer nanoparticles. *Reprinted with permission from Meziani et al. [50]. Copyright 2006, American Chemical Society.* (For color version of this figure, the reader is referred to the online version of this chapter.)

Ibuprofen NP has been also prepared by a similar approach [52]. Rapid expansion of $scCO_2$ containing polymers gave rise to monodispersed polymer NPs.

High molecular weight aromatic polycarbonate (PC) NPs (30–140 nm) have been produced using trans-esterification between bisphenol-A and diphenyl carbonate in supercritical CO_2[53]. Poly(ethylene oxide)–poly(propylene oxide) tri-block copolymer (PPO–PEO–PPO), NaOH and heptyl alcohol are used as stabilizer, initiator, and solvent, respectively. Three process steps (dispersion, condensation, and separation) are required to synthesize PC NPs using this supercritical CO_2 approach. Varying processing conditions reduce the average size from 98 to 45 nm [53].

4 SOLVOTHERMAL AND HYDROTHERMAL

Solvothermal is a heterogeneous reaction process of precursors dissolved in a solvent under high pressure and high temperature. To some extent, solvothermal is similar to sonochemical processes, although there is one important difference. In the case of sonochemical synthesis, reaction is often undertaken in an open reactor, whereas in the case of solvothermal synthesis, it is undertaken in a closed reactor. Often, reacting precursors are relatively insoluble under ordinary conditions [43]. For example, higher temperature increases solubility. Another difference is related to solvothermal operation temperature and pressure. Indeed, temperature and pressure values could be below the critical point. Two main factors are identified as affecting the characteristics of the synthesized NPs [54]:

a. chemical factors, including nature of the solvent, precursors, and mixture methodology;
b. thermodynamic factors, including temperature and pressure.

The role of these two factors could be monitored using specially designed reactors. A solvothermal reactor vessel, often referred to as autoclave, should resist the effects of the corrosive solvent at high temperature and pressure. Teflon and stainless steel are often used. Material choice for the autoclave will thus depend on the temperature, pressure, solvent, pH, and the overall solution composition.

The type of organic solvent could have some effect on the characteristics of synthesized NPs. Indeed, iron oxide NPs prepared in 1,4-butanediol have a higher thermal stability than those prepared in toluene, although both solvents give rise to single-crystal maghemite NPs [55]. A solvothermal approach using ethylenediamine as solvent has been used to synthesize $Cu(In,Ga)Se_2$ NPs [56].

When water is used as solvent, the solvothermal process is referred to as hydrothermal. Hydrothermal processing may involve three main steps: hydrolysis and hydrogen bonding, polymerization, and precipitation [57]. Surface-chelating agents are used to control size. Low cost and reduced environmental impact are the two advantages of the hydrothermal process. However, achieving NP monodispersity using water as solvent is very challenging.

Continuous synthesis of size-tunable NPs has been achieved using the hydrothermal system under supercritical conditions [40]. The solvent (here water) dielectric constant is reduced from 80 to less than 5 by increasing the temperature and/or pressure. Besides precursor solution and water pumping tube, this system includes the reactor with appropriate precursor delivery and product extraction, a cooling device, filtering, and a container for the final product. Metal oxide salt is mixed with heated water before introduction into the reactor. High dielectric value is often required to significantly increase the solubility properties of water. Higher dielectric constant could be obtained through continuous tuning of environmental pressure and temperature of the starting solution. Dielectric constant tunability above 20 is achieved by controlling the reaction temperature. Size tunable ceria NPs from 200 down to 20 nm have been obtained using this approach [58]. It is important to distinguish subcritical and supercritical hydrothermal conditions. In the case of CeO_2 particles, the diameter size is reduced from about 200 to about 20 nm, when the temperature is increased from 300 to about 370 °C [40]. This temperature range delineates the subcritical and the supercritical regimes [40]. This indicates that supercritical hydrothermal conditions are suitable for NP synthesis.

A recurring problem with hydrothermal reactions at supercritical conditions is the relatively wide range of size, shape, and composition distributions. This distribution is due to the time distribution it takes for the precursors to reach the supercritical conditions inside the reactor. Adschiri et al. developed a rapid and continuous supercritical hydrothermal synthesis of oxide microparticles and NPs [58]. This technique allows the starting materials to rapidly reach supercritical conditions, thus minimizing intermediary product formation [58].

The hydrothermal process has been used to prepare monodispersed anatase TiO_2 NPs [59]. First, the precursor is conditioned to include Ti-isopropoxide modification with acetic acid, hydrolysis, and peptization using HNO_3. The resulting solution is autoclaved (hydrothermal) by heating at 230 °C for 12 h. NPs with an average size of 20 nm are obtained following solvent evaporation, centrifugation, milling, and annealing [59]. This process allows size control by varying the hydrothermal processing temperature. Before utilization in screen printing film formation, these NPs are mixed with other compounds and homogenized to form a paste. These NPs are used for sunscreen, dye solar cells, and other applications. The world market for TiO_2 NPs is expected to surpass 1 million tons per year in the near future. Hydrothermal techniques could capture some significant share of this market.

Hydrothermal treatment of magnetic NPs synthesized by co-precipitation allowed reaching the critical size set at around 40 nm [60]. Monodisperse magnetic NPs ranging from 20 to 100 nm are more difficult to obtain with other techniques, although significant effort has been made, considering their potential applications.

For some applications such UV blocking and photo-oxidation, it is important to have a multiple band gap coating. This could be achieved by the combination of two or more oxide NPs with different but complementation energy bandgap values. Different energy bandgaps allow optimal solar radiation absorption over a wider range. A supercritical hydrothermal setup has been used to synthesize ZnO–TiO_2 nanocomposite [61]. These nanocomposites, consisting of TiO_2 with diameter in the 20–50 nm range, were inserted between the larger (200–300 nm) ZnO_2 particles. This insertion minimizes the agglomeration of the ZnO_2 particles.

5 MICROWAVE

Utilizing microwave heating for cooking food was first demonstrated and patented by Raytheon Company in the 1940s. Domestic applications took off in the 1970s, to become a basic kitchen item in most households. Microwave dielectric heating is now also a well-established commercial process for food, rubber, and paper processing. Until recently, microwave heating in chemical synthesis received very little attention [62,63]. Microwave heating has been used for synthesis and postsynthesis processing for decades. Dry and wet conditions are used. Microwave synthesis has been successfully used to synthesize a variety of nanomaterials [64]. Today, microwave systems for research, with precise power and temperature controllers, are commercially available. However, most data found in several scientific publications are still obtained using home-based microwave, with limited control over reaction temperature.

Microwave heating attributes have been used in nanostructured material synthesis [64–66]. Precise control over the heating process and uniform heating are two important attributes of microwave. Microwave synthesis of inorganic materials is fast [67] and the reaction occurs at relatively lower temperature [68]. Furthermore, in some cases, microwave synthesis is more efficient, with a lower amount of by-product waste. Microwave heating-based synthesis provides advantages over conventional heating in numerous cases [64–69].

Fast heating and green attributes make microwave heating a potential commercial process for producing quality NPs. Specific and nonspecific microwave interactions with reactants allow a higher heating rate. Selective and remote reaction is possible. Solvents with high dielectric constant (ε) are often required. Water ($\varepsilon=80$) and methanol ($\varepsilon=34$) are preferred solvents, whereas benzene ($\varepsilon=2.3$) and chloroform ($\varepsilon=4.8$) should be avoided [69]. Depending on the media, applied microwave power per unit volume is given by the following equations [70]:

- In the case of ionic conduction: $P_v=E^2qn\mu$

where E, q, n, μ are the electric field intensity, ion charge, ion density, ion mobility, respectively.

- In the case of dipolar rotation: $P_v=k\,E^2\omega\,\varepsilon''$

where k is a constant related to the unit of measurement used, ω is the applied angular frequency, E is the electric field intensity, and ε'' is the effective dielectric loss factor of the material as defined by *the* complex dielectric constant, $\varepsilon=\varepsilon_o(\varepsilon_r{}'-j\,\varepsilon_{r''})$, *with* ε_o *representing the* absolute dielectric constant.

Penetration depth (D_p) of the microwaves is a critical parameter, as it defines the effective volume of reaction and thus the optimal size of the reactor vessel. D_p is provided by the following equation [71]:

$$D_p(\text{cm})=\lambda_0\varepsilon'^{1/2}/2\pi\varepsilon''$$

where λ_0 is the excitation wavelength of the microwave system. Typically, commercial microwave systems use an excitation wavelength of 2.45 GHz. For example, at 95 and 25 °C the penetration depth of water is 5.7 and 1.4 cm, respectively [71]. The penetration depth of epoxy resin is around 4100 cm [71].

Microwave heating ability of a solvent is determined by loss tangent ($\tan\delta=\varepsilon''/\varepsilon'$) [62,63]. This parameter is a more representative figure of merit than the dielectric constant. High microwave absorbing solvents include ethylene glycol ($\tan\delta=1.35$) and ethanol ($\tan\delta=0.941$). Low microwave absorbing solvents include hexane ($\tan=0.02$) and toluene ($\tan\delta=0.04$). Water with a $\tan\delta=0.123$ is considered a medium absorbing solvent.

Microwave synthesis can also be used without a solvent or with a minimum amount of solvent. This provides opportunities to minimize waste and develop a clean synthesis process [72]. This issue is critical for reducing the environmental impact of NP manufacturing. There are already numerous reports on using microwave synthesis for large-scale process with both continuous flow and stop-flow techniques [73]. Dry microwave processing has been used mostly for materials post processing, rather than NP synthesis.

Reflux synthesis of NPs using conventional heating and polyol solvents has been widely used. This so-called polyol method is also amenable to microwave heating. In a polyol method, a diol acts as solvent, reducing agent, and surfactant. Typically, precursor compounds such as oxides, nitrates, and acetates are dissolved in a diol (ethylene or diethylene glucol family), followed by refluxing around the boiling temperature of the solvent. A polyol process and a variable frequency microwave (VFM) oven have been used to synthesize silver NPs [74]. This VFM was operated with an average frequency, bandwidth, and sweeping time set at 6.425 GHz, 1.15 GHz, and 0.1 s, respectively. Microwave heating using theses parameters provided a much faster reaction than conventional thermal heating. Indeed, temperature, reaction time, and precursor concentration seem to have a direct effect on NP characteristics [74].

Microwave heating has been used to synthesize metallic [75], semiconductor [76] and oxide [77] NPs. As shown in Fig. 2.8, semiconductor QDs are formed within the first minute of microwave heating, using cyclohexanone as susceptor and NP stabilizer. Varying the reaction from 30 to 300 s allows tuning of particle size from 2 nm to 6 nm, respectively. These NPs are easily integrated into conductive polymer matrix for photovoltaic devices without complicated work-up [76].

Kundu et al. showed that about 60 s are needed to synthesize gold NPs in the presence of poly(*N*-vinyl-2-pyrrolidone) (PVP) [78]. Besides directly integrating NPs into polymeric matrix, this technique uses aqueous solutions, eliminating the need for organic solvents. Particle sizes ranging from 6 to 22 nm have been obtained by increasing PVP molecular weight from 10 to 360 kDa. These gold NPs showed good catalytic activity for the reduction of aromatic nitro compounds by $NaBH_4$[78].

Various microwave synthesis parameters have been recently shown to affect specifically the NP size and quantum yield [79]. Indeed, applied power, reaction time, and reaction temperature are found to affect the NP size and quantum yield directly (Fig. 2.9). Reaction time has a more dramatic effect on both particle size and QY. Depending on the specific applications and also on the need to reduce energy cost, experimental conditions can be adjusted. High QY yield is achieved at reactor temperature of around 220 °C (Fig. 2.9C).

Carbon-coated $LiFePO_4$ NPs have been synthesized in aqueous solution, using relatively safe starting materials [80]. Electrochemical characterization

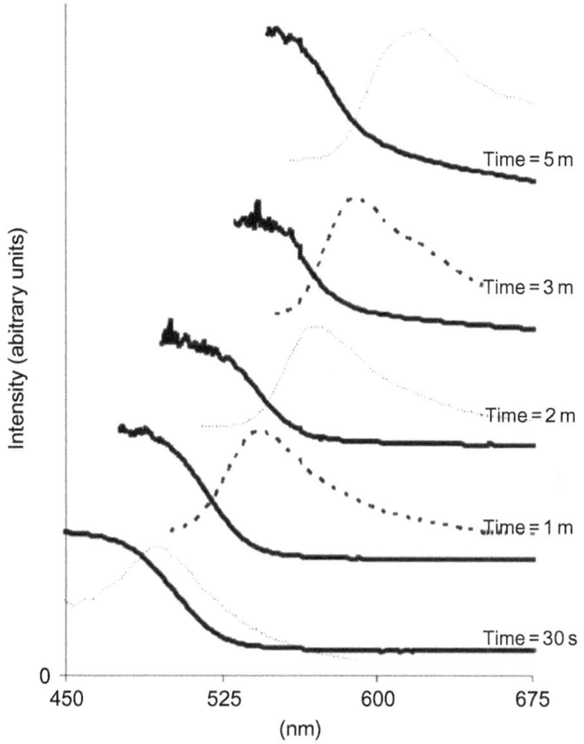

FIGURE 2.8 Absorption and (straight lines) and photoluminescence (dotted lines) spectra of CdSe nanocrystals obtained under different microwave heating time. *Reproduced from Ref. [76] with permission from The Royal Society of Chemistry.*

has showed promising charging and discharging capacity data. Microwave could be a cost-effective alternative for mass production of cathode materials like $LiFePO_4/C$ battery cars. Surfactant-free polystyrene NPs have been prepared by microwave heating, using potassium persulfate as the initiator in water [81]. NP size is controlled by varying the monomer-to-initiator weight ratio. Uniform heating allowed simultaneous initiation within the whole volume. In addition, water molecule stirring due to variable electromagnetic field contributed to uniform particle distribution.

Microwave heating has also been adapted to gas phase synthesis. Using silane precursor concentration of 380–2530 ppm, pressure of 20–30 mbar, and microwave powers of 120–300 W, silicon NPs with an average size of 6 nm have been obtained [82]. A quartz glass reactor, transparent to microwave, including a mean to feed Ar/H_2 mixture is used. This continuous process could be effective in producing high-quality silicon NPs. Furthermore, the relatively pure NP could be integrated into devices with minimal contamination. Gas phase NP synthesis using microwave may address issues related to quality control critical in several application areas.

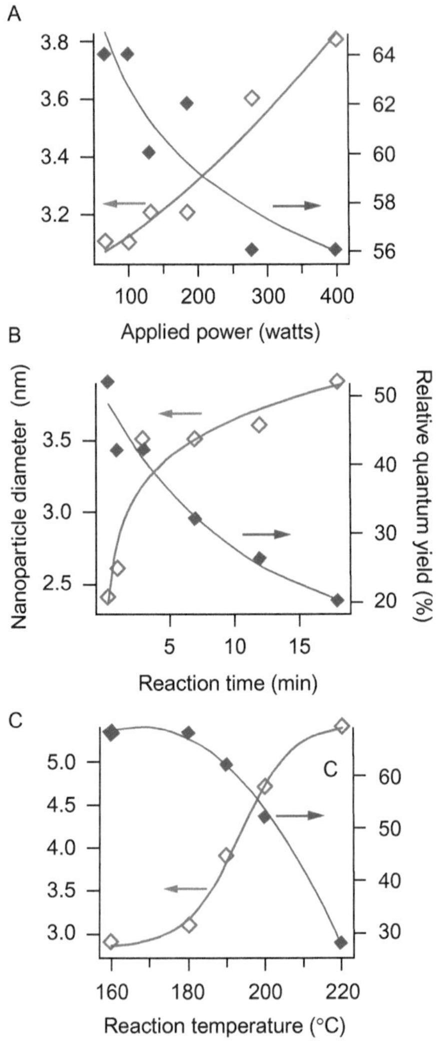

FIGURE 2.9 Variation of the nanoparticle size and quantum yield as a function of applied power, reaction time, and reaction temperature. *Reprinted with permission from Gerbec et al. [79]. Copyright 2005, American Chemical Society.* (For color version of this figure, the reader is referred to the online version of this chapter.)

6 ELECTRICALLY INDUCED SYNTHESIS

NP synthesis involving electrically induced charge transfer occurs under different forms. Electrochemical reaction process is the most widely used and versatile electrical-induced synthesis. Electrospray, a more scalable although less versatile technique, is also used for NP synthesis.

6.1 Electrochemical Techniques

Electrochemical techniques have been used to synthesize a wide variety of metallic, semiconducting, and oxide NPs [83–88]. Electrochemical synthesis is potentially low cost and has high production yield. Furthermore, electrochemical synthesis techniques provide the possibility of both batch and continuous processes. A schematic layout for an electrochemical synthesis of metallic NPs stabilized with tetraalkylammonium ligands is shown in Fig. 2.10. Reactions occurring at each electrode are summarized below [83]:

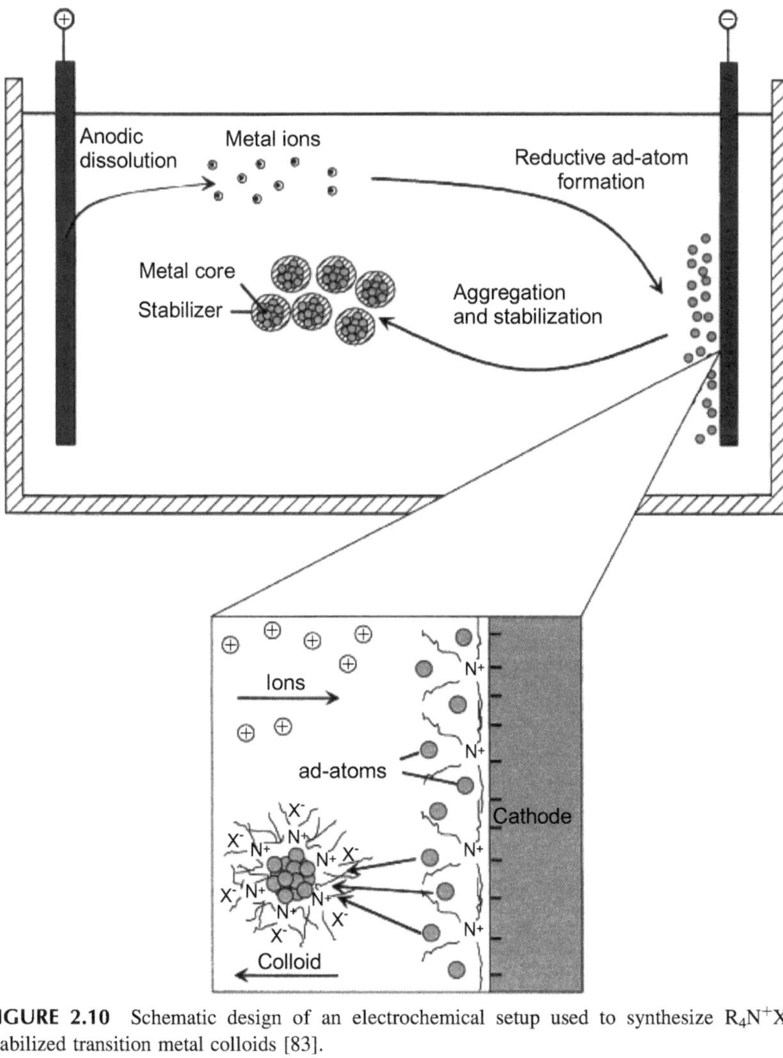

FIGURE 2.10 Schematic design of an electrochemical setup used to synthesize $R_4N^+X^-$-stabilized transition metal colloids [83].

- *Anode*: $M_{bulk} \rightarrow M^{n+} + ne^-$
- *Cathode*: $M^{n+} + ne^- + \text{stabilizer} \rightarrow M_{np}\text{-stablizer}$

where M_{bulk} and M_{np} refer to bulk and NP metals, respectively. In this process, the tetraalkylammonium salt ($R_4N^+X^-$), where X = Cl, Br, or I and $R = n\text{-}C_mH_{2m+1}$) is used as a NP stabilizer.

Numerous parameters, including solvent polarity, current density, charge flow, distance between electrodes, and temperature, are critical for controlling particle size selectivity of the electrochemical synthesis process. In the case of Pd, NP size from about 1 up to 5 nm is tuned using temperature, charge flow, and current density variable parameters [83].

This approach has also been used to synthesize bimetallic NPs including Ni–Pd, Fe–Co, and Fe–Ni. Cobalt NPs with a tunable average particle size from 2 to 7 nm have been synthesized electrochemically in the presence of tetraalkylammonium salts [87]. Lowest average particle size is obtained using a current density of about 3 mA/cm^2 or more [87]. Reducing current density increases the average particle size. The yield and scalability of this technique are not quantified. However, this electrochemical technique is also used to deposit these NPs into the substrate to form a film with tunable properties. Numerous metallic NPs with different size ranges are prepared using an inexpensive pulsed electrodeposition setup [88]. Shape of the current pulses, organic additives, bath temperature, and pH allow control of the physical properties of these NPs.

CdS NPs with an average size of 5 nm have been synthesized by a one-step electrochemical [84]. Using CdNO, elemental sulfur, ethylene glycol, and KNO_3 as the supporting electrolyte, stable NPs are obtained after a 2-h reaction at around 70 °C. Silver NPs with a size of about 20 nm have been synthesized by both potentiostatic and galvanostatic methods in ethanol solution [85].

Mono, binary, and ternary alloyed magnetic NPs have been synthesized using an ultrasound-assisted electropulse technique [86]. A working electrode is attached to the titanium horn of an ultrasound element. The reference electrode and the counter electrode are linked to a pulsing generator. Two approaches are used to produce NPs. In the first approach, the three-electrode configuration is used to control the anodic/cathodic potential of the tip of the horn versus a reference electrode. In the second approach, the anodic/cathodic current density passing through the tip surface is varied. Electrochemical and ultrasound parameters allow control of the NP characteristics. The precision of the electrochemical synthesis helps in obtaining high-quality NPs. Crystalline NPs without dentrite formation are obtained. Furthermore, the composition of the final product is similar to the starting material, confirming that this electropulse technique will provide high production yield. This electropulse technique has been used by other groups to synthesize metallic and semiconducting NPs [86].

A scalable and cost-effective electrochemical process for the production of titanium particles with an average size of around 100 nm from TiO_2 pellets in the solid state has been reported [90]. Known as the FFC (the first letters of the inventors' names) process, this technique is believed to produce metals and alloys from ores more efficiently than the conventional Knoll process. The FFC process is based on reducing solid TiO_2 powder, after combining with molten $CaCl_2$, to form the cathode (titanium) [91].

It has been shown that nanocrystal electrodeposition gives rise to nanocomposites with improved thermomechanical properties [92,93]. For example, hardness improves by a factor of 6 when reducing the average grain size to about 13 nm [93]. Using electrodeposition, high-density nanostructured films are achieved even prior to posttreatment. Thus, improvement of the product properties does not come at higher cost. Pulsed electrodeposition has been used to synthesize NPs directly integrated into an all-inorganic film nanocomposite (Fig. 2.11). In this case, electrodeposition yielding nanosized grains is possible when the experimental conditions are set to favor the nucleation of new grains instead of the growth of existing grains.

Metal NP films are formed directly on graphite or H-terminated silicon surfaces using the electrochemical process [95]. Combining chemical and electrochemical processes, QD are directly deposited to form a single layer on a graphite substrate (Fig. 2.12). This electrochemical method has been used to synthesize CuO, CdS, and ZnO NPs, directly assembled on graphite substrate (Fig. 2.12).

To produce semiconducting NPs (MX) from a generic semiconducting salt, metal NPs (M°) are first electrochemically obtained from metal ions (M^{n+}). This first step is followed by electrochemical oxidation of these metal particles for oxide species ($MO_{n/2}$). Finally, reduction of the metal oxide $MO_{n/2}$ using HX yields the semiconducting (MX) NPs. This technique has a clear advantage by yielding integrated QD into a solid state substrate in a single step. This method could be faster and cheaper than the more widely used molecular beam epitaxy.

Ni_3Sn_4 NPs with an average size of 50 nm have been obtained by electrodeposition on Cu nanorods [97]. There is no coalescence between the intermetallic NPs even after electrochemical (charging/discharging) cycling. Using copper as current collector, deposited NPs provide potentially high capacity and stable alternative electrode design for lithium batteries.

6.2 Electrospray

Electrospray synthesis, also called electro-hydro-dynamic atomization, uses an electrical field to break up a liquid jet into droplets [98]. A more detailed description of different nonelectric induced spray techniques is provided in the next paragraph. Synthesis of NPs and microparticles using electrospray

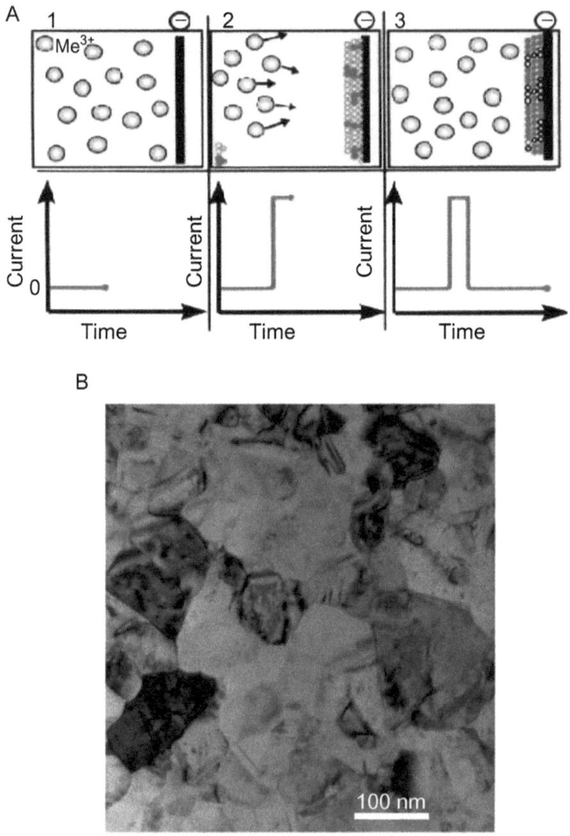

FIGURE 2.11 Schematic setup of a pulsed electrodeposition setup for synthesizing nanocrystalline materials (A) and an electronic image of nanoparticle-based nickel nanocomposite film using pulsed electrodeposition (B) [94]. (For color version of this figure, the reader is referred to the online version of this chapter.)

has been reviewed recently [99]. A typical electrospray setup used to fabricate fine powder is shown in Fig 2.13. Electrospray produces smaller monodisperse droplets in comparison to other spray processes. To obtain narrow size distribution, it is important to avoid aerosol explosion before evaporation takes place [100]. Using a cone-jet electrospray mode including a drying gas, high-voltage source, and an induction electrode, highly charged droplets are formed and discharged in a radioactive neutralizer. Monodisperse NPs between 4 and 180 nm have been thus obtained by combining electrospray and radioactive neutralizer. Electrospray synthesis from a solution containing precursors is based on droplet generation and electrical charging. Droplet size can be varied from 40 nm to 1.8 μm [100]. This range of droplet size could facilitate synthesis of tunable NPs. In spite of

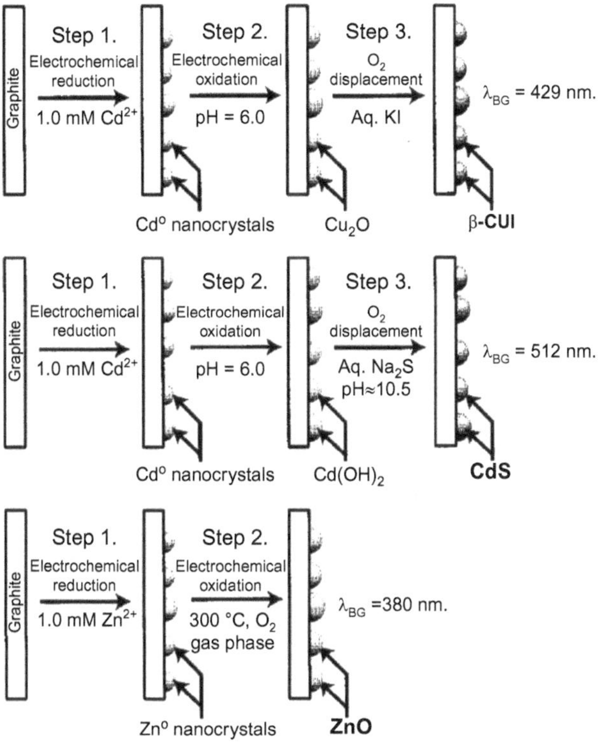

FIGURE 2.12 Schematic representation of a combined electrochemical–chemical fabrication process of different semiconducting nanoparticles on graphite. *Reprinted with permission from Penner [96]. Copyright 2000, American Chemical Society.*

all its advantages, it is difficult to use electrospray for large-scale production of NPs [101].

ZnS NPs with an average diameter ranging from 20 to 40 nm have been prepared by electrospray pyrolysis of ethanol solution containing zinc nitrate ($Zn(NO_3)_2$) and thiourea $SC(NH_2)_2$[102]. An electrospray process coupled with a CVD reactor was used to synthesize nonagglomerated SiO_2, TiO_2, and ZrO_2 NPs [103]. Size distribution of these NPs could be adjusted by varying the initial precursor concentrations. Indeed, SiO_2 particle size increases from 12 up to 30 nm when the precursor concentration is increased from 0.3×10^{-6} to 5×10^{-6} mol/L or more [103]. Although it is scalable and greener, the relative size distribution obtained using electrospray is much larger when compared to other wet chemical techniques.

Electrostatic forces increase interparticle repulsion. Thus, higher surface charge density of the electrospray jet leads to smaller droplets [104]. One approach allowing increase in surface charge density is to provide capillary effects. This could be achieved by using a sharp needle connected to a high-voltage supply [105]. This concept has been validated and successfully used

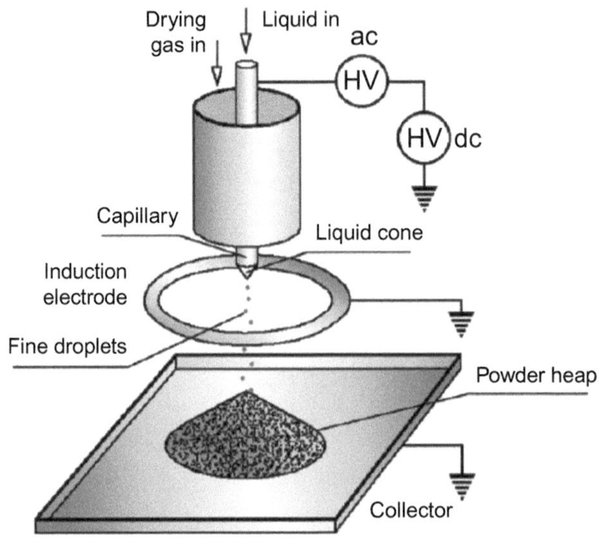

FIGURE 2.13 Schematic outline of an electrospray fabrication setup [99].

to synthesize 30- to 100-nm-sized SiO_2 particles by spraying a 10 vol % TEOS solution in ethanol. A continuous electrospray system consisting of two different sections has been developed [105]. One section of the electrospray setup corresponds to the droplet formation. The second section includes a furnace where decomposition allows NP formation. This technique is often referred as spray pyrolysis, described in more detail in the next paragraph.

7 SPRAY PYROLYSIS

Depending on the composition of the starting solution, spray techniques can be used to synthesize NPs, to produce film from NP precursors, or both. In this paragraph, we focus on producing NPs from a solution containing precursors. The various spraying techniques differ on the type and origin of the energy input required for NP formation. Synthesized NPs could be either in an aerosol state, dry powder, or NP embedded in a solid matrix.

The suitability of spray pyrolysis for continuous large-scale production of NPs has been demonstrated [106–108]. First, a pumping system is used to inject the solution precursors in the aerosol generator. Aerosols are generated using pneumatic, ultrasonic, or electrostatic atomization techniques. This is followed by their decomposition at high temperatures in one or several reactors (Fig. 2.14). Reactor temperature is usually higher than the melting temperature. A large-scale production of nickel particles using this approach has been recently reported [108]. Two reactors set at two different

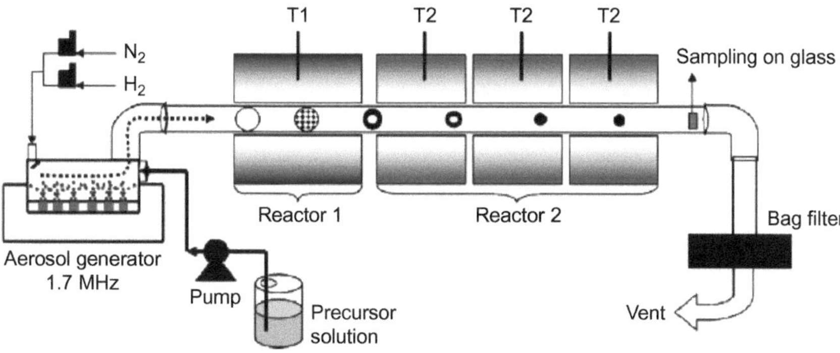

FIGURE 2.14 Schematic representation of a spray pyrolysis setup used in a large-scale nickel nanoparticle production [108]. (For color version of this figure, the reader is referred to the online version of this chapter.)

temperatures (T1 and T2) are used to decompose the aerosol droplets into solid particles. Finally, a combination of sampling and filtering steps is used to collect the formed particles. Numerous processing parameters have been used to optimize size distribution. For example, to obtain particle size below 200 nm, salt concentration should be less than 0.01 M. This setup has been successfully used in large-scale nickel NP production.

Spray hydrolysis process has been already used in pre-commercial and commercial manufacturing processes [109]. Following spray hydrolysis, the hydrated product is calcinated in the presence of additives. This is followed by milling, spray drying, and size reduction steps. The size reduction step followed by spray drying has been used to mass produce different oxide NPs. The main production steps used in oxide NPs using spray hydrolysis is shown in Fig. 2.15. Anatase TiO_2 NPs, with size ranging from 20 to 100 nm, have been produced from $TiCl_4$ dissolved in aqueous HCl solution.

Pyrolysis of ultrasonically atomized precursors has been used to fabricate silver NPs [110]. It is also possible to synthesize core–shell NPs using this technique. This technique is used to synthesize NPs with an average size of about 20 nm, using silver nitrate. Pyrolysis temperature of around 650 °C has been used for this. Increasing atomizer feed concentration and power allowed tuning of the average particle size up to 300 nm [110].

Monodispersed and crystalline $BaTiO_3$ NPs with tunable size from 30 to 360 nm have been prepared, using salt-assisted spray pyrolysis [111]. The particle size is controlled by varying the furnace temperature between 700 and 1000 °C. Up to around 100 nm, crystallite size and geometrical size values measured by XRD and TEM, respectively, are quite similar. This confirms the monocrystallinity of these NPs. However, above the nominal size of 100 nm, geometrical sizes are larger than the crystalline size [111].

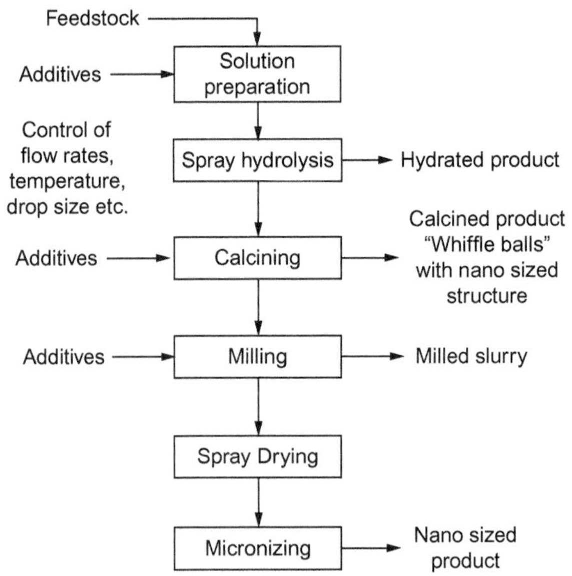

FIGURE 2.15 Process flowchart for oxide nanoparticle manufacturing [109].

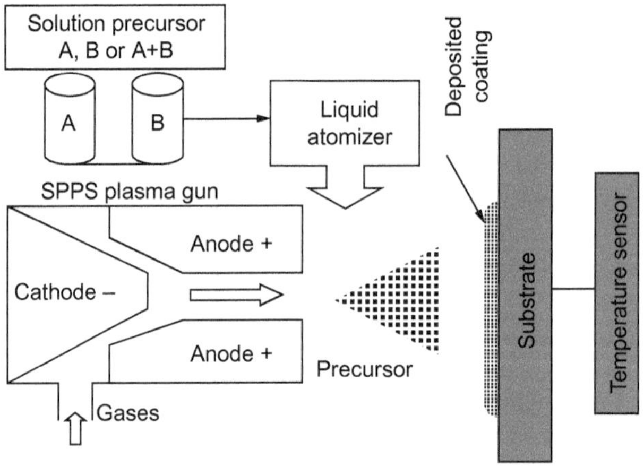

FIGURE 2.16 Schematic representation of the solution precursor plasma sprays (SPPS) technique [112].

Solution precursor plasma spray (SPPS) has been used to produce NPs directly integrated into the film (Fig. 2.16). This 2-in-1 approach for the fabrication of NP-based films could have a significant advantage for some applications. Besides reducing the number of manufacturing steps, this technique provides higher material yield with reduced energy requirements. This process also gives rise to higher quality films with a higher adhesion coefficient to the substrate. However, there is less control over the particle size distribution.

Indeed, large size distributions are obtained with significant contributions from micro-size particles.

A flame-based spray deposition, often referred to as flame-assisted ultrasonic spray pyrolysis (FAUSP), has been used to produce ceramic NPs with tunable size. The FAUSP setup consists of three parts, including the nebulizer with a carrying gas delivery and the spraying chamber connected by a tubing system [113]. Different NP collection plates are used [113]. By varying temperature and deposition time, NPs with size ranging from around 30 to more than 60 nm are obtained [113].

8 EMULSION

Emulsion droplets, liposomes, vesicles, micelles, and capsules are used as synthetic reactors for micro and submicron particles. For example, micelles and inverse micelles are used as templates for NP synthesis through either hydrolysis or reactive precursors. Micelles and inverse micelles form aqueous and nonaqueous (oil) dispersions, respectively [114]. In micelles, the hydrophobic group of the surfactant is miscible with the oil based core. In inverse micelles, the aqueous phase forms the core and the hydrophobic group of the surfactant is directed to the surrounding oil phase. These different emulsion droplets are used in pharmaceutical, consumer, and agrifood products.

It is important to differentiate the various micelles from emulsion (microemulsion, macroemulsion, and miniemulsion) [115]. The literature is sometimes quite confusing in this regard. Emulsions are obtained when amphiphiles (also called emulsifier or surfactant) are added to two immiscible liquids (usually water and oil). Depending on the relative ratio of water (W) to oil (O) concentration, either oil-in-water (O/W) or water-in-oil (W/O) emulsion droplets are obtained. Surfactants play an important role during the emulsion process. Different types of surfactants are used: water-in-oil or w/o, oil-in-water or o/w, and water in supercritical CO_2 or w/sc-CO_2[116].

Micron- and submicron-sized emulsions are often referred to as microemulsion. Microemulsion is a distinct, transparent, low viscosity, and thermodynamically stable phase. Larger sized emulsion is referred to as macroemulsion. Macroemulsion is thermodynamically unstable and highly viscous, requiring a co-emulsifier to improve stability. Heterophase consisting of nanodroplets dispersed in a continuous second phase are referred to as miniemulsions (also referred as nanoemulsion). Miniemulsion is a metastable phase. Given their size (200 nm or less), miniemulsions are transparent or at least transluent. In the polymerization process, one can distinguish conventional emulsion polymerization (macroemulsion) from nanoemulsion and microemulsion (Fig. 2.17). Conventional emulsion polymerization starts from an aqueous emulsion containing the monomer with an excess of surfactant that induces micelles. High-energy homogenization allows droplet formation

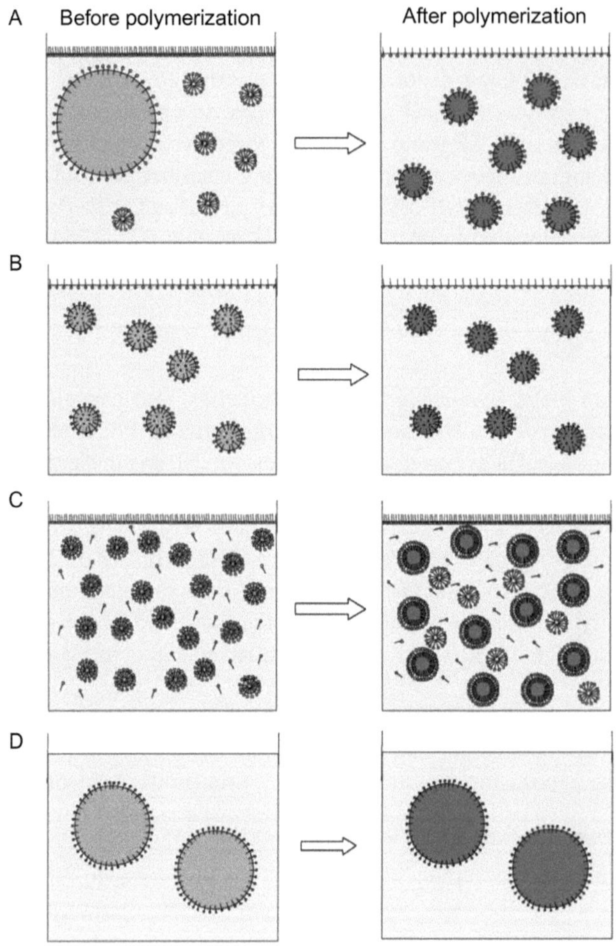

FIGURE 2.17 Schematic representation of different emulsion processes: (A) emulsion polymerization or macroemulsion, (B) miniemulsion polymerization, (C) microemulsion polymerization, and (D) suspension polymerization. The structures before (left) and after (right) polymerization are shown [117].

with polymerization mostly at the interface or continuous phase. Colloidal structures are thus obtained. Microemulsions are used to induce polymerization on a large scale through tight control of monomers, radical initiators, and co-surfactants distribution between the aqueous and organic phases.

Organic NP synthesis in aqueous phase is attracting a lot of interest due to its application in pharmaceuticals, neutraceuticals, cosmetics, printing inks and toners. Starting from a molecularly dispersed solution of hydrophobic organic compound (HOC), three different emulsion methods are distinguished [89]. This dichotomy depends mostly on the type of solvent used. With lipophilic solvents, o/w emulsion is obtained following mechanical homogenization in the presence of additives. In this case, HOC pseudo-latex or HOC hydrosol is obtained, depending on whether polymers are used. When amphiphilic solvents are used, HOC loaded pseudo-latex NPs are formed, following a spontaneous emulsification in an aqueous solution containing polymers. When a hydrophilic solvent is used, a spontaneous particle occurs in the presence of the stabilizer. In this case, HOC loaded pseudo-latex or hydrosol are also formed depending on whether a polymer is added. The formation of hydrosol in a lipophilic solvent is illustrated in Fig. 2.18.

A continuous process has been developed to produce hydrosol NPs by precipitation from a hydrophilic solution [89]. Monodispersed β-carotene hydrosol NPs have been obtained following a turbulent mixing at around 200 °C. Online measurement of the forming NPs using a laser optic device

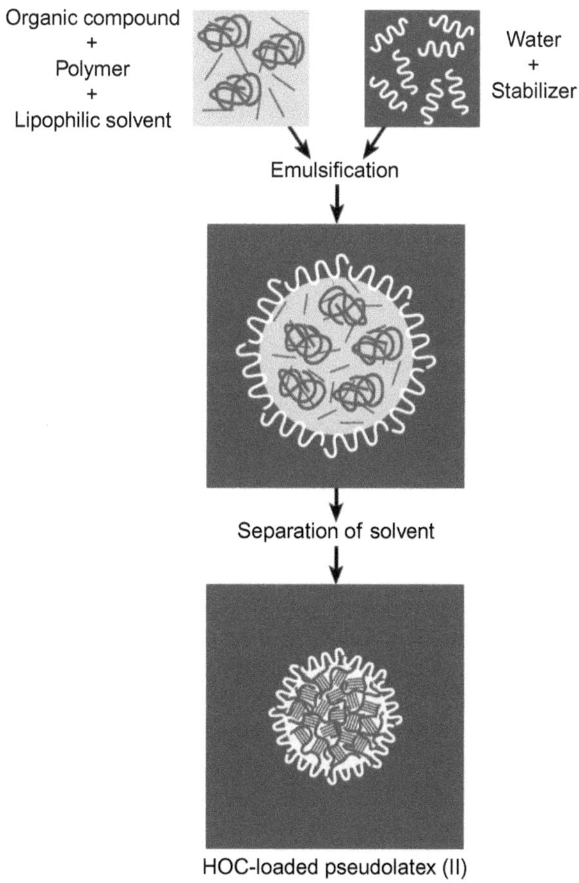

FIGURE 2.18 Simplified preparation process of active-compound hydrosol nanoparticles by the emulsification and evaporation processes [89]. (For color version of this figure, the reader is referred to the online version of this chapter.)

is possible. TEM data confirmed that the majority of NPs are less than 40 nm large [89].

8.1 Microemulsion

Water-in-oil remains the most used microemulsion technique [118]. Three basic steps leading to the formation of metal NPs by using microemulsion are distinguished [118]. First, two separate microemulsion aqueous phases consisting of metal salts (e.g., $FeCl_3$, $FeCl_2$, and $CuCl_2$) and reducing agent (NH_4OH, $NaBH_4$) are prepared. These two solutions are mixed, leading to percolation. Finally, following chemical reaction, metal and/or metal oxide precipitates are obtained. To optimize preparation of monodisperse Pd NPs using w/o microemulsion, a semi-batch reactor has been used under atmospheric nitrogen.

A 200-mL glass tank including four baffles and a four-pitched blade turbine impeller was used. A stirring rate was used to optimize NP properties. It was found that precharging the reactor with the microemulsion containing the reducing agent and the constant feeding of the Pd salts gives rise to monodisperse NP distribution. Two microemuslisons containing Pd salt and reducing agent are mixed under appropriate conditions that include defined uniform mixing conditions and stirring rate. Monodispersity of Pd is obtained by precharging the reactor with the microemulsion containing the reducing agent, followed by constant feed rate addition of the Pd microemulsion [119]. With optimized preparation parameters, monodisperse NPs with an average diameter of 5 nm have been achieved. It is quite difficult to synthesize multicomponent NPs such as YBCO-123 superconductor using standard co-precipitation. Microemulsion has been shown to provide YBCO-123 NPs with an average diameter of about 100 nm [120].

8.2 Miniemulsion

Also referred to as nanoemulsions, miniemulsion particles typically have a size of 20–200 nm [121–126]. Depending on the composition and size, these miniemulsions are transparent or at least transluent. Miniemulsions are obtained in aqueous and organic solution, using nonpolar phase with hydrophobe and polar phase with a lipophobe, respectively [123]. In spite of some similarities, miniemulsion is different from microemulsion. Miniemulsion is formed by mechanical shear, whereas microemulsion is formed by self-assembly [121]. It requires much more surfactant (5%) than macroemulsion (1%), although much less than oil/water microemulsion (20%) [122]. Miniemulsion uses less surfactant than microemulsion. Because of this, miniemulsions are not stable over a long period. Both microemulsion and miniemulsion have been used in radical polymerization [127].

A combination of modified emulsion templating and freeze-drying is used to synthesize organic NPs containing water-insoluble compounds [128]. First, surfactant-stabilized oil-in-water emulsion and polymer are mixed and then a freezing step is applied. This two-step process leads to uniform surfactant- and polymer-stabilized organic NPs. These NPs showed enhanced water solubility as demonstrated in the case of triclosan [128].

9 SOL–GEL

The sol–gel method was used for the fabrication of ceramics, monolithic glasses, submicron particles, and aerogel materials, at a relatively low temperature. Sol–gel consists of a solid skeleton (gel) filled with a solution of solid particles or chemical reagents (sol). It has been used to synthesize NPs in colloidal and film form for decades [129–131]. Controlling the pH value of the starting precursor solution is critical to obtaining the appropriate size and shape of the NP. There are two types of sol–gel structures and processes [131]. In one case, a gel structure with identifiable colloidal sols is used. In this case, sols consisting of suspensions of fine powder are obtained. In the second case, a polymeric gel with a solution of metal-organic compounds is used. In this case, 0D (NP) structures are often not achieved. In both cases, the drying process of the gel gives rise to a relatively dense product.

On a molecular level, sol–gel-based NP formation can be summarized as follows. Starting from the monomer precursor, dimers and oligomers, and then NPs, are successively formed. Using controlled pH value (base), NP growth is controlled in the absence of salt. In the presence of salt, a gelation of the dispersed sol begins to occur. Using the classical Stöber process [132], monodisperse colloidal silica with size ranging from 20 to 100 nm has been obtained.

In the colloidal-based sol–gel process, suspensions of fine particles (sol) are first formed. In this step, hydrolysis of organometallic precursors gives rise to a hydrous oxide. When the concentration of the sol reaches a critical level, they begin to link together to form a 3D gel. Although sol–gel is endowed with numerous advantages, mostly related to film formation (see Chapter 4), they suffer from one significant drawback related to precursors. Indeed, hydrolysable organometallic precursors are not easy to synthesize. This increases the overall cost of the sol–gel materials.

Metal (M = Si, Ti, Al, Zr) alkoxide solution is the most widely used sol–gel, giving rise to silica, alumina, titania, zirconia, and other metal oxides. The preparation includes three main process steps:

1. hydrolysis: $M{-}O{-}H + H_2O \rightarrow M{-}OH + R{-}OH$;
2. water condensation: $M{-}OH + HO{-}M \rightarrow M{-}O{-}M + H_2O$;
3. alcohol condensation: $M{-}O{-}R + HO{-}M \rightarrow M{-}O{-}M + R{-}OH$.

Tetraethyl orthosilicate (TEOS) or $Si(OC_2H_5)_4$, tetramethyl orthosilicate (TMOS) or $(Si(OCH3)_4$, Zr(IV)-propoxide, and Ti(IV)-butoxide are typical alkoxides. These compounds have been used to produce sol–gel NPs and microparticles.

In the colloidal sol routes, agglomeration of colloid NPs is induced by the pH and/or colloid concentration changes. Both aqueous and nonaqueous sol–gel processes have been developed. In both cases, preventing agglomeration during synthesis remains the main challenge. The aqueous sol–gel process provides an economical advantage with minimum environmental impact. Indeed, sol–gel processing has several attributes [133]:

a. efficient and cost-effective process to produce NPs and films;
b. most materials can be obtained at low temperatures;
c. possibility of greater control over material chemistry and homogeneity;
d. sol–gel powders can be consolidated at much lower pressures and temperatures;
e. high-density materials without the need for low-temperature binders;
f. thermomechanical processing of films and materials can be accomplished at significantly lower processing conditions;
g. uniform infiltration is achievable because of the continuous 3D structure of the sol–gel-based NP system.

Extraction of the pore fluid from a wet gel may lead to different materials characteristics. The characteristics of the final product may depend on the relative strength of the capillary forces. Three different gel forms can be distinguished [134]. These three gel states (xerogel, ambigel, and aerogel) are obtained from sol–gel materials following different postsynthesis processing. Xerogel and ambigel are often obtained by drying from sol–gel solvent (H_2O, ethanol) and alkane (minimal capillary pressure), respectively [134]. Aerogel is often obtained via supercritical CO_2 drying (no capillary pressure) [134]. Aerogel retains the low-density framework of the wet gel followed by residual solvent removal. Aerogels are receiving a lot of interest because of their low mass density.

During postprocessing of sol–gel materials, it is important to minimize particle agglomeration. Indeed, agglomeration is deleterious in forming nanosize grain and controlling film characteristics. To prevent agglomeration, surfactants are used, either during the preparation or after the preparation of the NPs. In the case of alkoxy-precipitated TiO_2, several surfactants can be used [135]. The ideal surfactant should provide optimum characteristics to control solubility and stability. Furthermore, reversible adsorption and stability during synthesis are also important. Hydroxypropyl (HPC) has been identified as the optimum polymer surfactant [135].

A wide variety of sol–gel-based NPs has been reported in the literature [135–139,13,140–143]. Preparation of II–VI semiconductor NPs embedded in the solid matrix using sol–gel is of particular interest. Using the sol–gel process, crystalline, photostable, and monodisperse QD with high PL efficiency are easy to obtain. Chemically stable CdTe NPs embedded in a glass pan with a narrow size distribution have also been reported [136]. The sol–gel process has also been used to synthesize γ-Fe_2O_3 NPs of different sizes in a SiO_2 matrix [139]. Using a modification of the so-called Pechini method [137], rare earth (Gd, Yb)-doped tungstate $K(WO_4)_2$ NPs have been obtained by sol–gel processing [138]. These NPs (also denominated as KREW) are considered a more cost-effective alternative to Nd:YAG materials. During mixing of reactant solutions (RE_2O_3, K_2CO_3, and $(NH_4)_{10}W_{12}O_{41}{\cdot}7H_2O$ in concentrated HNO_3), the nitrate solution is evaporated to form a precipitate. Appropriate amounts of organic complexing agent and ethylene glycol are added. The resulting polymeric resin with chelated cations is calcinated to form KREW NPs [138]. The concentration of cation metals (C_M=[CA]/[METAL]) and ethylene glycol (C_E=[CA]/[EG]), relative to the complexing agent, play a critical role in size distribution of the resulting NPs. Uniform and small NPs are obtained using $C_M=1$ and $C_E=2$ values [138]. The Pechini method [137,13] has been used to synthesize relatively monodisperse lithium manganate ($LiMn_{2-x}Cr_xO_4$) NPs using PVA sol–gel [140]. Even after calcination at about 800 °C, particle size of about 90 nm is obtained. This particle size distribution provides very good discharge capacity. A microemulsion templated sol–gel synthesis has been conducted in a one-pot approach [141].

Highly dispersed Pd–Ag/SiO_2 co-gelled xerogel catalysts are prepared from alkoxides, forming a chelate with palladium and/or silver cations [142]. For example, these catalysts are found to be very active and selective for hydro-dechlorination of 1,2-dichloroethane into ethylene. Little effect has been observed when using different reagent grades and mixing steps. No difference was found when an aero-sol–gel technique in which sol–gel reactions are carried out within aerosol droplets was used to produce aggregated NPs [143].

10 TEMPLATE SYNTHESIS

Numerous natural and synthetic templates are used for NP production and integration. *In situ* NP synthesis is discussed in Chapter 4. The impregnation of precursors inside a mesoporous structure is widely used to produce NP-based catalysts for the petrochemical and automotive industries. The impregnation method consists in wetting a porous support using solution-based precursors (salts). Following solvent evaporation and sample drying, calcination and reduction are applied to form NPs within the 3D structure.

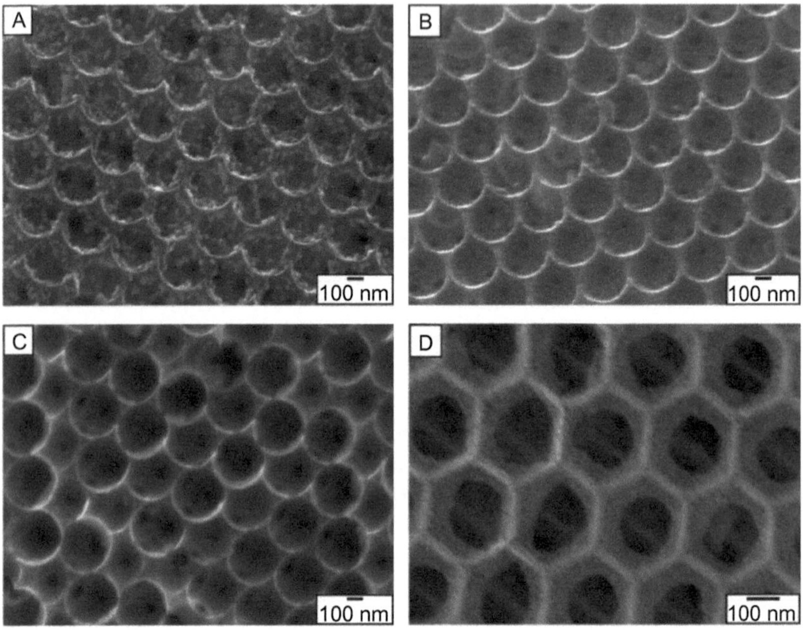

FIGURE 2.19 SEM images showing the evolution of the 3D-ordered macroporous (3DOM) consisting of (A) carbon impregnated with $SnSO_4$, (B) after decomposition of $SnSO_4$ to form SnO_2, (C) after 30 cycles, and (D) SnO_2 obtained after burning the carbon of the sample shown in Fig. 35 at 900 °C in air [144].

3D-ordered macroporous (3DOM) technique has been developed to provide a predictable catalyst support structure. After impregnating a 3DOM monolith support with an aqueous solution of $SnSO_4$ and providing thermal treatment, evenly dispersed SnO_2 NPs with diameters of 10–40 nm were obtained (Fig. 2.19). After further thermal treatment to remove carbon content, a SnO_2-based 3DOM is obtained. This 3D-ordered macroporous material consisting of interconnected pores is suitable for electrochemical-based devices, including batteries and capacitors. Other metal- and oxide-based nanocomposites are obtained using a similar approach [145].

Sub-100 nm transition metal (Cr, Mn, Fe, Co, Ni, Cu, and Zn) oxides are obtained by a wet chemical top-down approach. This method also called template-directed synthesis is a two-step process using an organic solvent (Fig. 2.20). First, a 3DOM structure is built via colloidal crystal templating. The second step consists of a spontaneous disassembly of these structures into particulate building blocks assisted by the introduction of amphiphilic surfactants. The last step, phosphate leaching, is performed using a NaOH etching process.

Micelles and inverse micelles are widely used as templates to form polymeric NPs [147]. First, monomers are introduced inside the spherical micelles. Following an *in situ* polymerization, surfactants are removed. This technique

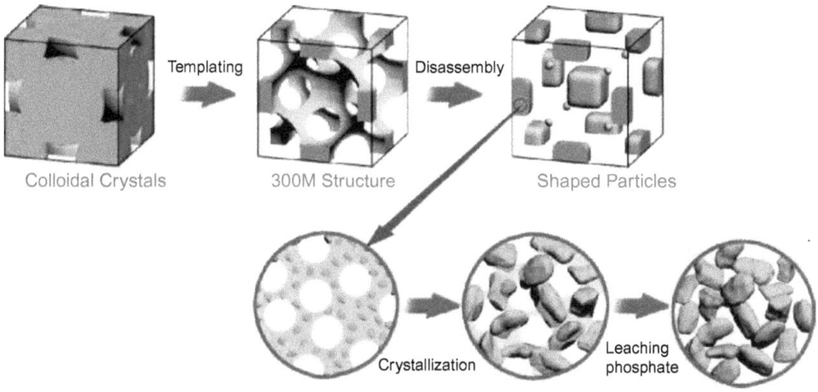

FIGURE 2.20 Schematic of nanoparticle synthesis using the colloidal crystal template and disassembly of a 3DOM structure. Crystalline and pure metal oxide phase is obtained by calcination. Phosphate contamination is removed by NaOH etching. *Reprinted with permission from Li et al. [146]. Copyright 2010, American Chemical Society.* (For color version of this figure, the reader is referred to the online version of this chapter.)

gives rise to highly monodisperse NP distribution. However, due to the large amounts of solvent and surfactant used in this method, the application is often limited to high value products.

Nucleic acid and nucleotide-templated NP synthesis with tunable properties are achievable using various constituent nucleotide or nucleotide-triphosphate building blocks [148,149]. Each chemical moiety of a nucleotide could play a specific role during the template synthesis of NPs. The different chemical moieties of a guanosine triphosphate nucleotide have been used in template synthesis of PbS NPs [149]. The triphosphate group and N7 group are used for the modulation of growth and emission properties, respectively [149]. The N2 group is used as stabilizer and/or for passivation [149]. The phosphate group can be used specifically as a binding site for Li^+, Na^+ and K^+ ions, whereas the base could be a binding site for Ag^+, Hg^{2+}, and Pt^{2+} ions.

Protein cages have also been used as size-constrained reaction vessels [150]. A core–shell structure, with protein cage-shaped as the shell, has been developed. Several metallic oxide (Co, Ni, Cr) and semiconductor (CdSe, ZnSe) NPs having an average diameter of 7 nm have been obtained [151]. Cobalt NPs are obtained via cobalt hydroxide oxidation using H_2O_2. CdS NPs are obtained by incrementally adding cadmium ions followed by sulfuric ions in the apoferritin cavity. A floating nanodot gate memory (FNGM) with a monolayer of Co or Fe nanodots has been fabricated using ferritin protein on a SiO_2 layer [151]. This FNGM is similar to a MOS-FET but has NPs embedded in the control SiO_2 gate layer.

Preformed voids and templates can also be used to form NPs and microparticles of different shapes [152]. Champion et al. used two schemes to

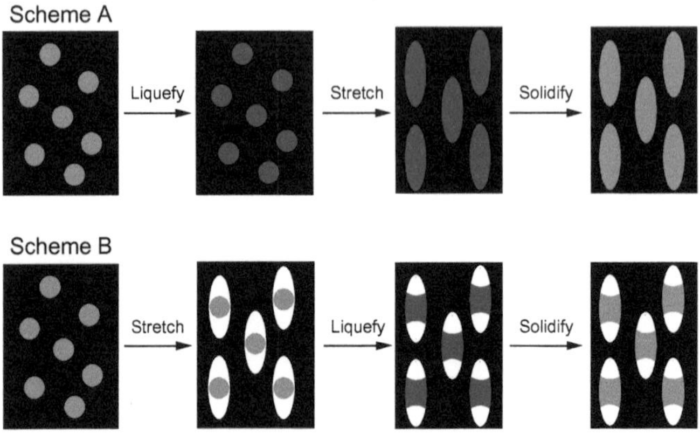

FIGURE 2.21 Schematic representation of two processes producing shaped structures from nanoparticles. The film containing liquefied nanoparticles is stretched in one or two dimensions, followed with a solidification step (A). The film is first stretched to introduce air between the particle and the matrix material, followed by liquefaction and then a solidification step (B) [152]. (For color version of this figure, the reader is referred to the online version of this chapter.)

fabricate NPs with more than 20 different shapes and sizes [152]. Particulates with size ranging from 60 nm to 30 μm have been obtained by a combination of stretching, liquefaction, and solidification (Fig. 2.21). In one approach, polystyrene NPs are liquefied first before stretching and solidification. In this approach, rod, circular and elliptical discs, and worms are obtained, depending on the viscosity of the starting materials. In a second approach, air is introduced within the pore to create voids before liquefaction using heat or toluene. The solidification gives rise to new and more complex shapes.

11 GREEN CHEMISTRY

Applying green chemistry principles to NP production should receive very high priority [153]. This is not currently the case. Current synthesis approaches use large amounts of toxic and or organic materials. Furthermore, the actual yield of high-quality NPs is relatively low, contributing to large amounts of waste.

Developing green processes for monodisperse NP manufacture is one of the major challenges of NP technology. It is quite difficult to have a strict definition of green chemistry. In this book, we limit our definition to the following principles:

a. optimize utilization of bio-based precursors;
b. minimize utilization of organic solvents during the synthesis and post-synthesis processes;
c. avoid utilization of toxic chemicals.

High surface area silica particles are widely used for catalysis, sensing, membrane, and insulation. However, the silicon alkoxide precursors used in current processes are expensive and have a significant environmental impact. Nanostructured silica is present in renewable biomass, including microalgae and rice husk (RH). Depending on the RH variety, hydrated amorphous silica represents up to 20% of the overall dry mass [154]. After boiling in HCl, rinsing, and drying, the RH is preheated to 700 °C for 2 h, followed by ultrasonication, stirring, filtration, and drying. Biogenic silica NPs (ca. 4.2 nm in diameter) are thus obtained [154]. These small NPs tend to agglomerate, forming large particles with an average diameter of 25–30 nm. This clustering provides a highly porous structure. Semicrystalline meso/macroporous frameworks with tunable pore size are obtained following doping with K^+ cations under specific pyrolysis conditions [154].

Supercritical CO_2 has been used as solvent for NP synthesis [155]. However, there are issues with isolation and collection of NPs [156], as well as toxicity of starting materials [157]. Aqueous-based synthesis provides numerous advantages over organic solvents. In the case of gold NP synthesis, the following advantages have been identified [158]:

a. water is a better solvent for metal ions;
b. there is no need for an additional antiaggregation agent, as it is readily achieved via electrostatic repulsion induced by the presence of surface (citrate and chloroaurate) ions;
c. there is the possibility of LBL (layer-by-layer) assembly due to the combination of negative and positive surface charging;
d. there is the possibility of shape, size, and bioconjugation control.

In the case of gold NP synthesis, there are a few drawbacks to the aqueous-based synthesis processes [158]. For example, dilute metal NP concentration is required to minimize ionic interaction. Furthermore, it is quite difficult to obtain monodisperse gold NPs with a close-packed hexagonal arrangement.

Aqueous synthesis involving biological media provides a sustainable fabrication process [159,160]. As shown in Table 2.1, a variety of microorganisms can be used for different NP synthesis. In the majority of cases, the synthesis occurs in the extracellular sites. Metal and semiconductor compound and magnetic NPs are obtained using different microorganisms. CdS, Au, and Ag NPs remain the most widely studied systems. Depending on the microorganism, tunable size distribution up to 500 nm has been reported (Table 2.1).

Material synthesis using biological molecules, also known as biomineralization, is an area rich in possibilities [161]. Bioinspired processes have been used to synthesize a variety of nanomaterials, including NPs [162–164]. In the case of silica, living organisms (diatoms, sponges, and grasses) provide an efficient platform to produce a variety of structures, including nanomaterials,

TABLE 2.1 Examples of Metallic Nanoparticle Synthesis Using Different Microorganisms [160]

Microorganism	Type of nanoparticle	Location	Size range (nm)
(A) Bacteria			
Pseudomonas stutzeri	Ag	Intracellular	~200
Morganella sp.	Ag	Extracellular	20–30
Lactobacillus strains	Ag and Au	Intracellular	—
Plectonema boryanum (Cyanobacteria)	Ag	Intracellular	1–10
			1–100
Escherichia coli	CdS	Intracellular	2–5
Clostridium thermoaceticum	CdS	Intracellular and extracellular	—
Actinobacter spp.	Magnetite	Extracellular	10–40
Shewanella algae	Au	Intracellular, pH 7	10–20
		Extracellular, pH 1	50–500
Rhodopseudomonas capsulata	Au	Extracellular, pH 7	10–20
		Extracellular, pH 4	50–400
Escherichia coli DH5α	Au	Intracellular	25–33
Thermomonospora sp.	Au	Extracellular	8
Rhodococcus sp.	Au	Intracellular	5–15
Klebsiella pneumoniae	Ag	Extracellular	5–32
Pseudomonas aeruginosa	Au	Extracellular	15–30
Shewanella oneidensis	Uranium (IV)	Extracellular	—
(B) Yeast			
MKY3	Ag	Extracellular	2–5
Candida glabrata and *Schizosaccharomyces pombe*	CdS	Intracellular	200
(C) Fungi			
Phoma sp. 3.2883	Ag	Extracellular	

TABLE 2.1 Examples of Metallic Nanoparticle Synthesis Using Different Microorganisms [160]—Cont'd

Microorganism	Type of nanoparticle	Location	Size range (nm)
			71.06–74.46
Fusarium oxysporum	Au	Extracellular	20–40
Verticillium	Ag	Intracellular	25 ± 12
Aspergillus fumigatus	Ag	Extracellular	5–25
Trichoderma asperellum	Ag	Extracellular	13–18
Phaenerochaete chrysosporium	Ag	Extracellular	50–200
Fusarium oxysporum and *Verticillium* sp.	Magnetite	Extracellular	20–50
(D) Plant and plant extracts			
Azadirachta indica (Neem)	Ag, Au, and Ag/Au bimetallic	Extracellular	50–100
Geranium leaves plant extract	Ag	—	16–40
Lemongrass plant extract	Au	—	200–500
Avena sativa (Oat)	Au	Extracellular	5–85
Alfalfa sprouts	Ag	Intracellular	2–20
Aloe vera	Au	Extracellular	50–350
Cinnamomum camphora	Au and Ag	Extracellular	55–80
(E) Algae			
Sargassum wightii	Au	Extracellular	8–12
Chlorella vulgaris	Au	—	9–20

under mild physiological conditions. Aquatic organisms and biomacromolecules give rise to ornate biomineral structures, using the biomineralization process. Understanding the biomineral formation process may help in biomimetic design strategies. For example, adsorption and storage of silicic acid from the environment by aquatic organisms can be controlled by modifying the attributes of the organism. However, one needs to be careful when using the insight gained from *in vitro* studies for *in vivo* processes [165]. Therefore, more detailed investigation is required.

Formation of many mineral-based NPs has been reported. Magnetite (Fe_3O_4) NPs have been obtained with the magnetospirillum gryphiswaldense cell [166]. Other examples of metallic and semiconductor NP synthesis in the presence of bacteria, fungi, and algae have been reported [163]. Silica NPs have been obtained using poly-L-lysine during the silicification process [165]. *In vitro* biomineralization processes, using different peptides, have also been developed. This approach has allowed better size control of CdS NPs [167]. Gold NPs with tunable size distribution are obtained using chitosan as a reducing/stabilizing agent [168]. The gelation of chitosan in the presence of tripolyphosphate polyanions allows control over the shape and size of gold NPs.

Soluble starch has been used as both the reducing and stabilizing agent for the synthesis of silver NPs from silver nitrates in an autoclave at 15 psi and 121 °C for 5 min [169]. Silver NPs with an average diameter of 5 nm have been obtained using water as solvent and starch as the reducing and protecting agent [156].

In the case of CuO NPs, numerous wet-based chemical processes have been used in the past (Table 2.2). The majority of the processes used to synthesize CuO NPs are not green. There are very few reported preparation methods using water as solvent and benign chemicals. The proposed CuO NP process provides not only small size distribution (10–20 nm) but will also reduce cost and environmental impact [72].

Hollow cobalt oxide formation using a biotemplating process in an aqueous solution containing ferritin has been reported [171]. After the introduction of the precursors in the ferritin protein shell, the nucleation step is initiated [171]. Tunability of the NP properties is achieved by controlling the cobalt atom concentration within the template.

Biological processes may give rise to biocompatible NPs. This is of interest not only for their excellent blood compatibility but also for their recycling capability. As discussed above (template synthesis), the biomineralization process has been used to synthesize metallic and semiconductor NPs in a cage-shaped protein supramolecule [151]. For example, ferritin NPs obtained using this route can be used for biological applications. Following stirring of a mixture of aqueous solution (15 mL) containing copper acetate monohydrate (0.02 g) and 0.23 g of β-cyclodextrin at room temperature, a filtration and drying step was applied [170]. The mixture containing blue copper acetate encapsulated with β-cyclodextrin compound was calcinated in the furnace at 500 °C for 1 h, to yield spherical and monodisperse 10–20 nm CuO NPs [170].

Recently, a rapid and inexpensive silver NP synthesis from aqueous Ag^+ solution containing culture supernatants of *Klebsiella pneumoniae* (Enterobacteriacae) has been reported [172]. Using incubation temperature of 25–35 °C, silver NPs with an average particle size of about 6 nm have been obtained.

TABLE 2.2 Comparaison of Various CuO Nanoparticle Preparation Techniques [170]

Preparation method	Solvent and chemicals used	Particle size (nm)
Sol–gel method	Ethanol and NaOH	5–55
Quick precipitation	Ethanol, glacial acetic acid, and NaOH	6
Sonochemical	Water, 10% water–DMF, and ethanol	20
Electrochemical	Tetraoctylammonium bromide in acetonitrile and tetrahydrofuran	4–6
Solid-state reaction, one-step	Ethanol and NaOH	12
Alcothermal synthesis	Ethanol	3–9
Reverse-micellar route	Isooctane	25–30
	n-Octane	70–80
	Methanol, chloroform, and *n*-butanol	
Microwave irradiation	Ethanol and NaOH	4
Reduction	Absolute ethanol, liquid ammonia solution with sodium metal and NaOH	5–15
Green approach (this study)	Water	10–20

CdS NPs with an average size of 10 nm have also been synthesized by stirring for 5 h and aging for 12 h in an aqueous mixture of cadmium nitrate and starch [173]. Biological organisms including bacteria, yeast, and fungi have been used to synthesize gold NP [174]. Other metal NPs with relatively wide size distribution have been synthesized using bacteria, fungi, and plants [175–178]. Monodispersed NPs have been synthesized in subcellular systems using a virus-based toolkit [179].

12 OTHERS

Intense interest in NPs during the last two decades has led to the development of dozens of new synthesis techniques. The methods that provide the best potential for low cost and scalability are described.

12.1 High-Temperature Coordinating Solvent

The so-called TOPO method developed by the Bawendi group [1] is widely used for small scale synthesis. This technique includes reacting appropriate metallic or organometallic (Zn, Cd, or Hg) and chalcogen (S, Se, or Te) precursors in a coordinating solvent at high temperatures. Although difficult to scale up, the TOPO process has received tremendous success in preparing highly monodisperse QDs.

Different versions of the original TOPO recipe have been reported [180–182]. Park et al. reported a general synthesis approach for large-scale production of monodispersed NPs in a single reactor batch [182]. Around 40 g of magnetite NPs with an average size of 12 nm have been obtained by mixing 36 g of iron–oleate complex and 5.7 g of oleic acid in 200 g of 1-octadecene at room temperature, followed by heating at 320 °C for 30 min. After cooling to room temperature, 500 mL of ethanol was added and the nanocrystals were separated by centrifugation [182]. They have also reported that using solvents with different boiling temperatures allowed tuning the particle size between 5 and 22 nm. Particle size distribution of less than 4% has been obtained. Thus, quality NPs are achieved because of the low-temperature synthesis [182].

12.2 Jet Aerosol

Production and application of aerosols have been practiced since the late nineteenth century [183]. Although liquid-based aerosol particles are usually larger than 1 μm, they may carry solid or liquid NPs inside. Two designs for the reaction chamber allowing scalable NP production using jet aerosol have been reported [184]. Operation pressure dictates the design choice [184]. Often, NP size increases with the gas pressure. Indeed, average Bi NPs of 1, 5, and 10 nm are obtained with gas pressure of 0.04, 0.52, and 1 kPa, respectively.

12.3 Photoreduction

Photon-based synthesis could provide another avenue to reduce utilization of solvents and other chemicals. A photochemically induced metal NP in the presence of polymer matrix involves a polymer–metal ion complexation followed by metal ion reduction and then finally metallic NP growth (Fig. 2.22). In the case of gold, particle size is easily controlled through variation of light incubation and metal-to-polymer relative concentration [185].

Gold NPs embedded in TiO_2 have been obtained by photochemical decomposition of $HAuCl_4$ precursor using a 125 W high-pressure mercury lamp [186]. A two-step process was used to produce tunable gold NPs from 5 up to 110 nm [187]. In the first step, seed particles from 5 to 20 nm were

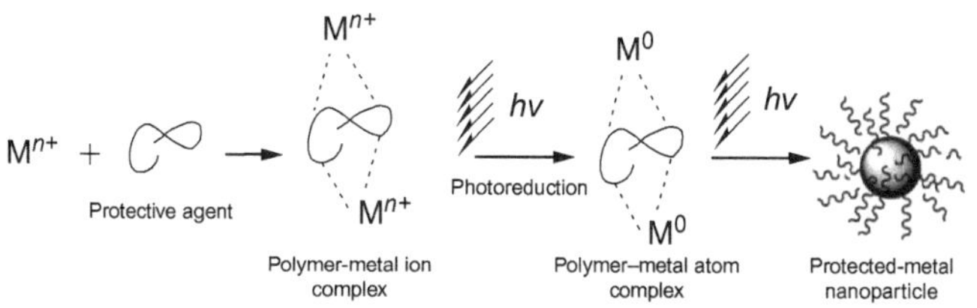

FIGURE 2.22 Schematic illustration of the main steps leading to the photoinduced metal nanoparticle synthesis [185].

fabricated by UV irradiation of a variable gold precursor to stabilizer/reductant ratio. In the second step, fresh Au(III) ions were reduced on the seed particle surface by ascorbic acid [187].

12.4 Laser Processing

For decades, laser energy source has been extensively used under vacuum conditions to prepare materials and films. More recently, it has also been applied to liquid media under atmospheric conditions. Single and alloyed metal NPs are obtained by laser ablation of a target immersed in water. In some cases, laser-induced NP synthesis has also been obtained from a solution of metal salts. Monodisperse silver, gold, and silver–gold alloy NPs are thus obtained readily, using this technique, from an appropriate mixture of $AgNO_3$ and $HAuCl_4$ salts [188]. The ablated silicon target provides a stabilizing and protective matrix for the NPs during their formation.

12.5 Thermal Decomposition

Also called thermolysis, this technique takes advantage of the low decomposition temperature of organometallic compounds. This method has been used to fabricate Fe and Ni NPs from their corresponding metal carbonyl. Heating $Ni(CO)_4$ to 200–300 °C inside an autoclave leads to nickel metal and carbon monoxide [189]. Fe–Cr–Mo–V–C steel has also been obtained by thermal decomposition of a solution containing $Fe(CO)_5$, $Cr(CO)_6$, Mo $(CO)_6$, and $V(CO)_6$, at 150 °C. Alternatively, laser could be also used as heat source.

There are two advantages in using the carbonyl thermal decomposition process to synthesize metal NPs. One is related to the relatively low operation temperature. Often, temperatures below 400 °C are used. Besides low energy cost, this relatively low synthesis temperature leads to narrow size distribution. The second advantage is related to the presence of CO as a by-product, preventing the oxidation of the metal NPs after synthesis. With a boiling temperature of 102 °C, iron pentacarbonyl, $Fe(CO)_5$, decomposes to form metallic Fe clusters around 160–180 °C [190]. Tunable particle size from 4.8 to

10.9 nm has been achieved [190]. A facile large-scale synthesis of monodisperse iron NPs has also been reported, using a similar approach [191]. Monodispersed Fe NPs with an average size of 9.3 nm are obtained using this approach [191]. Other precursors have also been used for thermolysis. Bimetallic Pd–Cu/CuzO colloids are prepared by thermal decomposition of acetate compounds [192].

To minimize toxicity due to CO handling, Ni NPs are produced using a two-step flow type process [189]. First, nickel carbonyl is formed by reacting CO at the surface of nickel raw materials. The temperature inset for this reaction is around 80 °C with a maximum around 100 °C, depending on the CO flow rate [189]. In the second step, carbonyl nitrogen and CO are fed into a hot wall reactor heated to 250–350 °C. In this process, CO can be recycled. Varying CO concentration and residence time allows tuning the particle size from 5 up to 65 nm or more [189].

12.6 Seeding Technique

Seeding technique has been developed to provide spherical gold NPs in the 30–100 nm range [193]. Gold NPs with a 2.6- and 12-nm diameter (seeds) obtained by BH_4^- reduction are grown to larger sizes by the addition of a boiling solution of $HAuCl_4$ and citrate [194]. Often, these seed NPs are obtained via micelle techniques. The polydispersity of the NP product depends on the polydispersity of the seed particles [193]. Shape- and size-tunable gold polyhedral NPs are obtained using a facile seeding technique [195]. Shape evolution from octahedral to trisoctahedral structures was achieved by changing the reductant concentration, with particle size up to 150 nm.

Slow co-injection of a metal-organic atom source and a reducing agent into a stirred solution containing seed NPs is used to deposit Ag on Au and Au on Ag [196]. This technique allows a fine tuning of the shell thickness over a wide range.

12.7 Photographic Process

A wet development process referred to as the develop-out process (DOP) has been used to produce Pt and Pd NPs [197]. This photographic process allows printed Pt and Pd NPs. Ferric oxalate was used as a UV sensitizer for Pt ion reduction in the presence of a developer (ammonium citrate). An alternative photographic technique referred to as the print-out process (POP) has been developed without the need of any further processing [198]. The POP technique allows wider and finer particle size tunability, from 30 to 250 nm. The metal NP deposition efficiency of POP is higher than that of DOP by a factor of 6–22 times, depending on the type of substrate used.

12.8 Single Source Compound NP

Compound NPs are challenging because of the requirement of multisource precursors, with different reactivities to environmental conditions. One way to solve this issue is to use a single source molecule containing all the elements of the compound NP. Ternary $Cu(In,Ga)(Se,S)_2$ is receiving considerable attention, given its optimal photovoltaic properties. As discussed in Chapter 5, $Cu(In,Ga)(Se,S)_2$ or CIGS gives rise to the highest power efficiency among thin film-based solar cells. The presence of all five elements is required to optimize the energy bandgap. Although current commercial CIGS photovoltaic films are obtained using the physical deposition technique, CIGS NPs are currently being developed for wet deposition. CIGS NPs with optimal chemistry are critical [199]. $CuInS_2$ NPs are obtained by irradiating the molecular precursor $[(TOP)_2CuIn(SR)_4]$ solution, where TOP=(octyl)3P and R=*n*-Pr (1), *t*-Bu (2) [200]. Dioctyl phthalate, benzene, toluene, and pentane are used as the organic solvent in this approach. Using single source precursor, Castro et al. [201] prepared colloidal $CuInS_2$ NPs between 2.7 and 4 nm.

12.9 Electroless Deposition

Noble metal thin film depositions on different substrates are readily obtained via electroless deposition. This technique, often referred to as electroless plating, involves spontaneous reduction of ionic metals in the absence of electric current. Electroless plating is easy to operate, providing a cost effective and scalable alternative to conventional metal plating [202]. Of late, electroless processes are used to form noble metal NPs directly integrated to surfaces. Various NP-based electroless processes have been reported [203]. In one process, metal ion reduction is catalyzed by the substrate. In another process, referred to as autocatalytic, the noble metal on the substrate surface catalyses other metal ion salts using an external reducing agent. Film deposition stops once the surface is completely covered. In the third process, referred to as galvanic displacement, the substrate surface is used as both the reducing agent and electron source for reduction of the metal ion. In the last case, thick layers are obtained following reduction, whereas the first process allows only a thin film as reduction stops once the surface is covered [203]. Contrary to the previous two processes, the autocatalytic process does not require a reducing agent as the substrate is used as both electron source and reducing agent. Solution concentration, temperature, and immersion time allow control over surface morphology and deposition rate.

Combining electroless deposition with soft-lithography patterning, NPs are directly formed onto pre-patterned surfaces. Three patterning approaches (photolithography, micro-contact printing, and dip-Pen nanolithography) are successfully used to produce high surface area and ordered metallic structures on different substrates including Ge(100), Cu, Zn, and Sn [203]. Galvanic displacement from aqueous metal salt solutions seems to provide the best results.

12.10 Metal-Steam Reforming

Using different techniques, coarse oxide scale is converted to zero-valent iron NPs [204]. Around 20 million tons of mill-scale is produced as a by-product of steel manufacturing. This mill-scale ends up mostly in landfill mines. Metal-steam reforming power by solar-energy is a sustainable approach to produce an unlimited amount of iron NPs from mill-scale.

12.11 Polyol Method

The polyol method involves suspending the metal precursor in a glycol solvent and subsequently heating the solution to a refluxing temperature. This technique has been used to synthesize metallic, oxide, and semiconductor NPs. Mono-metallic and metallic alloy NPs have been synthesized with this technique (Table 2.3). Often, a polyol is used as solvent, reducing agent, and ligand, to prevent NP agglomeration. The choice of polyol type is dictated by the optimum reflux temperature. Often, several hours of heating are required to produce NPs when convective heating is used. As discussed elsewhere in this chapter, microwave heating not only reduces heating time, but also improves the quality of NPs. Microwave heating using polyol solvent was successfully used in the past to produce metal and metal oxide NPs.

12.12 Hydrogen Reduction

Monodispersed crystalline silver NPs are obtained by reducing supersaturated Ag_2O solution. Using a Pyrex round-bottomed flask, the reduction was carried out at 70 °C in hydrogen gas pressurized at 10 psi above atmosphere [206]. As shown in Fig. 2.23, particle size is easily tunable (15–200 nm) by slow addition of silver ions. Low Ag_2O solubility dictates low dosing. Postsynthesis filtration and centrifugation allow suspensions of narrow size distribution and relatively high concentration (10^{12}–10^{14} particles/mL). This method is thus potentially efficient and easily scalable.

12.13 *In Situ* Reduction

In situ reduction of metal ions is widely used for the formation of NP-based catalysts. Catalyst support is first wetted with ion precursor solution, followed by an *in situ* reduction. CdS, ZnS:Cu, and Cu_2S/CdS/ZnS NPs are obtained by absorbing corresponding ions into sulfonated polymer networks (Fig. 2.24). NPs are obtained after reaction with H_2S. This approach allows high-quality monodispersed NPs to be directly and easily integrated.

TABLE 2.3 Synthesis Parameters and Products of Different Polyol Reactions

Material	Precursors	Concentration range used (mol/L)	Average crystalline size (nm) of powder	Average crystalline size (nm) of coating	Reaction time (h)
Fe	Iron(II) acetate	0.01–0.20	20		2
Co	Cobalt(II) acetate tetrahydrate; cobalt(II) chloride hexahydrate	0.05–0.20	12.1 14	15 (K) 14 (P) 23 (T)	2
Ni	Nickel(II) acetate tetrahydrate	0.02–0.20	20	9 (K) 30 (K) 15 (P)	1 1
Cu	Copper(II) acetate tetrahydrate	0.02–0.025	10–80	12 (AIN) 43 (K)	2
Ru	Ruthenium(III) chloride	0.021	5		1
Rh	Rhodium(III) chloride	0.01		9 (P)	1
Pd	Palladium(II) chloride	0.02–0.15	10	18 (K) 22 (P)	1
Ag	Silver nitrate	0.05–0.2	40	34 (T) 43 (K) 50 (P)	1
Sn	Tin(II) oxide	0.01-0.03	36		2
Re	Rhenium(III) chloride	0.02		14 (P)	1
W	Tungstic acid Sodium tungstate	0.012–0.20 0.03	8 10	12 (P) 12 (P)	3 3

Continued

TABLE 2.3 Synthesis Parameters and Products of Different Polyol Reactions—Cont'd

Material	Precursors	Concentration range used (mol/L)	Average crystalline size (nm) of powder	Average crystalline size (nm) of coating	Reaction time (h)
Pt	Potassium hexachloroplatinate(IV)	0.01–0.20	2	10 (K) 12 (T) 14 (GF) 15 (AF) (SF)	1
Au	Gold(III) chloride	0.01–0.20	28	32 (P)	2
Fe–Cu	Iron(II) acetate Copper(II) acetate tetrahydrate	0.016–0.16 0.018–0.14	27–47		2
Co–Cu	Cobalt(II) acetate tetrahydrate Copper(II) acetate tetrahydrate	0.01–0.20	17–35		2
Ni–Cu	Nickel(II) acetate tetrahydrate Copper(II) acetate tetrahydrate	0.0321 0.0963	8		1

K, kapton; P, pyrex; T, teflon; G, graphite; A, alumina; S, sapphire; F, fiber.
The range of crystallite size is given when it is concentration dependent [205].

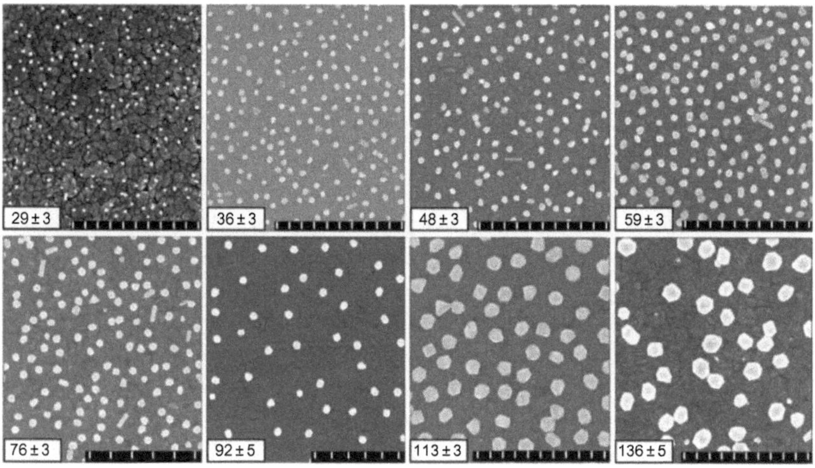

FIGURE 2.23 SEM micrograph with average nanoparticle size (nm) obtained by hydrogen reduction [207].

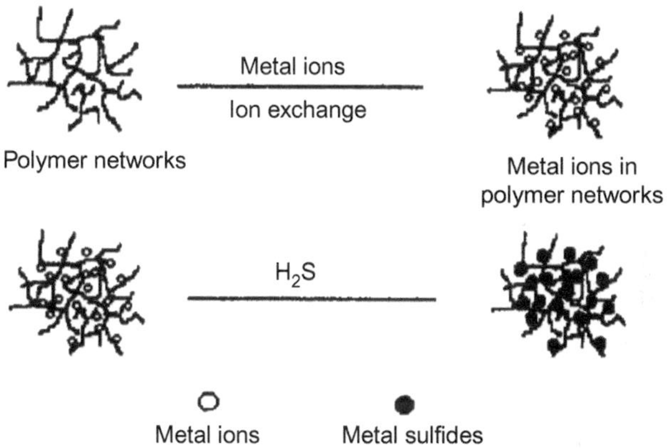

FIGURE 2.24 Schematic representation of the metal sulfide nanoparticle formation in a polymeric matrix [208].

13 CONTINUOUS FLOW REACTOR AND SCALE-UP

For most bench-scale NP synthesis methods, the only way to increase the production scale cost effectively is through the use of a continuous flow reactor. The reactor size is often constrained. Production of NPs using various continuous microreactors has been recently reviewed [209]. The development of the one-pot synthesis process [6] facilitates the utilization of a continuous flow reactor in the process of monodisperse core–shell QD production.

13.1 Conventional Continuous Flow Reactor

Kawa et al. reported stable production of about 13 g/h of highly luminescent and monodispersed CdSe nanocrystals, using TOPO [210]. They used a simple layout, including a continuous flow reactor. Precursor solution was prepared just before reaction and kept under nitrogen in a dark place. TOPO heated to 370 °C and precursor stock solution to 80 °C were injected into the reactor (350 °C) through the tubing system, with a flow rate of 42.4 and 17.6 mL/min, respectively. The collected solution was cleaned with a repetitive set of centrifugation and redispersion. Vacuum Metallurgical Co. Ltd. has developed a different continuous process that could be adapted to various wet-based NP syntheses [211].

There are numerous advantages in carrying out microfluidic synthesis of NPs [10]. Microfluidic reactors combine the advantage of a continuous and precise mixing process and precise time interval when compared to bulk synthesis. Small and monodisperse PLGA–PEG NPs have been synthesized using PLGA–PEG copolymer precipitation (Fig. 2.25). The three fabrication stages involving nucleation, growth, and aggregation are controlled by precise and rapid mixing within the microfluidic device. The fast precursor mixing and NP formation are due to diffusion of the solvent out of the focused stream and diffusion of water into the focused stream [212]. Inorganic milli- and microfluidic reactors have also been used to fabricate colloidal particles. In a recent design of microfluidic synthesis systems, factors contributing to NP size broadening have been identified and corrected [213].

13.2 Flow Tube Reactor

Controlling the experimental parameters influencing NP characteristics remains the biggest challenge to synthesis process scale up. The segemented flow tubular reactor (SFTR) has been developed to address this issue [214]. It consists of a reagent mixer and segmenter, a tubular section consisting of similar droplets or reactors, and a separator that separates the segmenting fluid (for recycling) from the product [214]. Barium titanate NPs with an average size of 30 nm were obtained using the SFTR technique [209].

Microfluidic devices are also used in continuous flow reactors. This allows rapid, precise, and reproducible changes in reaction parameters without significant local variation [215–217]. Local variation in the preparation conditions often leads to large dispersion in composition and size. The microfluidic reactor has been developed recently for a more efficient size and phase control of NP-based products. Face-centered cubic (β) cobalt NPs have been prepared using a polymeric microfluidic reactor under high flow rate [216]. The setup includes a flow controller, a microfluidic reactor, and a quenching stage (Fig. 2.26). The continuous flow reactor, based on microfluidic devices, has

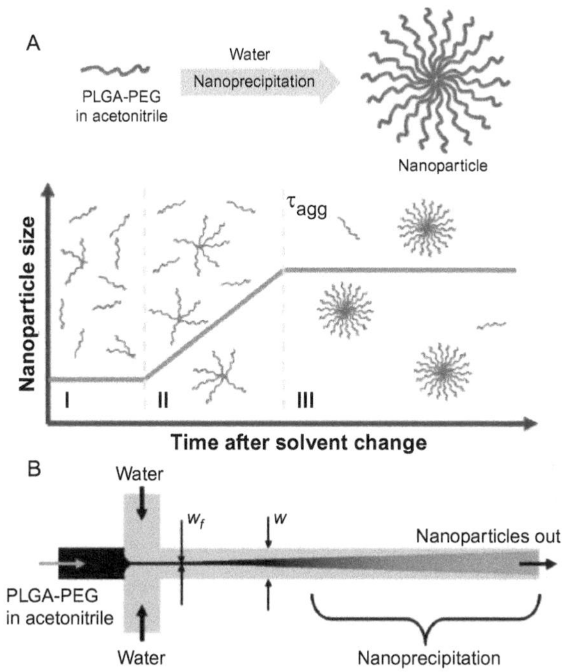

FIGURE 2.25 PLGA–PEG diblock copolymer nanoparticle fabrication using the nanoprecipitation process. (A) Nanoparticle formation and size variation in the three fabrication stages (I, II and III). PLGA–PEG is first dispersed in acetonitrile, followed by water addition. In this second stage, poorly soluble PLGA block initiates nanoparticle nucleation of nanoparticles. In the third stage, nanoparticle growth is kinetically locked after a characteristic aggregation timescale τ_{agg}. (B) Schematic representation of the mixing process in which the polymer stream is focused into a thin stream between two water streams. *Reprinted with permission from Karnik et al. [212]. Copyright 2008, American Chemical Society.* (For color version of this figure, the reader is referred to the online version of this chapter.)

been used to prepare thiol stabilized NPs [217]. Resulting gold NPs showed better monodispersity than those produced by the bulk synthetic method.

13.3 Continuous Bioinspired

Several synthesis techniques mimicking biological and natural processes have been reported. These techniques include biomineralization, molecular self-assembly, and others. However, these bioinspired synthesis techniques are often limited by their low throughput and scalability [218]. A simple and robust bioinspired synthesis technique for $BaTiO_3$ NPs has been recently reported [218]. Reagent delivery is achieved using ultrasonic/chromatographic nebulization. Concentric heating coils are used to improve the aging process [218].

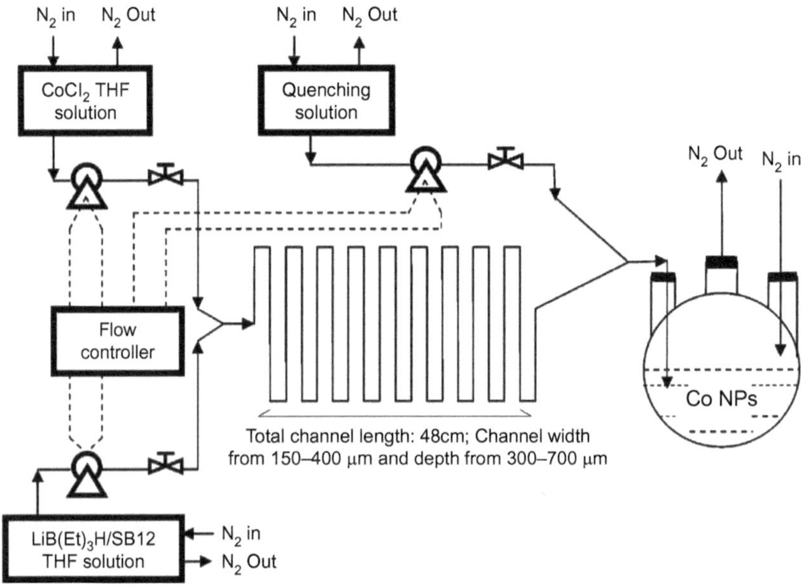

FIGURE 2.26 Schematic outline of a microfluidic reactor used in synthesis of phase-controlled cobalt nanoparticles. *Reprinted with permission from Song et al. [216]. Copyright 2006, American Chemical Society.*

13.4 Spray-Drying

Liquid solutions, including suspensions, slurries, emulsions, and melts are cost effectively transformed to dry powder on the industrial scale. The resulting powder is often large and agglomerated. The company Büchi has started marketing a lab-scale spray drier, nanospray dryer B-90, to prepare submicron powder particles [219]. This technique is particularly suitable for poorly soluble drugs as it redisperses flowable dried powder.

14 PURIFICATION, SIZE-SELECTIVE AND OTHER POSTTREATMENT PROCESSES

Quality control is the weak link in most NP-based value propositions. Purification and size selection of NPs are critical steps. The quality of the product is directly related to the NP properties [220]. In the early years of the twentieth century, several purification methods were already in use [221]. This includes dialysis, electrodialysis, diafiltration, and ultrafiltration. A second solvent in which the stabilized NPs are not soluble but miscible with the first solvent (often referred as nonsolvent) is often used for purification. This will result in aggregation that is often followed by sedimentation or phase separation. Dispersion and chemical properties have direct consequences on the final product functionalities. For example, surface modification will affect interparticle interactions. These interactions will define the macroscopic properties, including stability, rheology, and conductivity. Thus the interaction forces involving the different components will define the relationship between the microscopic surface properties on the one hand and the macroscopic properties on the other.

As shown in the case of CdSe NPs, purification has an important effect on optical properties [222]. Self-assembling of gold NPs into 1D and 2D arrays is affected by their purity[223]. Some of these purification processes may take more than 10 L of solvent (mostly organic) per gram and up to several days of processing.

There are numerous published reports claiming production of highly monodisperse NPs. However, these processes suffer from low throughput and difficulty in scaling up cost effectively. Three postsynthetic processes are generally used to produce monodispersed NPs. Centrifugation, filtration, and dialysis are quite straightforward, although they do not always provide highly monodispersed size distribution. Electrophoresis and chromatography are also used. Size-selective precipitation provides the best results through fractionation. However, specific postprocessing steps are required for specific NP applications.

Combination of solvent–antisolvent pairs with centrifugation remains the most widely used approach to separate, purify, and obtain monodisperse NPs [1]. This approach is widely used for small-scale demonstration, taking advantage of the inherent interparticle attractions leading to precipitation. Using an appropriate surfactant, these NPs could be redispersed in a solvent. Addition of antisolvents to dispersed NPs helps accelerate the precipitation. With centrifugation, larger particles precipitate first. Repeating these two steps could help enhance the monodispersity of the synthesized NP. Another alternative is to use a direct atom addition on the seed NP [224]. However, this atomic scale process is relatively slow and not easy to adapt to large-scale manufacturing.

An elegant and easy approach to size selection has been reported by McLeod et al. using the pressure tunable solvent properties of CO_2-expanded liquids [225]. As shown in Fig. 2.27, size-selective precipitation of silver NPs at controlled locations is obtained by pressure and expansion of an organic solvent using carbon dioxide gas. Indeed, varying the fraction of CO_2 within the organic solvent at different pressures allows the tunability of the initial solvent strength. Indeed, there is a correlation between the pressure and the UV absorbance intensity from various silver QDs.

Diafiltration has been successfully used for purification and size selection of various NPs [226,23]. As shown in Fig. 2.28, in the case of gold NPs, diafiltration provides significant advantages as an efficient and convenient purification and size separation process of aqueous NP solutions. Besides eliminating the organic solvent, Sweeney et al. claim that the diafiltration process requires just 15-min process time [23].

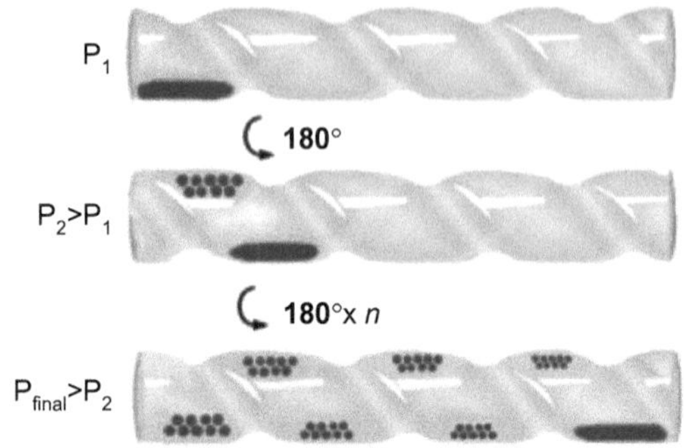

FIGURE 2.27 Simplified setup used for nanoparticle size selection apparatus. Organic solvent volume is expanded using CO_2 gas, altering solvent strength (solubility). The 180° tube rotation provides locations for the multiple size-selected populations. *Reprinted with permission from McLeod et al. [225]. Copyright 2005, American Chemical Society.*

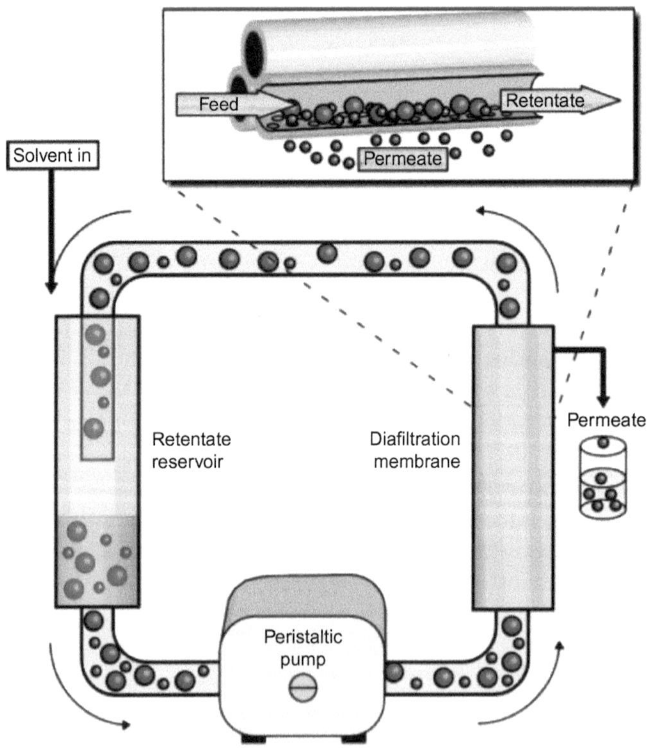

FIGURE 2.28 Illustration of the continuous diafiltration used for gold nanoparticle filtration and size separation. After introducing the solution containing nanoparticles into the retentate reservoir, it is pumped through the diafiltration membrane. Only small molecule impurities or small nanoparticles (blue) are eluted, while the large nanoparticles (purple) remain with the circulating fluid. *Reprinted with permission from Sweeney et al. [23]. Copyright 2006, American Chemical Society.* (For color version of this figure, the reader is referred to the online version of this chapter.)

Electrophoresis [227] and centrifugation [228,229] are two potentially efficient, scalable filtration techniques with minimum solvent use. Polydispersed PVP-coated gold NPs are separated into different fractions containing monodispersed NPs using electrophoresis [229]. In this case, the viscosity gradient is used to separate the NPs of different sizes, instead of gravity.

Numerous wet-based surface functionalization techniques of NPs have also been developed to reduce agglomeration. Bonding, coordinative, electrostatic, and hydrogen bonding are often used. These techniques have been applied during preparation or after preparation. Numerous bifunctional surface modifiers have been used for NP stabilization. One functional group is used for anchoring to the surface of the NP by physical, ionic, or covalent bonding. A second functional group interacts with other particles or matrix. In the case of metallic (e.g., Au) and chalcogenide (e.g., CdSe) NPs, thiol (R-SH)-based bifunctional molecules provide efficient polar and nonpolar functionalization. Grafting techniques are also used to functionalize NPs [230]. Depending on the composition of the NP and on the application, RSi $(OMe)_3$, $R—NH_2$, —COOH, $—SO_2OH$, $—PO(OH)_2$ functional groups can also provide efficient stabilization [230].

REFERENCES

[1] C.B. Murray, D.B. Norris, M.G. Bawendi, J. Am. Chem. Soc. 115 (1993) 8706–8715.

[2] K.J. Klabunde, Free Atoms, Clusters, and Nanoscale Particles, Academic Press, San Diego, 1994.

[3] C. Hayashi, R. Uyeda, A. Tasaki (Eds.), Ultra-Fine Particles: Exploratory Science and Technology, Noyes Publications, Westwood, NJ, 1997.

[4] G. Schmid (Ed.), Nanoparticles: From Theory to Application, Wiley-VCH, Weinheim, 2004.

[5] C.M. Donega, P. Liljeroth, D. Vanmaekelbergh, Small 1 (2005) 1152–1162.

[6] J.I. Kim, J.-K. Lee, Adv. Funct. Mater. 16 (2006) 2077–2082.

[7] L.A. Juhaiman, L. Scoles, D. Kingston, B. Patarachao, D. Wang, F. Bensebaa, Green Chem. 12 (2010) 1248–1252.

[8] N. Toshima, T. Yonezawa, New J. Chem. 11 (1998) 1179–1201.

[9] S. Yu, C.-J. Sun, G.M. Chow, in: Carl C. Koch (Ed.), Nanostructured Materials, 2nd Edition: Processing, Properties and Applications, second ed, William Andrew Publishing, Norwich, USA, 2007, pp. 173–234.

[10] J. deMello, A. deMello, Lab Chip 4 (2004) 11–15.

[11] A. Camenzinda, W.R. Caserib, S.E. Pratsinis, Nano Today 5 (2010) 48–65.

[12] V.V. Srdic, M. Winterer, J. Eur. Ceram. Soc. 26 (2006) 3145–3151.

[13] B.L. Cushing, V.L. Kolesnichenko, C.J. O'Connor, Chem. Rev. 104 (2004) 3893–3946.

[14] B.V. Enustun, J. Turkevich, J. Am. Chem. Soc. 85 (1963) 3317–3328.

[15] J. Kimling, M. Maier, B. Okenve, V. Kotaidis, H. Ballot, A. Plech, J. Phys. Chem. B110 (2006) 15700–15707.

[16] H. Goesmann, C. Feldmann, Angew. Chem. Int. Ed. Engl. 49 (2010) 1362–1395.

[17] Y. Xiong, Y. Xia, Adv. Mater. 19 (2007) 3385–3391.

[18] Y. Xia, Y. Xiong, B. Lim, S.E. Skrabalak, Angew. Chem. Int. Ed. Engl. 48 (2009) 60–103.
[19] A.C. Shi, R.I. Masel, J. Catal. 120 (1989) 421–431.
[20] M. Yoshimura, W. Suchanek, K. Byrappa, Mater. Res. Soc. Bull. 25 (2000) 17–24.
[21] T. Yonezawa, M. Sutoh, T. Kunitake, Chem. Lett. 7 (1997) 619–620.
[22] S.L. Cumberland, M.G. Berrettini, A. Javier, G.F. Strouse, Chem. Mater. 15 (2003) 1047–1056.
[23] S.F. Sweeney, G.H. Woehrle, J.E. Hutchison, J. Am. Chem. Soc. 128 (2006) 3190–3197.
[24] J.D.N. Cheeke, Fundamentals and Applications of Ultrasonic Waves, CRC Press LLC, Boca Raton, FL, 2002.
[25] K.S. Suslick, Science 247 (1990) 1439–1445.
[26] Y. Mastai, A. Gedanken, in: C.N.R. Rao, A. Muller, A.K. Cheetham (Eds.), The Chemistry of Nanomaterials, Wiley-VCH, Weiheim, 2004, pp. 113–169.
[27] K.S. Suslick, G.J. Price, Annu. Rev. Mater. Sci. 29 (1999) 295–326.
[28] A. Gedanken, Ultrason. Sonochem. 11 (2004) 47–55.
[29] K.S. Suslick, MRS Bull. 20 (1995) 29–34.
[30] F. Grieser, Muthupandian Ashokkumar, in: F. Caruso (Ed.), Colloids and Colloid Assemblies: Synthesis, Modification, Organization and Utilization of Colloid Particles, Wiley-VCH, Weinheim, 2004, pp. 120–149.
[31] N.A. Dhas, A. Gedanken, Appl. Phys. Lett. 72 (1998) 2514–2516.
[32] H. Kim, K. Cho, H. Song, J.-H. Kim, J.W. Lee, J.S. Lee, B. Min, S.H. Kim, S. Kim, Key Eng. Mater. 277–279 (2005) 961–965.
[33] C.G. Hu, Y. Li, J.P. Liu, Y.Y. Zhang, G. Bao, B. Buchine, Z.L. Wang, Chem. Phys. Lett. 428 (2006) 343–347.
[34] S. Zhu, H. Zhou, M. Hibino, I. Honma, M. Ichihara, Adv. Funct. Mater. 15 (2005) 381–386.
[35] K. Okitsu, M. Ashokkumar, F. Grieser, J. Phys. Chem. 109 (2005) 20673–20675.
[36] K. Vinodgopal, Y. He, M. Ashokkumar, F. Grieser, J. Phys. Chem. 110 (2006) 3849–3852.
[37] M. Bradley, F. Grieser, J. Colloid Interface Sci. 251 (2002) 78–84.
[38] H. Xia, Q. Wang, J. Nanopart. Res. 3 (2001) 401–411.
[39] P.S. Shah, T. Hanrath, K.P. Johnston, B.A. Korgel, J. Phys. Chem. B 108 (2004) 9574–9587.
[40] T. Adschiri, Y. Hakuta, K. Sue, K. Arai, J. Nanopart. Res. 3 (2001) 227–235.
[41] J. Eastoe, M.J. Hollamby, L. Hudson, J. Eastoe, M.J. Hollamby, L. Hudson, Adv. Colloid Interface Sci. 128–130 (2006) 5–15.
[42] Y. Zhang, C. Erkey, J. Supercrit. Fluids 38 (2006) 252–267.
[43] K. Byrappa, S. Ohara, T. Adschiri, Adv. Drug Deliv. Rev. 60 (2008) 299–327.
[44] J. Jung, M. Perrut, J. Supercrit. Fluids 20 (2001) 179–219.
[45] C. Aymonier, A. Loppinet-Serani, H. Reveron, Y. Garrabos, F. Cansell, J. Supercrit. Fluids 38 (2006) 242–251.
[46] Z. Knez, E. Weidner, Curr. Opin. Solid State Mater. Sci. 7 (2003) 353–361.
[47] E. Alonso, I. Montequi, S. Lucas, M.J. Cocero, J. Supercrit. Fluids 39 (2007) 453–461.
[48] X.L. Kirk, J. Ziegler, A. Ghezelbash, K.P. Johnston, B.A. Korgel, Nano Lett. 4 (2004) 969–974.
[49] C.A. Fernandez, C.M. Wai, Chem. Eur. J. 13 (2006) 5838–5844.
[50] M.J. Meziani, P. Pathak, T. Desai, Y.-P. Sun, Ind. Eng. Chem. Res. 45 (2006) 3420–3424.
[51] S.M. Pourmortazavi, S.S. Hajimirsadeghi, Ind. Eng. Chem. Res. 44 (2005) 6523–6533.
[52] P. Pathak, M.J. Meziani, T. Desai, Y.P. Sun, J. Supercrit. Fluids 37 (2006) 279–286.
[53] J.-Y. Lee, C.-H. Song, J.-I. Kim, J.-H. Kim, J. Nanopart. Res. 2 (2002) 53–59.
[54] G. Demazeau, J. Mater. Sci. 43 (2008) 2104–2114.
[55] S. Chaianansutcharit, O. Mekasuwandumrong, P. Praserthdam, Cryst. Growth Des. 6 (2006) 40–45.

[56] Y.-G. Chun, K.-H. Kim, K.-H. Yoon, Thin Solid Films 480–481 (2005) 46–49.
[57] M. Yoshimura, A.K. Byrapp, J. Mater. Sci. 43 (2008) 2085–2103.
[58] T. Adschiri, K. Kanazawa, K. Arai, J. Am. Ceram. Soc. 75 (1992) 1019–1022.
[59] A. McEvoy, M. Grätzel, in: J. Garcia-Martinez (Ed.), Nanotechnology for the Energy Challenge, Wiley-VCH, Weinheim, 2009, pp. 33–45.
[60] T.J. Daou, G. Pourroy, S. Begin-Colin, J.M. Greneche, C. Ulhaq-Bouillet, P. Legare, P. Bernhardt, C. Leuvrey, G. Rogez, Chem. Mater. 18 (2006) 4399–4404.
[61] R. Viswanathan, G.D. Lilly, W.F. Gale, R.B. Gupta, Ind. Eng. Chem. Res. 42 (2003) 5535–5540.
[62] B.L. Hayes, Microwave Synthesis: Chemistry at the Speed of Light, CEM Publishing, Matthews, 2002.
[63] C.O. Kappe, D. Dallinger, S.S. Murphree, Practical Microwave Synthesis for Organic Chemists, Wiley-VCH, Weinheim, 2009.
[64] I. Bilecka, M. Niederberger, Nanoscale 2 (2010) 1358–1374.
[65] G.A. Tompsett, W.C. Conner, K.S. Yngvesson, Chemphyschem 7 (2006) 296–319.
[66] K.T.J. Loones, B.U.W. Maes, G. Rombouts, S. Hostyn, G. Diels, Tetrahedron 61 (2005) 10338–10348.
[67] K.J. Rao, B. Vaidhyanathan, M. Ganguli, P.A. Ramakrishnan, Chem. Mater. 11 (1999) 882–895.
[68] W. Huang, Y.M. Xia, H. Gao, Y.J. Fang, Y. Wang, Y. Fang, J. Mol. Catal. B Enzym. 35 (2005) 113–116.
[69] C. Gabriel, S. Gabriel, E.H. Grant, B.S.J. Halstead, D.M.P. Mingos, Chem. Soc. Rev. 27 (1998) 213–223.
[70] R.F. Schiffmann, in: A.S. Mujumdar (Ed.), Handbook of Industrial Drying, third ed, Taylor & Francis, Boca Raton (FL), 2006 (Chapter 12).
[71] www.pueschner.com/basics/eindringtiefe_en.php (accessed 10.08.11).
[72] C.R. Strauss, R.S. Varma, Top. Curr. Chem. 266 (2006) 199–231.
[73] J.M. Kremsner, A. Stadler, C. Oliver Kappe, Top. Curr. Chem. 266 (2006) 233–278.
[74] H. Jiang, K.-S. Moon, Z. Zhang, S. Pothukuchi, C.P. Wong, J. Nanopart. Res. 8 (2006) 117–124.
[75] D. Li, S. Komarneni, J. Am. Ceram. Soc. 89 (2006) 1510–1517.
[76] A.V. Firth, Y. Tao, D. Wang, J. Ding, F. Farid Bensebaa, J. Mater. Chem. 15 (2005) 4367–4372.
[77] F. Bensebaa, F. Zavaliche, P. L'Ecuyer, R.W. Cochrane, T. Veres, J. Colloid Interface Sci. 277 (2004) 104–110.
[78] S. Kundu, K. Wang, H. Liang, J. Phys. Chem. C 113 (2009) 5157–5163.
[79] J.A. Gerbec, D. Magana, A. Washington, G.F. Strouse, J. Am. Chem. Soc. 127 (2005) 15791–15800.
[80] S. Beninati, L. Damen, M. Mastragostino, J. Power Sour. 180 (2008) 875–879.
[81] W. Zhang, J. Gao, C. Wu, Macromolecules 30 (1997) 6388–6390.
[82] B. Giesen, H. Wiggers, A. Kowalik, P. Roth, J. Nanopart. Res. 7 (2005) 29–41.
[83] M.T. Reetz, M. Winter, R. Breinbauer, T. Thurn-Albrecht, W. Vogel, Chem. Eur. J. 7 (2001) 1084–1094.
[84] Y.J. Yang, L.Y. He, H. Xiang, Russ. J. Electrochem. 42 (2006) 954–958.
[85] M. Starowicz, B. Stypuła, J. Banas, Electrochem. Commun. 8 (2006) 227–230.
[86] J.L. Delplancke, in: M.I. Baraton (Ed.), Synthesis, Functionalisation and Surface Treatment of Nanoparticles, American Scientific Publishers, Stevenson Ranch, CA, 2003, pp. 17–29.

[87] A. Ledo-Suárez, L. Rodríguez-Sánchez, M.C. Blanco, M.A. López-Quintela, Phys. Stat. Sol. (a) 203 (2006) 1234–1240.
[88] H. Natter, R. Hempelmann, J. Phys. Chem. 100 (1996) 19525–19532.
[89] D. Horn, J. Rieger, Angew. Chem. Int. Ed. Engl. 40 (2001) 4330–4361.
[90] G.Z. Chen, D.J. Fray, T.W. Farthing, Nature 407 (2000) 361–364.
[91] D.J. Fray, Int. Mater. Rev. 53 (2008) 317–325.
[92] U. Erb, Nanostruct. Mater. 6 (1995) 533–538.
[93] I. Gurrappa, L. Binder, Sci. Technol. Adv. Mater. 9 (2008) 1–11.
[94] M.A. Meyers, A. Mishra, D.J. Benson, Prog. Mater. Sci. 51 (2006) 427–556.
[95] R.M. Penner, J. Phys. Chem. B 106 (2002) 3339–3353.
[96] R.M. Penner, Acc. Chem. Res. 33 (2000) 78–86.
[97] P.G. Bruce, B. Scrosati, J.-M. Tarascon, Angew. Chem. Int. Ed. Engl. 47 (2008) 2930–2946.
[98] T. Ciach, K.B. Geerse, J.C.M. Marijnissen, in: Ph. Knauth, J. Schoonman (Eds.), Nanostructured Materials, Kluwer Academic Publishers, New York, 2002, pp. 43–71.
[99] A. Jaworek, Powder Technol. 176 (2007) 18–35.
[100] D.-R. Chen, D.Y.H. Pui, S.L. Kaufmar, J. Aerosol Sci. 26 (1995) 963–977.
[101] O.V. Salata, Curr. Nanosci. 1 (2005) 25–33.
[102] I.W. Lenggoro, K. Okuyama, F.J. de la Mora, N.J. Tohge, Aerosol Sci. 31 (2000) 121–136.
[103] K. Nakaso, B. Han, K.H. Ahn, M. Choi, K. Okuyam, Aerosol Sci. 34 (2003) 869–881.
[104] K. Kim, C.K. Rye, NanoStruct. Mater. 4 (1994) 597–602.
[105] J.C.M. Marijnissen, J. van Erven, K.-J. Jeon, in: W. Sigmund, H. El-Shall, D.O. Shah, B.M. Moudgil (Eds.), Particulate Systems in Nano- and Biotechnologies, Taylor & Francis, 2009, pp. 15–26 Boca Raton, FL.
[106] G.L. Messing, S.-C. Zhang, G.V. Jayanthi, J. Am. Ceram. Soc. 76 (1993) 2707–2726.
[107] K. Okuyama, I.W. Lenggoro, Chem. Eng. Sci. 58 (2003) 537–547.
[108] K.Y. Jung, J.H. Lee, H.Y. Koo, Y.Ch. Kang, S.B. Park, Mater. Sci. Eng. B 137 (2007) 10–19.
[109] D. Verhulst, B.J. Sabacky, T.M. Spitler, J. Prochazka, A New Process for the Production of Nanosized TiO_2 and Other Ceramic Oxides by Spray Hydrolysis, Altair Nanomaterials Inc., Reno, Nevada http://www.b2i.cc/Document/546/A2003-02-25Verhulst.pdf (accessed 22.09.09).
[110] K.C. Pingali, D.A. Rockstraw, S. Deng, Aerosol Sci. Technol. 39 (2005) 1010–1014.
[111] Y. Itoh, I.W. Lenggoro, K. Okuyama, L. Madler, S.E. Pratsinis, J. Nanopart. Res. 5 (2003) 191–198.
[112] V. Viswanathan, R. Filmalter, S. Patil, S. Deshpande, S. Seal, J. Am. Ceram. Soc. 90 (2007) 870–877.
[113] J. Schoonman, Solid State Ion. 135 (2000) 5–19.
[114] R.M. Richards, in: C.I. Contescu, K. Putyera (Eds.), Dekker Encyclopedia of Nanoscience and Nanotechnology, Taylor & Francis Group, LLC, New York, 2009, pp. 2126–2139.
[115] Y.S. Lee, Self-Assembly and Nanotechnology: A Force Balance Approach, Wiley, New Jersey, 2008, p. 93.
[116] J. Eastoe, M.J. Hollamby, L. Hudson, Adv. Colloid Interface Sci. 128–130 (2006) 5–15.
[117] M. Antonietti, K. Landfester, Prog. Polym. Sci. 27 (2002) 689–757.
[118] I. Capek, Adv. Colloid Interface Sci. 110 (2004) 49–74.
[119] J. Schmidt, C. Guesdon, R. Schomacker, J. Nanopart. Res. 1 (1999) 267–276.
[120] F. Li, C. Vipulanandan, IEEE Trans. Appl. Supercond. 13 (2003) 3196–3198.
[121] T.G. Mason, J.N. Wilking, K. Meleson, C.B. Chang, S.M. Graves, J. Phys. Condens. Matter 18 (2006) R635–R666.
[122] J.F. Salager, A. Forgiarini, L. Marquez, Nanoemulsion, FIRP Booklet # E237-A.

[123] K. Landfester, Annu. Rev. Phys. Res. 36 (2006) 231–279.
[124] M.S. El-Aasser, C.D. Lack, J.W. Vanderhoff, F.M. Fowkess, Colloids Surf. 29 (1988) 103–1188.
[125] D. Morales, J.M. Gutierrez, M.J. Garcıa-Celma, Y.C. Solans, Langmuir 19 (2003) 7196–7200.
[126] P. Fernandeza, V. Andre, J. Rieger, A. Kuhnle, Eng. Asp. 251 (2004) 53–58.
[127] I. Capek, in: Advances in Polymer Science, vol. 145, Spinger, Berlin, 1999, pp. 1–55.
[128] H. Zhang, D. Wang, R. Butler, N.L. Campbell, J. Long, B. Tan, D.J. Duncalf, A.J. Foster, A. Hopkinson, D. Taylor, D. Angus, A.I. Cooper, S.P. Rannard, Nat. Nanotechnol. 3 (2008) 506–511.
[129] C.J. Brinker, G.W. Scherer, Sol-Gel Science, Academic Press, New York, 1990.
[130] H. Schmidt, Macromol. Symp. 159 (2000) 43–55.
[131] M.N. Rahaman, Ceramic Processing and Sintering, second ed, New York, Marcel Dekker, 2003 (Chapter 1).
[132] W. Stöber, A. Fink, E. Bohn, et al., J. Colloid Interface Sci. 26 (1968) 62–69.
[133] P.K. Sharma, W. Miao, A. Giri, S. Raghunathan, in: C.I. Contescu, K. Putyera (Eds.), Nanomaterials: Manufacturing, Processing, and Applications: Dekker Encyclopedia of Nanoscience and Nanotechnology, Taylor & Francis Group, LLC, 2009, pp. 2799–2814.
[134] D.R. Rolison, B. Dunn, J. Mater. Chem. 11 (2001) 963–980.
[135] T.E. Mates, T.A. Ring, Colloids and Surfaces 24 (1987) 299–313.
[136] C. Li, Norio Murase, Langmuir 20 (2004) 76–81.
[137] M.P. Pechini, US Patent No. 3.330.697 July 1, 1967.
[138] M. Galceran, M.C. Pujol, M. Aguilo, F. Dıaz, J. Sol-Gel Sci. Technol. 42 (2007) 79–88.
[139] S. Solina, G. Piccaluga, M.P. Morales, C.J. Serna, Acta Mater. 49 (2001) 2805–2811.
[140] C.H. Lu, Y. Lin, H.C. Wang, J. Mater. Sci. Lett. 22 (2003) 615–620.
[141] M. Kirchhoff, U. Specht, G. Veser, Nanotechnology 16 (2005) S401–S408.
[142] S. Lambert, Fabrice Ferauche, B. Heinrichs, N. Tcherkassova, J.P. Pirard, C. Alie, J. Non-Cryst. Solids 352 (2006) 2751–2762.
[143] G. Beaucage, J. Hyeon-Lee, D.J. Kohls, S.E. Pratsinis, J. Nanopart. Res. 1 (1999) 379–392.
[144] K.T. Lee, J.C. Lytle, N.S. Ergang, S.M. Oh, A. Stein, Adv. Funct. Mater. 15 (2005) 547–556.
[145] A. Stein, Z. Wang, M.A. Fierke, Adv. Mater. 21 (2009) 265–293.
[146] F. Li, Y. Qian, A. Stein, Chem. Mater. 22 (2010) 3226–3235.
[147] J. Jang, Adv. Polym. Sci. 199 (2006) 189–259.
[148] H. Yoshimura, Colloids Surf. A Physicochem. Eng. Asp. 282–283 (2006) 464–470.
[149] L. Berti, G.A. Burley, Nat. Nanotechnol. 3 (2008) 81–87.
[150] M.A. Allen, M.M. Prissel, M.J. Young, T. Douglas, in: P. Behrens, E. Bauerlein (Eds.), Handbook of Biomineralization, Wiley-VCH, 2007, pp. 311–334.
[151] K. Iwahori, I. Yamashita, J. Cluster Sci. 18 (2007) 358–370.
[152] J.A. Champion, Y.K. Katare, S. Mitragotri, PNAS 104 (2007) 11901–11904.
[153] A. Dahl, B.L. Maddux, S.E. Hutchison, Chem. Rev. 107 (2007) 2228–2269.
[154] W. Wang, J.C. Martin, X. Fan, A. Han, Z. Luo, L. Sun, ACS Appl. Mater. Interfaces 4 (2012) 977–981.
[155] P.S. Shah, J.D. Holmes, R.C. Doty, K.P. Johnston, B.A. Korgel, J. Am. Chem. Soc. 122 (2000) 4245–4246.
[156] P. Raveendran, Jie Fu, Scott L. Wallen, J. Am. Chem. Soc. 125 (2003) 13940–13941.
[157] K.P. Johnston, T. Randolph, F. Bright, S. Howdle, Science 272 (1996) 1726.
[158] M. Sastry, Curr. Sci. 85 (2003) 1735–1745.
[159] N. Krumov, I. Perner-Nochta, S. Oder, V. Gotcheva, A. Angelov, C. Posten, Chem. Eng. Technol. 32 (No. 7) (2009) 1026–1035.

[160] K.N. Thakkar, S.S. Mhatre, R.Y. Parikh, Nanomed. Nanotechnol. Biol. Med. 6 (2010) 257–262.
[161] E. Bauerlein, in: E. Bauerlein (Ed.), Handbook of Biomineralization, Wiley-VCH Verlag GmbH & Co, Weinheim, 2007.
[162] E. Ruiz-Hitzky, K. Ariga, Y.M. Lvov (Eds.), Bio-Inorganic Hybrid Nanomaterials, Wiley-VCH Verlag GmbH & Co. KGaA, Weinheim, 2008.
[163] J. Wendy, J. Crookes-Goodson, J.M. Slocik, R.R. Naik, Chem. Soc. Rev. 37 (2008) 2403–2412.
[164] J.F. Banfield, H. Zhang, in: Nanoparticles and the Environment, J.F. Banfield, A. Navrostsky (Eds.), Mineralogical Society of America, Washington D.C., 2001, pp. 1–58.
[165] S.V. Patwardhan, N. Mukherjee, M. Steinitz-Kannan, S.J. Clarson, Chem. Commun. (2003) 1122–1123.
[166] D. Schüler, R.B. Frankel, Appl. Microbiol. Biotechnol. 52 (1999) 464–473.
[167] E.D. Spoerke, J.A. Voigt, Adv. Funct. Mater. 17 (2007) 2031–2037.
[168] H. Huang, X. Yang, Biomacromolecules 5 (2004) 2340–2346.
[169] N. Vigneshwaran, R.P. Nachane, R.H. Balasubramanya, P.V. Varadarajan, Carbohydr. Res. 341 (2006) 2012–2018.
[170] T. Premkumar, Kurt E. Geckeler, J. Phys. Chem. Solid 67 (2006) 1451–1456.
[171] J.W. Kim, S.H. Choi, P.T. Lillehei, S.H. Chu, G.C. King, G.D. Watt, Chem. Commun. (2005) 4101–4103.
[172] A.R. Shahverdi, S. Minaeian, H.R. Shahverdi, H. Jamalifar, A.-A. Nohi, Process Biochem. 42 (2007) 919–923.
[173] Q. Wei, S. Kang, J. Mu, Colloids Surf. A Physicochem. Eng. Asp. 247 (2004) 125–127.
[174] M. Gericke, A. Pinches, Hydrometallurgy 83 (2006) 132–140.
[175] J.L. Gardea-Toresdy, J.G. Parsons, E. Gomez, J. Peralta-Videa, H.E. Trojani, P. Santiago, Nano Lett. 2 (2002) 397–401.
[176] X. Bansai, D. Rautaray, A. Ahmad, M. Sastry, J. Mater. Chem. 14 (2004) 3303–3305.
[177] D. Bhattacharaya, R.K. Gupta, Crit. Rev. Biotechnol. 25 (2005) 199–204.
[178] D. Mandal, M.E. Bolander, D. Mukhopadhyay, G. Sarkar, P. Mukherjee, Appl. Microbiol. Biotechnol. 69 (2005) 485–492.
[179] C.D. Mao, D.J. Sols, B.D. Reiss, S.T. Kottman, R.Y. Sweeny, A. Haynurst, Science 303 (2004) 213–217.
[180] C. de Mello Doneg, P. Liljeroth, D. Vanmaekelbergh, Small 1 (2005) 1152–1162.
[181] R.E. Bailey, A.M. Smith, S. Nie, Physica E 25 (2004) 1–12.
[182] J. Park, K. Aa, Y. Hwang, J.-G. Park, H.J. Noh, J.-Y. Kim, J.-H. Park, N.-M. Hwang, T. Hyeon, Nat. Mater. 3 (2004) 891–895.
[183] S.C. Lock, Aerosols: In Theory and Practice, Aerosols Ltd, London, 1947.
[184] S. Tohno, M. Itoh, S. Aono, H. Takana, J. Colloid Interface Sci. 180 (1996) 574–577.
[185] S.-A. Dong, S.-P. Zhou, Mater. Sci. Eng. B 140 (2007) 153–159.
[186] P. He, M. Zhang, D. Yang, J. Yang, Surf. Rev. Lett. 13 (2006) 51–55.
[187] T.K. Sau, A. Pal, N.R. Jana, Z.L. Wang, T. Pal, J. Nanopart. Res. 3 (2001) 257–261.
[188] E. Jimenez, K. Abderrafi, R. Abargues, J.L. Valdes, J.P. Martınez-Pastor, Langmuir 26 (2010) 7458–7463.
[189] E. Kauffeldt, T. Kauffeldt, J. Nanopart. Res. 8 (2006) 477–488.
[190] H. Shao, H.S. Lee, Y. Huang, I.Y. Ko, C.O. Kim, IEEE Trans. Magnet. 41 (2005) 3388–3390.
[191] H. Yang, F. Ito, D. Hasegawa, T. Ogawa, M. Takahashi, J. Appl. Phys. (2007) 09J112-1–09J112-3.
[192] K. Esumi, T. Tano, K. Torigoe, K. Meguro, Chem. Mater. 2 (1990) 564–567.

[193] K.R. Brown, D.G. Walter, M.J. Natan, Chem. Mater. 12 (2000) 306–313.
[194] E.G. Schutt, Eur. Patent Application 90317671.4, filed September 25, 1990.
[195] M. Eguchi, D. Mitsui, H.-L. Wu, R. Sato, T. Teranishi, Langmuir 28 (2012) 9021–9026.
[196] J.P. Wilcoxon, P.P. Provencio, J. Am. Chem. Soc. 126 (2004) 6402–6408.
[197] J. Jiang, T.D. Hall, L. Tsagalas, D.A. Hill, A.E. Miller, J. Power Sour. 162 (2006) 977–984.
[198] J. Jiang, A.E. Miller, J. Power Sour. 173 (2007) 867–876.
[199] L. Al Juhaiman, L. Scoles, D. Kingston, B. Patarachao, D. Wang, F. Bensebaa, Green Chem. 12 (2010) 1248–1252.
[200] J.J. Nairn, P.J. Shapiro, B. Twamley, T. Pounds, R. Wandruszka, T. Rick Fletcher, M. Williams, C. Wang, M.G. Norton, Nano Lett. 6 (2006) 1218–1223.
[201] S.L. Castro, S.G. Bailey, R.P. Raffaelle, K.K. Banger, A.F. Hepp, J. Phys. Chem. B 108 (2004) 12429–12435.
[202] Y. Okinaka, M. Hoshino, Gold Bull. 31 (1998) 3–13.
[203] L.A. Porter Jr., H.C. Choi, A.E. Ribbe, J.M. Buriak, Nano Lett. 2 (2002) 1067–1071.
[204] A.M. Azad, S. Kesavan, S. Al-Batty, Key Eng. Mater. (2008) 229–255.
[205] L.K. Kurihara, G.M. Chow, P.E. Schoen, Nanostruct. Mater. 5 (1995) 607–613.
[206] D.D. Evanoff Jr., G. Chumanov, J. Phys. Chem. B 108 (2004) 13948–13956.
[207] D.D. Evanoff Jr., G. Chumanov, Chemphyschem 6 (2005) 1221–1231.
[208] J.M. Huang, Y. Yang, B. Yang, S.Y. Liu, J.C. Shen, Thin Solid Films 327–329 (1998) 536–540.
[209] R. Schenk, V. Hessel, N. Jongen, V. Buscaglia, S. Guillemet-Fritsch, A.G. Jones, in: H.S. Nalwa (Ed.), Encyclopedia of Nanoscience and Nanotechnology, vol. 7, American Scientific Publishers, New York, 2004, pp. 287–296.
[210] M. Kawa, H. Morii, A. Ioku, S. Saita, K. Okuyama, J. Nanopart. Res. 5 (2003) 81–85.
[211] G.B. Segeev, NanoChemsitry, Elsevier, Amsterdam, 2006, p. 28.
[212] R. Karnik, F. Gu, P. Basto, C. Cannizzaro, L. Dean, W. Kyei-Manu, R. Langer, O.C. Farokhzad, Nano Lett. 8 (2008) 2906–2912.
[213] J.D. Winterton, D.R. Myers, J.M. Lippmann, A.P. Pisano, F.M. Doyle, J. Nanopart. Res. 10 (2008) 893–905.
[214] A. Aimable, T. Strachowski, E. Wolska, W. Lojkowski, P. Bowen, Process. Appl. Ceram. 4 (2010) 107–114.
[215] B.K.H. Yen, A. Gunther, M.A. Schmidt, K.F. Jensen, M.G. Bawendi, Angew. Chem. Int. Ed. Engl. 44 (2005) 5447–5451.
[216] Y. Song, H. Modrow, L.L. Henry, C.K. Saw, E.E. Doomes, V. Palshin, J. Hormes, C.S.S.R. Kumar, Chem. Mater. 18 (2006) 2817–2827.
[217] D. Shalom, R.C.R. Wootton, R.F. Winkle, B.F. Cottam, R. Vilar, A. deMello, C.P. Wilde, Mater. Lett. 61 (2007) 1146–1150.
[218] T. Ould-Ely, M. Luger, L. Kaplan-Reinig, K. Niesz, M. Doherty, D.E. Morse, Nat. Protoc. 6 (2011) 97–104.
[219] X. Li, N. Anton, C. Arpagaus, F. Belleteix, T.F. Vandamme, J. Control. Release 147 (2010) 304–310.
[220] W. Peukert, H.C. Schwarzer, F. Stenger, Chem. Eng. Process. 44 (2005) 245–252.
[221] H.N. Holmes, Laboratory Manual of Colloid Chemistry, third ed, Johm Wiley & Sons, London, 1934.
[222] G. Kalyuzhny, R.W. Murray, J. Phys. Chem. B 109 (2005) 7012–7021.
[223] E.W. Foster, G.J. Kearns, S. Goto, J.E. Hutchison, Adv. Mater. 17 (2005) 1542–1545.
[224] J. Park, E. Lee, N.-M. Hwang, M. Kang, S.Ch. Kim, Y. Hwang, J.G. Park, H.J. Noh, J.-Y. Kim, J.H. Park, T. Hyeon, Angew. Chem. Int. Ed. Engl. 44 (2005) 2872–2877.

[225] M.C. McLeod, M. Anand, C.L. Kitchens, C.B. Roberts, Nano Lett. 5 (2005) 461–465.
[226] I. Limayem, C. Charcosset, H. Fessi, Sep. Purif. Technol. 38 (2004) 1–9.
[227] M. Hanauer, S. Pierrat, I. Zins, A. Lotz, C. Sonnichsen, Nano Lett. 7 (2007) 2881–2885.
[228] G. Chen, Y. Wang, L.H. Tan, M.X. Yang, L.S. Tan, Y. Chen, H.Y. Chen, J. Am. Chem. Soc. 131 (2009) 4218–4219.
[229] P. Qiu, C. Mao, Adv. Mater. 23 (2011) 4880–4885.
[230] H. Althues, J. Henle, S. Kaskel, Chem. Soc. Rev. 36 (2007) 1454–1465.

Chapter 3

Dry Production Methods

1 INTRODUCTION

Nanoparticle (NP) manufacturing under dry conditions has seen relatively more commercial success compared to the wet manufacturing process [1–4]. Dry NP processes are often easier to scale up cost effectively. However, monodispersity and surface chemistry control remain a challenge in the case of dry processes. Physical (e.g., spray) and chemical (e.g., precipitation) methods have been described in the previous chapter as wet methods. Two large groups of dry techniques can also be distinguished. The first group consists of the physical methods including metal evaporation in a flux of inert gas followed by condensation at a lower temperature. The second group consists of the chemical methods such as precursor decomposition followed by recombination at atomic scale. Physical methods used to prepare nanoparticles (NPs) consist of both top-down (e.g., ball milling) and bottom-up (chemical vapor condensation) processes.

Physical methods used in dry NP synthesis include molecular beam, gas evaporation, ball milling, microwave plasma, laser ablation, sputtering, shock waves, combustion flame, electro-explosion, laser induced electro-dispersion, and supersonic jet [5]. Mechanical attrition and gas-phase methods remain the two most viable commercial processes for large-scale manufacturing. Besides particle characteristics, cost and scalability are factors providing an edge to dry synthesis processes. Surface fuctionalization is easily achievable via wet methods, particularly those involving organic solvents. The main drawback of nonaqueous based wet methods is the requirement for high volume of solvent, which contributes to higher NP cost.

Other critical factors include energy requirement and environmental impact. In general, electrically powered dry methods consume relatively more energy than wet methods. This is particularly so in the synthesis of light oxide NPs such as TiO_2 [6]. Dry and wet fabrications of TiO_2 NPs comprise a similar number of processing steps [6]. The dry process provides NPs of relatively higher purity when compared to wet processing. At present, it is difficult to provide a more quantitative comparison between the two techniques based on efficiency and production cost. This depends on product characteristics and other requirements.

Interface Science and Technology, Vol. 19. http://dx.doi.org/10.1016/B978-0-12-369550-5.00003-3

Although dry methods provide several advantages, the relatively large particle size distribution combined with difficulty in surface functionalization [7] reduces the scope of their applications. Posttreatments that include size selection are often required to reach monodispersity. Surface functionalizaton of dry process NPs is also possible. Posttreatment based on solution processing of aerosol NPs has been reported [8]. In this case, organic passivation allowed enhancement of PL properties of silicon nanocrystals.

Combining attributes of wet and dry synthesis techniques could provide opportunities for near-ideal manufacturing processes. Andres et al. have reported a continuous process combining physical evaporation and surfactant-based stabilization [9]. This preparation is conducted in four steps [9]. In the first step, following gold atom evaporation from a carbon crucible, a room temperature He stream is used to condense the atoms into small clusters. In the next two steps, the formed clusters are melted and recrystallized while still in the gas phase. In the last step, by contact with an organic solvent mist containing 1-dodecanethiol, these gas-phase clusters are collected as a stable colloidal suspension. The resulting NPs are scrubbed from the gas phase using a solvent mist in the presence of surfactants. The presence of surfactants prevents NP growth and agglomeration. Using this hybrid and continuous fabrication process, high-quality monodispersed gold NPs are obtained. This technique could be easily applied to other chemical compositions.

In this chapter, mainly two categories of dry techniques are discussed: mechanical and gas phases. Other techniques, including cryomelting, laser pyrolysis, reactive plasma arc evaporation, microwave plasma process, epitaxial growth, rapid quenching solidification, ion-beam sputtering, self-propagating high-temperature synthesis (SHS), and hydrogen reduction, are also discussed. Similar to the approach taken in the preceding chapter, dry methods that have (or will have) significant potential for large-scale manufacturing are discussed more extensively.

2 MECHANICAL SYNTHESIS

Size reduction using crushing, grinding, or milling machines has been known for centuries. These size reduction techniques are often referred to as comminution. Micrometer- and millimeter-sized particles have been obtained using ball mills, rod mills, and sag mills. These techniques are widely used in the mining industry and to a lesser extent in the manufacturing industry. More recently, several mechanical methods have been reported that allow submicron particle fabrication [10]. Mechanical techniques are suitable for large-scale NP fabrication. However, the low purity and inhomogeneity of the particles limit the application of these techniques. Purely mechanical techniques are further limited by the relatively large size distribution. Particles with sizes less than 100 nm have been obtained using mechanochemical techniques.

Typically, mechanical attrition occurs via either smashing (jet mills and vibratory mills) or rubbing and smearing (sigma-blade kneader) [11]. Some mechanical techniques such as ball milling combine smashing and smearing. Mechanical grinding belongs to the "top-down" group of techniques [12]. These milling (or attrition) approaches use mechanical power to provide energy for size reduction of coarse particles. Different types of grinding processes have been used including impact (disc mill), attrition, mortar, jet mill, and ball mill (tumbling, vibration).

It is important to differentiate mechanical alloying from mechanical attrition. In the case of mechanical alloying, the starting elements are mixed in a defined proportion. In the case of mechanical attrition, mechanical refining of the micro- or macrostructured materials down to NP size has been carried out [13]. Both techniques are referred to as ball milling. Ball milling gives the lowest average particle size when compared to vibration milling, jet milling, and roll milling [14] and is the most widely used mechanical technique.

Purely mechanical techniques often produce a relatively larger particle distribution, rarely in the submicron size range. Mechanochemical synthesis has been successfully used to produce NPs. The chemical reaction often occurs at the surface of the starting materials. Small particles formed on the surface of larger particles are subsequently removed from the solid surface. Often, NPs obtained using mechanical or mechanochemical techniques contain a large amount of amorphous phase, requiring heat treatment and sintering for conversion to the crystalline phase.

The minimum grain size obtained by ball milling is inversely proportional to the melting temperature of the ball-milled elements [13]. Grain sizes down to a few nanometers are obtained when materials with a low melting temperature are used [13]. It is worth noting that NPs are also obtained naturally due mostly to mechanical forces. For example, sliding wear and erosion processes generate NPs and microparticles. However, the size distribution of these particles is quite large.

2.1 Ball Milling

A variety of ball milling techniques (e.g., vibratory, rotary, planetary, shaker, and tumble) have been used to produce submicron-sized particles. Typically, large reactant powder particles (50 μm) are introduced in a sealed vial with hardened steel- or WC-coated balls. Ball milling is performed by shaking and violent agitation (Fig. 3.1). The vial is filled with gas (often inert). The milling process includes powder splitting, exposing highly reactive surfaces. The ensuing plastic deformation leads to nanometer-scale structures of the powder. High-energy mechanical attrition raises the temperature to around 150 °C. The presence of highly reactive surfaces promotes coalescence allowing formation of homogeneous particles [10,15].

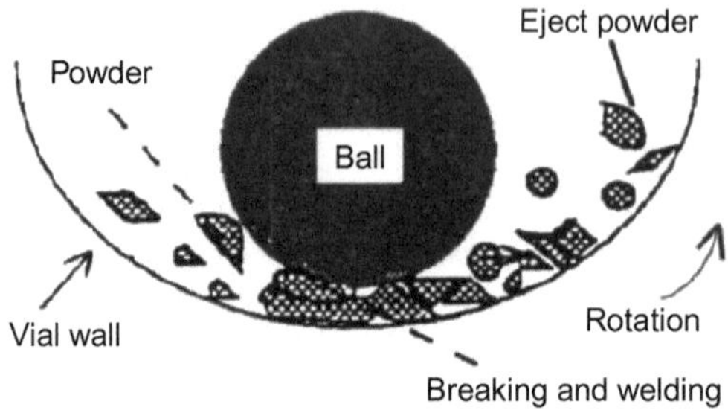

FIGURE 3.1 Schematic representation of the ball milling process used for particle size reduction [15].

Ball milling has several attributes when compared to other methods of synthesis of NPs. It is simple, relatively inexpensive, and can be applied to any material and composition. However, in the case of a conventional mechanical milling technique, the average size and size distribution of the produced NPs are relatively large. This drawback can limit utilization of the ball milling technique when monodisperse NPs are required. [16]. GaN NPs with optical bandgap consistent with that of the bulk material have been synthesized using the mechanical method. Particle size distribution depends on both bead diameter and processing time [17]. Only a few minutes of processing time are required to obtain particles of 100 nm or less using 100 μm beads. Using 30 and 60 μm beads, sub-100 nm particles of less than 100 nm are obtained quite instantaneously. However, with beads larger than 100 μm, particles of several hundred nm are obtained after more prolonged processing times.

Extended processing times lasting several hours are often required to achieve a smaller NP distribution [18,19]. Beyond a few hours of milling, the average size of iron powder reaches a minimum between 20 and 40 nm [19]. Cyclic operation has also been shown to affect the particle size distribution [18]. It is likely that ball milling facilitates sintering of fcc metals to form larger particles [2].

Modified ball milling methods have been used to address some of the shortcomings of the conventional approach. Vibratory mill has been used to synthesize MS (where M = Pb, Zn, Cd, Cu) nanocrystalline particles on a large scale at room temperature [20]. Recently, a modified ball milling method using elemental precursors as a reactant instead of large powder particles has been developed. Reactive ball milling, ball milling with reactive fluid (gas and liquid), has extended the range of applications of normal ball milling. Mechanochemical ball milling combining milling and chemical reaction is discussed later.

Small graphite particles are used to increase the Li/C ratio of carbon intercalated graphite (CIG) in lithium-ion battery anodes. Ball milling often leads to amorphous graphite and defects in addition to CIG phases. Using water as a fluid, it has been shown that these shortcomings could be minimized [21]. Production of defect-free NPs using ball milling (Fig. 3.2) may involve a significant amount of water during particle size reduction [22]. This milling process can be operated in a batch or recirculation mode. Addition of fluid inside the ball milling reactor could moderate the mechanical shock and minimize the amorphization of the material (e.g., graphite) and reagent [23].

Embedded Fe–C alloy NPs with an average size of 50 nm were obtained by ball milling using preheated steel cut chips [24]. Pd NPs with an average size of 5.5 nm have been obtained using a commercial planetary ball mill system [25]. Note that the starting material consists of 400 μm Pd particles. In terms of small sizes and monodispersity, this technique provides the best quality NPs using mechanical milling [25]. Ball milling has also been used to produce ceramic [26,27] and polymeric [28] NPs. Often, the produced particles showed a relatively large size distribution. Posttreatment yields higher quality NPs and also allows tuning of particle properties [29]. In the case of ZnO, the average size can be controlled from 28 up to 57 nm by varying posttreatment temperature using this approach [29].

Ball milling has been successfully used for the synthesis of alloys, nanocomposites, and intermetallic materials. Mechanical alloying starting from dissimilar powders such as Ti and Al lead to the formation of $Ti_{25}Al_{75}$ NPs

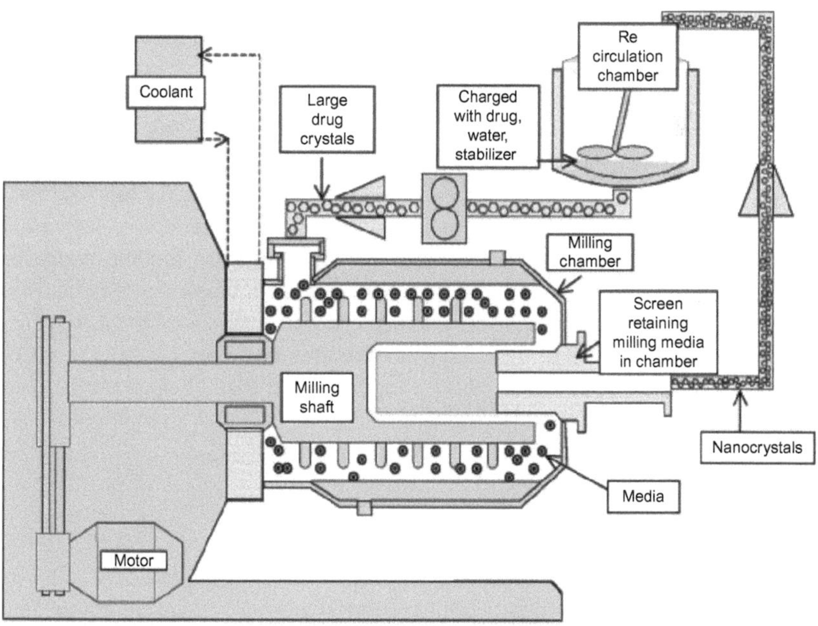

FIGURE 3.2 Schematic representation of the media milling process used to manufacture microparticles and NPs [22].

with crystalline size in the range of 10–30 nm [30]. A gradual phase transformation is observed during the mechanical alloying of the multicomponent $Zr_{60}Al_{10}Ni_9Cu_{18}Co_3$ powder mixture. Nanoparticles (NPs) with a single crystalline phase are clearly demonstrated using XRD after 11 h of milling [31]. Below 10 h of milling, only large crystalline particles are observed. Transformation from the initially crystalline powder mixture to an amorphous glass-like structure is also observed.

Nanocomposites consisting of NPs embedded in a compositionally different matrix provide numerous advantages. For example, they could be more stable than ultra fine grained monolithic materials. Mg–Ti–C TiC nanocomposites and NPs embedded in a magnesium matrix to form Mg–TiC nanocomposites exhibit good ductility and compressive strength properties [32]. The mechanical process is also quite suitable for the preparation of alloyed Al–Fe NPs with an average size of about 10 nm [33]. Starting with Al and Fe microparticles, it was established that the processing time has a direct effect on the degree of alloying. The lattice parameter is used to measure the Al/Fe alloying ratio [33]. Indeed, the lattice parameter is reduced from 4.05 to below 4.04 Å after a processing time of several hundred hours [33].

Besides processing time and temperature, material properties may affect the final grain and particle sizes. In particular, the minimum grain size is found to diminish with increasing melting temperature of the starting materials [34]. This variation is much steeper for lower melting point metals. A minimum grain size of 50 and 40 nm is observed with Sn and Mg powders having melting points between 800 and 1000 °C [34]. In the case of Fe, PdNi, Co, and Cr that have a melting temperature of about 2000 °C and an average minimum grain size of 10 nm has been reported [34]. For C with a melting temperature of 4000 °C, the minimum grain size is just few nanometers [34]. Furthermore, the phase may also play a significant role. Indeed, it appears that the minimum grain size (d) varies as follows: $d(\text{fcc}) < d(\text{bcc}) < d(\text{hcp})$ [34].

Contamination from the vial and grinding media remain two important issues with ball milling. Other sources of contamination include additives and by-products. The nature and concentration of these contaminations depend on the milling conditions. As in the case of all contaminations, they may hinder applications of these high surface area powders. Metallic contamination of lithium-based active nanostructured electrode battery materials has been singled out as a possible explanation for laptop battery explosion.

2.2 Mechanochemical method

A mechanochemical (also referred as mechanosynthesis) method was successfully used to synthesize intermetallics, solid solutions, oxides, hydrides, and organic compounds [35]. These NPs are ill defined, having different phases and shapes, and relatively wide size distribution. Sub-100 nm metallic and oxide NPs have been reported by McCormick's group since the early 1990s [32,36–38]. Mechanochemical synthesis is often carried out in a sealed container, starting with powder precursors. The precursor reactant sizes are reduced significantly to provide intimate phase mixing and enhance diffusion and chemical reactivity. For example, relatively contamination-free silicon quantum dots have been synthesized by silica reduction in a ball-milling reactor containing graphite and SiO_2 powder [39]. GaN NPs with irregular shapes ranging from 4 to 20 nm have been synthesized by reacting Ga_2O_3 with Mg_3N_2 [16].

SnO_2 NPs provide a significant improvement in storage capacity per unit mass in lithium-ion batteries [40]. Simply mixing molten salts (lithium nitrate, lithium hydroxide monohydrate) and Sn(II) chloride dehydrate in an agate mortar, followed by heat treatment up to 300 °C, gives rise to porous SnO_2 NPs [40]. These NPs with an average size of less than 10 nm are incorporated into matrix materials to form the anode.

Co_3O_4 NPs in the range of 15–20 nm have been prepared by mechanochemical reactions involving cobalt carbonate, sodium oxide, and sodium carbonate [41]. Three main process steps are involved during the mechanochemical synthesis of these Co_3O_4 NPs [41]. The reaction is completed after three days of milling. First, the reactants (Na_2O, $CoCO_3$, and Na_2CO_3) are mixed to form smaller particles, allowing intimate mixing. After 12 h, cobalt oxide is formed. After another 60 h of milling, single-phase Co_3O_4 NPs are obtained.

A relatively similar approach has been used by Balaz et al. to synthesize ZnS, CdS, and PbS NPs with diameters in the range of 5–18 nm [41]. In this case, the reaction kinetics is much faster [42]. Using metal acetate as precursor and Na_2S, the reaction during the milling process proceeds as follows [42]:

$$(CH_3COO)_2Me + Na_2S \rightarrow MeS + 2CH_3COONa$$

An industrial vibratory mill system has been used to synthesize MeS (Me = Pb, Zn, Cd, Cu) NPs with an average size range of 4–18 nm [20]. This technique is suitable for large-scale production. Soft magnetic NPs, such as $Fe_{73.5}Nb_3Cu_1Si_{13.5}B_9$, often referred to as Finemet, have been developed for potential industrial applications [43]. Finemet iron NPs have been mechanochemically synthesized using a planetary milling system [44]. Properties of TiO_2 NPs obtained by mechanochemical synthesis have been compared to those obtained by sol–gel and by combustion [45]. The mechanochemical route yields an average particle size of 20 nm after 12 h of milling. Sol–gel and combustion yield particle sizes of 14 and 50 nm, respectively. There are also some microstructural differences between the NPs obtained by these three techniques.

Applying High-temperature self-propagating synthesis (SHS) concepts in mechanosynthesis allows finely dispersed NPs [35,46]. Although developed for ceramic and intermetallic powders, SHS could also be used for NP preparation. SHS takes advantage of the generated exothermic heat to complete the reaction through self-sustaining combustion propagation. After adding the heated precursors inside the reaction zone, energy is released, which leads to increased temperature and reactivity. Material conversion and temperature profiles show a similar trend. Both values increase before reaching a plateau. The maximum temperature is related to combustion temperature (T_C). In the case of TiC, combustion temperature increases when particle size is reduced [46]. Sub-100 ZrB_2 NPs have been synthesized by SHS using a planetary ball mill containing zirconia, magnesium, and boron oxide or boric acid. A similar approach has been used to prepare sub-50 nm TiB_2 NPs. Addition of NaCl lowers the particle size from around 50 (no NaCl) down to 20 nm (20 wt% NaCl) [46].

2.3 Severe Plastic Deformation

Synthesis of ultrafine grains using plastic deformation has been known for decades. Using mechanical deformation stress, microstructured materials are broken down to nanoscale structures including NPs [47]. In the cases of SPD (severe plastic deformation), the deformation step can be obtained by extrusion, rolling, or other means.

SPD has been used to synthesize pure metal and alloy materials with grain sizes of about 100 nm or less [48,49]. Under SPD conditions, microstructured materials are subdivided into progressively finer grains. Numerous bulk materials and films owe their strength to the presence of ultrafine grains. Two types of SPD have been developed: high-pressure torsion (HPT) and equal-channel angular pressing (ECAP) [49]. In the case of ECAP, a piston is used to apply a strain on a sample machined to fit in a die. The two channels intersecting at an angle around 90° facilitate the plastic deformation. ECAP is the most popular SPD technique for the production of nanocomposite materials containing sub-100 nm grains. In the case of HTP, the sample is linearly compressed between two anvils. A shear stress is applied because of the rotation of one of the two anvils. HPT provides the possibility to independently control the cumulative strain, applied pressure, and strain rate [49].

2.4 Cryomilling

Cryomilling is a mechanical attrition approach to synthesize powders in a cryogenic liquid medium [50]. Developed first by Exxon Research [51], cryomilling has been used by several other groups to reduce the size of powders down to nanometer range [50]. Cryomilling systems (Fig. 3.3) with a capacity of 1–35 kg have been used in laboratory and commercial systems.

Different metal (Al, Fe, Ni) and alloy (AlRu, AlMg, inconel) NPs have been prepared using cryomilling systems. Size is controlled by varying

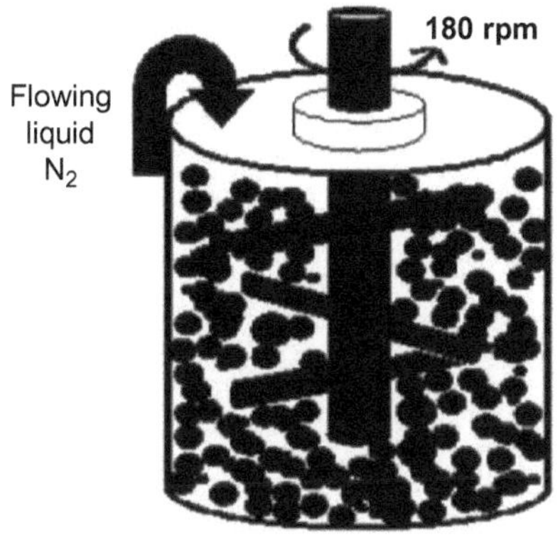

FIGURE 3.3 Schematic representation of the cryomilling technique [50].

milling time and temperature [50]. One interesting point regarding cryomilled NPs is the potential improvement in their mechanical properties once they are consolidated.

3 GAS PHASE

Although gas-phase synthesis of small particles could be traced back to ancient China [52], it is only during the last three decades of the last century that significant developments have been reported [1,53–56]. Several sources and methods are used to evaporate the precursor. Thermal, laser, sputtering, and electron beams have been used in the past. There are three main process stages in gas-phase synthesis. This includes nucleation initiation, nucleation growth to form NPs, and collection following collision with a cold finger or molecules. There are differences related to the initiation of the nucleation and collection stages.

The gas-phase fabrication process of NPs revolves mostly around the nucleation stage once supersaturation is reached. Following the nucleation stage, the conditions leading to supersaturation are altered to prevent further nucleation [56]. Size distribution depends mostly on the ability to control the supersaturation state before and after nucleation is initiated. Indeed, a high degree of supersaturation is required to obtain high nucleation density, and at the same time, the source of supersaturation should be slowed to avoid uncontrolled particle growth. This supersaturation level is adjusted continuously or quasi-continuously within a timeframe of milliseconds to seconds [56]. Following NP formation, coalescence and coagulation should be minimized to avoid the formation of large agglomerated particles with a wide size distribution. Particle coalescence is slowed or eliminated by lowering temperature. However, some degree of agglomeration is difficult to eliminate completely [56].

Gas-phase processing techniques are viable for commercial production. When compared to other fabrication processes, NP synthesis using gas-phase techniques provides the following advantages [54]:

a. Higher product purity is obtained when compared to nonvacuum methods;
b. Aerosol behaves like a reactor allowing complex chemical structures and control over product characteristics;
c. It is a cheap alternative to expensive vacuum synthesis techniques;
d. A higher deposition rate is obtained by combining fabrication and integration of NPs into a single step;
e. It is a continuous fabrication process.

A few drawbacks hinder the wider utilization of gas-phase techniques. Issues related to wide size distribution, surface composition, and yield should be addressed as high-priority issues. High precursor concentrations are required for a practical and scalable gas-phase fabrication process. This gives rise to pore size distributions, and crystalline phases and compositions.

Gas-phase synthesis of NPs could be divided into two categories. This division depends on the nature of the precursor and the energy source in the dispersion step [56]. The first category consists of solid precursors used in laser ablation, spark discharge, and inert-gas condensation to fabricate NPs. In the second category, liquid and vapor precursors are used in conjunction with spray and laser pyrolysis, flame pyrolysis, or plasma processing [56]. In the following paragraphs, the different gas-phase techniques are discussed according to the process of gasification. In some cases, the same technique can use solid and nonsolid precursors.

3.1 Inert-Gas Condensation

Subsequent to the early work of Granqvist and Buhrman [1], gas-phase condensation was successfully used to synthesize numerous metallic, compound, and oxide NPs [57]. Following metal evaporation in a gas-filled (e.g., Ar) chamber, NPs are formed and deposited on a water-cooled finger in the form of fractals (Fig. 3.4). See Ref. [1] for more details. Modifications of this technique have been developed [57]. Inert-gas condensation (IGC) allows geometrical and structural properties. This is achieved by adjusting the rate of energy supply, energy removal of hot atoms, and the rate of cluster/particle removal from the supersaturated gas [34]. To achieve this control, several processing parameters can be used. This includes inert-gas type, pressure, flow rate, evaporation rate, and temperature [34].

To increase the average particle size, inert-gas pressure, inert-gas molecular weight, crucible temperature, and growth distance and/or evaporation rate need to be increased. Decreasing inert-gas flow rate or inert-gas temperature also results in increased average particle size [34]. IGC is a continuous process providing flexibility in material range and composition.

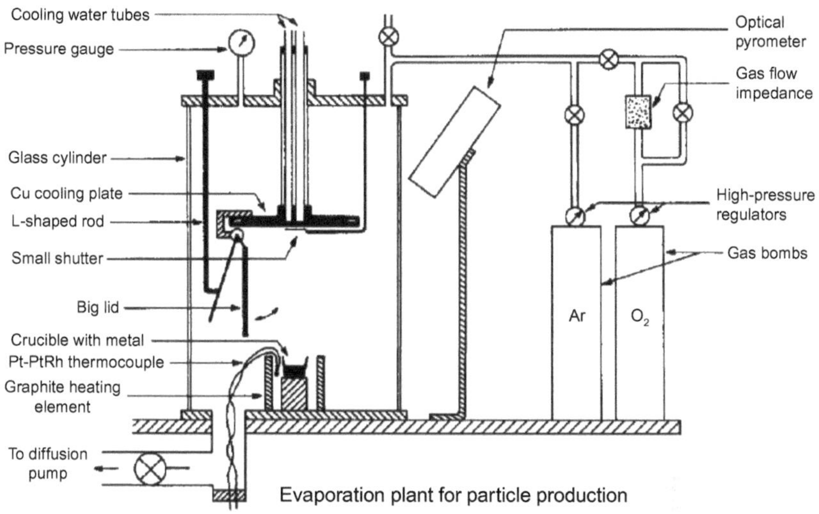

FIGURE 3.4 Schematic representation of the inert-gas condensation (IGC) setup for NP synthesis [1].

High-quality NPs (in purity and size distribution) are easily achievable with some postprocessing. However, this is achieved at the expense of relatively high production costs. Agglomeration occurs nonetheless when these NPs are consolidated. So far, it has not been possible to synthesize NPs of a few compounds such as GaAs and organic-based materials.

3.2 Aerosol Spray Pyrolysis

Aerosol processing has been used to synthesize particles with a wide range of sizes, shapes, and compositions [58]. Aerosol spray pyrolysis can be used to synthesize a variety of NP compositions with controlled size and porosity structure (Fig. 3.5).

Aerosol could be formed via pneumatic nebulization, ultrasonic nebulization, or electrospray techniques [60]. In some cases, these various atomization techniques could have some effect on the particle characteristics [60,61]. The pneumatic nebulizer has been used to spray heated solutions on heated substrates to form uniform CdS NPs in the range of 3–6 nm [62]. In this case, large CdS powders are first dissolved in methanol for use as precursors.

An aerosol generator has been coupled to a fluidized bed reactor (FBR) to synthesize metallic zinc NPs using ZnO powder as raw material [63]. Highly stable and defect-free anatase titania NPs have been obtained from propanolic titanium tetra-isopropoxide aerosol. The temperature inside the reactor is at least 600 °C [64]. Unlike other synthesis processes, no anatase-to-rutile phase transition is observed.

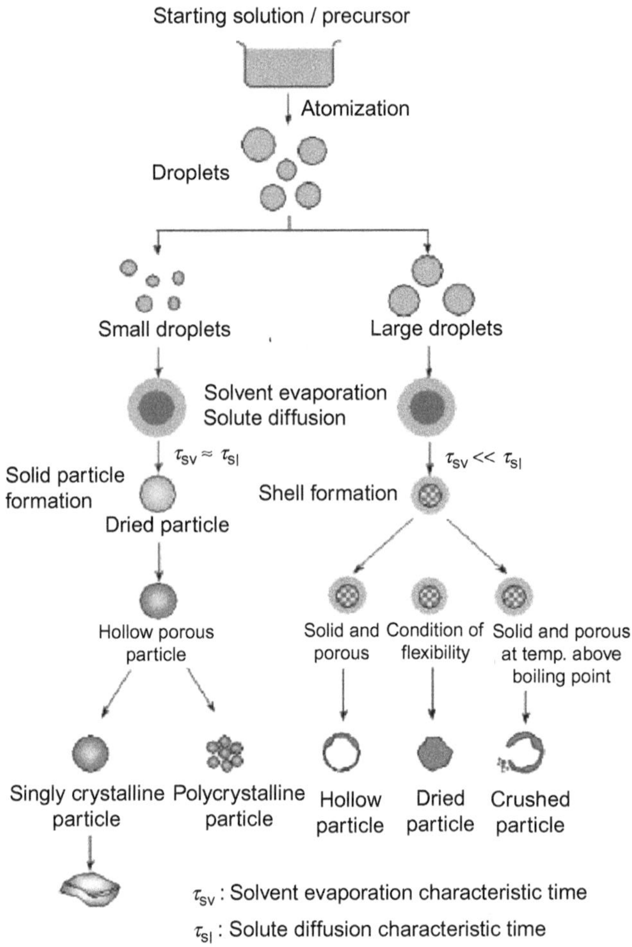

FIGURE 3.5 Process flowchart showing the small and large droplet approaches leading to NPs with different structures [59].

3.3 Combustion Flame Synthesis

Combustion synthesis is widely used to synthesize small oxide particles from redox oxides [65]. Compared to other synthesis techniques, exothermic combustion processes are more energy efficient because they can sustain high process temperatures. Depending on the state of the reactants, one can distinguish between solid state combustion and solution combustion (SC) [66]. Although developed only recently, SC has been successfully used for synthesis of NPs of various oxides [66,67].

In general, the precursors consist of redox mixtures. The reducing agent (fuel) is an organic compound such as urea, citric acid, or PVA (polyvinyl alcohol) polymer. The oxidizing agent is a metal salt such as nitrate. After ignition using an external source, the precursor mixture burns via an exothermic redox reaction to form the product. The presence of a large volume of gas formed after the combustion induces disintegration of large particles into NPs [68].

One advantage of this technique is the low energy input, while the disadvantage is the presence of a relatively large amount of carbon in the final product. However, even when a polymer is used, most of the organic and other volatiles are removed at relatively high temperatures (350 °C). This is the case of Co_3O_4 NP combustion synthesis process [69]. Furthermore, for some applications, the presence of a carbon coating on the surface of the resulting NPs can be beneficial. In the case of ceramic materials used for anode and cathode films in lithium batteries, interconnected carbon shell layers are used as a pathway for electric charge transport. The presence of a carbon layer will reduce particle growth and agglomeration/aggregation.

Gas-phase synthesis using the flame combustion process has been widely used [54,55,70]. Flame heat can be used to initiate nucleation of both aerosol and "non-aerosol" precursors. Agglomerated and aggregated NPs are obtained via aerosol flame synthesis (FS) using either liquid-fed or vapor-fed precursors (Fig. 3.6). Starting with the liquid or vapor precursors, a nucleation of monomers/atoms gives rise to nucleate, which forms the seed NPs (Fig. 3.6). Agglomeration occurs following coalescence of seed NPs. Aggregated NPs are obtained from coagulated/sintered seed NPs.

A couple of million tons of carbon black, silica, and titania are produced annually [55]. Flame synthesis has been used for large-scale production of ceramic micropowder commodities for a long time [72]. Depending on particle composition and characteristics, and production requirements, different flame reactor designs are used [73]. The first two designs (co-flow diffusion flame and premixed flat flame) are simple and widely used. The third one (counterflow diffusion flame) is more complicated although it can provide some advantage in terms of quality control.

In the case of a co-flow design, there are some key flame reactor design parameters that allow the optimization of the NP fabrication process [74,75]. Flame characteristics such as length and shape are often used to tune particle characteristics. The flame length can be controlled by increasing the oxidant feeding rates into the reactor. The laminar flame is induced by turbulence caused by high oxidant flow rates. Shorter flames caused by faster combustion (high temperature and steep cooling rates) reduce particle residence times. Depending on the detailed particle temperature history, shorter flames give rise to smaller particles with various degrees of agglomeration [75]. Oxygen flow allows control of the flame's height and shape for the production of Mn/SiO_2 NPs [75]. Emissions from manganese-containing species in the flame lead to the green color [75].

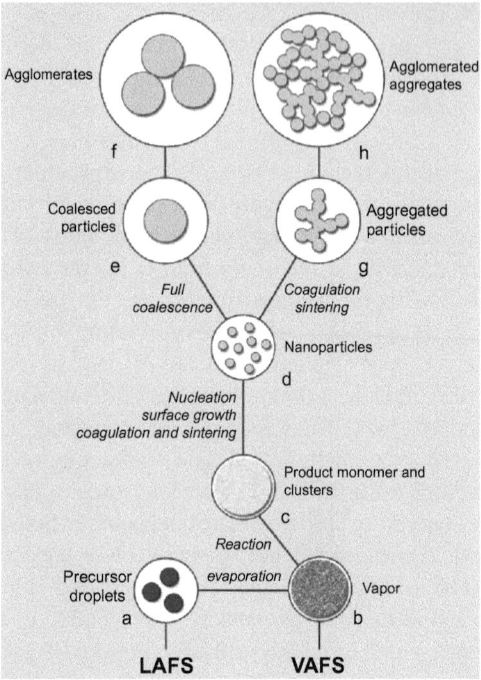

FIGURE 3.6 Liquid-fed (LAFS) and vapor-fed (VAFS) flame aerosol synthesis of agglomerated and aggregated NPs [71]. (For color version of this figure, the reader is referred to the online version of this chapter.)

Scalability of flame spray pyrolysis has been recently demonstrated with a pilot plant production rate of 1.1 kg/h of SiO_2 NPs. Furthermore, this technique allowed size distribution control in the range of 10–75 nm [76]. Size control remains a critical factor, particularly in the case of large-scale production [76–78]. The setup used for continuous SiO_2 NP production is shown in Fig. 3.7. Agglomerated TiO_2, SiO_2, and SnO_2 NPs are easily prepared from $TiCl_4$, $SiCl_4$, and $SnCl_4$, respectively, in a CH_4/O_2 flame. To prevent agglomeration and oxidation, NP growth is confined in larger particles that are later removed. This has been recently demonstrated in the case of metallic titanium NPs [77]. Metallic NPs dispersed on TiO_2 support have been achieved using FS [78].

Silica NPs have been successfully obtained using a commercial setup based on the FS process (Fig. 3.8). The product of silica materials consists mostly of agglomerated clusters. Individual NPs are loosely bonded through hydrogen forces. The aggregated product may occur through strong chemical bonds. The product is commercialized under the brand name of Aerosil, with an average diameter varying from 10 up to around 40 nm [80]. A similar approach has also been used to produce other oxides such TiO_2 and Al_2O_3.

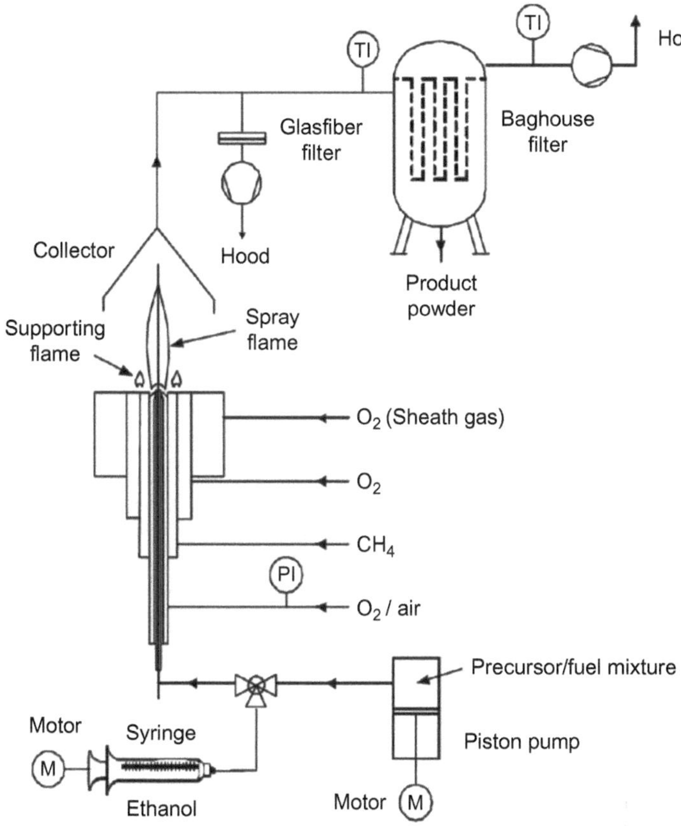

FIGURE 3.7 Schematic of flame spray pyrolysis (FSP) setup used to fabricate silica NPs [76].

Liquid flame spray has been used to produce tunable Ag and Ag-TiO_2 NPs [81]. Depending on the distance between the flame torch and the NP-collecting device, the size of the sample silver NP could vary between 2 and 10 nm. The same trend was observed for Ag and Ag-TiO_2 systems. This behavior is directly related to the ambient temperature at which the particles are collected. Two different options can be adopted to control the size distribution of the final product. In one option, a system allowing collection of NPs at different distances from the burner has been used. A second option may consist in varying the physical size of the flame.

Size-tunable silica (SiO_2) NPs from 15 to 170 nm are obtained by hexamethyldisiloxane oxidation in a flame reactor at atmospheric pressure [82] and a production rate of 130 g/h has been achieved. Proper design of the diffusion burner helps to eliminate particle deposition on the burner face [82]. NP agglomeration/aggregation increases with distance separation from the burner

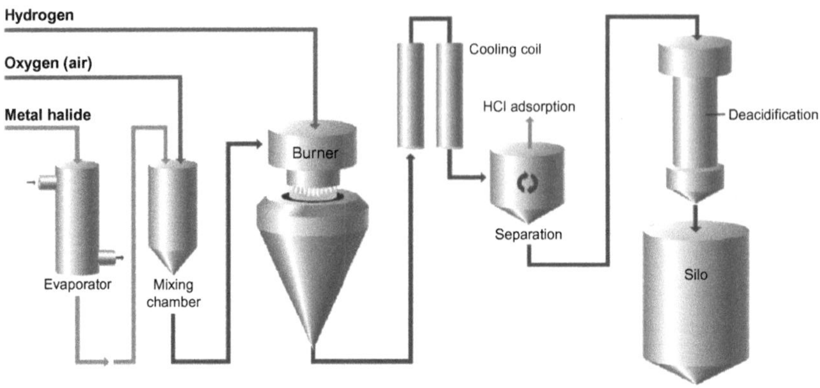

FIGURE 3.8 Schematic setup for the process used to produce silica NPs [79].

[81]. From around 2500 °C near the burner, the temperature is reduced to about 1300, 800, and 600 °C at a distance of 10, 15, and 20 cm, respectively. The average particle size often increases at higher temperatures. This has been confirmed in Ag, TiO_2, and Ag-TiO_2 NPs, showing nucleation, surface growth, coagulation, sintering, and agglomeration as they travel away from the burner [81]. Alternatively, high-voltage DC is used to generate a corona electric discharge. Velocity of ionic particles allowing size differentiation is controlled [83].

A continuous flame pyrolysis process was used to produce indium tin oxide (ITO) NPs with an average size ranging from 11 to 20 nm [76]. The flame aerosol apparatus consists of an ultrasonic atomizer and a diffusion flame burner (Fig. 3.9). Particle size is critical to the characteristics of ITO films. Precursor aerosols are first generated and then fed into the flame. This allows precise control of the molar concentration of the precursor, which directly affects the average particle size. Increasing flame temperature by controlling the H_2 flow rate gives rise to a larger average particle size. Particle size control is critical for producing highly transparent and conductive ITO films. Indeed, with a larger average NP size, the transparency and the surface resistance is decreased [85]. High optical transparency and low resistivity are two important figures of merit of the utilization of ITO films in optoelectronic devices.

Large size distribution is often obtained under high feed concentration. When high-concentration feed/production is required, a high-power CO_2 laser beam has been used to control the size and morphology of the produced silica [86] and titania [87] NPs. Rapid temperature increase when using the laser beam during FS prevents aggregation formation. In the case of SiO_2 NP manufacturing, the temperature of synthesized particles by hydrolysis and oxidation of $SiCl_4$ in a flame quickly reaches optimal sintering conditions. Using

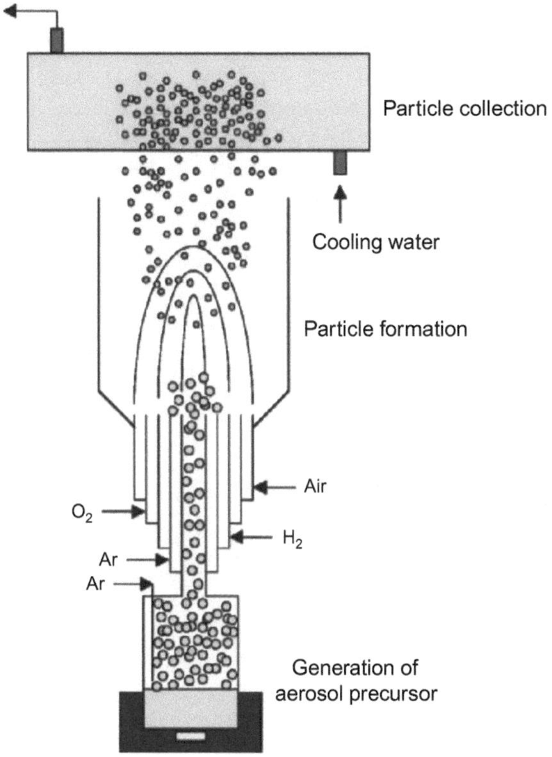

FIGURE 3.9 Flame aerosol pyrolysis setup used for ITO NPs synthesis [85]. (For color version of this figure, the reader is referred to the online version of this chapter.)

high-power laser provides size and coalescence control during FS [73]. This technique has been termed coalescence-enhanced synthesis.

Because of their potential industrial applications, binary NP systems (TiO_2/SiO_2 and SnO_2/SiO_2) have been investigated. This allows direct integration of NPs with bifunctional properties. Single SnO_2 and binary SnO_2/SiO_2 mixtures have been obtained using the atmospheric pressure flame aerosol process [88]. Discrete SnO_2 NPs have been successfully formed on larger SiO_2 particles using this technique [88]. The flame spray technique has also been successfully used to produce In- and Mg-doped ZnO NPs [89]. Flame synthesis using Zn and Al precursors gives rise to zinc aluminate ($ZnAl_2O_4$) NPs when appropriate conditions are applied [90]. Bandgaps of doped $Mg_xZn_{1-x}O$ and $In_yZn_{1-y}O$ NPs respond differently to the doping level. In the case of indium doping, the bandgap decreases with increasing doping. The opposite happens in the case of Mg doping. The ability to change the bandgap is critical in some energy, environmental, biomedical, and electronic applications.

Tungsten trioxide (WO_3) NPs have been recently reported using a combustion process of an oxidant (peroxopolytungstic acid derivative) and a fuel (glycine, urea, or thiourea) mixture [91]. The average particle size of these NPs is lower than that of the NPs commercially available. Furthermore, the average optical bandgap (2.52–2.56 eV) is also lower when compared to that of the commercial WO_3. Geometrical and optical parameters of these NPs could be of interest in photocatalysis applications.

The gas-phase combustion flame process has been developed to produce salt-encapsulated high-purity Ta NPs [92]. Encapsulation allows control of size and morphology. This prevents contamination and oxidation of the core particle. The bulk of the salt is removed by sublimation, which is followed by high-temperature vacuuming of the synthesized product.

3.4 Plasma Spray (or Thermal Spray)

Thermal spray techniques have been used to produce coatings and powder particles for decades. Thermal spray techniques are often conducted using fluid (solution, gas) or solid media (Fig. 3.10). Gas- and solid-phase thermal spraying are already being used commercially. Solution precursor plasma spray (SPPS) and suspension plasma spray (SPS) use dissolved precursors in a solution and a suspension, respectively [93,94]. SPPS and SPS techniques are currently attracting a lot of interest to address the shortcomings of conventional plasma spray techniques.

Numerous plasma processes have been used to fabricate NPs. Often, plasma frequency and temperature are critical parameters allowing tunability of material properties [95]. NP synthesis using thermal spray involves four stages [96]:

a. introduction of solid precursors in the plasma reactor using a gas carrier;
b. heating and melting, followed by dissociation into gaseous species;

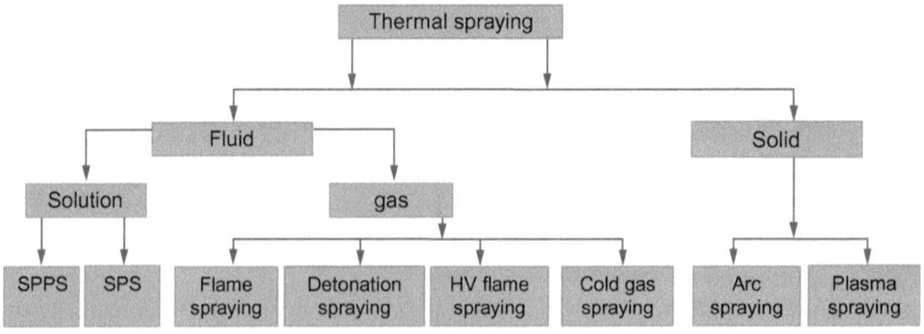

FIGURE 3.10 Simplified classification of the different thermal spray techniques. (For color version of this figure, the reader is referred to the online version of this chapter.)

c. reaction of the gaseous species to form a new seed cluster;
d. nucleation and growth to form NPs.

Each step involves several parameters allowing control of NP characteristics. Steps related to the precursor solution, spray gun, and collection system (Fig. 3.11) allow optimization of the overall production process.

Inductively coupled thermal radio frequency plasma (ICP) has been used to produce microparticles [98] and NPs [99]. ICP could be used to fabricate small particles either via vaporization and condensation of larger particles or via direct reaction of the precursor liquid solution in the plasma [100]. Sub-10 nm titania NPs have been obtained from a solution of Ti-isopropylate [100]. Silicon NPs with a size range of 2–10 nm have been synthesized using the low-pressure plasma process [101]. ICP has been used to manufacture TiO_2 NPs [100]. This direct synthesis technique of NPs from liquid precursors is facilitated by a chemical reaction in the plasma.

Fe_3Al intermetallic NPs have been produced using the hydrogen arc plasma method [102]. The average NP size is about 40 nm. Amorphous carbon and iron NPs are produced by atmospheric plasma jet glow discharge at near ambient temperature [103]. The production rate is increased when precursor concentration or applied power is increased, although often at the expense of reduction in NP monodispersity.

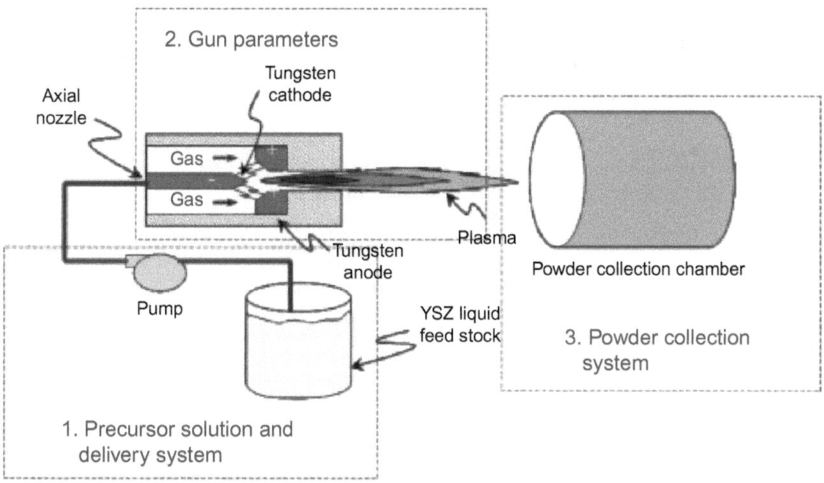

FIGURE 3.11 Schematic representation of the three main steps of NP fabrication using plasma spray processing [97]. (For color version of this figure, the reader is referred to the online version of this chapter.)

3.5 Chemical Vapor Deposition—Fluidized Bed Reactor

Chemical vapor deposition (CVD) is widely used in commercial powder and thin film manufacturing. Size distribution of powder materials obtained by CVD is relatively wide owing to Brownian motion-induced coagulation following gas reaction and nucleation [104]. Charging reactant gases induced by ionization prevents agglomeration following nucleation. This approach has been applied to CVD reactors using a sonic-jet corona discharger [104]. Using this approach, monodisperse SiO_2 NPs have been prepared from tetraethylorthosilicate precursors.

Aerosol-assisted CVD (AACVD) is based on a slight alteration of the conventional CVD [105]. AACVD is used to prepare NPs and nanocomposite films (Fig. 3.12). AACVD has been developed for powder synthesis and deposition and involves the use of a starting precursor solution followed by its

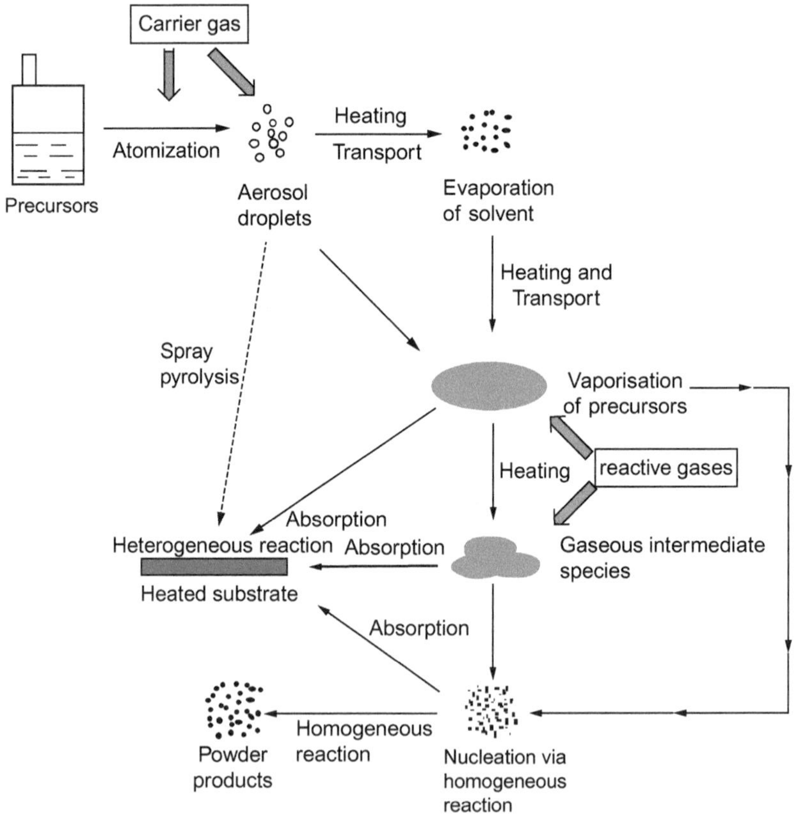

FIGURE 3.12 Schematic diagram of the AACVD process used for powder fabrication and deposition using liquid precursors [105].

atomization [105]. Using a carrier gas, aerosols are formed from the precursor solution. Afterward, NPs are obtained either by homogenous reaction as free powder following vaporization or via heterogeneous reaction on a heated substrate (Fig. 3.12).

Powder processing using CVD allows large-scale and continuous processing. Several gas–solid reactor designs have been developed including fixed bed, rotary cylinder, flat hearth, and fluidized bed. Larger scale CVD fabrication of powder material in FBRs is a widely used process. Gas–solid fluidization in FBR-CVD provides numerous advantages: (i) high rates of physical mixing with uniform temperature; (ii) high bed turnover frequency; and (iii) high heat transfer coefficients and high mass-transfer rates. Continuous CVD processes involving catalyst NP formation within heated porous supports yield improved functional properties [106]. For example, molybdenum carbide on alumina support was shown to provide two orders of magnitude higher activities than the same catalyst obtained by impregnation [107]. This catalyst has been obtained by CVD from a vapor mixture of $MoCl_5$, C_6H_6, and H_2.

FBR technique could also involve wet chemistry processing in some cases, although more often FBR uses gas-phase reactions [106–108]. FBR-CVD is currently used to produce micro-sized particles (20–500 μm) for use in various industrial processes. It is relatively difficult to use submicron particles in an FBR reactor. Often, NPs are synthesized in an FBR reactor by using porous particles as support media. TiN NPs are formed by CVD reaction of $TiCl_4$ and NH_3 at 973 K on α-Al_2O_3 [109]. The average particle size has been estimated at about 20 nm. In comparison with conventional mechanical synthesis processes, the FBR-CVD nanocomposite provides higher electroconductivity [109].

Rh NPs dispersed in 100–200 μm silica particles for use in FBRs have been obtained by CVD [110]. The particle sizes are about 1–1.7 nm depending on the preparation conditions. Dispersion of Pt NPs in a carbon matrix using a similar approach has also been reported [111]. A typical setup used for FBR-CVD synthesis of supported NPs is shown in Fig. 3.13. This setup has been used to synthesize Rh NPs on 100 and 200 μm silica support particles [110]. These Rh NPs, although relatively aggregated, are relatively small (1–2 nm). A similar setup has also been used to synthesize Pt NPs (3–6 nm) on submicron carbon particles [111].

Gas-phase impregnation and decomposition (GPI-D) has also been used to produce low loading dispersed catalysts. GPI-D shares some similarities with CVD. Both techniques involve gas-phase precursors followed by high-temperature decomposition. Like CVD, GPI-D has been used to fabricate nanostructured catalysts on oxide supports [112]. GPI-D differs from CVD techniques, as it involves two distinct process steps: first, gas-phase precursor absorption, followed by thermal treatment. In the case of CVD, the two steps occur simultaneously. GPI-D could also involve repeating these two steps more than once. However, GPI-D requires highly volatile precursors.

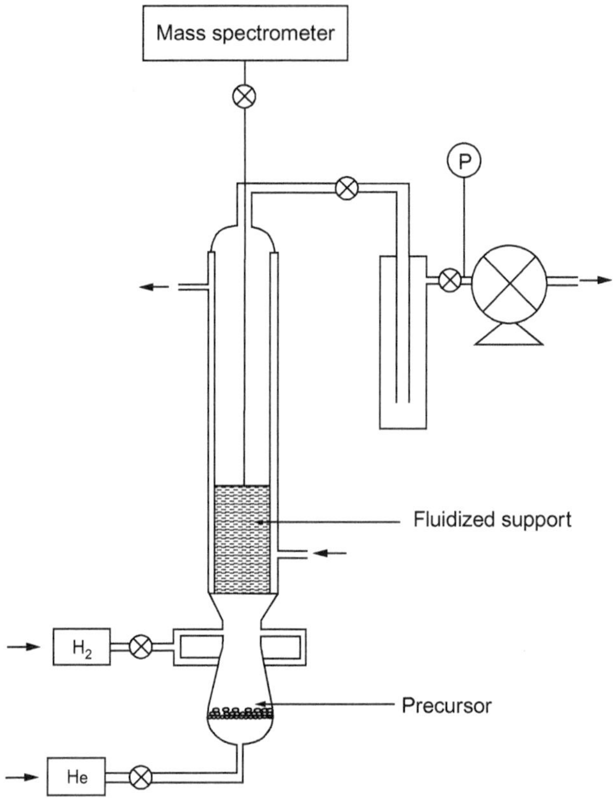

FIGURE 3.13 Schematic setup of a fluidized bed reactor (FBR) technique used to synthesize supported NPs [110].

FBRs are also used in postprocessing particulates. For example, FBR has been recently used to functionalize various submicron particles by ALD (atomic layer deposition) using reactive gases [113]. The hydrodynamic behavior of NP-based catalysts within FBR is critical to take advantage of their larger surface area [114].

CVD on porous substrates has been used to prepare supported NPs [112]. The production rate is limited by vapor pressure and slow mass-transfer kinetics [114]. This technique has been used to prepare monodispersed gold NPs on different organic and inorganic supports [115,116]. A rapid method for growth of metal NPs on nanowires has been developed using the plasma-enhanced CVD (PE-CVD) technique [117]. Ni NPs with sizes varying from 2 to 6 nm have been obtained depending on the nanowire substrate temperature. Hot-wire CVD (HW-CVD) has been successfully used to fabricate crystalline WO_3 NPs [118].

3.6 Chemical Vapor Condensation

Although it is somewhat analogous to CVD [119], chemical vapor condensation (CVC), sometimes also referred to as chemical vapor synthesis, is relatively different. As schematically outlined in Fig. 3.14, CVD is mostly used for thin film deposition on solid substrates, seed particles or simply material supports. CVC is used to prepare nanosized materials without any support. CVC provides more flexibility in the product composition given the relatively larger database of precursor chemistries developed for CVD-based processes. There is significant confusion in the literature because several CVC processes are reported as CVD.

Chemical reactions induced by high temperature inside furnace and plasma reactors have been used for various oxide NPs [73,53,120,121]. Furnace reactors have been used to synthesize Fe(Co), TiO_2, SnO_2, and ZnO NPs from iron carbonyl (cobalt carbonyl), titanium-tetra-isopropoxide (TTIP), tetra-*n*-butyl tin (TNBT), and zinc acetate, respectively [73].

Combustion flame-CVC (CF-CVC) has been used for producing ceramic NPs and microparticles [122]. This technique is relatively easy to scale. Low values of the O_2/He ratio produce NP ratios with minimum agglomeration. CF-CVC involves pyrolysis of the precursors. Different process parameters of CF-CVC are used to control particle size distribution and the degree of crystallinity [123]. Various oxides with an average particle size of 100 nm and an average size of 25 nm for the primary particles have been obtained [124].

NEI Corporation has developed a large-scale flame combustion process using liquid solution precursors. Following pyrolysis of metalorganic precursors, oxide NPs are obtained using CF-CVC [123]. Combination of flame geometry and rapid condensation results in a narrow distribution of NPs.

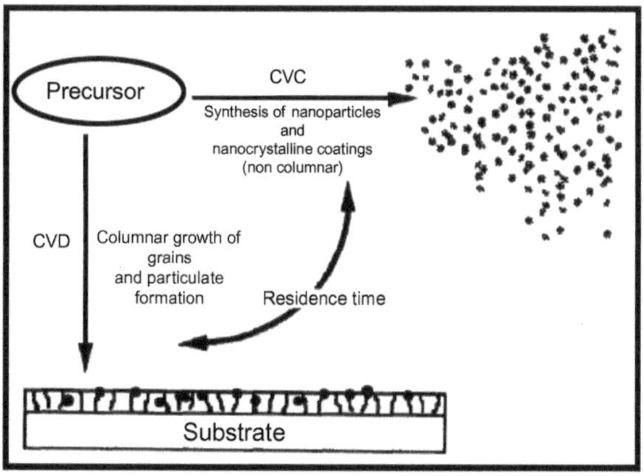

FIGURE 3.14 Simplified representation of CVD for thin film deposition and CVC for nanosized particle synthesis [53].

3.7 Evaporation Techniques

Numerous evaporation techniques have been used, including thermal, plasma, sputtering, electric discharge, and laser processing. Evaporation techniques provide many incentives for large-scale synthesis of NPs [124]. They provide high-purity, crystalline NPs with relatively cleaner surfaces. These physical techniques include thermal evaporation, sputtering, arc discharge, laser ablation, and exploding wire. In the case of thermal evaporation, the most widely used approach, numerous heat sources are used.

As summarized in Fig. 3.15, the physical vapor synthesis (PVS) consists of three main steps. First, atomic vapor is formed following solid precursor bombardment or thermal heating. Then clusters are formed when reactive additives are provided. Finally, NPs are formed following a cooling or condensation process. Although the excitation source may vary, the downstream processing steps are often similar.

Various sources of thermal jet heating processes have been used to synthesize NPs. RF induction heating, plasma heating, and laser processes have been used in the case of solid precursors [54,124–129]. A crucible containing a solid material precursor is placed in a heated flow of inert carrier gas. NP formation by heating to evaporation of metal in a crucible has been known for several decades [1]. NP synthesis using evaporation followed by cryocondensation has been used for numerous transition metals. The production rate, which is mostly controlled by vapor pressure, varies significantly depending on the metal precursor [125]. It has been estimated that in the case of copper,

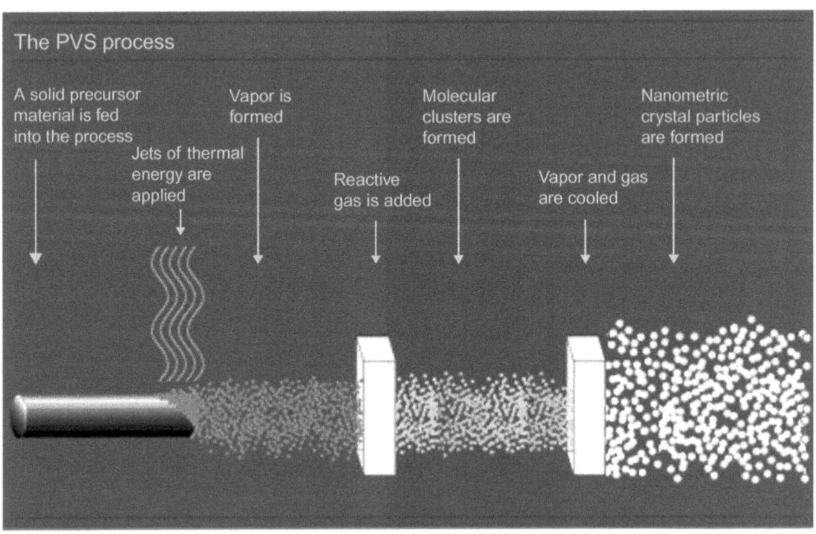

FIGURE 3.15 Schematic of the physical vapor synthesis (PVS) process for NP fabrication [124]. (For color version of this figure, the reader is referred to the online version of this chapter.)

the precursor should be heated to around 2000–2200 K under high inert-gas pressure to produce 1 g/min of Cu NPs [125]. Similar production rates have been obtained for Fe, Ni, and Co. However, in the case of Al, production is about an order of magnitude lower.

Al–Mn quasicrystal (QC) NPs have also been prepared by thermally stimulated evaporation heating of Al and Mg metals in argon gas [126]. It is now well established that QCs give rise to hard films with lower thermal and electronic diffusivity [128]. Helium gas pressure and the temperature of the heating boat are critical to the final size distribution of the synthesized particles [128].

NP collection and stabilization represent important steps in the case of evaporation techniques. To prevent surface degradation and agglomeration, a fluid-based collection process has been used. Fe-, Co-, or Ni-based ferrofluids have been prepared using a metal evaporator in a drum-shaped vacuum chamber [129]. As the chamber rotates, the hydrocarbon oil containing surfactants is drawn into contact with freshly formed NPs.

Tube-furnaces involving a sublimation and condensation approach have been used to synthesize crystalline, quasi-spherical, and monodisperse PbS NPs with tunable sizes between 2 and 50 nm [130]. Local heating is often used to reduce space requirements, energy consumption, and also the time required to stabilize temperature [131]. A possible process pathway to achieve these requirements has been proposed [132]. A small ceramic heater that is capable of heating up to 1500 °C in about 10 s an area of $5 \times 10\ mm^2$ has been used. The temperature is precisely controlled with an applied voltage between 60 and 120 V. In the case of silver, the average NP size varies between 6 nm at 1000 °C and 22 nm at 1350 °C.

Laser and, to some extent, electron beams are also used as the energy carriers to replace the ion beams of sputtering. Electron beams have been used to synthesize sub-10 nm TiN and AlN NPs by evaporation of Ti and Al in an NH_3 or N_2 atmosphere [133]. Laser ablation of soot particles with an average diameter of 250 nm gives rise to carbon NPs of 20–50 nm [134]. This technique is particularly of interest given applications of high surface area carbon and the availability of soot material as a “waste” for recycling. Depending on the distance between the collector and the nozzle, the particle size obtained using laser-assisted production (Fig. 3.16) varies between 40 and 150 nm.

Laser ablation and sputtering are the most widely used PVS techniques [136–140]. Sputtering and laser ablation offer numerous advantages when compared to thermal processes [54,139]:

a. thermal heating processes are prone to oxidation and contamination, mostly from the crucible;
b. sputtering and laser ablation can be used for a wide range of compositions including metals, semiconductors and insulators;

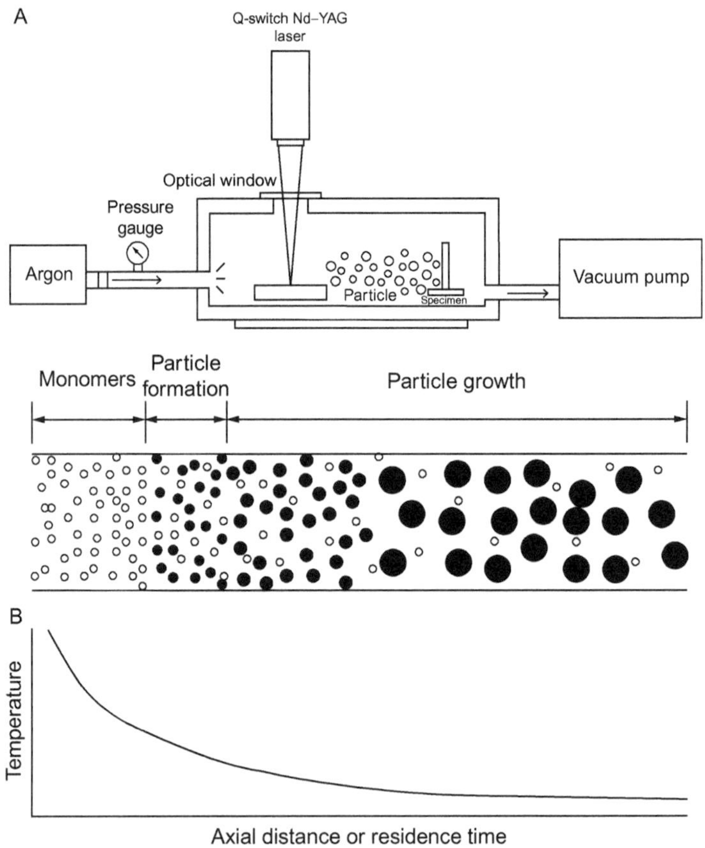

FIGURE 3.16 Simplified setup for a laser-assisted gas-phase NP synthesis (A) and the interparticle collision process (B) [135].

c. sputtering and laser ablation provide more synthesis precision for the NP characteristics;

d. sputtering and laser ablation are characterized by lower energy consumption.

However, production rates using laser ablation and sputtering are usually low. Therefore, applications of the two techniques are quite limited.

Various metals and oxides are obtained by laser ablation [141]. Laser ablation provides high production rates, good particle size control, and good monodispersity. Laser ablation allows control over material properties using process parameters including laser tuning and multi-target precursors. Furthermore, for some applications, pulsed laser ablation can also be used to vaporize materials that cannot be readily evaporated [136]. However, laser ablation may give rise to significant aggregation/agglomeration.

Chromium sputtering and deposition on amorphous silica allows precise control of particle size [137]. Well-dispersed Au, Ag, Cu, and GaAs NP layers, ranging from 1 to 15 nm in diameter, were obtained by sputtering on amorphous silica substrates [138]. Furthermore, it is also possible to control the interparticle distance (surface coverage) by varying the deposition time. Direct-current (DC) magnetron sputtering has been successfully used to produce Al, Mo, CuMn, AlTi, and ZnO_2 NPs with diameters in the 7–50 nm range [140]. DC magnetron sputtering consists of a gas inlet and pumping system, a magnetron sputtering source, and a cold finger and scraper for NP collection [140]. Average NP sizes are tuned by varying the sputtering power, the substrate temperature, the pressure, and the sputtering gas [139]. In the case of silver NPs, the particle size is increased by lowering the argon gas pressure and by increasing substrate temperature [139].

Metallic NPs suitable for pyrotechnic, explosive and propellant applications are currently produced by the electrical exploding wire (EEW) method [142–147]. A schematic outline of the electric circuit used for EEW is provided in Fig. 3.17. A wire is fed automatically to a gas-tight explosion chamber. Following repetitive short pulses (1–10 μs), the produced NPs are collected by a working gas (Fig. 3.17). At the power consumption of about 5 kW, production rates of 50–100 and 100–220 g/h have been reported in the case of metal nanopowders and oxide nanopowders, respectively [143]. An average particle size of 20–100 nm has been obtained with aluminum and copper. Argonide Inc. is marketing numerous metal and oxide NPs with an average diameter of 50 and 100 nm obtained using EEW [145].

The EEW technique yields relatively pure NPs. Since the reactor is operating under pressurized inert gas, the capital and operation costs are lower when compared to high vacuum techniques such as evaporation. The size distribution of metal NPs obtained using EEW is often in the high size range,

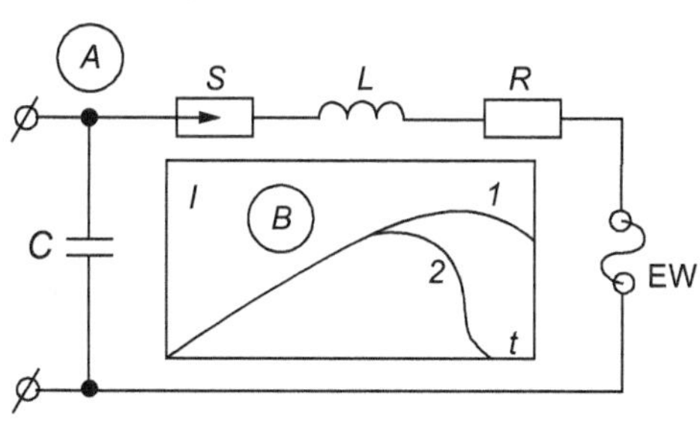

FIGURE 3.17 (A) Schematic outline of the electric circuit of the setup used for the exploding-wire technique. The inset (B) provides a diagram of the current oscillograms for shorting (1) and during the electrical explosion (2) stages [142].

typically between 50 and 100 nm [146]. Monodisperse NP distribution is difficult to achieve with EEW [147].

Nonmetallic NPs obtained using EEW have been recently reported [148]. Indeed, ZnO NPs with a median average ranging from 9 to 22 nm have been obtained using pure zinc plates. Alumina NPs with an average size of 22 nm have also been reported with the use of this technique [145].

Oxide (Al_2O_3, Fe_2O_3, TiO_2, ZrO_2) NPs are readily obtained using EEW, with a rate output of up to 200 g/h and power consumption of less than 50 kWh/kg [145]. The energy efficiency of this technique is much better than that of spark discharge, which requires 200 kWh/kg of NPs with a relatively similar setup [149]. Nickel ferrite ($NiFe_2O_4$) NPs with tunable sizes from 15 to 45 nm have been obtained using the pulsed wire discharge method in an oxygen-filled chamber [150]. Starting with Ni/Fe wires, pulse widths of 10–20 μs and voltage of 4–6 kV were used.

A few shortcomings related to oxidation and alloying have been addressed here. The issue of oxidation of metallic NPs prepared by the electro-exploding wire technique has been addressed by passivation using nickel, fluoropolymer, boron, and stearic and oleic acid followed by suspension in kerosene and ethanol [151]. Electrodeposition has been used to fabricate alloys and intermetallic wires [152]. Explosion of electrodeposited Al–Cu wires gives rise to sub-100 nm Al–Cu NPs. Regular arrays of gold NPs have been obtained directly on a solid substrate using vacuum evaporation [128]. It is worth noting that the interparticle distance is less than 40 nm.

3.8 Atmospheric Pressure Electrical Discharges

Atmospheric pressure electrical discharge (APED) leads to condensable vapors from plasma–surface interaction or from plasma-induced gas-phase reactions [153]. The APED technique is quite similar to the electrical wire discharge method described in the previous paragraph. Nonthermal APED methods for NP synthesis are under development. APED NP production consists of three main steps. Following the production using photon, thermal, or ionic excitation, nucleation leads to sub-10 nm NP formation. Agglomeration leads to particle growth. These agglomerated particles are later collected from the reactor. Particle collection may include integration in bulk composite materials or film coating.

4 OTHERS

Numerous other techniques used for NP production have not been discussed in the previous paragraphs. Some of these techniques are difficult to classify within the different categories discussed earlier. Only techniques that are potentially scalable for commercial applications have been considered.

4.1 Cryomelting

Cryomelting, sometimes also called cryocondensation, is a continuous process used to generate Al, Cu, Fe, Ni, Co, Sn, Fe–Ni, and Fe–Co from their corresponding solid bars [125,153]. It shares some similarities with the conventional condensation method. However, cryomelting may provide a better control over size distribution [84,125].

There are very few reports of cryomelting-based NP fabrication. This technique is highly scalable and potentially cost effective. High production rates and yields of aluminum NPs, with 60% of them having a size of less than 70 nm, have been obtained [84]. Cryomelting is basically a two-step process (Fig. 3.18). Metallic gas is first obtained by rapid overheating of a metal using the radio frequency heating process. In the second step, NPs are formed following rapid condensation using cryogenic fluid. In this second step, nucleation is achieved following the condensation when the atom gas is in contact with argon gas.

Laser vaporization is also used as a source of atomic gas generation [154]. Diffusion and convection help to remove formed NPs away from the nucleation zone. This step prevents the formation of larger particles. This technique can control the size of the condensing particles precisely through a combination of temperature gradient, reactor pressure, and laser power generation [154]. FeAl NPs with diameters between 6 and 9 nm have been obtained using this technique [154].

4.2 Laser Pyrolysis

Rapid and localized heating using infrared laser has been used to synthesize various NPs [155–158]. Laser energy is transferred either directly to the

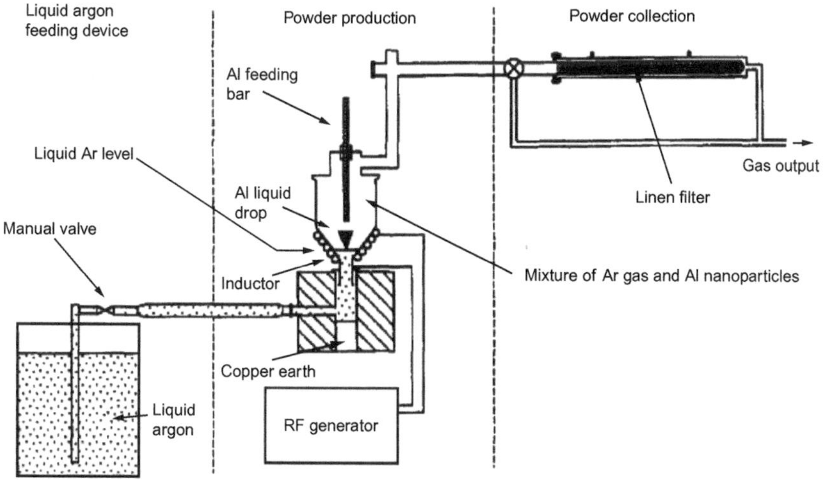

FIGURE 3.18 Experimental cryomelting set up used for high-rate Al NP synthesis [84].

precursors or indirectly through an inert photosensitizer. Laser pyrolysis can be used in the production of NPs on the basis of a continuous process. In this case, the laser pyrolysis setup consists of a continuous laser pyrolysis system using (1) laser beam, (2) ZnSe window, (3) water-refrigerated aluminum target, (4) nozzle, (5) pressure gauge, (6) ultrasonic bath, (7) iron pentacarbonyl solution in isopropanol, (8) nonreturn valve, (9) ball valve, (10) three-way ball valve, (11) argon rotameter, (12) mass controller of air flux, (13) stainless steel filter to collect the produced powders, (14) heating resistance, (15) pressure controller valve, (16) rotary vacuum pump, and (17) filter to capture oil mist [157]. Using silane and acetylene as precursors, this technique allows a production rate of more than 1 kg per hour of SIC NPs [157]. The laser intensity was found to affect the size and structure of synthesized SiC NPs.

4.3 Reactive Plasma Arc Evaporation

Reactive plasma arc evaporation has been used to synthesize metal NPs at rates of up to 1.2 g/min under an atmospheric pressure of H_2 or N_2 [158]. Hydrogen molecules first dissociate into two separate atoms in the arc plasma. These hydrogen atoms dissolve in the bulk of the molten metal. Carbon NPs have also been produced using atmospheric pressure cold plasma at a frequency of 76 kHz [159]. The same technique has also been used to synthesize iron NPs coated with a carbon layer.

4.4 Microwave Plasma Process

Ceramic NPs have been synthesized inside a tube within a resonant microwave cavity [160]. Following microwave plasma ignition, reactants are dissociated and ionized at a relatively low temperature (400–750 K) to form seed particles. Electrostatic repulsion attributed to the unipolar charged particles prevents further particle growth.

Small Pd NPs have been prepared by microwave plasma following deposition of the palladium source on a modified polymeric surface [161]. High-pressure microwave plasma minimizes solvent utilization while allowing higher operating temperatures. Silicon NPs in the range of 5–8 nm have been obtained by microwave heating of silane (SiH_4) in a quartz reactor containing an Ar/H_2 gas mixture [162].

4.5 Epitaxial Growth

Epitaxial growth is broadly defined as the condensation of gas precursors to form a film on a substrate. Liquid precursors are also used, although the vapor phase from molecular beams is more in use. Vapor precursors are obtained by CVD and laser ablation. Controlling the chemical environment and keeping the precursor concentration low at and near the surface are critical to obtain high-quality NPs. Epitaxial growth of NPs has been studied for decades, mostly in the context of heterogeneous catalysis [163]. More recently, growth of QD using molecular beam epitaxy (MBE) and metalorganic vapor phase epitaxy (MOVPE) has attracted interest with potential applications in optoelectronic devices [164,165].

One clear advantage of epitaxial growth is the fact that monodispersed NP synthesis and integration are performed simultaneously. In the case of MBE, atom deposition leads to integrated nanostructure films based on QD formation. Furthermore, the resulting films do not require any posttreatment to improve their properties. MBE has been used to directly form an ultrathin layer of InAs (higher lattice constant) on a GaAs (lower lattice constant) substrate. Because of the strain imposed by the lattice mismatch, QD dots are formed to relieve the stress (Fig. 3.19).

Monodisperse InAS quantum dots are formed on GaAs(100) substrates when the critical coverage is reached [165]. These QDs (or islands) start to form with the coverage above 1.5 monolayer (ML) [164,166]. InAs QDs with a bimodal distribution have been grown by MBE on GaAs (001) substrates [167]. Combining lithographic nucleation sites with self-assembled dot growth allowed assembling of sophisticated structures [168]. Varying the incidence angle of indium beams allowed control of the 3D QD configuration.

4.6 Rapid Quenching Solidification

Crystallization of amorphous mixture melts allows the growth of NPs within a solid matrix. The mixture melt is often amorphous and can be obtained by dry as well as wet techniques. Controlled rapid solidification allows porosity-free film and bulk material. Furthermore, rapid solidification allows tuning of NP size distribution [169]. This approach was first used to synthesize the so-called FINEMET (FeCuNBSiB) materials and other alloy nanostructures [43].

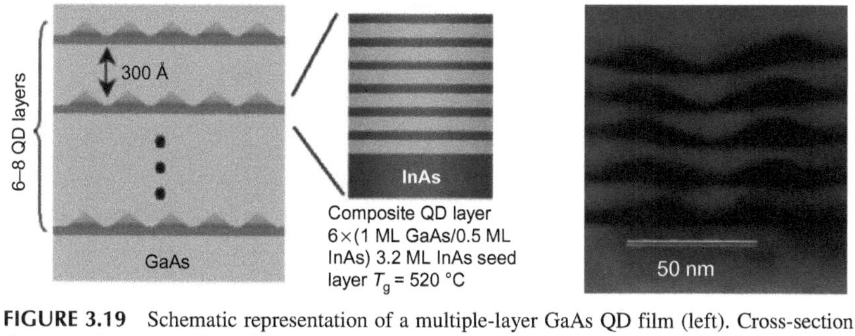

FIGURE 3.19 Schematic representation of a multiple-layer GaAs QD film (left). Cross-section electron micrograph of a multilayer GaAs film (right) [166].

4.7 Ion Beam Sputtering

Controlled nucleation and growth in ion beam sputtering allow the synthesis of NPs of tunable size directly integrated into solid substrates [170,171]. NPs directly embedded in solid matrices are obtained in a two-step process. After ion implantation, annealing is applied, which gives rise to well-defined nanocrystals dispersed in the solid matrix [171]. Similar to conventional NP synthesis processes, nucleation occurs once the implanted ions reach a critical concentration (supersaturation). Typically, supersaturation is reached with a surface concentration of around 10^{16} ions/cm^2. Beyond this concentration, the nuclei size grows to form NPs. Depending on the ionic dose, Ostwald ripening and coalescence may give rise to wide size distribution. Postprocessing, which includes annealing, often helps to control this size distribution. Several II–VI, III–V, and IV–VI compound semiconductor NPs are obtained using this approach, with potential application in optoelectronic devices.

4.8 Self-Propagating High-Temperature Synthesis

The SHS technique has been developed for powder production [172]. This combustion technique operates at high temperatures allowing synthesis of refractive materials powder. SHS is a scalable and cost-effective method. Tungsten NPs have been obtained using this method [173]. It has also been shown that addition of NaCl lowers the operation temperature and gives rise to a more uniform size distribution. The operating temperature could be used to control the final size distribution.

4.9 Hydrogen Reduction

Hydrogen reduction of supersaturated AgO_2 solution at 70 °C gives rise to tunable and crystalline silver NPs [174]. Indeed, average particle sizes of 29, 36, 48, 59, 76, 92, 113, and 136 nm are obtained by varying the reaction time. Due to low AgO_2 solubility, silver ion supply is low, allowing for the slow growth of the NPs. Furthermore, the constant concentration of H_2 allowed a more uniform size distribution.

4.10 NP Formation on Grapheme

NPs integrated into grapheme oxide substrates are easy to integrate into a variety of devices [175]. Ti, Fe, and Au metal NPs are obtained by sputtering on chemically modified grapheme. In the case of gold, tunable NP sizes starting from less than 1.5 nm and coverage up to 5×10^4 particles/μm^2 have been achieved. In the case of Ti, a uniform film has been obtained.

Different structures have been observed with different metals. This difference is ascribed to the different surface diffusion energy barriers of metal adatoms on grapheme. Furthermore, the relative metal–metal cohesion energy versus the metal–graphene binding energy is also critical [175]. Ti and Fe form relatively dense films due to their strong binding to graphene and slow adatom diffusion. The weak binding and fast diffusion of Au and Ag atoms allow the formation of NPs on graphene.

REFERENCES

[1] C.G. Granqvist, R.A. Buhrman, J. Appl. Phys. 47 (1976) 2200–2221.
[2] H. Gleiter, Prog. Mater. Sci. 33 (1989) 223–315.
[3] F. Einar Kruis, H. Fissan, A. Peled, J. Aerosol Sci. 29 (1998) 511–535.
[4] U. Backman, Studies on Nanoparticle Synthesis via Gas-to-Particle Conversion, VTT Publications 562, ESPOO 2005.
[5] G.B. Segeev, NanoChemistry, Elsevier, Amsterdam, 2006, p. 30.
[6] N. Osterwalder, C. Capello, K. Hungerbuhler, W.J. Stark, J. Nanopart. Res. 8 (2006) 1–9.
[7] W. Peukert, H.C. Schwarzer, F. Stenger, Chem. Eng. Proc. 44 (2005) 245–252.
[8] L. Mangolini, D. Jurbergs, E. Rogojina, U. Kortshagen, Phys. Stat. Sol. C 3 (2006) 3975–3978.
[9] R.P. Andres, J.D. Bielefeld, J.I. Henderson, D.B. Janes, V.R. Kolagunta, C.P. Kubiak, W.J. Mahoney, R.G. Osifchin, Science 273 (1996) 1690–1693.
[10] C.L. De Castro, Nanoparticles from mechanical attrition, in: M.I. Baraton (Ed.), Synthesis, Functionalization and Surface Treatment of Nanoparticles, American Scientific Publishers, Stevenson Ranch (CA), 2002, pp. 1–15.
[11] H. Mollet, A. Grubenmann, Formulation Technology, Wiley-VCH, Weinheim, 2001, p. 137.
[12] J. Cho, M.A. Waters, R. Hogg, Int. J. Miner. Proc. 44–45 (1996) 607–615.
[13] C.C. Koch, Nanostruct. Mater. 2 (1993) 109–129.
[14] M.N. Rahaman, Ceramic Processing and Sintering, second ed, Marcel Dekker, New York, 2003, pp. 125–180.
[15] R. Janot, D. Guerard, Prog. Mater. Sci. 50 (2005) 1–92.
[16] S. Cai, T. Tsuzuki, T.A. Fisher, B.D. Nener, J.M. Dell, P.G. McCormick, J. Nanopart. Res. 4 (2002) 367–371.
[17] H.W. Way, Grinding and Dispersing Nanoparticles, Netzsch Fine Particle Technology, Exton, PA, 2004.
[18] Y.D. Kim, J.Y. Chung, H. Jeon, Mater. Sci. Eng. A 291 (2001) 17–21.
[19] I. Borner, J. Eckert, Mater. Sci. Eng. A226–228 (1997) 541–545.
[20] E. Godočíková, P. Baláž, E. Gock, W. Sik Choi, B. Sun Kim, Powder Technol. 164 (2006) 147–152.
[21] D. Guerard, R. Janot, J. Metastable Nanocryst. Mater. 20–21 (2004) 311–312.
[22] E. Merisko-Liversidge, G.G. Liversidge, E.R. Cooper, Eur. J. Pharm. Sci. 18 (2003) 113–120.
[23] D. Guerard, R. Janot, J. Ghanbaja, P. Delcroix, J. Alloys Compd. 434–435 (2007) 410–414.
[24] M. Umemoto, Z.G. Liu, K. Masuyama, X.J. Hao, K. Tsuchiya, Scr. Mater. 44 (2001) 1741–1745.
[25] B. Sampedro, T.C. Rojas, A. Fernandez, A. Hernando, Phys. Stat. Sol. A 203 (2006) 1201–1205.

[26] S. Indris, D. Bork, P. Heitjans, J. Mater. Synth. Process. 8 (2000) 245–250.
[27] L.B. Kong, J. Maa, H.T. Huang, W. Zhu, O.K. Tan, Mater. Lett. 50 (2001) 129–133.
[28] A.P. Smith, H. Ade, C.M. Balik, C.C. Koch, St.D. Smith, R.J. Spontak, Macromolecules 33 (2000) 2595–2604.
[29] A.C. Dodd, A.J. McKinley, M. Saunders, T. Tsuzuki, J. Nanopart. Res. 8 (2006) 43–51.
[30] T. Klassen, M. Oehring, R.T. Bormann, J. Mater. Res. 9 (1994) 47–52.
[31] H.-J. Fecht, C.I. Contescu, K. Putyera (Eds.), Dekker Encyclopedia of Nanoscience and Nanotechnology, Taylor & Francis Group, LLC, New York, 2009, pp. 3030–3039.
[32] S. Hwang, C. Nishimura, P.G. McCormick, Scr. Mater. 44 (2001) 2457–2462.
[33] B. Huang, K.N. Ishihara, P.H. Shingu, Mater. Sci. Eng. A231 (1997) 72–79.
[34] C.C. Koch, C.I. Contescu, K. Putyera (Eds.), Dekker Encyclopedia of Nanoscience and Nanotechnology, Taylor & Francis Group, LLC, New York, 2009, pp. 3019–3029.
[35] T.F. Grigorieva, A.P. Barinova, N.Z. Lyakhov, J. Nanopart. Res. 5 (2003) 439–453.
[36] P.G. McOrmick, Mater. Trans. Jpn. Inst. Metals 36 (1995) 161–169.
[37] H. Yang, G. Nguyen, P.G. McCormick, Scr. Metall. Mater. 32 (1995) 681–684.
[38] H. Yang, P.G. McCormick, J. Solid State Chem. 110 (1994) 136–141.
[39] C. Lam, Y.F. Zhang, Y.H. Tang, C.S. Lee, I. Bello, S.T. Lee, J. Cryst. Growth 220 (2000) 466–470.
[40] Z.P. Guo, G.D. Du, Y. Nuli, M. Faiz Hassan, H.K. Liu, J. Mater. Chem. 19 (2009) 3253–3257.
[41] G. Garcıa-Pacheco, J.G. Cabanas-Moreno1, H. Yee-Madeiral, F. Cruz-Gandarilla, Nanotechnology 17 (2006) 2528–2535.
[42] P. Balaz, E. Boldizarova, E. Godocıkova, J. Briancin, Mater. Lett. 57 (2003) 1585–1589.
[43] Y. Yoshizawa, S. Oguma, K. Yamauchi, J. Appl. Phys. 64 (1988) 6044–6046.
[44] J.H. Wu, L.B. Kong, Y.K. Kim, Solid State Phenomena 124–126 (2007) 851–854.
[45] T.V. Anuradha, Bull. Mater. Sci. 30 (2007) 263–269.
[46] S.K. Mishra, L.C. Pathak, Key Eng. Mater. 395 (2009) 15–38.
[47] T.G. Langdon, Rev. Adv. Mater. Sci. 13 (2006) 6–14.
[48] T.C. Lowe, R.Z. Valiev, JOM 56 (2004) 64–68.
[49] V. Viswanathan, T. Laha, K. Balani, A. Agarwal, S. Seal, Mater. Sci. Eng. R 54 (2006) 121–285.
[50] D.B. Witkin, E.J. Lavernia, Prog. Mater. Sci. 51 (2006) 1–60.
[51] R. Petkovic-Luton, J. Vallone, USPTO 4619699, 1986.
[52] G.D. Uhich, Chem. Eng. News 6 (1984) 22–29.
[53] H. Hahn, Nanostruct. Mater. 9 (1997) 3–12.
[54] F.E. Kruis, H. Fissan, A. Peled, J. Aerosol Sci. 29 (1998) 511–535.
[55] M.S. Wooldridge, Prog. Energy Combust. Sci. 24 (1998) 63–87.
[56] M.T. Stewart, Curr. Opin. Colloid Interface Sci. 8 (2003) 127–133.
[57] R.W. Sigel, MRS Bull. 15 (1990) 60–67.
[58] A. Gurav, T. Kodas, T. Pluym, Y. Xiong, Aerosol Sci. Technol. 19 (1993) 411–452.
[59] K. Okuyama, I.W. Lenggoro, Chem. Eng. Sci. 58 (2003) 537–547.
[60] W. Lenggoro, K. Okuyama, J.F. de la Mora, N. Tohge, J. Aerosol Sci. 31 (2000) 121–129.
[61] M. Eslamian, N. Ashgriz, ASME J. Eng. Mater. Technol. 129 (2007) 130–142.
[62] L. Amirav, A. Amirav, E. Lifshitz, J. Phys. Chem. 109 (2005) 9857–9860.
[63] H.H. Funke, H. Diaz, X. Liang, C.S. Carney, A.W. Wimer, P. Li, Int. J. Hydrogen Energy 33 (2008) 1127–1134.
[64] P.P. Ahonen, A. Moisala, U. Tapper, D.P. Brown, J.K. Jokiniemi, E.I. Kauppinen, J. Nanopart. Res. 4 (2002) 43–52.

[65] K.C. Patil, S.T. Aruna, S. Ekambaram, Curr. Opin. Solid State Mater. Sci. 2 (1997) 158–165.
[66] K.C. Patil, S.T. Aruna, T. Mimani, Curr. Opin. Solid State Mater. Sci. 6 (2002) 507–512.
[67] A.S. Mukasyan, P. Epstein, P. Dink, Proc. Combust. Inst. 31 (2007) 1789–1795.
[68] K.P. Jayadevan, T.Y. Tseng, N.S. Nalwa (Ed.), Encyclopedia of Nanoscience and Nanotechnology, vol. 8, American Scientific Publishers, Valencia, CA, 2004, pp. 333–376.
[69] J. Jiu, Y. Ge, X. Li, L. Nie, Mater. Lett. 54 (2002) 260–263.
[70] G.D. Ulrich, Chem. Eng. News (1984) 23–32.
[71] A. Camenzinda, W.R. Caserib, S.E. Pratsinis, Nano Today 5 (2010) 48–65.
[72] S.E. Pratsinis, Prog. Energy Combust. Sci 24 (1998) 197–219.
[73] M. Choi, J. Nanopart. Res. 3 (2001) 201–211.
[74] K. Wegner, S.E. Pratsinis, Powder Technol. 150 (2005) 117–122.
[75] W.J. Stark, S.E. Pratsinis, Powder Technol. 126 (2002) 103–108.
[76] D.E. Rosner, Ind. Eng. Chem. Res. 44 (2005) 6045–6055.
[77] D.P. Dufoux, R.L. Axelbaum, Combust. Flame 100 (1995) 350–358.
[78] T. Johannessen, J.R. Jensen, M. Mosleh, J. Johansen, U. Quaade, H. Livbjerg, Chem. Eng. Res. Des. 82 (2004) 1444–1452.
[79] C.S. Isfort, M. Rochnia, Toxicol. Lett. 186 (2009) 148–151.
[80] Evonik Degussa GmbH, www.evonik.com
[81] H. Keskinen, J.M. Makela, M. Aromaa, J. Ristimaki, T. Kanerva, E. Levanen, T. Mantyla, J. Keskinen, J. Nanopart. Res. 9 (2007) 569–588.
[82] H.K. Kammler, S.E. Pratsinis, J. Nanopart. Res. 1 (1999) 467–477.
[83] S. Vemury, S.E. Pratsinis, Appl. Phys. Lett. 66 (1995) 3275–3277.
[84] Y. Champion, J. Bigot, Nanostruct. Mater. 10 (1999) 1097–1110.
[85] H.D. Jang, C.M. Seong, H.K. Chang, H.C. Kim, Curr. Appl. Phys. 6 (2006) 1044–1047.
[86] D. Lee, M. Choi, J. Aerosol Sci. 31 (2000) 1145–1163.
[87] D. Lee, M. Choi, J. Aerosol Sci. 31 (2000) S632–S633.
[88] K.K. Akurati, R. Dittmann, A. Vital, U. Klotz, P. Hug, T. Graule, M. Winterer, J. Nanopart. Res. 8 (2006) 379–393.
[89] H. Li, Y. Zhang, X. Pan, H. Zhang, T. Wang, E. Xie, J. Nanopart. Res. 11 (2009) 917–921.
[90] J.R. Jensen, T. Johannessen, S. Wedel, H. Livbjerg, J. Nanopart. Res. 2 (2000) 363–373.
[91] W. Morales, M. Cason, O. Aina, N.R. de Tacconi, K. Rajeshwar, J. Am. Chem. Soc. 130 (2008) 6318–6319.
[92] J.L. Barr, R.L. Axelbaum, M.E. Macias, J. Nanopart. Res. 8 (2006) 11–22.
[93] P. Fauchais, JTTEE5 17 (2008) 1–3.
[94] Adapted from: Thermal spraying for power generation components/Klaus Erich Schneider, Wiley-VCH and John Wiley, Weinheim and Chichester, 2006, p. 6.
[95] D. Vollath, J. Nanopart. Res. 10 (2008) 39–57.
[96] L. Tong, R.G. Reddy, JOM (2006) 62–66.
[97] R. Hui, Zh. Wang, O. Kesler, L. Rose, J. Jankovic, S. Yick, R. Maric, D. Ghosh, J. Power Sour. 170 (2007) 308–323.
[98] P.C. Kong, Y.C. Lau, Pure Appl. Chem. 62 (1990) 1809–1816.
[99] M. Shigeta, H. Nishiyama, J. Heat Transfer 127 (2005) 1222–1230.
[100] B. Dzur, Metal Powder Rep. 63 (2008) 12–15.
[101] A. Bapat, C. Anderson, C.R. Perrey, C.B. Carter, S.A. Campbell, U. Kortshagen, Plasma Phys. Control. Fusion 46 (2004) B97–B109.
[102] H. Chuncheng, C. Zuolin, Y. Yansheng, Z. Zhikun, J. Nanopart. Res. 4 (2002) 107–110.
[103] M.D. Barankin, Y. Creyghton, A. Schmidt-Ott, J. Nanopart. Res. 8 (2006) 511–517.
[104] M. Adachi, S. Tsukui, K. Okuyama, J. Nanopart. Res. 5 (2003) 31–37.

[105] X. Hou, K.-L. Choy, Chem. Vap. Deposit. 12 (2006) 583–596.
[106] C. Vahlas, B. Caussatb, P. Serp, G.N. Angelopoulos, Mater. Sci. Eng. R 53 (2006) 1–72.
[107] T. Miyao, I. Shishikura, M. Matsuoka, M. Nagai, Chem. Lett. (1996) 561–562.
[108] D. Geldart, Powder Technol. 7 (1973) 285–292.
[109] K. Tsugeki, T. Kato, Y. Koyanagi, K. Kusakabe, S. Morooka, J. Mater. Sci. 28 (1993) 3168–3172.
[110] P. Serp, R. Feurer, R. Morancho, P. Kalck, J. Catal. 157 (1995) 294–300.
[111] P. Serp, R. Feurer, Y. Kihn, P. Kalck, J.L. Faria, J.L. Figueiredo, J. Mater. Chem. 11 (2001) 1980–1981.
[112] P. Serp, P. Kalck, R. Feurer, Chem. Rev. 102 (2002) 3085–3128.
[113] D.M. King, J.A. Spencer, X. Liang, L.F. Hakim, A.W. Weimer, Surf. Coat. Technol. 201 (2007) 9163–9171.
[114] D.P. Long, J.M. Blackburn, J.J. Watkins, Adv. Mater. 12 (2000) 913–915.
[115] M. Okumura, S. Nakamura, S. Tsubota, T. Nakamura, M. Azuma, M. Haruta, Catal. Lett. 51 (1998) 53–58.
[116] T. Ishida, M. Nagaoka, T. Akita, M. Haruta, Chem. Eur. J. 14 (2008) 8456–8460.
[117] A.D. LaLonde, M.G. Norton, D. Zhang, D. Gangadean, A. Alkhateeb, R. Padmanabhan, D.N. McIlroy, J. Nanopart. Res. 8 (2006) 99–104.
[118] S.H. Lee, R. Deshpande, P.A. Parilla, K.M. Jones, B. To, A.H. Mahan, A.C. Dillon, Adv. Mater. 18 (2006) 763–766.
[119] K.L. Choy, Prog. Mater. Sci. 48 (2003) 57–170.
[120] B.K. Kim, C.J. Cjoi, Scr. Mater. 18 (2001) 2161–2164.
[121] B.K. Kim, G.G. Lee, H.M. Park, N.J. Kim, Nanostruct. Mater. 12 (1999) 637–640.
[122] A. Singhal, G. Skandan, N. Glumac, B.H. Kear, Scr. Mater. 44 (2001) 2203–2207.
[123] G. Skandan, A. Singhal, C.I. Contescu, K. Putyera (Eds.), Dekker Encyclopedia of Nanoscience and Nanotechnology, Taylor & Francis Group, LLC, New York, 2009, pp. 2788–2798.
[124] Nanophase Technologies, http://www.nanophase.com, May 11th 2008.
[125] Y. Champion, Ann. Chim. Sci. Materiaux 31 (2006) 281–294.
[126] Y. Saito, K. Mihama, H.S. Chen, Phys. Rev. B 35 (1987) 4085–4091.
[127] Z. Minevski, C.L. Tennakoon, K.c. Anderson, C.J. Nelson, F.C. Burns, D.J. Sordelet, C.W. Haering, D.W. Pickard, Mater. Res. Soc. Symp. Proc. 805 (2004) 345–350.
[128] R. Uyeda, Prog. Mater. Sci. 35 (1991) 1–96.
[129] M. Nakatani, T. Furubayashi, T. Takahashi, H. Hanaoka, J. Magn. Magn. Mater. 65 (1987) 261–264.
[130] F.E. Kruis, H. Fissan, B. Rellinghaus, Mater. Sci. Eng. B 69–70 (2000) 329–334.
[131] J.H. Jung, H.C. Oh, H.S. Noh, J.H. Ji, S.S. Kim, Aerosol Sci. 37 (2006) 1662–1670.
[132] M.B. Attoui, S. Deluche, A. Schmidt-Ott, J. Aerosol Sci. 35 (2004) 773–774.
[133] S. Iwama, K. Hayakawa, T. Arizumi, J. Cryst. Growth 56 (1982) 265–269.
[134] C.B. Stipe, J.-H. Choi, D. Lucas, C.P. Koshland, R.F. Sawyer, J. Nanopart. Res. 6 (2004) 467–477.
[135] J.-H. Jang, J. Lin, Surf. Coat. Technol. 202 (2008) 6136–6141.
[136] W. Marine, L. Patrone, B. Luk'yanchuk, M. Sentis, Appl. Surf. Sci. 154–55 (2000) 345–352.
[137] M. Hu, S. Noda, H. Komiyama, J. Appl. Phys. 93 (2003) 9336–9344.
[138] H. Shirakawa, H. Komiyama, J. Nanopart. Res. 1 (1999) 17–30.
[139] P. Ayyub, R. Chandra, P. Taneja, A.K. Sharma, R. Pinto, Appl. Phys. A 73 (2001) 67–73.

[140] H. Hahn, R.S. Averback, J. Appl. Phys. 67 (1990) 1113–1115.
[141] G.F. Gaertner, H. Lydtin, Nanostruct. Mater. 4 (1994) 559–568.
[142] Y.F. Ivanov, M.N. Osmonoliev, V.S. Sedoi, V.A. Arkhipov, S.S. Bondarchuk, A.B. Vorozhtsov, A.G. Korotkikh, V.T. Kuznetsov, Propellants Explosives Pyrotechnics 28 (2003) 319–333.
[143] Y.A. Kotov, J. Nanopart. Res. 5 (2003) 539–550.
[144] Nanoparticle technology: production and application development, Report of a DTI Global Watch Mission, September 2005.
[145] http://www.argonide.com/nanopowders.html (accessed 02.02.08).
[146] C. Rinaldi, T. Franklin, M. Zahn, T. Cader, C.I. Contescu, K. Putyera (Eds.), Dekker Encyclopedia of Nanoscience and Nanotechnology, Taylor & Francis Group, LLC, New York, 2009, pp. 1922–1938.
[147] L. Kaledin, F. Tepper, C.I. Contescu, K. Putyera (Eds.), Dekker Encyclopedia of Nanoscience and Nanotechnology, Taylor & Francis Group, LLC, New York, 2009, pp. 2165–2174.
[148] N.K. Singh, A. Alqudami, S. Annapoorni, Phys. Stat. Sol. A 207 (2010) 2153–2158.
[149] N.S. Tabrizi, M. Ullmann, V.A. Vons, U. Lafont, A. Schmidt-Ott, J. Nanopart. Res. 11 (2009) 315–332.
[150] P.Y. Lee, K. Ishizaka, H. Suematsu, W. Jiang, K. Yatsui, J. Nanopart. Res. 8 (2006) 29–35.
[151] A.A. Gromov, U. Förter-Barth, U. Teipel, Powder Technol. 164 (2006) 111–115.
[152] W. Kim, J.-S. Park, C.-Y. Suh, H. Chang, J.-C. Lee, Mater. Lett. 61 (2007) 4259–4261.
[153] J.P. Borra, J. Phys. D Appl. Phys. 39 (2006) R19–R54.
[154] Y.B. Pithawalla, M.S. El-Shall, S.C. Deevi, V. Strom, K.V. Rao, J. Phys. Chem. B 105 (2001) 2085–2090.
[155] W.R. Cannon, S.C. Danforth, J.H. Flint, J.S. Haggerty, R.A. Marra, J. Am. Ceram. Soc. 65 (1982) 330–335.
[156] J.D. Casey, J.S. Haggerty, J. Mater. Sci. 22 (1987) 4307–4312.
[157] A. Reau, B. Guizard, C. Mengeot, L. Boulanger, F. Ténégal, Mater. Sci. Forum 534–536 (2007) 85–88.
[158] T. Yokoyama, M. Hosokawa, K. Nogi, M. Naito (Eds.), Nanoparticle Technology Handbook, Elsevier, Amsterdam, 2007, pp. 428–431.
[159] V. Vons, Y. Creyghton, A. Schmidt-Ott, J. Nanopart. Res. 8 (2006) 721–728.
[160] D. Vollath, D. Vinga Szabo, J. Nanopart. Res. 8 (2006) 417–428.
[161] P. Korovchenko, A. Renken, L. Kiwi-Minsker, Catal. Today 102–103 (2005) 133–141.
[162] B. Giesen, H. Wiggers, A. Kowalik, P. Roth, J. Nanopart. Res. 7 (2005) 29–41.
[163] C. Chapon, C.R. Henry, A. Chemam, Surf. Sci. 162 (1985) 747–754.
[164] P. Petroff, Top. Appl. Phys. 90 (2003) 1–25.
[165] P. Bhattacharya, S. Ghosh, A.D. Stiff-Roberts, Annu. Rev. Mater. Res. 34 (2004) 1–40.
[166] M. Dinu, R. Rapaport, G. Chen, H.R. Stuart, R. Giles, Bell Labs Tech. J. 10 (2005) 215–234.
[167] S.-F. Tang, S.-Y. Lin, S.-C. Lee, J. Nanopart. Res. 3 (2001) 489–492.
[168] M.K. Yakes, C.D. Cress, J.G. Tischler, A.S. Bracker, ACS Nano 4 (2010) 3877–3882.
[169] C. Syryanarayana, Int. Mater. Rev. 40 (1995) 41–64.
[170] S. Xu, B.L. Evans, J. Materials, Science 27 (1992) 3108–3117.
[171] D. Litvinov, S. Khizroev, C.I. Contescu, K. Putyera (Eds.), Dekker Encyclopedia of Nanoscience and Nanotechnology, Taylor & Francis Group, LLC, New York, 2009, pp. 2825–2835.

[172] Z.A. Munir, U. Anselmi-Tamburini, Mater. Sci. Rep. 3 (1989) 277–375.
[173] H.H. Nersisyan, J.H. Leeb, C.W. Won, Combust. Flame 142 (2005) 241–248.
[174] D.D. Evanoff Jr., G. Chumanov, Chem. Phys. Chem. 6 (2005) 1221–1231.
[175] P.A. Pandey, G.R. Bell, J.P. Rourke, A.M. Sanchez, M.D. Elkin, B.J. Hickey, N.R. Wilson, Small 7 (2011) 3202–3210.

Chapter 4

Nanoparticle Assembling and System Integration

1 INTRODUCTION

Currently, most man-made objects are obtained by subtractive, modifying, and additive methods. The first two methods are used in the production of high-resolution and precise micro- and nanostructures. The last method including self-assembly is a cost-effective manufacturing technique using building blocks such as nanoparticle (NP). Incorporation of nanoparticles (NPs) into finished goods requires intermediary processing steps that include substrate preparation, NP deposition, and posttreatments. NPs can be used to make various structures by means of different laboratory-scale processes [1]. Integration of NPs into 1D, 2D, and 3D structures has been achieved. As such, NPs can fill the gap between the top-down microfabrication techniques and the bottom-up chemical synthesis techniques. In this chapter, the focus is mostly on scalable processes for NP integration, although other newly developed laboratory-scale techniques are also described.

The organization of colloidal particles (including NPs) has been studied for decades. Given its complexity, it remains a very challenging subject for both experimentalists and theoreticians [2]. Several complex structures are possible using colloidal assembly. Predicting and controlling the 2D and 3D structures of organized NPs are challenging given the variety of forces and the wide range of the strengths of these forces. The equilibrium status of the assembled NP system, with and without an external field, may give rise to different phases. At least six different geometrical complexes have been identified [2].

NPs can be formed in gas, liquid, or solid state. Often, utilization of NPs in the solid state requires their integration into thin films. These films are either used directly (coating) or they are part of a device (fuel cell and solar cells). Depending on the application, NP-based films are required to be dense, amorphous, crystalline, porous, reticular, grainy, or fractal-like [3]. Thus, consideration of the deposition technique should be based not only on the nature of the particle and matrix but also on other factors such as film and product requirements.

Interface Science and Technology, Vol. 19. http://dx.doi.org/10.1016/B978-0-12-369550-5.00004-5

Creating patterns and their duplication is critical for NP integration. Patterning or lithography, consisting of designing a specific feature and its implementation onto a substrate, is a well-established field in the microelectronic industry. Current requirements for high spatial resolution (below 90 nm) in the microelectronic industry are almost at the fundamental limit. Furthermore, the high capital cost of photolithography has pushed the industry toward alternative patterning processes. Soft lithography, nanoskiving, and self-assembly have the potential to manufacture smaller and more complex features at a lower cost [4]. However, these techniques have not led to reliable and cost-effective devices.

Several interparticle forces are involved in the process of film formation. These forces are somewhat similar to those leading to NP aggregation. Indeed, aggregation (or film formation) occurs when attractive forces are larger than repulsive forces. Electrostatic, steric, and van der Waals forces are the dominant pair-wise attractive and repulsive interactions. Figure 4.1 shows the variation of different pair-wise forces as a fu. Non-pair-wise forces include hydrodynamic forces (drag forces), Brownian motion as a function of the distance between two particle surfaces, and buoyancy and gravity forces may also contribute [6]. The last two forces (buoyancy and gravity) are negligible in the case of NPs. The total Brownian force is also negligible in most cases. This is not the case with the hydrodynamic forces. The hydrodynamic forces are expressed as:

$$f_{\mathrm{hydr}} = 6\pi\mu R v,$$

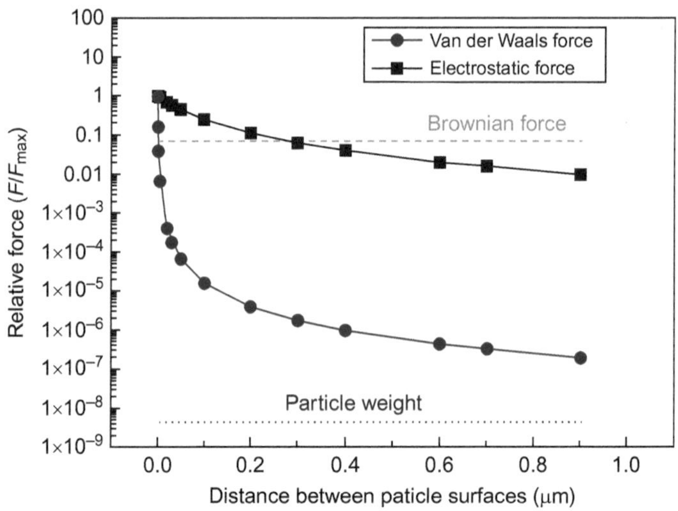

FIGURE 4.1 The main colloidal forces and their respective ranges as a function of the interparticle distance [5]. (For color version of this figure, the reader is referred to the online version of this chapter.)

where R corresponds to the particle radius, and μ and v correspond to the relative velocity between liquid and particle, respectively [6].

The distance range of these forces is also critical, not only during the synthesis step but also during shipping and utilization. The forces can be divided into three categories. The first category is represented by the attractive capillary forces with a range of about 1 mm. The second one consists of long-range forces including electric double-layer and Van der Waals forces. The range of these two forces may extend up to 100 nm or more. The third category consists of forces with a distance range of about 10 nm or less. In this last category fall the repulsive steric forces with a distance range of about 1–2 nm.

NP–substrate and/or NP–medium interactions may also play a role in film formation. The simultaneous occurrence and the magnitude range of these interactions make any attempt to control the film formation process quite challenging. Obtaining specific film properties using NP deposition requires the control of parameters defining the strength of the different forces. For example, the amplitudes of some forces need to be reduced while increasing the amplitudes of others. Core–shell structure, surface charging, NP concentration, solvent media, NP size, temperature, and surface substrate structure are a few experimental conditions that allow controlling the relative strengths of different forces.

In the case of colloidal NPs, several forces contribute to the self-assembly processes. These include electrostatic, van der Waals, Brownian, and electrosteric forces [7]. Other forces related to depletion, biological processes, and solvation may also play a role depending on the composition. The strength of these forces depends on numerous factors including particle size, interparticle distances, and the composition of the liquid medium [7]. Note that the Brownian forces may play a significant role, whereas the contribution of particle weight is relatively negligible (Fig. 4.1). Furthermore, the relative amplitude of weight and Brownian forces is independent of the interparticle distance (Fig. 4.1).

The deposition of NPs from aqueous solution is of interest. This process could be described within the context of the classical Derjaguin–Landau–Verwey–Overbeek theory [8]. Minimizing organic solvent can help reduce the financial and environmental costs of NP-based technology development. Besides the interparticle interactions, particle–substrate interactions are also important. The propensity of NPs to aggregate can affect film characteristics. In particular, aggregation may lead to settling even before deposition. Aggregation timescale, defined as the initial particle population to form aggregates (t_c), is critical as it relates to the NP film formation. Above 0.2 μm particle diameter, the aggregation timescale (t_c) is constant, and it is controlled mostly by mechanical stirring [8]. Below this critical size, the dominant Brownian motion gives rise to a reduction in the aggregation time.

Besides size distribution and concentration, energy input, temperature, and pressure during deposition are three important parameters used to control film characteristics [9]. In the case of vacuum-based techniques, a wide range of film structures can be obtained depending on the specific value of these three parameters. Although possible, tunability in film density and structure can be challenging using nonvacuum processes.

Posttreatment following deposition is an integral part of NP integration. For example, the drying process is critical for obtaining the required film characteristics [10–12]. Indeed, when high-density film is needed, the drying process is often carried out in two steps. In the first step, a slow and low-temperature heating to remove the solvent is used. This is followed by a second step consisting of fast and high-temperature annealing. Solvent evaporation leads to the formation of films of higher density. Further densification is achieved at higher temperatures [10].

Today, chemical and physical deposition techniques are used in different industries to manufacture thin films. These coatings are used to provide either the main functional properties of the overall system (e.g., photovoltaic thin film layer) or a protective and handling packaging solution (e.g., surface finish). Often, molecular precursors are used in chemical deposition, whereas atomic precursors are used in physical depositions. Sol–gel, plating, and chemical vapor deposition (CVD), atmospheric pressure chemical vapor deposition (APCVD), Low Pressure Chemical Vapor Deposition (LPCVD), Plasma-enhanced chemical vapor deposition (PECVD), and metal-organic chemical vapor deposition (MOCVD) techniques are the main chemical deposition methods. Evaporation (thermal, electron beam, molecular beam, and laser ablating) and sputtering (DC, RF, and magnetron) are the main physical deposition techniques. When NPs are used instead as precursors, these different techniques may or may not work.

Physical vapor deposition (PVD) allows high-quality films with precise control over the morphology at the atomic level [9], quite easy to reproduce when NPs are used as building blocks. However, numerous industrial applications do not require such high-level control and sophistication. Vacuum-based deposition techniques are relatively expensive. The power requirement of PVD techniques is about one to two orders of magnitude higher [13]. In particular, the use of an e-beam system to develop nanosized features is quite prohibitive for some applications, besides the fact that it is relatively slow. A drop casting requires almost no power, whereas an e-beam evaporator requires 10 kW. Between these two extremes, a spin coater and a sputter coater require 40 and 180 W, respectively [13]. In comparison, a laptop requires around 30 W.

Nanomaterials in general and NPs in particular are considered as the future building blocks of advanced materials and devices that will contribute to fulfilling the numerous promises and potential of nanotechnology. In addition to their unique properties, NP-based products provide specific advantages, particularly from the manufacturing point of view. Their parallel deposition, low temperature, and vacuumless processes can address numerous issues facing the conventional industries. In particular, NP integration could address issues related to high capital cost and difficulty in handling small and/or fragile components,

To reach this goal, improved approaches to assembling NPs into stable, cost-effective working films and devices are critical. Indeed, using NPs as building blocks of functional devices requires robust and precise methodologies. Self-assembling is the most natural and obvious process of integrating NPs into films. Self-assembling is, for example, limited by the lack of control over the process of integration into a final product [14]. Thus, for precise control over the film properties, it is critical to develop assembling procedures that will allow the trapping of the building blocks into a spatiotemporal arrangement. Higher-order structures obtained by self-assembly using NPs as building blocks have been achieved through numerous approaches. 2D and 3D self-assembling from colloidal solution have been widely studied because of their simplicity and versatility. Langmuir–Blodgett (LB), self-assembling monolayer (SAM), layer-by-layer (LBL), electrophoretic (EPD), and evaporation-driven self-assembly are important NP integration methods [15]. Assembling and deposition techniques take advantage of the different NP–NP and NP–substrate forces. Often, the choice of the integration methodology for a specific NP system can be dictated by several factors. Long-range order, film density, and rate of deposition are the most important. As shown in Table 4.1, the deposition rate may vary by up to two orders of magnitude depending on the integration method.

Until recently, most nanostructured films and devices were obtained using the top-down approach, in which the size of the bulk material is reduced using

TABLE 4.1 Approximate Fabrication Rate Estimation and Typical Thickness of Different Integration Techniques (Including Data from Ref. [15])

Nanoparticle integration technique	Deposition rate (cm^2/h)	Thickness range (μm)
Spin coating	>100	0.1
Langmuir–Blodgett (LB)	10	1
Self-assembling monolayer (SAM)	1	0.1
Layer-by-layer (LBL)	1	0.1
Electrophoretic deposition (EPD)	100	10
Evaporation-driven self-assembly (EDSA)	10	1

physical or chemical means. A bottom-up approach is now used more, although mostly at the laboratory or pilot plant stages. Nanostructured films and devices could be produced from atom, molecule, and/or nanomaterials. In this chapter, the focus is on the bottom-up approach using NPs as the main building blocks.

Numerous confinement methods have been used to produce quantum dot (QD)-based film and devices [16]. This includes lithography-defined QD, epitaxial self-assembled QD, and colloidal QD. The last category is the main focus of this chapter. When direct assembling is difficult, organic or inorganic templating has been used to provide a solid matrix framework for NP integration. Various inorganic structures can be used depending on the size of the NPs and their applications [17]. Microporous (around 1 nm or less), mesoporous (2–50 nm), and macroporous (above 50 nm) structures are obtained in zeolite and clays, MCM and mesoporous silicate, and photonic crystals, respectively [17]. Organic templates such as carbon nanotubes, polymers, and functionalized porous surfaces have been developed.

There are a lot of similarities between molecular-based organic systems [18] and NP-based building blocks. Both systems are constructed using bottom-up processes. Similar to molecular organic systems, existing deposition processes need to be adapted to fit the physicochemical characteristics of NPs. Brownian motion is the most important hurdle to overcome in NP integration into controlled structures. Stronger driving forces should contribute to overcome the random and diffusional nature of these NP movements. Covalent, van der Waals, electrostatic, hydrophobic, and capillary forces enable 1D, 2D, or 3D NP structure formation. Several review articles [15,16,19] and books [20–22] have been written on this subject. Indeed, NPs are used to fabricate thin films with tunable optical, magnetic, and electronic properties. They are also used to protect materials and devices against degradation. Vacuum deposition techniques remain the dominant approach in the fabrication of numerous functional thin films. However, new requirements in various industries call for alternative, cost-effective, and sustainable film methods that are more flexible and reliable.

2 TEMPLATE AND NANOREACTOR

NPs do not register to a defined position if no additional driving forces are added to help nullify the Brownian motion effect. The position change increases with temperature, surrounding fluid viscosity, and reduced particle size. Templating is probably the simplest approach toward assembling NPs into macroscopic objects. In particular, nanoporous materials offer a variety of structures, shapes, and dimensions to choose from. Both natural (biological systems and minerals) and man-made (in purpose or by-products) nanostructured systems have been described in the literature. Geometrical confinement of NPs is critical in some applications [23]. Synthetic porous materials play and will continue to play a vital role in industrial processing [24,25]. Porous support plays a vital role in the petrochemical industry. Applications include separation, catalysis, and energy storage [26–28].

Catalysis is the main field of applications. There are numerous examples in the literature showing significant potential for future applications of nanostructured templates. Close-packed, ordered gold NPs on a DNA matrix through electrostatic interaction have been reported [29]. Novel materials have been recently developed providing more flexibility and advantages to enhance templating properties. For example, dendrimers have been recently developed as porous matrix stabilizers for the preparation of highly monodisperse NPs.

According to IUPAC classification, three categories of porous materials are distinguished: microporous with pore diameters of 2 nm or less, mesoporous with pore diameters of 2–50 nm, and macroporous with pores greater than 50 nm. Another term, nanoporous, is often used in the literature and it is related to porous materials with pore diameters between 1 and 100 nm, corresponding to porosity (void volume) higher than 0.4. Catalysis is the main application of NPs integrated into a nanotemplate. A three-level porosity reactor is used to optimize interface properties between the reacting liquid gas and the deposited NPs [30]. This includes the first level of porosity consisting of micropores within the particles. The second level of porosity is that of mesopores consisting of interparticle spacing within a single array. The last level has the macropores consisting of the space between the arrays.

A monolithic reactor (Fig. 4.2), sometimes referred to as a honeycomb structure, is widely used as an industrial catalyst support for NP and microparticle catalysts. The fabrication of these catalyst supports involves bulk-like chemical reactions starting from inorganic and organometallic precursors [31]. Alternative preparations involving NPs allow better control over the porosity. This includes 3D NP assembling followed by posttreatments. NP

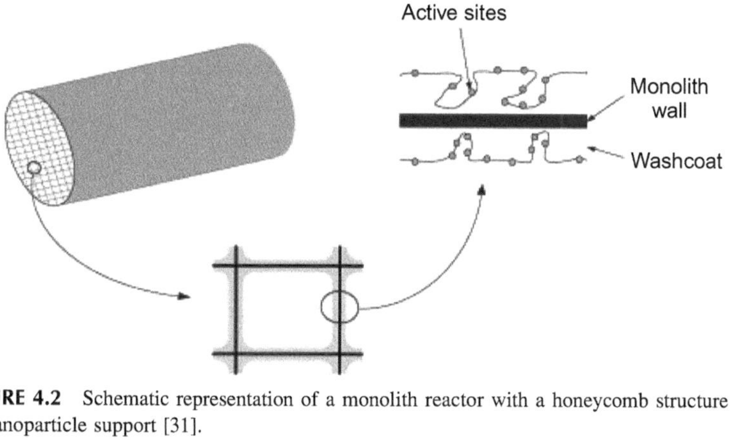

FIGURE 4.2 Schematic representation of a monolith reactor with a honeycomb structure used for nanoparticle support [31].

loading and stability depend on the porosity and surface composition of the support.

Well-ordered nanoporous templates are critical for achieving optimized material and device characteristics. Seven different nanoporous templating modes are reported [30–33]:

a. Emulsion templating;
b. Biomimetic templating;
c. Surfactant templating;
d. Aerosol spray;
e. Inverse opal micromolding;
f. Colloidal templating;
g. Double templating.

The choice of the templating mode can be dictated by several factors. Depending on the starting materials and specifications, these different template modes provide some flexibility. Significant progress has been made toward controlling the size and shape of the pores as well as the activity and distribution of the sites. As shown in Fig. 4.3, a wide choice of nanoporous materials and size ranges are available. Most zeolites are endowed with uniform and small pore sizes. For numerous catalytic applications, relatively larger pore sizes are needed. Several mesoporous materials (FSM, MCM, and SBA) with pore sizes up to 50 nm or more have been reported [34]. In the case of microporous materials, a VPI-5 aluminophosphate template has been developed.

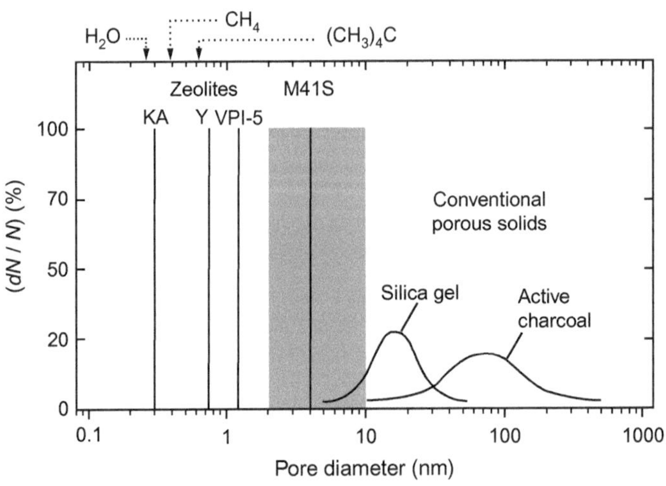

FIGURE 4.3 Typical pore diameter distributions of porous solids (KA is the K form of zeolite A; VPI-5 is a microporous aluminophosphate with 18-membered-ring pores; M41S is a family of mesoporous MCM materials). For comparison, the kinetic diameters of water, methane, and neopentane are also indicated [34].

The double and surfactant templating modes provide versatility and potential applications. Other nanoporous templating modes are also described. One of the most important attributes of nanoporous materials is their ability to select molecules based on their size and shape. This property is particularly appealing for NP assembling. Furthermore, the presence of these confined areas of specific size and shape is also used as a nanoreactor site. Materials choice for the templating is based on several factors. Porosity, permeability, and cost along with the chemical, mechanical, and thermal characteristics are the most important factors. Carbon and oxides are the most widely used materials, although metallic templates have also been reported. Technico-economic factors of specific applications will dictate materials composition and type of templates.

The majority of nanoporous templates are based on organic and oxide compositions. Recently, several nanoporous metal foams (NMFs) have been developed [35]. These NMFs combine the thermomechanical and catalytic properties of metal and high surface area, ultralow density, and high strength-to-weight ratio of alumina. It has been argued that the nature of the support has a direct effect on platinum particle size distribution [36]. The difference in particle size distribution has been assigned to the strength of metal–support interaction.

Mesoporous and microporous templates have been produced from polymers, carbon, glass, aluminosilicates, oxides, and metals. The optimal materials choice will depend on several parameters. Pore size, stability, and cost are the most important parameters used to classify these porous materials. Aluminosilicates are endowed with high chemical and thermal stability combined with high surface area and low cost. Theses aluminosilicates are probably the best choice [37]. Other material characteristics are sometimes important for some specific applications. They include permeability, conductivity, strength, and lifetime [37]. In such cases, glass and metal and, to a lesser extent, advanced carbon materials could be considered [37].

2.1 Zeolites and Mesoporous Materials

Natural and synthetic zeolites are used for numerous applications. There are numerous synthetic zeolites [28]. Although about three quarters of mass-produced zeolites are used for ion exchange applications, zeolites have also been used as porous support, particularly for fine chemical manufacturing and petrochemical refining. Growth and applications of metallic clusters in a zeolite matrix have been known for a long time [38]. The use of these cluster-supported catalysts is well documented. The main role of the zeolite support is to prevent NP sintering following heat treatment before and during their use.

Mesoporous materials are widely used as ion exchange in detergent formulation. Other applications include purification and catalysis. Support and catalytic properties of zeolites are widely used in the chemical industry due to the following attributes [39]:

a. High surface area combined with high absorption capacity;
b. Tunable absorption properties (hydrophobic and hydrophilic);
c. Active catalytic sites with tailored strength and concentration;
d. Tunable channel and cavity sizes and shapes within a suitable range;
e. Good mechanical and thermal properties allowing stable operation in industrial processes.

A wide variety of mesoporous material compositions have been reported. Even in the case of zeolite, there is a wide range of compositions. The chemical composition of a zeolite is represented by the general formula:

$$M_{x/m}{\cdot}Al_xSi_{2-x}O_4{\cdot}nH_2O,$$

where m is the valence of the cation M; n, the water content; and $0 \leq x \leq 1$. The pore size is tunable from about 0.5 to 8 nm and is obtained using aluminosilicates and MCM-41 [40]. Numerous applications involving incorporation of metal and oxide NPs in zeolites have been studied [40]. In the case of metal NPs, reductive and oxidative catalysis of organic reactions are reported. Typical metal size values are between 0.5 and 1.5 nm.

When acid zeolites and zeolites with oxidizing properties are used in combination with noble metal NPs, bifunctional catalyst systems are obtained [41]. They are used in downstream petrochemical processes. Oxide NPs including Na_2O, Cs_2O, MgO, CaO, and $CsLaO_2$ have been grown inside zeolite structures. Although they provide advantages in shape-selective catalysis, their industrial application is quite limited. Crystalline molecular sieves, particularly zeolites, are limited by their internal dimensions [42]. Indeed, in zeolites with pore size of 1.5 nm or less, large molecules are denied access to active sites within the pores. Recently, Corma et al. synthesized catalytical active molecular sieves with 18- and 10-member rings using high-throughput synthesis methods [43].

Using submicron particle morphology, zeolites may provide even higher catalysis efficiency. Indeed, zeolite NPs may provide high external as well as internal surface areas. Zeolite NPs are obtained using various techniques [44]. Figure 4.4 illustrates the synthesis process used to produce 40–50 nm FAU-type zeolite NPs. The process includes gel formation followed by crystallization via propagation and agglomeration. Besides catalysis, these faujasite-type zeolites have potential applications in optoelectronics, chemical sensing, and medicine.

Hierarchically structured ceramic monoliths with adjustable porosity have been obtained using aluminum titanium oxide (AlTi), amorphous silica (Si),

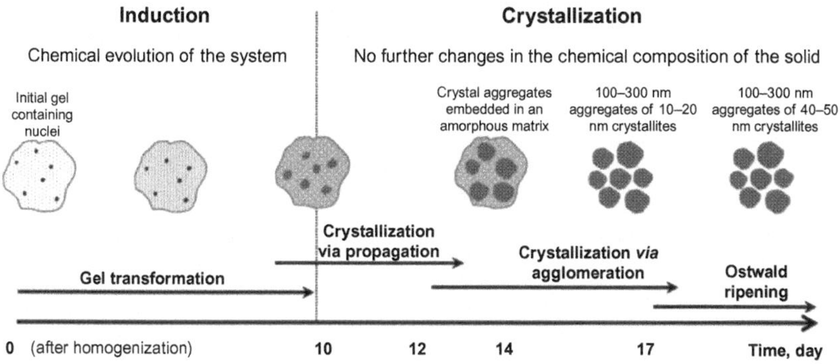

FIGURE 4.4 Schematic illustration of the synthetic process of FAU-type zeolite nanoparticles. *Reprinted with permission from Valtchev and Bozhilov [45]. Copyright 2004, American Chemical Society.*

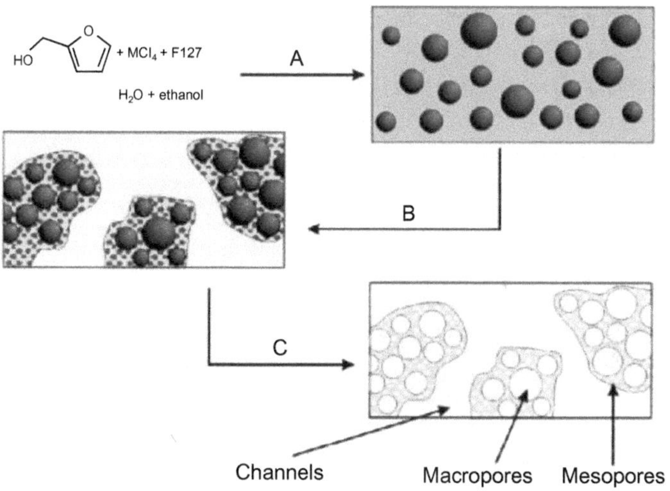

FIGURE 4.5 Schematic representation of a three-step process using one pot for the production of hierarchically porous metal oxide monoliths. First (A) polyfurfuryl alcohol (PFA) nanoparticles are synthesized (black sphere). Second (B) evaporation of the solvents leads to condensation and formation of a mesostructured network around the PFA particles. Third (C) a calcination is carried out to remove PFA, leaving behind a bimodal structure [46]. *Reprinted with permission from Drisko et al. [46]. Copyright 2010, American Chemical Society.* (For color version of this figure, the reader is referred to the online version of this chapter.)

titania (Ti), amorphous titaniumzirconiumoxide (TiZr), yttriumzirconiumoxide (YZr), and zirconia (Zr). A single-pot process has been recently reported (Fig. 4.5). A bimodal porous structure has been obtained using this method. Indeed, macropores and mesopores are obtained following polyfurfuryl alcohol NP calcinations.

2.2 Inorganic Porous Membrane and Porous Silicon

Nanoporous alumina membranes have been fabricated using controlled anodization of aluminum surfaces in aqueous acids [47]. The diameter of the parallel running pores is tunable from 5 to 300 nm [47–49]. Various engineered inorganic structures with potential application in catalysis have been developed [50,51]. Nanoporous alumina membranes have been recently used to grow uniform Pd and Ru NPs [47,52]. As shown in Fig. 4.6, small (around 3 nm) and highly uniform Ru NPs are fabricated inside 20 nm wide porous channels. Various transition metal NPs embedded in porous alumina are successfully used in gas-phase hydrogenation [52].

The possibility of controlling pore size is an additional parameter allowing the tuning of NP properties [53]. For example, varying applied voltage and acid type during oxalic anodic oxidation of alumina allowed a tunable diameter between 10 and 200 nm. Porous silicon-containing silver NPs with an average diameter of 3 nm have been successfully prepared [54]. The modified porous silicon switches reversibly between transparent and opaque, due to changes in the outer silver layer oxidation state. Gold NPs have been used to fabricate porous metallic film [55].

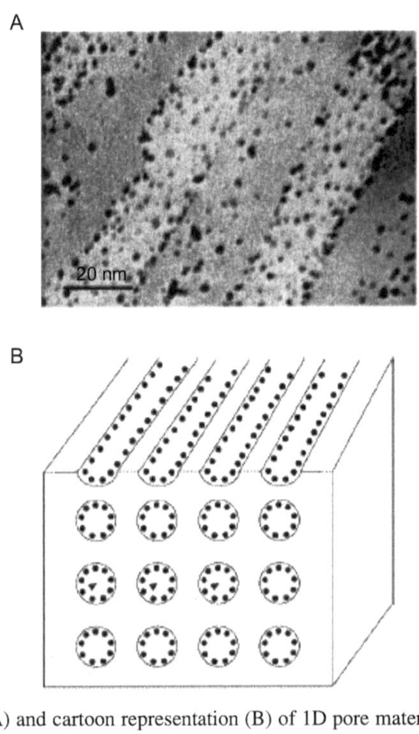

FIGURE 4.6 TEM (A) and cartoon representation (B) of 1D pore materials filled with nanoparticles [52].

RuO_2 NPs have been successfully incorporated into MFI-type zeolites [56]. Incipient wetness impregnation has been used to synthesize iron oxide NPs inside MCM-41 zeolite [57]. Substrates with coarse pores are used to draw NP-laden gas exiting a flame reactor forming a finer porous layer [58]. Such a system could be used in numerous applications including SOFC and filtration.

Uniform and high surface porosity films have been obtained using alumina NPs deposited onto a porous substrate. Ceramic α-Al_2O_3 tubes with average pore diameters of 450 and 2700 nm are used for this purpose [58]. Gas-phase NPs provide higher diffusion rates and adhesion coefficients to the substrate walls when compared to solution-based NPs. Indeed, NPs obtained by flame synthesis could be used directly following their formation. The presence of surfactant in colloidal solution inhibits NP coalescence and growth.

2.3 Activated Carbon and Carbon Nanotube

Activated carbon forms disordered and ill-defined porous structures with an inhomogeneous surface chemistry [59]. Overall surface area is estimated to be 500–1500 m^2/g and up to 50% higher than ordered and well-defined zeolites. Charcoal is widely used to support NPs for catalysis applications. Often, charcoal is impregnated by stirring in the suspension containing the NPs. Before use, charcoal is treated (activated) to increase its porosity and optimize its interactions with NPs.

Porous carbon prepared from natural precursors (e.g., coconut shell) or synthetic precursors (e.g., phenolic resin) provides limited control over porosity [60]. Mesoporous carbons synthesized by carbonizing sucrose inside the pores of the self-ordered mesoporous silica template give rise to tunable pore size in the range of 2–4 nm [61]. Porous carbon is widely used in batteries and supercapacitors. Higher porosity provides high storage capacity. Carbon nanotubes have provided a good model system with large pores and high conductivity. This leads to large power densities but low energy density. High production cost precludes industrial applications of carbon nanotubes at present. Metal NPs dispersed on the surface of carbon nanotubes have been achieved [62]. Carbon nanotube filling using fluorescent NPs has been reported [63]. A high filling factor has been obtained using this approach (Fig. 4.7). There are numerous other reports demonstrating the feasibility of oxides, and metallic and semiconducting NPs on single and multiwall carbon nanotubes for different applications.

NP attachment to organic-based porous materials is facilitated by surface functionalization. Numerous approaches have been developed for porous surface functionalizations. These include surface oxidation, KOH activation, CO_2 activation, sulfonation, halogenation, and grafting [64]. Electrostatic attraction between TiO_2 NPs and modified 3-DOM carbon leads to

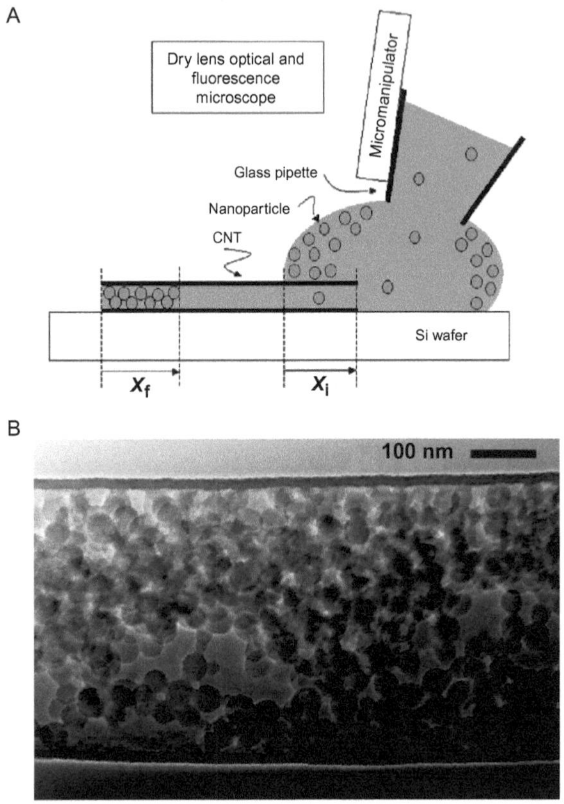

FIGURE 4.7 (A) Schematic representation of the filling process of nanoparticles inside a nanotube; (B) TEM image of the nanoparticle-filled nanotube. *Reprinted with permission from Kim et al. [63]. Copyright 2005, American Chemical Society.* (For color version of this figure, the reader is referred to the online version of this chapter.)

well-dispersed NPs [65]. Polymer surface coating could also enhance surface area and surface chemistry for NP deposition [64].

2.4 Polymers

Polymeric nanocomposites offer numerous options for NP integration. NPs can be either integrated into preformed porous structures or combined with a polymer before forming. Polymer–NP nanocomposites offer possibilities to control the final morphology and structure [66,67]. Two general approaches allowing the control of NP distribution within a polymeric matrix are pursued (Fig. 4.8). The first approach consists in using an externally applied force (external-in) to form a pattern. This is referred to as directed patterning of NP dispersions (DPND). The phase behavior of this pattern may be modulated by uniform changes in the intrinsic (pressure or temperature) or extrinsic

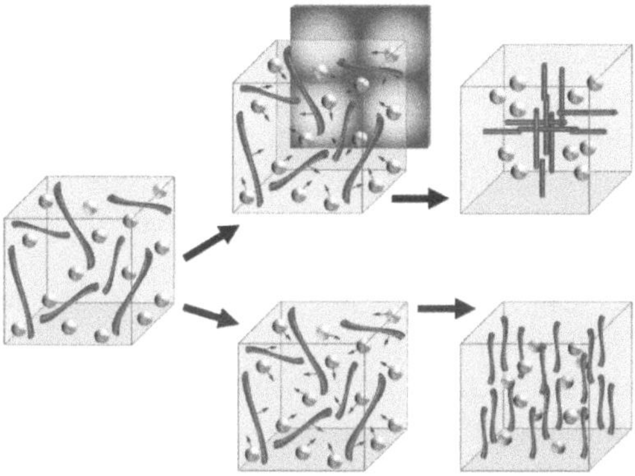

FIGURE 4.8 Schematic representation of two approaches to controlling morphology and structure of nanoparticle–polymer nanocomposites. Directed patterning of nanoparticle dispersions (DPND) (top) and mesophase assembly of nanoparticles MANP (bottom) are obtained using external means (external-in) and internal interaction (internal-out), respectively. *Reprinted with permission from Vaia and Maguire [66]. Copyright 2007, American Chemical Society.* (For color version of this figure, the reader is referred to the online version of this chapter.)

(number density or entropy) thermodynamic parameters. A random NP distribution is thus transformed to an ordered structure following NP mass flow (arrows), minimizing the overall energy of the system. In the second approach, a mesophase assembly of NPs (MANP) is induced by tailored inter-NP and NP–matrix interactions (internal-out). In this case, uniform changes in the system's environment will affect the NP organization in MANP [66]. Practical application of each approach will depend on the requirements of the final product. The first approach could be used when specific ordering patterns are needed.

Nanoporous polymers refer to porous polymeric structures with pore size in the range of 1–100 nm [68]. Several examples of polymer-stabilized metal NPs have been reported [69]. Indeed, NP dispersion has been obtained in water-soluble polymers, functionalized resin, inside a polymer particle, and on polymer surfaces. The first two approaches could be used to produce bulk materials and film using printing techniques. The last two nanoporous polymer structures can be used for carrying and delivering NPs through liquid-containing systems.

Microporous polymer structures have also been prepared using zeolite templates [70]. There are three main steps for replica microporous polymer production [70]. Monomers are first introduced within the zeolite channels. Thermoset phenol formaldehyde resins were cured at around 900 °C within the channel networks of different zeolites. Following HF etching, a relatively well-structured zeolite replica polymer is obtained. Various parameters could be used to control the size distribution of these replicas. Macroporous and mesoporous structures of the zeolite are the main parameters.

Thermoplastic polymeric matrices made from poly(aryl ethers) incorporating an oxadiazole moiety are used to fix silver ions as seed sites for NP growth [71]. These polymer-ions are first integrated into a thin film. A transition to a nanocomposite film is obtained using a chemical reducing agent. A spherical 3D wormhole-like magnetite mesostructure has been obtained in a four-step process [72]. First, triblock copolymers are self-assembled into micelles in ethanol solution. Following a mixture with magnetite NPs, a robust aggregated magnetite mesostructure is formed. The last step consists in washing and calcination at 500 °C.

Molecularly imprinted polymer nanocomposites have been prepared and integrated into biosensing chips [73]. Voicu et al. have produced a polymer-based nanotemplate using microcontact imprinting [74]. The porous surface structure can accommodate molecules and small NPs. A relatively simple approach has been developed to produce amphiphilic NPs at room temperature within a polyester matrix [75]. Following Ag^+ incorporation into PEG sebacate domains, silver NPs are formed in contact with PEG-reducing fragments [75]. Dispersed polyester particles containing Ag NPs are thus obtained. The hydrophobicity of these NPs will depend on the type of solvent used as a dispersion medium [75].

2.5 Gel Structures

Sol–gel materials are readily amenable to forming 2D and 3D structures. The sol–gel film can be obtained by a dip coating process consisting of substrate dipping in the solution followed by a slow pulling out from the solution. The gelation takes place during contact with the atmosphere at room or relatively high temperature. Postprocessing steps are often used to produce optimized structures.

Polymeric nanogels and microgels are 3D interconnected polymeric NP and microparticle gels, respectively [76]. Colloidal gel has been used to fabricate 3D networks with some control over macroscopic and microscopic features [77]. Shear forces are used to control the gel structure of PLGA–PVAm and PLGA–PEMA NP mixtures (Fig. 4.9). With excessive shear forces, the sol–gel network could be destroyed.

Nanoporous aerogel is a light material (around four times air density) and often quite brittle. It is usually prepared using the sol–gel route. Aerogel silicon is prepared by etching silicon. Transition-metal NPs have been suspended in sterically stabilizing microgels [78]. Compared to dendrimers, for example, porosity distribution of microgels is relatively large. There is less control over size. Hence, microgels are not useful as a support for NPs when precise size distribution is required. Nanogels have been developed to address this shortcoming. Drug-loaded nanogel has been prepared using the reverse micelle method [79]. These cross-linked particles provide high surface area with tunable dimension in the sub-100 nm regime.

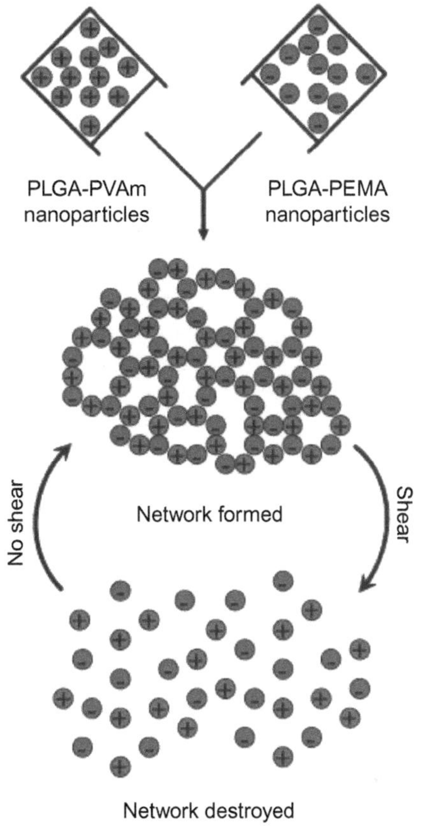

FIGURE 4.9 Schematic representation of a gel structure in the absence and presence of shear forces [77]. (For color version of this figure, the reader is referred to the online version of this chapter.)

It is important to distinguish sol–gel from aerogel and xerogel. The difference is in the degree of shrinkage between the wet sol–gel, aerogel obtained by supercritical-fluid extraction, and xerogel obtained by ambient evaporation. It is also worth noting the difference in porosity between the aerogel and xerogel. Low-weight materials using aerogels could have important implications in numerous industries. Since their discovery almost a century ago [80], aerogels have received a lot of interest because high surface area and strong monolith structure are achievable from most materials. Aerogels with fractal-like geometries [81] can be used as powder, monolith, or thin film. Silica [82] and carbon [83] aerogel materials have been developed. These materials can be used for different applications for their mechanical and catalytic attributes [84].

A basic sol–gel process could be modified to produce multicomponent and/or functional nanocomposites with different levels of porosity [82].

Addition of NPs gives rise to silica matrix-based nanocomposites (Fig. 4.10). The sol–gel approach to produce a NP–matrix composite is straightforward. Adding functionalization and improving the properties of the gel and/or aerogel include solution- or gas-phase modification, as indicated at process stages I, II, III, and IV (Fig. 4.10). If needed, the dry aerogel could be redispersed in solution. Oxidative aggregation of metal chalcogenide has also been used to produce II–VI or IV–VI semiconductor aerogels [86].

Carbon aerogel consisting of 3D network of interconnected particles gives rise to highly porous materials [87]. Porosity up to 99% has been achieved [87]. These carbon aerogels are often obtained from carbon aerogels produced using resorcinol and formaldehyde precursors. Organic aerogels are readily obtained by polycondensation of resorcinol (R) and formaldehyde (F), followed by CO_2 supercritical drying [87,88]. The porosity of these aerogels is directly related to the value of R/C mole ratio. A similar sol–gel approach has been used to produce porous chalcogenide aerogels [86,88]. Thiolate-capped NPs are first prepared, followed by a gelation step by controlled surface group stripping, and then a supercritical CO_2 drying. Monolith-based CdS and CdSe aerogels are obtained by this method. Tunable quantum confinement properties remain in the aerogel state. The presence of mesopore (2–50 nm) structures in colloidal-like CdS and CdS aerogels has been shown

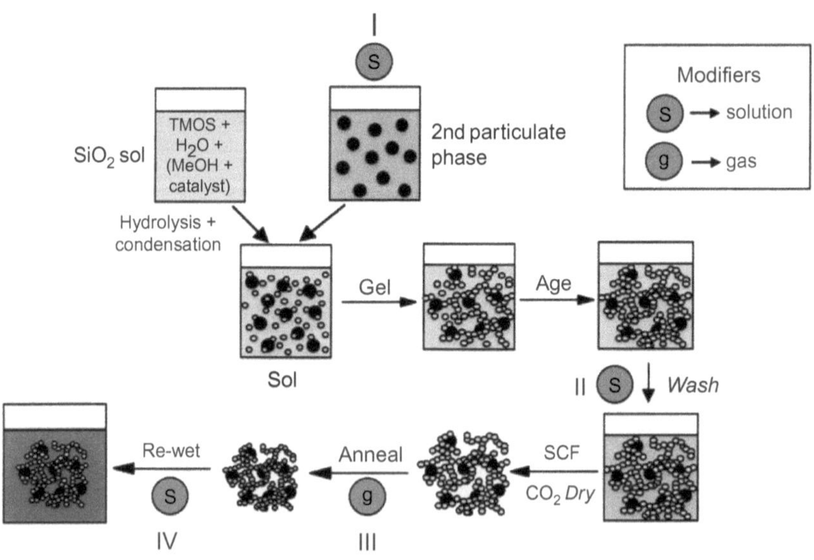

FIGURE 4.10 A modified sol–gel process for dry and wet aerogel nanocomposite preparation. Starting with SiO_2 and a second particulate phase, a sol mixture is prepared followed by gelation and aging. After supercritical-fluid (SCF) CO_2 drying, an annealing step is applied [82]. (For color version of this figure, the reader is referred to the online version of this chapter.)

[86]. Minimal volume loss is observed in the conversion of wet gel to aerogel. In the case of xerogel formation, a significant volume loss has been reported [86]. It is worth noting that modification of silica aerogel with some functional amine groups may increase the density and strength of the native aerogels by several orders of magnitude.

2.6 Dendrimers

Nitrogen-based dendrimers, such as poly(amidoamine) and poly(propylene imine), have been used as a template for the growth, stabilization, and solubility of NPs [89]. The size distribution range is usually limited to 1–5 nm. NP synthesis using this approach involves ion salt absorption followed by reduction. Dendrimer-embedded NPs are considered for catalysis and other applications. Dendrimer templating gives rise to highly monodisperse NPs. However, high dendrimer cost, synthesis complexity, and scalability will likely hinder utilization of this technique.

2.7 Opal and Inverse Opal

Photonic crystals have attracted tremendous interest because of their potential application in optoelectronics [90]. Gold NPs infiltrated into a photonic crystal using 640 nm beads showed unique behavior [91]. Optical properties of photonic crystals are modified with gold and semiconductor NPs. Preparation of InP NPs by atmospheric MOCVD gives rise to optical properties depending on the filling factor [92]. Well-defined pore structure has been obtained by electrophoretic deposition of Au and CdTe in polystyrene (PS) bead-based opal [93].

The electrochemical co-deposition technique was used to form a NP opal structure (Fig. 4.11). PS microparticles are first deposited on an ITO substrate followed by filling up of the remaining uncovered area using epoxy gluing. This modified ITO substrate was used as a working electrode in a solution containing appropriate pyrrole (Py) and gold (or CdTe) NP concentrations. After electropolymerization, THF was used to remove the PS template. A well-organized NP distribution in a polypyrrole-inverse opal is obtained (Fig. 4.11). Utilization of the glue may have improved the overall film integrity. However, glue residues may contaminate the final product. PS microparticles have also been used as a 3D template to prepare nanostructured titania. Monodispersed organic-layer-coated TiO_2 NPs are easily infiltrated to form compact packing with a 3D array structure [94]. After thermal decomposition of the organic template, a near-perfect inverse opal has been obtained. This inverse opal has been shown to have promising applications in TiO_2-based solar cells.

Heating rate has been used to control the porosity of the 3D colloidal crystal template [95]. A capillary-enhanced NP packing process is shown in

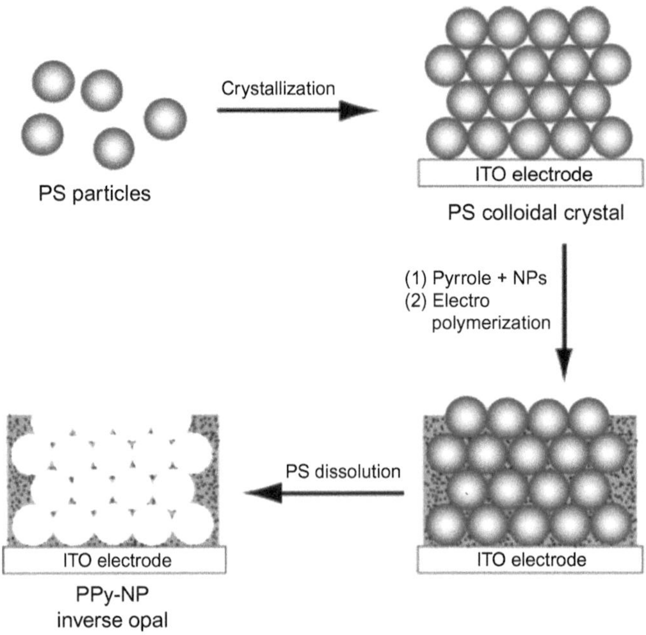

FIGURE 4.11 Schematic representation of nanoparticles embedded in polypyrrole (PPy)-inverse opals. *Reprinted with permission from Yu et al. [93] Copyright 2004, American Chemical Society.*

Fig. 4.12. The liquid interface of the capillary-enhanced process is used in poorly packed particles (Fig. 4.12A). Following NP deposition, a poorly packed film is obtained. Depending on the position of the liquid interface, the final packing will be reduced (Fig. 4.12A) or increased (Fig. 4.12B) following heat treatment. Indeed, when the liquid interface is below the substrate, only the secondary capillary and the gathering forces are effective within the gel-like bulk structure. Once there is no liquid interfacing, the primary capillary force is nullified. Thus, a highly packed film is obtained. If water is drained off from interstitial sites by evaporation, secondary forces (enhanced capillary) will help in densifying the opal structure (Fig. 4.12B).

2.8 Biological Template

Large 2D assemblies of protein building blocks are used as a template to grow regular structures of NPs [96,97]. Ni–Pd NPs have been synthesized and assembled into regular patterns in a single fabrication step (Fig. 4.20). This technique has been used by other groups also [99]. 2D arrays of iron and indium NPs have been obtained on carbon film and silicon wafer by taking advantage of the regular lattice arrangement formed after protein crystallization. NPs caged in Ferritin form well-ordered 2D structures

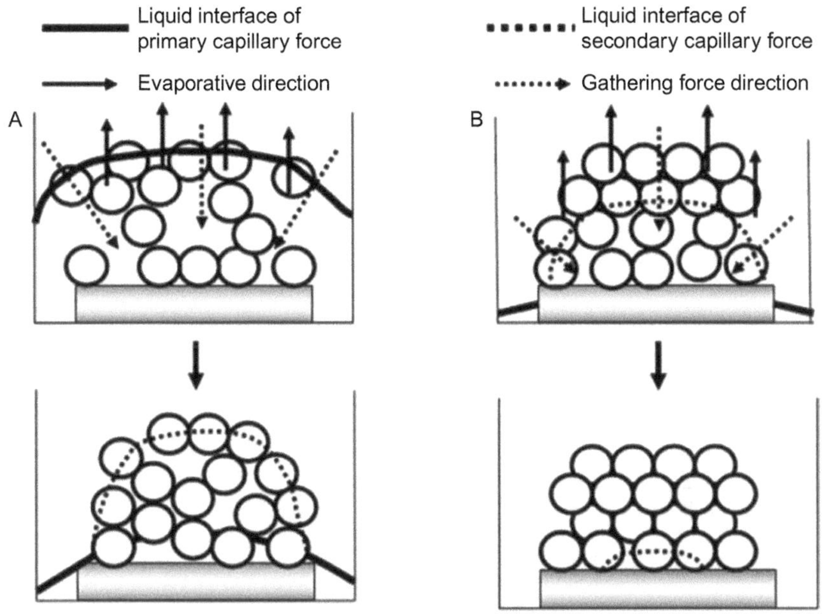

FIGURE 4.12 Schematic representation of nanoparticle packing density due to (A) primary capillary and (B) secondary capillary forces [95].

even after the protein template is removed using heat or chemical treatment [100].

NPs are very good precursor candidates for self-sustaining structures using a biomineralization process [101]. Differences in critical nucleus size and activation energy will affect the stability of the final phase. Indeed, a correlation has been established between increasing metastability and decreasing surface energy [101]. Similar trends in variation of the free energy of two different polymorphs as a function of particle size have been reported [101]. However, these polymorphs may have different free-energy scales [101].

Biomineralization of cellulose materials also gives rise to nanoporous silica structures [102]. Two production processes of mineralized silica frameworks from wood cellular templates have been reported [102]. In the first process, silica sol at low pH is infiltrated followed by drying and calcination. In the second process, silicate–surfactant mesophase infiltration and condensation at high pH are followed by washing/drying and calcination. The silicic acid is infused into wood cellular structures in order to obtain a structure mimicking the original hierarchically ordered silica. Ceramic (SiO_2), magnetic (Fe_3O_4), and semiconducting (CdS) NPs were successfully incorporated into supercellular swellable bacterial templates [103]. Particle charging has a direct effect on the resulting pattern characteristics.

2.9 Block Copolymer Template

Block copolymer templating is probably the most versatile process to fabricate porous structures. As schematically summarized in Fig. 4.13, a variety of 1D, 2D, and 3D porous structures are possible using block copolymer templates. We will not cover all possible structures. This area has been recently reviewed by Forster et al. [17,104].

A five-step process was used to produce $PbTiO_3$ NPs on a template with $Si/Pt/TiO_2$-multilayered surface [105]. First, gold-loaded poly(vinyl pyrrolidone) (PVP)-*b*-PS micelles were deposited on Si precoated with Pt, followed by gold reduction and reactive-ion-beam etching tempering, allowing gold seed particle formation. Then, a reactive-ion etching (RIE) was applied on exposed TiO_2. After eliminating the Au template by RIE, TiO_2 seeds are exposed. $PbTiO_3$ grains are thus grown on TiO_2 seed patterns. These patterns are used to grow $PbTiO_3$ submonolayer NPs. This bottom-up approach is a good alternative to the relatively more expensive and time-consuming top-down evaporation techniques.

Triblock copolymers have also been used as templates for the fabrication of anatase thin films with tunable pore sizes ranging from 8.3 to 14 nm [106]. Slowly adding $Ti(OBu(n))_4$ in HCl with amphiphilic triblock copolymer $HO(CH_2CH_2O)_{20}(CH_2CH(CH_3)O)_{70}(CH_2\text{-}CH_2O)_{20}H$ in ethanol produces

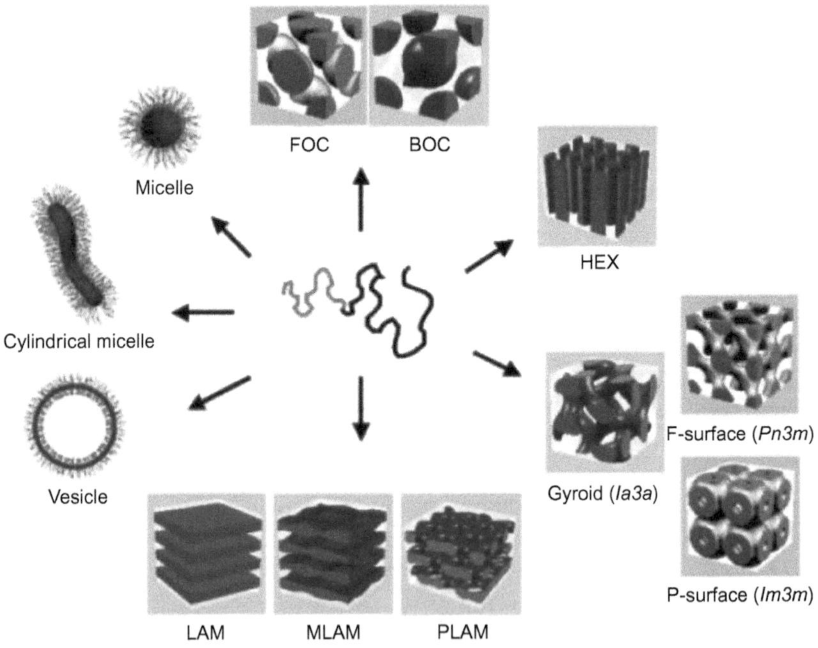

FIGURE 4.13 Template structures using block copolymers and surfactants [104].

colloidal sols. After film deposition on an ITO substrate and calcination, an anatase thin film with tunable pore diameters was obtained.

2.10 Double Templating

The double templating fabrication process is summarized in Fig. 4.14. Double templating consists of five process steps [107]:

a. Micron-sized silica particles are assembled into a 3D close-packed array;
b. Insertion of polymers in the interstices followed by polymerization allows complete particle encapsulation;
c. Selective chemical etching is applied to remove the silica particles. Micron-sized air cavities are obtained;
d. Polymer latex spheres are injected into the spherical cavities. The polymeric spheres assemble within the voids to form an ordered, close-packed structure;
e. An inorganic precursor is infiltrated into the interstices between the latex spheres and gelled to capture the ordered structure. Thermal decomposition of all the organic material produces the ordered nanoporous structures, also called photonic balls.

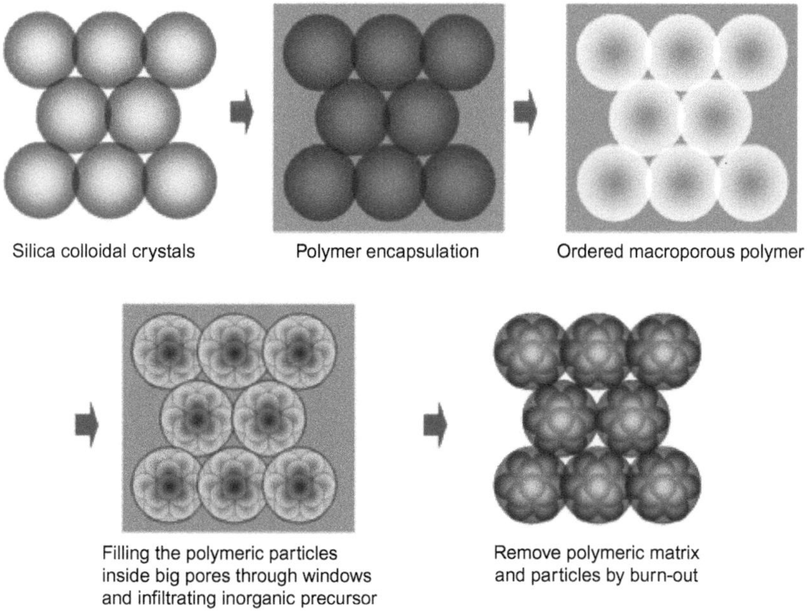

FIGURE 4.14 Fabrication process of an ordered mesoporous structure using the double templating method. *Reprinted with permission from Yi et al. [107]. Copyright 2002, American Chemical Society.* (For color version of this figure, the reader is referred to the online version of this chapter.)

This technique allows a sophisticated and ordered porous structure that is difficult to achieve with other techniques.

3 SELF-ASSEMBLING AND DIRECT ASSEMBLING

Many self-assembling approaches have been developed including vertical deposition (sedimentation), electric field-induced self-assembly, self-assembly at interface, physical confinement, air-pulse-driven self-assembly, and printing colloidal crystal arrays [108]. A strong interest in NP self-assembling emerged following the work of the IBM group in the 1990s, which showed good spatial ordering of CdSe NPs in the 1990s [109].

Velev and Gupta recently summarized the advantages and drawbacks of different colloidal assembling techniques [85]. Self-assembling has long-term potential for cost-effective bottom-up manufacturing [85]. As shown in Table 4.2, the different SAM techniques allow a variety of assembling types with different complexity and cost structures. Furthermore, some of these SAM techniques are potentially scalable. However, the majority of these

TABLE 4.2 Characteristics of Different Assembling Techniques [85]

Method of assembly	Schematics	Types of assembly	Scalability	Complexity/cost
Sedimentation	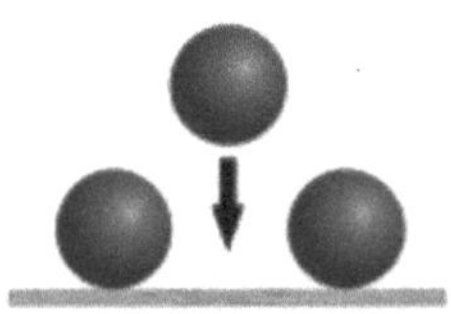	3D	Yes	Small
Evaporation	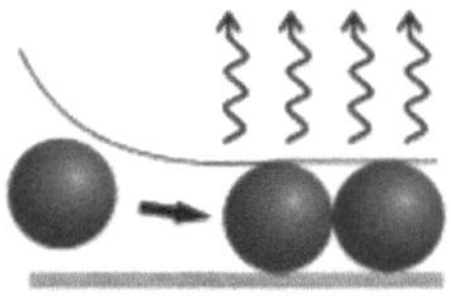	2D, 3D	Yes	Small
Adsorption/LBL	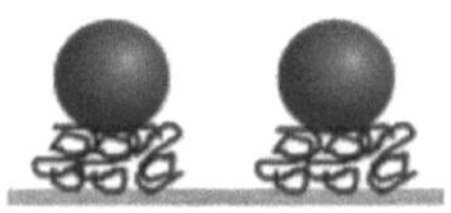	2D, 3D	Yes	Small
External force field		1D, 2D, 3D	Limited	Medium

TABLE 4.2 Characteristics of Different Assembling Techniques [85]—Cont'd

Method of assembly	Schematics	Types of assembly	Scalability	Complexity/cost
Biospecific		1D, 2D, 3D	Limited	High
Templated on surfaces		1D, 2D, 3D	Limited	Very high
Templated on droplets	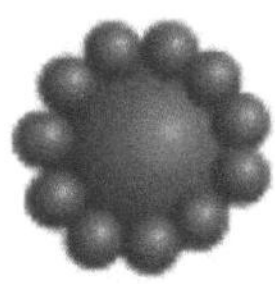	Supra	Yes	Medium

techniques are still at their infancy. They face serious fundamental and engineering challenges. Scalability is probably the most critical challenge due to low production yield, thermal and chemical gradients, and uniformity. Some of these challenges have already been addressed, although using a complicated, slow, and costly process. An attempt has been recently made to evaluate the cost of different colloidal assemblies. A direct relationship has been implied between the overall cost and the value added as it pertains to the complexity (Fig. 4.15). The relative positions of simple particles, complex particles, and microelectronic chip technologies are provided. There are some uncertainties in this classification. However, Fig. 4.15 provides some guidance when choosing alternative solutions.

Two important parameters govern the self-assembling processes [110]. The first one is related to the increase in the particle-binding energy with the particle diameter. For particles with a diameter above 2 nm, the binding energy is much larger than kT, which may lead to nonequilibrium systems. Second, surface diffusion occurs at a much shorter time scale in the case of NPs. Using the $\langle x\rangle^2 \sim \sqrt{Dt}$ relation where $\langle x\rangle$ is the mean-square distance, D is the diffusion coefficient, and t is the elapsed time, it would take 1 year and 30 s for a surface molecule to circle a particle of 10 μm and 10 nm, respectively.

A plethora of attractive (van der Waals, solvation, depletion, hydrophobic, π–π stacking, hydrogen bond) and repulsive (electric double layer, salvation, hydration, and steric) forces contribute to the self-assembling process [111].

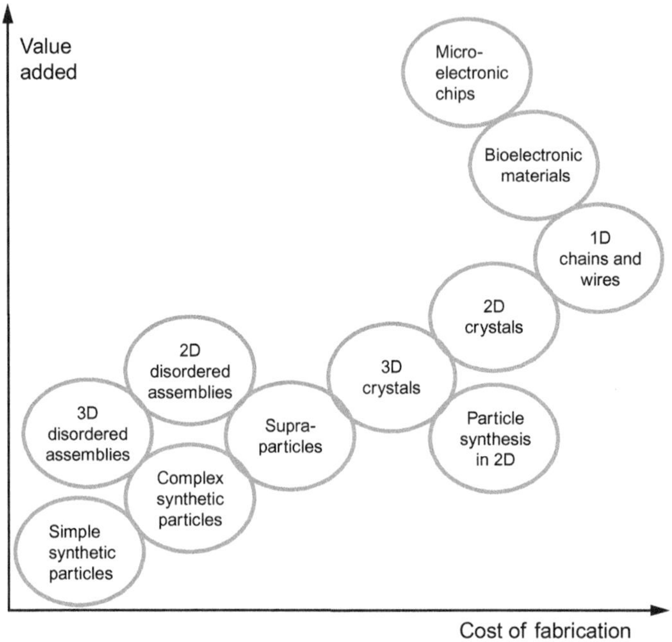

FIGURE 4.15 Hypothetical relationship between the fabrication cost and the value added of different particle-based self-assembling systems. For comparison purposes, the relative position of simple particles, complex particles, and microelectronic chip technologies are provided [85].

Stability of colloidal NPs and their self-assembling film is governed by the overall potential energy as reported by Motte et al. [112]:

$$U_{total} = U_{vdW} + U_{magn} + U_{elec} + U_{steric},$$

where U_{vdW}, U_{magn}, U_{elec}, and U_{steric} represent the van der Waals, magnetic, electrostatic, and steric interparticle interactions, respectively. Not included in the equation above are the specific forces that could play a critical role in the self-assembling process [110]. These include solvation [113], capillary [114,115], convective [114,116], and shear [117] forces.

Often, the van der Waals interactions are the most dominant and common forces in these systems. There are three groups of van der Waals interactions [118]:

a. Keesom interactions consisting of dipole–dipole forces between polar molecules;
b. Debye interactions consisting of dipole-induced dipole forces between polar and nonpolar molecules;
c. London interactions consisting of dispersion forces between nonpolar molecules.

Hamaker has estimated the potential energy of the van der Waals interactions in the case of solid particles with a diameter a using the following relation [118]:

$$U_{vdW} = -16Aa^6/9d^6,$$

where the Hamaker constant A is given by the following relation:

$$A = -0.75\left[h\nu(\alpha/4\pi\varepsilon_0)^2\pi^2\rho^2\right],$$

where $h\nu$, α, ε_0, ρ are first ionization potential, polarizability constant, permittivity of the free space, and mass density, respectively. The Hamaker constant is affected significantly by the intervening medium. Often the Hamaker constant of particles is much lower when they are dispersed in water than when they are in vacuum. For example, the Hamaker constants of amorphous TiO_2 are estimated to be 15.3×10^{-20} J and 5.35×10^{-20} in vacuum and water, respectively [118]. The amplitude of van der Waals forces is proportional to d^{-7} where d is the interparticle distance. Thus, their effect is reduced over large distances. In well-dispersed systems, van der Waals forces are nil.

Self-assembled NPs are governed by other forces including particle–solvent and particle–substrate interactions. Particle–solvent interactions defined by surface tension give rise to capillary forces that allow the formation of close-packed NP films. Quality and properties of the self-assembled films are also affected by the particle size and shape distribution, air pressure, humidity, moving speed of contact line, and by the solvent evaporation rate [108].

Nondirectional forces and insufficient mobility during assembly often lead to amorphous or short-range order self-assembling [119,120]. The mixture of two different-sized colloidal particles crystallizes into AB, AB_2, AB_5, or AB_{13} colloidal crystals [121]. Other self-assembling systems based on complex structures have been investigated [108]. They include binary colloidal crystals, non-close-packed colloidal crystals, patterned colloidal crystals, and non-spherical colloidal crystals.

The conventional self-assembling process of NPs is random and nonhierarchical. Several strategies have been devised to control the self-assembly process. Chemical functionalization is often used to regulate NP self-assembly. This includes patterned surfaces of NPs [122], antibody–antigen recognition [123], NP–matrix interaction [124], and NP anisotropy [125].

Self-assembly engineering could also take advantage of specific properties of different shapes and tethers [125]. Five main classes of tethered NPs are distinguished related to the (i) number of tethers, (ii) length of tether, (iii) number of NP vertices, (iv) relative location of tethers, and (v) tether specificity [125]. More complex anisotropies that could be used as building blocks can be described as combinations of these dimensions [125].

Directional assembling using external forces gives rise to a hierarchical structure [126]. These external forces could originate from gravity, fluid flow, applied electrical or magnetic field, and mechanical or capillary forces. The combination of interparticle interactions and externally applied forces provides possibilities for tuning SAM structures. Key to self-assembling of NPs is the ability to control size, shape, functionalization, and environmental factors. Non-stabilized NPs lead to unpredictable flocculated films. The presence of large size distribution often leads to imperfections, although bimodal NP distribution is sometimes used to obtain dense films. Numerous structures of NPs have been obtained at the laboratory scale by SAM [1,127,128]. However, very few large-scale productions have been reported so far.

3.1 One-Dimensional Assembly

Although rare, several NP-based 1D assembling systems have been reported [129–131]. Numerous approaches have been used to build nanorods from NPs and/or atoms as precursors. They include growth in 1D templates, directional catalyst-based growth, oriented attachment, growth by selective adhesion, growth under directional external field, and seeded growth [132]. There are still numerous fundamental and practical issues that need to be addressed before using these 1D assemblies reliably in commercial devices [130]. Using binding functionalities and peptide templates, a 1D nanofiber was assembled from modified gold NPs [133]. However, the most interesting application of 1D assembling of NPs is the fabrication of self-supporting and unified nanorods.

Reduction of CuO nanowires coated with Al_2O_3 shells results in 1D NP chains [134]. A three-step process including CuO nanowire, Al_2O_3 coating, and high-temperature reduction is used (Fig. 4.16). The morphological properties of these 1D structures depend on the reduction temperature (Fig. 4.16). Cu NP dimension and the overall length of the 1D chain are controlled by the initial CuO nanowire dimensions. CuO nanowire reduction is performed in H_2 at 400, 500, and 600 °C for 1 h [134]. Higher reduction temperature gives rise to well-defined Cu NPs within the nanotubes.

3.2 Two-Dimensional Assembly

The balance between NP–substrate and NP–NP interactions is critical to the formation of well-ordered 2D NPs using the self-assembling process [135]. These interactions are affected by the type of solvent, concentration, size, and surface functionalization. Well-controlled 2D NP ordering is critical in applications such as surface-enhanced Raman scattering (SERS) [136]. In this regard, SAM could provide advantages when compared to other NP array fabrication techniques [137,15].

A 2D-structured array formed by ordered and hexagonally close-packed NPs has been studied by several groups [138,139]. Indeed, SAM obtained

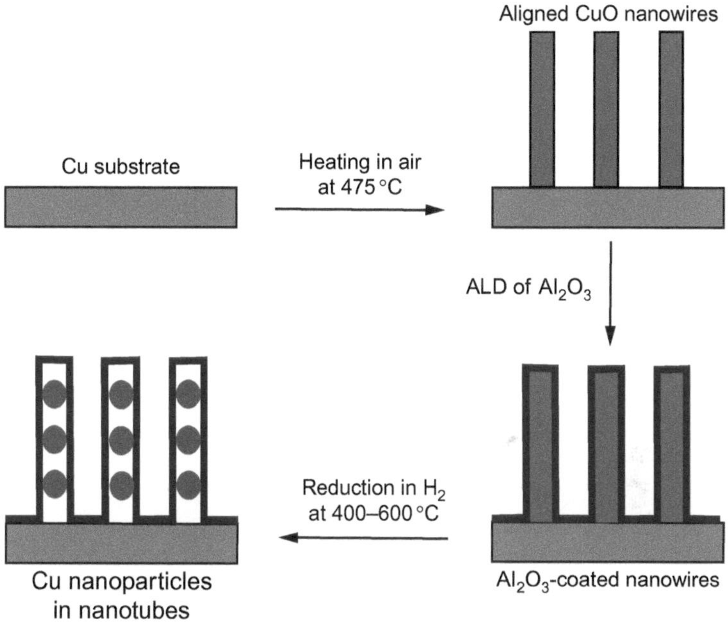

FIGURE 4.16 The three-step fabrication process of 1D structures using Cu nanoparticles as precursors. *Reprinted with permission from Qin et al. [134]. Copyright 2008, American Chemical Society.* (For color version of this figure, the reader is referred to the online version of this chapter.)

from a solution of monodispersed NPs can form close-packed arrays on a variety of substrates following evaporation of the dispersing solvent. Such long-range ordered NPs form a superlattice that is similar to the atomic ordering of bulk materials obtained following crystallization. However, long-range order remains one of the main issues of self-assembled NPs on a solid substrate. The lack of ordering is ascribed to fluid flow and solvent fluctuations during the drying process.

Long-range ordered 2D self-assembled gold NPs have been obtained by controlling solvent dewetting (Fig. 4.17). Increasing the concentration of nonvolatile dodecanethiol ligands helps to interlock NPs inside the superlattice. Furthermore, enhanced interaction between the NPs and silicon nitride substrate also helps increase the ordering range. Arrays obtained with NPs that have their surface saturated with ligands exhibit long-range ordering, when compared to arrays with NPs surrounded by an incomplete ligand shell [140].

Nonequilibrium states following the drying stages and the interparticle and particle–substrate interactions play pivotal roles in shaping 2D self-assembling. Concentration of the nonvolatile dodecanethiol ligands and the rapid dewetting of the volatile toluene solvent are minimized. This allows long-range ordering [140]. Long-range order of gold nanocrystals on silicon

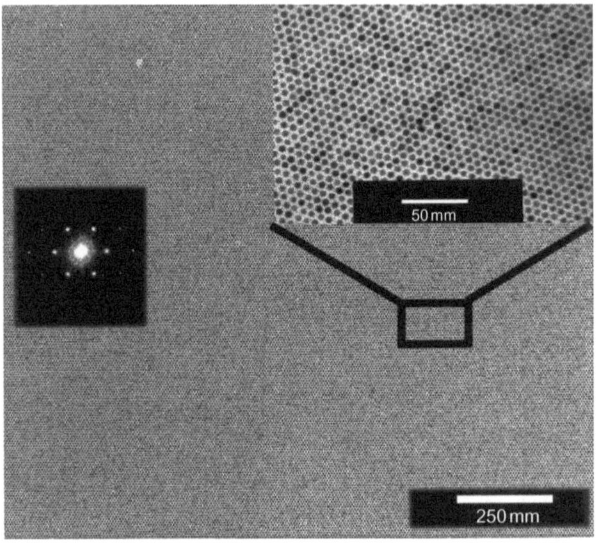

FIGURE 4.17 Long-range-ordered gold nanoparticle monolayer obtained from toluene colloidal solution. Detailed view of this 2D SAM is shown in the upper right inset. The diffraction pattern is also shown in the left side inset. *Reprinted with permission from Qin et al. [134]. Copyright 2008, American Chemical Society.*

nitride substrate over several micrometers has been reported [140]. Even longer-range order has been obtained using rapid early-stage drying [141]. Indeed, monodisperse gold NPs have been shown to self-assemble as an ordered monolayer across a 3 mm × 4 mm Si_3N_4 substrate. This is obtained using low concentration solution of monodispersed dodecanethiol-ligated gold nanocrystals in toluene.

A saturated 2D solution of NPs at the liquid–air interface is often required to obtain self-assembled NPs of long-range order [141]. For this to happen, rapid solvent evaporation is triggered. This allows high concentration of the NPs at the liquid–air interface. Under these conditions, an attractive interaction between the NPs and the liquid–air interface occurs [141].

Functionalized NP assembling on a modified substrate allows better control over the density and structure of the final SAM. Homogeneous 2D and 3D SAMs are achieved using metallic [142] and semiconducting [143] NPs. A simple two-step assembly process including surface modification followed by modified NPs has been used [144]. This approach allowed a stable monolayer of NPs [144].

Coe et al. demonstrated that spin coating of an aromatic organic host (e.g., TPD) and aliphatically (oleic acid) capped QD (e.g., PbSe) mixture in a solvent (e.g., chloroform) can give rise to long-range order hexagonally close-packed crystalline domains [145]. Solution concentration, solvent type, QD size distribution, and morphology are parameters that affect the size of the colloidal crystal domains. Some other parameters also affect the long-range order of the self-assembled NPs. Chain length of metal NP coatings has been shown to affect the ordering of 2D film [112,146]. Detailed characterizations of 2D SAM from Pd NP with diameter (d) coated with variable chain length (l) thiol molecule (C4–C16) have been reported [146]. It was concluded that 2D ordering occurs only for $1.5 < d/l < 3.8$ [146].

Crack patterns have been observed and reported [147]. This phenomenon could be quite detrimental for practical applications. Indeed, these cracks could lead to electric shorting. These cracks form on the macroscale and mesoscale levels [147]. Features of these cracks are affected by the film thickness and external fields. Substrate plays a significant role in building crack-free long-range order NP-based monolayers. Besides NP–NP and NP–substrate interactions, solvent–substrate interaction plays a significant role as reported in the case of Ag_2S NPs assembled on HOPG and MoS_2 [112].

Nonionic amphiphilic ligands have been used to form ordered 2D structures from high-concentration NPs in a variety of solvents [148]. The advantage of using nonionic ligands is related to the fact that a large concentration of nonagglomerated NPs could be stabilized in aqueous solution. Thus, only the interparticle interactions will be the dominant force, and the electric double-layer forces that change with ion/electrolyte concentrations during the solvent evaporation will not be a factor [148]. Water-dispersed thiol-terminated glycol-coated gold NPs are used to form hexagonally closed-packed 2D arrays on SiO_2 surfaces during the solvent evaporation [148].

Monodisperse gold NPs modified by alkanethiol self-assemble into ordered 2D arrays with three different types of structural organization:

a. Monodisperse NPs with two different average size distributions self-assemble into ordered bimodal arrays following an intimate mixing.
b. In the absence of an intimate mixing, size-segregated regions are occupied by hexagonal close-packed monodisperse NPs.
c. Polydisperse NPs occupy random positions in a pseudo-hexagonal lattice.

Long-range order has been obtained using bimodal NP distribution [149]. High-magnification TEM micrograph and electron diffraction pattern confirmed that ordered 2D gold NPs are achievable. Good ordering has been obtained using NPs with two distinct average sizes $R_B/R_A \approx 0.58$, where R_A and R_B correspond to the radius of large and small particles, respectively [149].

Another approach for obtaining long-range 2D ordering is to use specific surface functionalization [150]. Hydrophobic surfactants including alkanethiols improve NP (16–170 nm) self-assembling at the air–water interface. High-quality monoparticulate films have been obtained by a manual

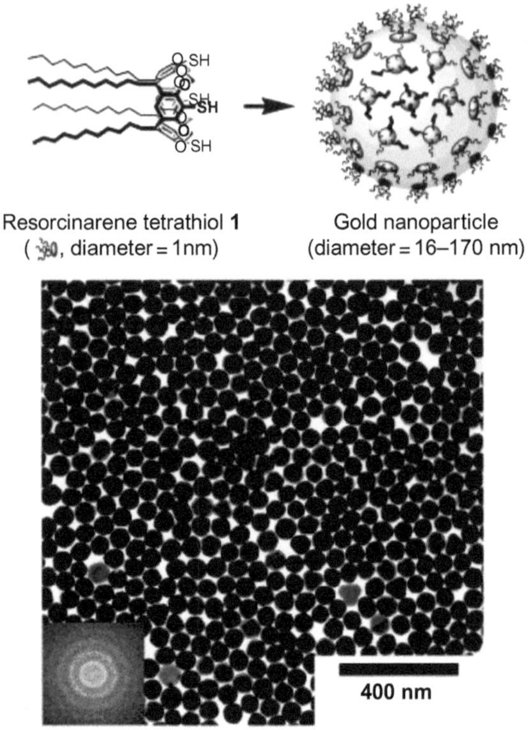

FIGURE 4.18 TEM micrograph of a 2D array of 70 nm gold nanoparticles modified by resorcinarene tetrathiol. The relative size of the surfactant (1 nm) and the gold nanoparticles are also shown *Reprinted with permission from Kim et al. [150]. Copyright 2001, American Chemical Society.*

Langmuir–Schaefer deposition technique in the absence of applied lateral compression (Fig. 4.18). It was found that the average interparticle spacing is reduced from 1 to about 0.4 nm when the particle size is increased from 20 to 110 nm, respectively [150]. These 2D structures are promising as substrates for SERS-based sensing [150]. Indeed, SERS signal has been shown to depend on the metallic NP size and the interparticle distances. With size-tunable NPs and surfactants, it is thus possible to optimize these nanoparticulate films to obtain the highest Raman signal-to-noise ratio.

2D and 3D nanocrystal superlattices have been obtained using simple evaporation techniques [151]. Ordered 2D SAM CdSe and cobalt NPs have been obtained [109]. Even more interesting is the ability to produce 3D binary superlattice NPs [152]. A binary Fe_2O_3:PbSe NP system with compatible surfactant stabilizers is used to obtain an ordered superlatice. An assembly of 11 nm Fe_2O_3 magnetic NPs (blue) with 6-nm semiconducting PbSe NPs (red) with a Fe_2O_3/PbSe ratio of 1:13 is used in this case [153].

This convective assembly process has been shown to produce well-ordered 2D monolayer and 3D multilayer systems using both micro- and NPs [116]. This technique has also been used for multilayer systems leading to a colloidal crystal. Monolayer and multilayer films have been achieved using 1.1 μm PS latex particles [116].

Other approaches producing bulk nanostructures starting from 2D layers have been reported. Controlled LBL 2D self-assembling of PS-*b*-polyferrocenyldimethylsilane block copolymer NPs on topographical templates has been reported [154]. Well-packed and ordered NP arrays have been obtained. A linear relationship between the confinement width and the number of monolayer rows is observed [154]. This template self-assembling combines top-down and bottom-up processing. It allows precise control of spatial arrangement in hybrid structure nanofabrication.

Combination of spontaneous self-assembling and continuous processes gives rise to an ordered and large-area monolayer [155]. The design setup of this coating technique is quite simple and scalable. It consists of a nozzle spraying on a horizontally moving substrate. Silica NPs with an average size of 25 nm are assembled on an octadecyltrichlorosilane-modified silicon substrate. Close-packed monolayers are obtained at high rates. Using the increasing capillary forces and controlled wetting properties of the substrate to advantage, different surface coverages ranging from 20% to 90% have been obtained [155].

3.3 Three-Dimensional Self-Assembly

Self-assembling of NPs forming bulk materials and superstructures has attracted less attention. However, a few successful attempts to produce these 3D colloidal crystals (also referred to as superlattices) have been reported. Superlattices consisting of periodically positioned NPs in 2D and 3D are achievable. Often, monodispersed NPs with potentially high van der Waals interactions are required.

DNA-based and synthetic polymer-based NP assemblies are not described in this section. Crystals of about 20 μm have been prepared using $CoPt_3$ NPs in a slow precipitation process (Fig. 4.19). 3D structures have also been reported using CdS, CdSe, and FePt NPs [156].

These 3D colloidal crystals are obtained using controlled oversaturation conditions in solutions (Fig. 4.20). CdSe crystals up to 100 μm in size have been obtained following the addition of a nonsolvent (methanol or methanol with propan-2-ol) to a toluene solution containing CdSe NPs inside a glass tube. The presence of propan-2-ol as a buffer layer allows control of the self-assembling kinetics. Fast growth (without buffer) gives rise to 3D structures with an irregular shape (Fig. 4.20B). Slow growth in the presence of a buffer layer allows perfectly shaped hexagonal platelets (Fig. 4.20C).

FIGURE 4.19 Three-dimensional images of crystals obtained by 3D SAM. *Reprinted with permission from Shevchenko et al. [156]. Copyright 2002, American Chemical Society.*

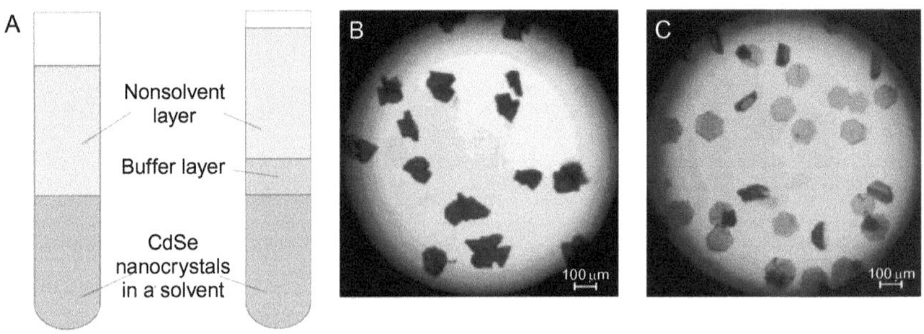

FIGURE 4.20 (A) Schematic illustration of the nucleation process allowing 3D colloidal crystals. A nonsolvent (methanol or methanol with propan-2-ol) is added to the CdSe nanoparticle toluene solution (left tube). Methanol diffuses directly into the colloidal solution. In the second case, a buffer layer of propane-2-ol slows down the diffusion of methanol (right tube); (B) and (C) show the optical micrographs of colloidal crystals CdSe [98]. (For color version of this figure, the reader is referred to the online version of this chapter.)

Well-ordered and crystalline 3D PbSe structures have been obtained using self-assembly of faceted NPs [153]. NPs with different shapes are first synthesized. Preferential growth on different facets provides opportunities to fabricate structure and shape with submicron resolution, which is difficult to obtain with conventional techniques. Indeed, 3D cubes, stars, nanowires, and nanorings are produced in this way. However, only 3D cubes give rise to dense films [153]. Deformable macrosize 3D assembly of interconnected metal NPs has also been recently reported [157].

3D self-assembling is obtained following solvent freeze and deposition of 5 nm gold NPs in toluene, 1-octanol, *n*-decane, and chloroform [158]. Structural characterization allowed comparisons of the 3D ordering that already exists in the solution and in the film. Detailed cryo-TEM showed some differences between frozen and deposited 3D assembling depending on the dispersing solvent [158]. *n*-Decane solvent provides the best overall solvency [158].

3D self-assembling of NPs could be simply obtained from a vertical growth of multiple 2D layers following a drop-cast and solvent evaporation [159]. However, it is quite difficult to obtain large-order range using this approach. There are several solutions to address this issue. One possible solution is to create specific and strong interactions between the NP and the substrate. Another option is adding an external force applied normally to the surface. Indeed, self-assembling of NPs is affected by applied electric and/or magnetic fields. When magnetic NPs are used, the intensity and the direction of the applied magnetic field directly affect the morphology of the 3D structure [160].

Two types of 3D structures are usually distinguished [161]. The first are crystalline colloidal arrays induced by interparticle (repulsive) electrostatic interactions. The second category consists of artificial opals induced by capillary and gravitational forces. Self-assembling of NPs on a prepatterned substrate allows the fabrication of mesostructure and macrostructure with nanometer-scale precision [162]. As shown in Fig. 4.21, variable film thickness was first formed by spin coating on a metal-coated silicon wafer. Particle size, NP concentration, spin-coating speed, and the number of spin-coating cycles are used to control NP film thickness.

Using photoresist (PR) and interferometric lithography (IL), patterns are defined. These patterns are obtained through RIE and PR removal. 2D patterns are obtained by self-assembling of silica NPs on silicon-based substrates [162]. SEM characterization confirmed that high-quality 3D silica-particle patterns on Si wafers can be achieved. Other structures (e.g., disc, diamond, and hole) have been obtained using a similar approach [162]. These high-quality, large-area, and controllable thickness patterns are obtained by simply combining spin-coating and post-deposition IL and RIE processes.

An alternative approach to building similar patterns is to start with a chemically patterned monolayer followed by preferential NP assembling. This approach combines bottom-up assembling with top-down patterning using ultraviolet (UV), X-ray, and electron-beam lithography. Chemical patterning is obtained by lithographic radiative techniques, including UV light, X-ray, and electron beam of a self-assembled monolayer of alkanethiolate. This approach needs improvement to overcome the high volume throughput required for its commercialization [137].

Quasicrystalline superlattices are obtained from a mixture of two different types of NPs [163]. Spontaneous formation of binary quasicrystals is

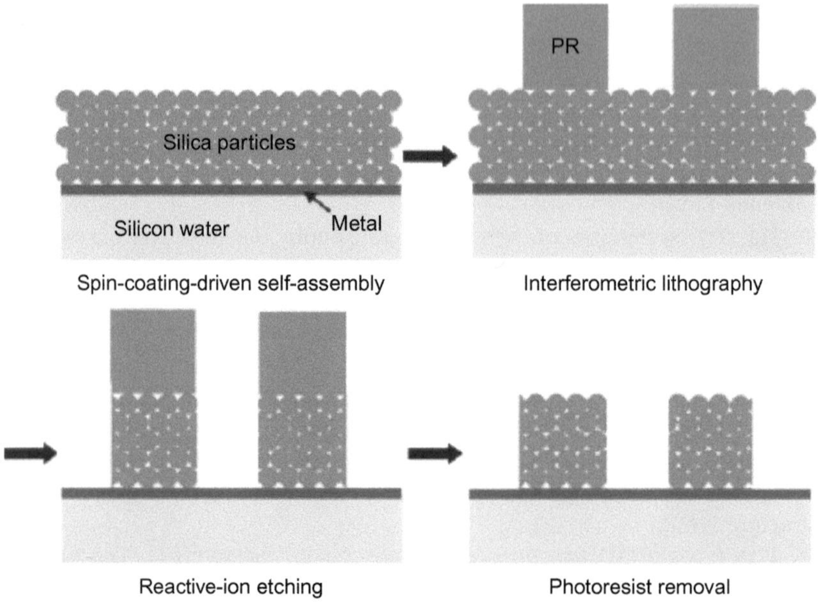

FIGURE 4.21 3D growth of a mesoscopic pattern using a combination of spin coating, interferometric lithography (IL) and reactive-ion etching (RIE) [162].

achieved from various combinations of different NPs. Indeed, by combining 13.4 nm Fe_2O_3 NPs with 5 nm gold NPs, or 9 nm PbS NPs with 3 nm Pd NPs, different scaffold structures are possible. The final structure does not depend on the NP composition [163]. Co-assembling of PbSe semiconductor QDs and Fe_2O_3 magnetic nanocrystals produces precisely ordered 3D superlattices governed by the surfactant oleic acid [152]. Different size ratios lead to AB_{13} or AB_2 superlattices with potentially tunable optical and magnetic properties [152]. Redl et al. showed that optimum AB_{13} and AB_2 superlattices have been obtained by a binary NP mixture with a diameter ratio $d_{PbSe}/d_{Fe_2O_3}$ of 0.55 [152]. The shortest distance between Fe_2O_3 NPs is about 1.8 nm.

Using surface-bound organic templates, 2D and 3D structures are obtained using NPs [164]. This method allows control over the surface density. These surface-bound organic templates were used to direct the spatial arrangement of gold NPs [164]. The 2D template is made of 3-aminopropyltriethoxysilane (APTES) with a concentration gradient along the solid substrate [164]. The 3D matrix comprises surface-anchored poly(acryl amide), whose molecular weight changes gradually [164]. SEM characterization showed clearly that the NP surface density changes as a function of the surface position following APTES surface concentration. NP coated with appropriate polymer and dispersed in the corresponding homopolymer matrix could self-assemble into specific anisotropic superstructures [165].

3.4 Layer-by-Layer

Introduced by Iler [166] and later developed by Decher et al. [167,168], the LBL deposition technique allows building a multilayer coating by depositing one layer at each step. This is accomplished through a Coulombic interaction of positively and negatively charged NPs. Sometimes, secondary interactions such as hydrogen bonding and π–π interactions may also play a role. Several groups have attempted to obtain a conformal film mostly using NPs coated with polyelectrolytes [167,168]. There are numerous applications of LBL as a high-precision manufacturing technique for fabrication of films [169]. The LBL concept can be combined with other soft-lithography techniques to provide controlled architectures and functions. Using charged and noncharged particles, molecular imprinting may allow fabrication of multilayered structures with selective patterning.

A multilayer LBL film is obtained by sequential deposition following alternate dipping of the substrate into cationic and anionic solutions. Using a modified gold substrate and citrate-stabilized gold colloids, Abdelrahman et al. showed that it is possible to build a multilayer film with precise thickness [170]. The gold substrate was first modified with 1,4-benzenedimethanethiol (BDMT) from ethanolic solution using the SAM approach. Using gold particles of 2.6 nm diameter, one-layer and multilayer NP films were built by soaking the modified substrate in a solution of NPs with a diameter of 2.6 nm (Fig. 4.22). TEM confirmed that after each step (comprising BDMT and gold colloid assembling), the film thickness increased by about 5–6 nm. An LBL film obtained by alternating modified CdS NP hydrosols and ionene (I-6-Me-BF_4) depositions has been reported [171]. Often, appropriate surface modification is needed before assembling. Improved film stability can be obtained via photo-cross-linking. This requires appropriate NP surface modification containing photo-sensitive moieties.

For some applications, 3D interpenetrating structures are required to increase the interface area [172]. As illustrated in Fig. 4.23, bi-continuous interpenetrating networks of the NP-based cathode and anode of a battery cell form a 3D interpenetrating network. Repulsive short-range (negative Hamaker constant, $A_{123} < 0$) and attractive ($A_{121} > 0$) forces allow electrochemical junction and percolating network formation, respectively. Interpenetrating anode and cathode materials form a bi-continuous percolating network with a repulsive bipolar junction (Fig. 4.23). This structure provides high-performance batteries, fuel cells, and solar cells.

With the goal of building an antireflecting and antifogging coating, Lee et al. used positively charged TiO_2 NPs (average size of about 7 nm) and

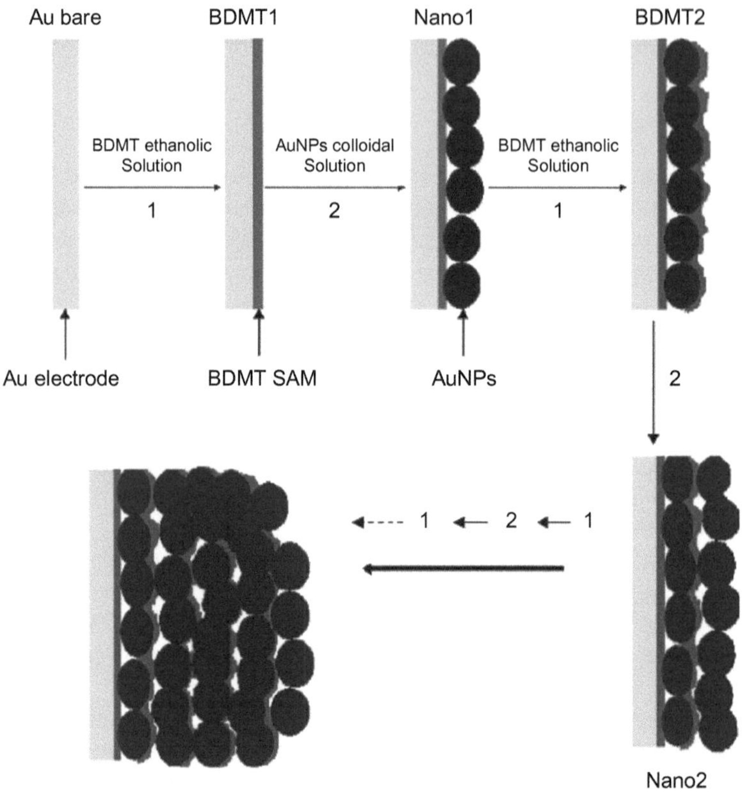

FIGURE 4.22 Simplified illustration of a multilayered nanoparticle film obtained through the combination of BDMT, SAM, and functionalized gold nanoparticles. Nano1, Nano2, ... correspond to 1, 2, ... layers, respectively. *Reprinted with permission from Abdelrahman et al. [170]. Copyright 2006, American Chemical Society.* (For color version of this figure, the reader is referred to the online version of this chapter.)

negatively charged SiO_2 NPs (average size of 22 nm) [173]. This LBL coating was obtained by using only NPs. The presence of nanoporous structures within these films provides opportunities for special optical and interfacial processes. It is possible to extend these results to other compositions and applications including magnetism and catalysis.

Dimensionally graded LBL structures are obtained using CdTe NPs with different sizes. CdTe NPs with an average size of 2–3 and 5–6 nm exhibit green (495 nm) and red luminescence (peak at 620 nm), respectively [174]. Intermediary particle sizes are used for yellow (peak at 550 nm) and orange luminescence (575 nm). NP integration forming a graded structure is obtained by depositing four bilayers of "green," "yellow," "orange," and "red" CdTe NPs [174]. Uniform and close-packed layers are obtained. Luminescence of

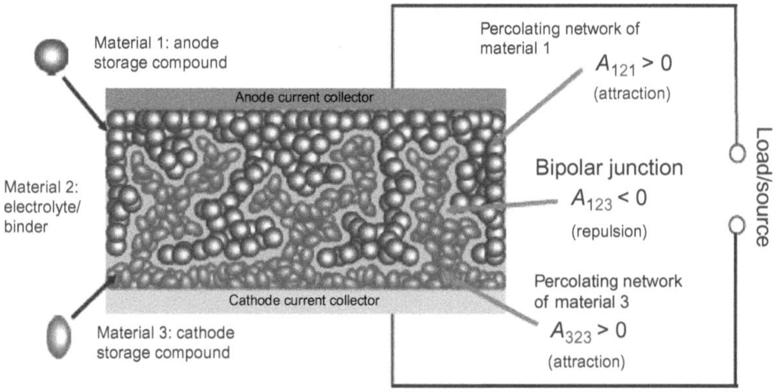

FIGURE 4.23 Schematic representation of bi-continuous percolating networks consisting of three different materials. Material 2 (electrolyte/binder) is sandwiched between two attracting materials (1 and 3) with positive Hamaker constants [172]. (For color version of this figure, the reader is referred to the online version of this chapter.)

the resulting film from large particles is broadened and red-shifted. The proposed sequential assembly provides opportunities for photonic devices.

3.5 Bio-Assembling and Bioinspired

Bioinspired materials self-assembling gives rise to a variety of structures, although the majority do not involve NPs. Spherical NPs do not facilitate assembling, although truncated NPs may. Indeed, bioinspired self-assembling often involves chiral and asymmetric structures as building blocks. Bioinspired self-assembling involves three main stages [175]. In the first stage, primary building blocks (NPs and other nanomaterials) form rod (dimer) and disc (oligomer) primary self-aggregates with dimensions of less than 100 nm. In the second stage, secondary self-aggregates are formed. These secondary self-aggregates have a wide range of structures (rod, hollow tube, ribbon, twisted tape, sheet, and helix) and length scale in the range of micrometers or less. Double-helix and fibers of more than 1 μm are typical tertiary self-aggregate structures.

Inspired by the self-assembly of the lipid membrane surrounding living cells, silver and gold NPs (with a micelle-like structure) are assembled into cylinder-shaped structures about 100–1000 nm long [176]. V-shaped amphiphiles of 2-nm diameter gold NPs stabilized by PS-b-poly(ethylene oxide) surfactants have also been synthesized [176].

Bioinspired robotic self-assembling of nanostructured materials is attracting a lot of interest [177]. There are numerous examples of naturally occurring and artificial self-assembling of bio-based nanomaterials [15,178]. Programmed array formation using the biomimetic antibody–antigen recognition approach has been achieved. Directed self-assembly of monometallic (Au) and bimetallic (Au and Ag) NPs has been obtained [15].

The highly specific recognition properties of antibodies and antigens allow programmed assembly of NPs in solution [179]. As shown in Fig. 4.24, cross-linked gold NPs are obtained by attaching anti-dinitrophenyl (DNP) IgE antibodies to the surface. On a macroscopic scale, densely packed anti-DNP IgE/Au and anti-biotin IgG/Au NPs are obtained (Fig. 4.24).

A biochemical computing concept has been recently used to build 3D SAM, taking advantage of signal-responsive NPs coupled with the enzyme-based logic gates [179]. This concept is based on pH-responsive NPs coupled with enzyme-based logic gates. The pH changes produced *in situ* by AND/OR enzyme logic gates induce protonation (low pH) and deprotonation (high pH). NPs coated with a stimuli-responsive shell layer allow the turning on and turning off of interparticle interactions. The on/off interaction leads to aggregation or dissociation, respectively. This concept has been demonstrated using 200 nm silica particles coated with poly-2-vinylpyridine (P2VP). This polymer is a weak polyelectrolyte responding

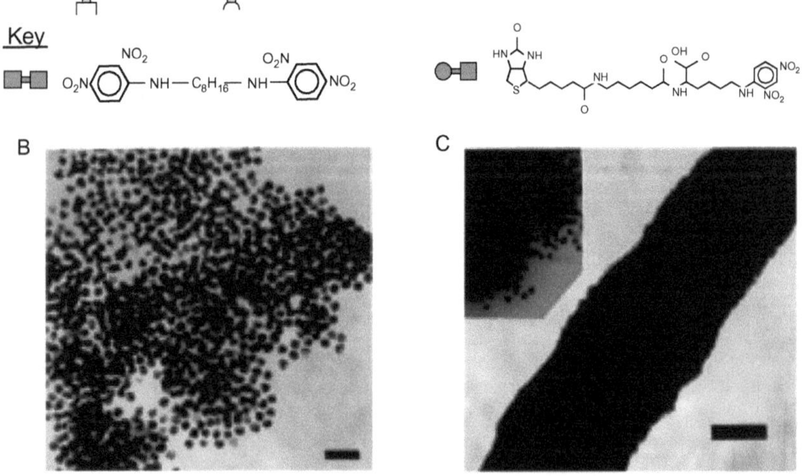

FIGURE 4.24 (A) Schematic representation of an idealized cross-linked nanoparticle structure. (B) TEM image of anti-DNP IgE/Au nanoparticles cross-linked by DNP–DNP antigen (scale bar = 50 nm). (C) Structure of a 3D filament consisting of densely packed anti-DNP IgE/Au and anti-biotin IgG/Au nanoparticles. Inset: edge of the filament (scale bar = 1 mm) [179].

to changes in the pH [180]. This is confirmed by scanning probe microscopy (SPM) images of films obtained from the respective solutions on a wafer substrate. Particle size distribution obtained using SPM and dynamic light scattering (DLS) showed the aggregation and dissociation states of the different combinations.

A genetically engineered M13 bacteriophage and ZnS NPs are used to fabricate ordered self-supporting hybrid domains. The main process steps used in this novel assembling technique include bioselectivity of peptide inserts with ZnS substrates, bacterial amplification following pH elution and repeat interaction, and liquid crystal superstructure formation following nanocrystal binding [181]. The scale of these domains ranges from the nanoscale up to the micrometer scale (72 μm) [181]. The genetic selection of a peptide recognition moiety allows continuous 3D structures over a centimeter-length scale obtained from different types of inorganic NPs.

Another bioinspired self-assembly route has been reported by McNally et al. [182]. They took advantage of the hybridization of single-stranded (ss) with double-stranded (ds) DNA and the strong affinity between the protein avidin and biotin to self-assemble gold NPs. DNA-encoded NPs are assembled on a solid support to form both symmetric and asymmetric nanostructures [183]. This stepwise surface-encoding protocol allows nanoarchitectures with tailored functions.

3.6 Magnetic Self-Assembling

Several techniques can be used to produce mono- and multilayer magnetic NPs on solid substrates. This includes dip coating, spin coating, LB deposition, and tape casting. The first two techniques are widely used and are described in this section. Other techniques are described elsewhere in this chapter. When using magnetic NPs, dipolar interactions dominate over other forces including van der Waals, Brownian, and steric forces. However, other factors such as particle size distribution and concentration, temperature, nature of solvent, and surface ligands may affect the relative importance of these different forces. Furthermore, chemical modification of the substrate and/or NPs allows controlling the quality and the structure of the NP film.

Magnetic NPs tend to self-assemble into a linear chain [184]. Other self-assembling structures have also been reported [185]. The chain-like organization of magnetic NPs occurs when the ratio of dipolar to thermal energy $\lambda = \mu^2/(4\pi\mu_0 d^3{}_{pp}kT)$ is above 4 [186]. Other groups have also described the magnetic-driven self-assembling process [186,187].

A polymer-mediated self-assembly of FePt magnetic NPs has been achieved by alternatively absorbing poly(ethylenimine) (PEI) and FePt NPs on a solid surface (Fig. 4.25). This technique allows control over thickness and dimension of the assembly.

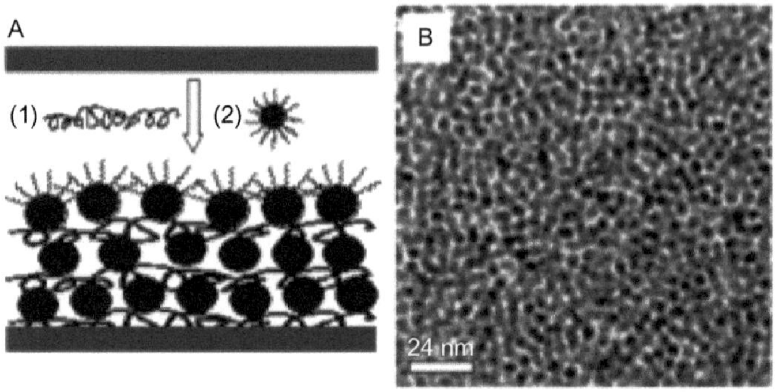

FIGURE 4.25 Schematic illustration (A) and TEM (B) of PEI-mediated self-assembled FePt nanoparticles. *Reprinted with permission from Sun et al. [188]. Copyright 2002, American Chemical Society.*

3.7 Electrostatic

Equally sized and functionalized gold and silver NPs self-assemble into a 3D diamond-like crystal [189]. In this case, gold and silver NPs are functionalized by $HS(CH_2)_{10}COOH$ (MUA) and $HS(CH_2)_{11}NMe_3{}^{3+}Cl^-$, respectively. These modified NPs are used for the formation of large, sphalerite (diamond-like) crystals. In these sphalerite crystals, each NP has four oppositely charged neighbors. Varying the gold/Ag NP ratio leads to different optical characteristics [189].

The relative importance of Brownian motion may prevent electrostatic force-based assembling. Fudouzi et al. showed that a dielectrophoretic (DEP) force of particles in a suspension is larger than disturbing forces such as Brownian movement [190]. Using 100 nm Al NPs dispersed in perfluorocarbon solvent, particle patterns are formed following electrostatic attraction between NPs and an electrified pattern [191]. This electrified pattern is obtained by focusing an e-beam on $CaTiO_3$ substrates. A simple model and simulation confirmed that DEP forces are responsible for the particle deposition [190,191]. DEP has been successfully used to assemble 50 nm PS NPs on gold nanowire templates.

3.8 Template and Confined Self-Assembly

Mono and binary colloid-guided deposition and clustering on patterned templates allow directed self-assembly [192,111]. A wide range of dimensions and complexities are obtained. Uniform aggregates could be obtained from a colloidal solution in an inclined fluidic cell (Fig. 4.26). The solution flow gives rise to three forces that could affect the structure of the aggregates. The gravitational (F_g) and the capillary (F_c) forces are often dominant. Depending on the signs of charges on the surfaces of particles and templates,

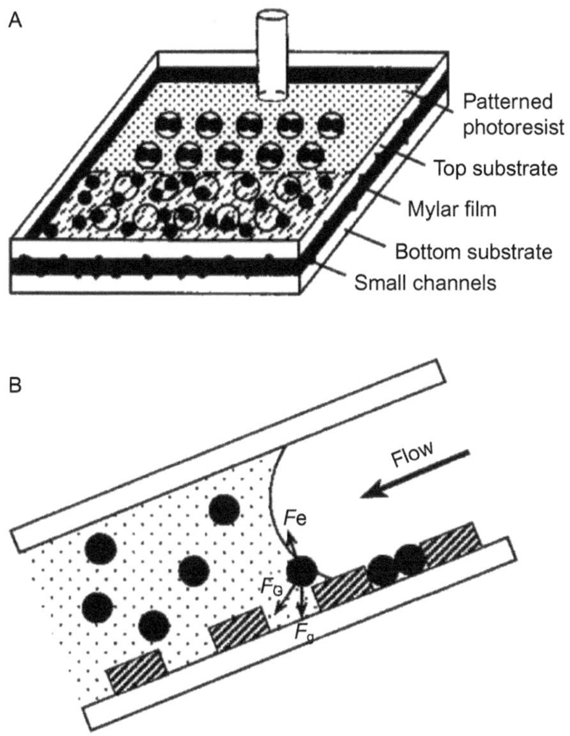

FIGURE 4.26 Schematic representation of (A) the top and (B) the cross-sectional views of the fluidic cells used to assemble colloidal particles into well-defined aggregates. Capillary (F_c), gravitational (F_g), and electrostatic (F_e) are the main forces affecting the assembling of the particles near the interface [193]. *Reprinted with permission from Yin et al. [193]. Copyright 2001, American Chemical Society.*

the electrostatic force (F_e) between the surfaces of particles might play a role. As illustrated in Fig. 4.26, and based on the surface charge sign the interparticle interactions are mostly repulsive.

Epitaxial growth of colloidal particles on a (100) fcc template created on a poly(methyl methacrylate) (PMMA)-modified gold-coated substrate allowed formation of bulk colloidal crystals [194]. Although these 3D structures are obtained using 200 nm particles, it is also feasible using sub-100 nm NPs. Micropatterned substrates are ideal structures for direct self-assembling of NPs. Using appropriate electric charges, substrates are chemically patterned using the microcontact printing process. This could result, for example, in well-defined anionic and cationic regions. Second, charged NPs are attached to the modified substrate. After a drying step, a complex ordered 2D structure is obtained under the influence of screened electrostatic and lateral capillary interactions.

Evaporative lithographic patterning is a simple assembly route allowing NP deposition without premodification of substrate surface [195]. A binary mixture consisting of high colloidal microsphere concentration and a dilute NP concentration has been used [195]. The patterned mask consists of a hexagonal holes array with defined diameter and center-to-center spacing. A specific gap height between the mask and the underlying binary colloidal suspension is also used. Following the evaporative step, dried film consisting of a periodic array of discrete NP features imbedded within a continuous colloidal network is obtained [195].

3.9 Other Forces

Besides electrical and magnetic fields, there are numerous other forces that could provide directional self-assembling [111]. These forces include capillary, flow, special confinements, and gravity. In some cases, capillary forces could play a significant role. Colloidal particles are subjected to capillary forces when the interfaces involve a liquid. This occurs in the case of liquid–gas, liquid–liquid, liquid–solid interfaces [196]. Capillary forces directed normally to the contact line are ascribed to either liquid-in-gas or gas-in-liquid capillary bridges. The lateral capillary forces are parallel to the contact line. This second group of capillary forces is due to the overlap of interfacial deformations created by separate colloidal particles [196]. Two different subclasses of lateral forces are distinguished: flotation and immersion. The deformations in the case of flotation are caused by the particle weight and buoyancy. Immersion includes three different types of deformation (infinite menisci, finite menisci, and capillary multimodes). These deformations are caused by the wetting properties of the particle surface and, in particular, the position and shape of the contact line.

4 NANOCOMPOSITE-BASED DEPOSITION TECHNIQUES

NPs are prone to agglomeration during processing, requiring particular care before and after synthesis. Nanocomposites based on NP incorporation in (or combined with) a second phase provide numerous advantages for handling and integration [197,198]. These nanocomposites can have numerous shapes and structures: surface or bulk dispersion, thin film, bulk material, porous material, agglomeration, or core-shell structure. Both the nano- and microlevel structures could provide particular properties to the final product. NPs integrated into films and sheeting materials improve the properties of the overall matrix. NPs embedded in a thermoplastic matrix are easier to integrate into devices. Integration of these nanocomposites is enabled by existing printing, coating, and shaping techniques.

Addition of nanosized reinforcements (nanofillers) to polymers instead of micro- and macrofillers may enhance resin strength and stiffness, while improving its processability and appearance. Furthermore, NP-based fillers may give rise to uniform distribution within the matrix. The majority of nanofiller products in the market are based on carbon and clay particulates [199]. NPs have been used as nanofillers to form superior composites [200,201]. Well-dispersed NPs up to 100 nm in diameter in a 3D matrix provide enhanced mechanical properties. Well-dispersed SiO_2 NPs in a PS matrix improve the glass-transition temperature [202]. NPs embedded in a polymer matrix are the most widely used nanocomposites, as they provide the possibility of combining the desirable properties of NPs and polymer matrix. Polymer-based nanocomposites have been used to enhance nanocatalyst performance, increase strength and toughness of polymers, and provide materials for electromagnetic interference shielding [198]. Usually the fabrication of nanocomposite-based films includes the following steps:

a. NP modification using surfactant, coupling agents, polymer encapsulation, and various grafting processes;
b. Formulation, mixing, and compounding;
c. Film deposition or integration into device;
d. Posttreatment.

Often, surface modification is a critical step to achieve the right formulation. Depending on the NP type and application, there is a wide range of functional and anchoring groups [203]. The choice will depend on the requirements for the anchoring group, the length (thickness) of the modifier, and the end group. In the case of silane-based groups, different functional groups are used to achieve polymerization, base/acid, nonreactive, or adhesive properties. Surface modification facilitates NP dispersion [204–206]. The rheological properties of these NPs are related to the degree of NP dispersion, which directly affects the film quality. For example, mechanically dispersed Ag NPs using a three-roller mill give rise to the highest film conductivity [207].

The mixing and compounding steps are critical. These conditioning steps will dictate not only the integration technique to be used but also the overall quality of the product. A wide variety of mixing systems are available and have been used in the past. They include agitator, colloid mill, roller and ball mill, toothed disc, high-pressure homogenizer, and kneader [204]. Both melt and solution mixings have been used successfully in the past to produce homogeneous nanocomposites [198].

Thermodynamically stable dispersion of NPs is critical [208]. Surface modifications are often the first step toward addressing this issue. Control of viscosity of the NP solution is critical for high-quality thin film. The chemical composition, molecular weight, relative concentration, and the solvent play a major role in controlling the viscosity [205]. The propensity of NPs to agglomerate during the various fabrication process steps is a critical hurdle in the quest for real-world applications. This issue has been studied and addressed in numerous cases [209]. Another issue related to the absence of cost-effective methods for producing and controlling the dispersion of the NPs in a polymeric matrix is slowing down their market development. Although there are numerous reports showing successful systems at the laboratory scale, difficulties appear when these processes are scaled up.

The majority of liquid nanocomponent synthesis processes can be grouped into three types [210]. In melt processing, polymer melt (mostly polyolefins) is used to disperse modified NPs. In casting methods, a solution containing polymer and hydrophobic NPs yields a solid nanocomposite following solvent evaporation. In the case of *in situ* polymerization, monomers are polymerized in the presence of well-dispersed NPs resulting in a nanocomposite. The choice of method depends mostly on the composition and on the requirement of the ensuing integration technique. More details are provided on the various manufacturing processes of these nanocomposites by Althues et al. [210].

Besides inorganic NPs in a polymer matrix and polymeric NPs embedded in an inorganic matrix, there are other nanocomposite types. Combining polymer NPs and inorganic NPs could provide some advantage [211]. Incorporation of NPs in a sol–gel structure has been used for homogeneous bulk dispersion [211]. Of particular interest is the possibility of integrating NPs into a conductive or semiconductive matrix. Zhou et al. [212] successfully prepared various monodisperse metal NPs (Pd, Au, and Pt) dispersed in π-conjugated poly(dithiafulvene) with strong electron-donating properties. Cross-linking of NPs involving molecules and polymers enhances the flexibility, versatility, and stability of the films [213–215].

4.1 Spin Coating and Dip Coating

Dip coating (also called free meniscus) is obtained by the immersion of a solid substrate in a solution. After removing the substrate at constant speed, the resulting film is subjected to various treatments. The properties of the coated film (e.g., thickness and density) depend on various parameters. Withdrawal rate of the substrate and heat treatment are critical parameters. Dip coating is not restricted by the shape or the complexity of the substrate. Indeed, both porous and cylindrical substrates have been used. For bulky substrates, dip coating is much easier and less expensive than vacuum-based techniques. Three main processing steps are distinguished [216]: first immersion, followed by deposition, and finally drainage. Additional drying and post-treatment steps are often required. These processing steps can be carried out in batches or continuously.

In the case of spin coating, targeted film thickness is obtained by varying concentration and rotation speed. Film thickness is proportional to the rotation speed process [217]:

$$d \approx 1/\sqrt{\omega}.$$

Drop casting followed by a controlled evaporation is the simplest fabrication process. This fabrication involves also a self-assembling step of the NPs. However, the multitude of forces involved makes control of the surface composition and structure quite difficult. A simple solution deposition followed by a controlled solvent dewetting yields long-range and ordered NP supperlattices [140]. The solution concentration of monodisperse gold NPs is a critical factor for long-range order on solid substrates. Indeed, high film coverage is obtained on silicon nitride substrates when the starting gold NP concentration in toluene solution is increased from 1.0×10^{12}, to 4.8×10^{12}, and then 1.2×10^{13} m/L [140]. Inorganic–organic hybrid nanocomposites were successfully spin-coated on a PMMA substrate [218]. The resulting coatings are dense, flexible, abrasion-resistant, and UV-absorbent. Spin coating has been successfully used to prepare large-area ordered semiconductor NP film [144]. A four-step process leading to large-area close-packed QD film has been reported [144]. Posttreatment of the spin-coated layer is critical as phase separation and low-density films may lead to defective devices. This QD film is used as an active layer in LED.

In the case of sol–gel, a wide range of film thickness can be achieved, although above a thickness of 1 μm, cracks start to appear [118]. Following deposition, the film is dried. After film shrinkage, a fluid flow is applied in the pores [118]. This step is critical as it could prevent the appearance of cracks in thick films. Supercritical drying, consisting of liquid removal from the pore above the critical temperature and pressure of the liquid, has been used [118]. Another way to increase film thickness limit is to incorporate NPs in the starting solution. Stable and monodisperse NPs are important to obtain a smooth film.

4.2 Tape Casting

Tape casting, often referred as doctor blading or knife coating, is widely used to form thin ceramic films mostly for capacitor and electronic packaging laminates. When combined with NP attributes, this technique could address current device miniaturization challenges including control of the green tape thickness. A scraping (or doctor) blade is used to remove excess deposited coating materials from a moving substrate. Film thicknesses from 5 μm to a few millimeters are currently obtained using tape casting [219]. Additional postprocessing steps are required to control film composition and properties. One clear advantage of tape casting is the possibility of using the continuous

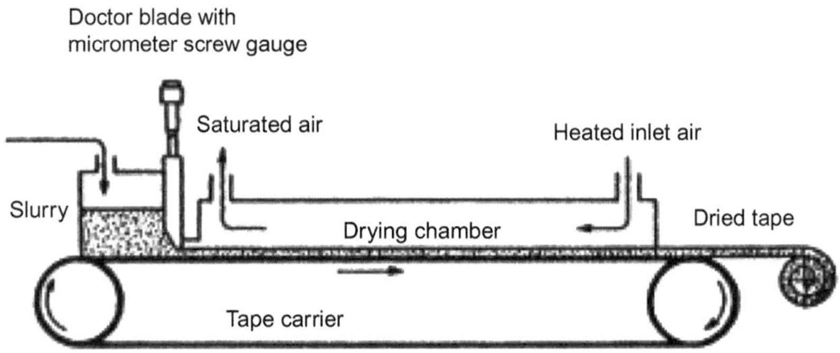

FIGURE 4.27 Simplified schematic outline of a continuous tape-casting setup [220].

process on flexible substrates (Fig. 4.27). Although tape casting is mostly based on nonaqueous solvent, aqueous-based tape-casting systems have also been used.

Higher ionic conductivity and improved mechanical properties are obtained by incorporating ceria gadolinia NPs as film in low-temperature solid oxide fuel cells (SOFC). Powder NPs (ceria gadolinia) were mixed with a solvent (butyl acetate), dispersant (sorbitane monooleate), plasticizer (polyethyleneglycol 300 and a bis(2-ethylhexyl) phthalate), and binder (polyvinylbutanal) in well-defined proportions and mixing steps [221]. The suspension was deposited onto a silicon-coated PET foil followed by a casting using a doctor blade with a slit width of 1 mm at a blade speed of 4 mm/s. After removal from the substrate foil, the casted film was heated up to 500 °C at a heating rate of 30 °C/h and held for 1 h to slowly remove all organic additives. It was then heated up at 30 °C/min to a final sintering step at 940 °C and 980 °C for 2 h [221]. Following sintering, grain size increased from 37 to about 200 nm, with >92% theoretical density.

For economical, toxicity, and environmental reasons, aqueous-based slurry for tape casting is becoming very attractive [222,223]. The aqueous-based fabrication process is quite similar to nonaqueous tape casting [222]. However, aqueous tape casting has some specific techniques such as slow drying, higher crack sensitivity, flocculation, and poor wetting. The high apparent density of the casted film requires relatively high inorganic and high plasticizer contents. Furthermore, appropriate milling of ceramic powder and dispersant in solvent is often required [222].

The relative NP content and the interparticle separation have a dramatic effect on the quality of aqueous tape-casted films [224]. Surface-to-surface separation distance (SSD) between NPs is used as the criterion to prepare crack-free NP-PVA (polyvinyl alcohol) films. Indeed, with the high Al_2O_3 NP content and small enough SSD, PVA binder is unable to bind the entire alumina NP inducing cracks within the casted film (Fig. 4.28). In spite of

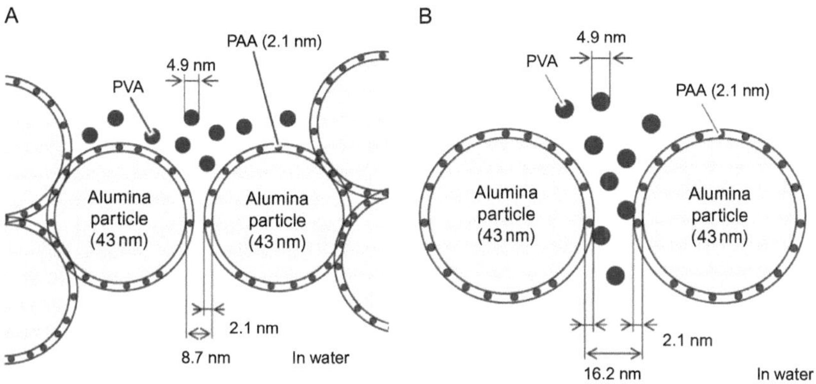

FIGURE 4.28 Schematic modelization of the effect of PAA-modified Al_2O_3 nanoparticle content on the interparticle separation [224]. (For color version of this figure, the reader is referred to the online version of this chapter.)

all the effort, aqueous tape casting remains uneconomical. Although organic solvents are more expensive, low yield and slow processing time make aqueous tape casting still more expensive.

Solvent type and dispersant are key components of tape-casting slurry [225]. Dense and smooth green lead zirconate titanate $Pb(Zr_xTi_{1-x})O_3$ tapes were obtained using xylene–ethanol as solvent and phosphate ester as dispersant. The average starting particle size is about 75 nm. The doctor-blading technique, consisting of mechanically spreading a NP-based paste within a well-defined boundary, has been widely used to prepare 5–10-μm-thick TiO_2 porous films [226]. Bodnarchuk et al. obtained square-inch assemblies consisting of ordered islands [227]. Indeed, wustite/cobalt ferrite superlattices are formed from magnetic 10–20 nm core/shell nanocrystals [227].

4.3 Screen Printing

Screen printing is used extensively in the printing and electronic industries. It's suitable for batch and continuous (roll-to-roll) processes. Screen printing of particles embedded in a matrix is often used to form relatively thick films (10–50 μm). Following paste deposition on the screen, pressure is applied to transfer the past through the screen to the substrate (printing). After completing this printing step, the screen is removed and the printed materials are allowed to settle. Often a drying step is required. Final dried film thickness is given by the relation [228,229]:

$$D = V_{\text{screen}} k_{\text{p}} (c/\rho),$$

where V_{screen} is the paste volume on the screen measured in cm^3/m^2; $k_{p,}$ the pick-out ratio; ρ, the density of the material in the final film in g/cm^3; and c, the concentration of the solid material in g/cm^3.

4.4 Inkjet Printing

Inkjet printing technology, consisting of directing fluid droplets (ink) onto a substrate, has been incorporated into printing systems at the industrial scale.

Inkjet printing is used to build 2D and 3D films and devices based on a wide variety of materials. This is obtained by moving the print head across the printing substrate or vice versa [230,231]. Four processing steps were used to build a plastic thin film transistor using one inkjet printer [232]. One conductor, one dielectric, two conductors, and one semiconductor layers are successively deposited. These processing steps allow minimization of the amount of waste, good layer-to-layer registration, high speed, and large-area coverage. Furthermore, these processing steps are amenable to a roll-to-roll (R2R) manufacturing process used for making flexible electronic devices [232,233].

The majority of today's inkjet printers are based on the drop-on-demand (DOD) methods [234]. After release from the nozzle, the ink comes in contact with the substrate. The ink is allowed time to wet the surface and the solvent to evaporate. A specific drying process is often required. Although thermal, piezoelectric, electrostatic, or acoustic processes are used to generate droplets, the majority of today's DOD printers use either thermal or piezoelectric print-heads [234]. Utilization of stable dispersion of metal NPs is quite challenging. Indeed, with large-scale requirement of high metal loading ink (20–80%), practical metal NP concentration is limited to about 10^{-2} mol/l or less [234]. However, there are a few reports where the lower loading limit has been achieved using silver NPs [235].

Submicron spatial resolution has been achieved using electrohydrodynamically induced fluid flows through fine microcapillary nozzle jet printing [236]. A 1-octanol solution of 3 nm diameter NPs is successfully printed using this method. The flow is reduced to form accurately positioned 13-μm-diameter equi-spaced droplet relics containing NPs [237]. In the case of NP-based ink, besides viscosity and surface tension, other properties may affect the quality of the printed features [238]. For example, small particle size and stability are critical as large particles may clog the printing nozzle. Formulation of the solution containing NPs is critical to the successful utilization of inkjet printing. Particle size, solvent type, and concentration seem to affect the resistivity of the metal film obtained by inkjet printing [239]. Formulation of the SnO_2 NP solution could involve several steps [240]. Ball milling and ultrasonic mixing of the $SnO_{2,}$ which includes acetylene black, dispersant, and binder additions, are the main process steps.

There are also system (size, curing, positioning, throughput, and print speed) and head (nozzle, ink reservoir) parameters that require special attention for any given NP formulation. For the printhead, two options are often used: continuous and impulse inkjet. In the case of impulse inkjet, a piezoelectric inkjet head is often used. A pressure impulse is induced within the ink chamber, expelling an ink droplet from the orifice [241]. Droplet control formation using an electric field allows fine-tuning of the geometrical features of the formed film [241]. In this case, an electrical signal is used to drive a deforming membrane using a piezoelectric crystal. Pressure impulse leads to the formation of a droplet through an orifice. Other approaches such as airbrush, electrostatic, and thermal are also used for the ink release control. Ink consists of stabilized NPs and solvent. Capping agents and other additives are used to control the rheological properties of the ink.

Recently, ink printing has been successfully used to print metallic NPs [239,242], oxide sols [243], and semiconductors [244]. An obvious advantage of inkjet printing is the possibility of surface patterning with resolution down to 20 μm [239]. Gold NPs with an average size of 5 nm capped with alkanethiolates are used to fabricate conductive lines using inkjet printing [244]. Sintering at 150 °C was sufficient to melt the NPs and remove the capping agent. Uncapped NPs in molten state allowed the formation of bulk-like gold lines albeit with a much higher processing temperature.

Laser sintering of inkjet-printed gold NPs has been used to create high-resolution metal patterning on a flexible substrate [245]. Following the inkjet deposition on the surface, a focused continuous laser is used to create a first pattern on the surface. Non-sintered NPs are easily removed from the surface. If needed, an additional feature could be built by repeating the same steps on the same substrate containing the first pattern.

A commercial drop-on-demand piezoelectric inkjet printing method has been used to produce patterned metal interconnects [246]. Annealing at 200 °C for half an hour is sufficient to obtain good electrical and structural properties [246].

4.5 Other Printing Technologies

For mass production of NP-based large-area thin film, several other printing technologies could be considered [15]. Printing technologies include gap coating, slot die coating, roll coating, and gravure coating [15]. Slot die coating (or slide coater) is probably the most promising to deliver cost-effective micron-thick films [247,248]. Indeed, slot die coating technology is currently a widely used method for the production of optical films with high precision. Slot die is usually limited by the lower thickness limit [249]. Indeed, in most cases, the lowest film thickness is about 5 μm. Other printing technologies that are receiving significant interest are microcontact techniques, microchannel printing, dip-pen lithography, and hot embossing, along with roll and gravure coating. In some cases, more than two printing processes are needed to provide high-quality NP-based films.

A TiO_2–PVA nanocomposite is often used as a light-diffusing backlit film produced by inkjet printing [250]. Slot die technology offers numerous advantages, as it allows wider, multilayer films under a single coating operation [251]. Slot die film of a nanocomposite also offers flexibility in varying parameters. In the case of NPs in PVA medium, NP type and pH along with molecular weight and degree of hydrolysis of PVA could be used to optimize the film properties [252]. Combination of the soft-lithography process and inkjet printing technique provides high-resolution patterns [253,254]. Recently, regularly patterned poly(*p*-phenylene vinylene) polymer, a semiconductor, has been reported [255]. These regular patterns could be used to obtain high-performance plastic-based high-efficiency OLED and solar cells.

Kraus et al. successfully inked 60 nm gold NPs directly onto a printing plate taking advantage of tailored adhesion forces [256]. A two-step process of high-resolution NP printing with templating grooves has been used. First, the NP dispersion fills the predefined topographical features. Then, dried particle assembly is printed from the plate onto a substrate through tailored adhesion [256]. SEM imaging showed 60-nm monodispersed gold NPs arranged in u-grooves of different widths [256]. The relative position of the NP is preserved during the transfer to the substrate from the pattern. An accuracy of 100 nm or better was achieved [256].

Dip-pen nanolithography has been used to fabricate spatial patterns of high resolution [257,258]. Dip-pen lithography could be considered a nanoscale version of bulk material deposition using inkjet printing. However, the dip-pen technique is limited to specific interactions and is a relatively slow process. Recently, Salaita et al. reported a parallel dip-pen printing technique that addresses some of these limitations [257].

NP arrangement in conventional self-assembling is stochastic [259]. In some processes, NPs are arranged deterministically on predefined structures. This leads to much higher precision in the deposited pattern with much fewer defects. [256]. Direct ink writing of colloidal inks has been used to build a 3D structure LBL [260].

4.6 Roll-to-Roll Deposition

Currently, newspaper and other media support are produced by R2R processes [228,261,262]. R2R uses a direct printing process to transfer an ink containing a functional layer from a patterned roll to a moving substrate (Fig. 4.29). A multilayer coating is also achievable using R2R [261,262]. R2R is used to produce storage batteries, flexible displays, lighting, and solar devices. Although current models predict lower cost when compared to conventional processes [261], to the best of our knowledge there is no commercial plant

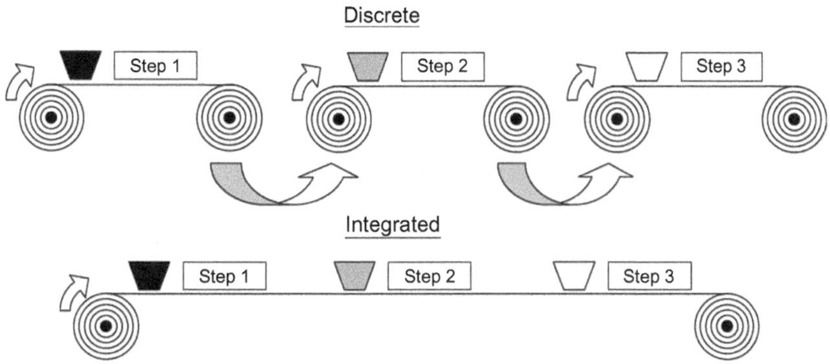

FIGURE 4.29 Schematic representation of a three-layer solution deposition using the roll-to-roll (R2R) process [228].

confirming this advantage. Contrarily to existing microelectronic processes, printed electronics are obtained by an additive process. Three R2R platforms are used [262]:

a. Mass printing: flexography, gravure, offset printing;
b. Sheet-fed printing: inkjet, screen printing;
c. Step-by-step printing: stamping, nano-imprint, UV lithography.

Although R2R is not a deposition technique per se, it could be considered as a platform for large-scale NP integration. It is particularly suitable for solution-based deposition techniques to produce coatings. Four R2R coating techniques have been commercially developed: knife-over-edge coating, slot die, gravure, and meniscus [228]. For simultaneous multilayer coating, a multi-slot feeding ink has been suggested [228].

A comparison between different coating technologies has been established in the context of polymeric film [228]. Ink waste, patterning, speed, ink formulation, wet thickness, and R2R compatibility are the main criteria used to rank 16 printing techniques [228]. Except for spin coating, all these techniques are compatible with R2R. Different printing techniques could be used to satisfy the requirements of each parameter. With a current throughput of several 10,000 m^2/h, flexographic, gravure, offset, and inkjet printing technologies are used for high-speed processes. This ranking still applies for NP-based ink formulations. NPs are easy to functionalize, providing interface properties similar to polymeric materials.

4.7 Mechanical and Thermomechanical Processing

Metal injection molding (MIM) is a process used in powder metallurgy to fabricate parts. This method of fabrication is inspired by the polymer injection process that is widely used to manufacture plastic parts and components.

Coarse particles of several microns or more are often used to fabricate theses parts. Ultrafine and spherical powders are often used to produce high-end products. Combining coarse particles with NPs could be beneficial for some applications. Film obtained from bimodal NP distribution, where small NPs are inserted at the interstice of large particles, could give rise to high-density film.

Micro-MIM of micro- and NPs within a sacrificial plastic mold is currently being evaluated for the fabrication of microstructured parts [263]. Figure 4.30 provides two possible flowcharts used to produce these microstructured parts. There is a direct relationship between the particle size and the surface roughness (R_y) of green compact materials [263]. From R_y=5–10 μm for NPs of less than 8 nm, R_y increases to 20 μm for NPs of 20 nm. This difference is even more drastic in the case of sintered parts [263]. Indeed, R_y=2.5 and 28 μm were obtained when using NPs of 8 and 20 nm, respectively.

Hot press of nanocomposites is also used to fabricate dense films [264]. A typical hot press technique consists of a base, a heating chamber (including the sample stage), and a weighing cylinder. In the case of polymer electrolytes for batteries and fuel cell, temperature and mechanical pressure allow precise control over the electrical properties [265]. Lamination has been recently used to form 3D interpenetration nanostructure from semiconductor NPs.

Sol–gel nanocomposites provide a lot of flexibility in terms of integration as films or as free-standing materials. Integration of sol–gel into films and parts involves transformation of the gel phase driven by the evaporation of the solvent. Depending on the drying process, xerogel, aerogel, film, fiber, or fine powder is obtained [266]. The bulk xerogel phase is obtained either

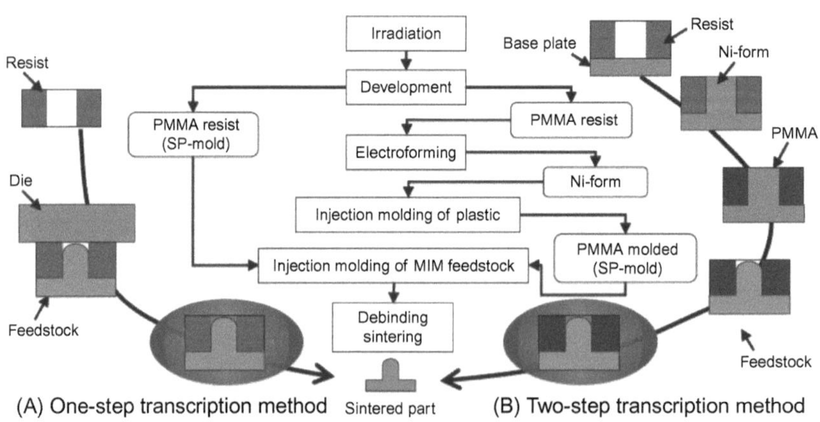

FIGURE 4.30 Two possible micro-sacrificial plastic mold insert metal injection molding (μ-SPiMIM) methods: (A) one-step transcription (B) and two-step transcription [263]. (For color version of this figure, the reader is referred to the online version of this chapter.)

directly in solution, or after the solution is spread over the surface of the solid substrate using dip coating, spin coating, or spraying. Heating the film removes the solvent and surfactants. Multilayer film is possible by repeating the deposition and heating steps. Aerogel is obtained using supercritical drying at high temperature. Fine powders are obtained under very slow evaporation of the solvent at ambient temperature. Sol–gel solutions are transformed into sheets, tubes, and fiber.

5 EXTERNAL FIELD

Various external forces and constraints have been used to assemble and improve 2D and 3D NP organizations. Indeed, unmediated assembling of NPs offers limited control over packing density and number of deposited layers. In some cases, sintering will depend not only on the composition and properties of NPs but also on external factors such as electromagnetic fields [267]. Like most electrical-based deposition techniques, EPD deposition is relatively easy to integrate into large-scale manufacturing [268].

5.1 Electrically Directed Assembly—Electrophoretic Deposition

Most electrically directed assemblies of NPs use solution-based NPs. EPD is widely used to deposit microparticles and NPs on solid substrates. Although there are a few reports of electrically directed gas-phase NP deposition onto a substrate [269,270], this chapter focuses mostly on liquid-phase EPD processes. Other electrically induced NP integration processes are also discussed later. They include electrostatic self-assembly (ESA), electroless, electrospinning, and electrodynamic techniques. Note that electrospray, involving droplet formation using high-voltage-induced shear, was discussed in Chapter 2 in the context of NP synthesis.

5.1.1 Electrophoretic Deposition

The potential of EPD for integrating particles to form high-quality film, bulk materials, and even devices was recognized following the observation of an electric field-induced movement of aqueous clay particles two centuries ago [271]. Today, EPD is used to deposit a variety of materials including borides, carbides, metals, nitrides, organics, oxides, and phosphors [272]. EPD is particularly well suited to deposit phosphorus micron and submicron particles for display applications [273]. EPD techniques allow preparation of films with thickness varying from less than 10 nm up to several 100 μm [273]. Furthermore, EPD provides the following specific advantages:

a. Rapidity and low deposition cost;
b. Scalability;

c. Not limited by the shape and size of the substrate;
d. Wide composition and thickness ranges.

Under an applied DC electric field, a colloidal solution consisting of charged particles is directed onto a conductive substrate (Fig. 4.31). The substrate could be of any shape. Particle coagulation on the substrate leads to the formation of film. EPD could be performed at either constant potential or constant current. The second approach is often preferred because at constant voltage the current decreases significantly once a thick layer is obtained. Often, posttreatment is required to remove trapped solvent and/or anneal the green film.

Electrolytic deposition (ELD), also called electroplating, has also been used to produce coatings. ELD uses ions in solution. EPD uses NPs in solution to provide a higher deposition rate, up to 1 mm/min [274]. However, one drawback of EPD is the lack of film uniformity when large and polydisperse particles are used. Both ELD and EPD are currently referred in the literature as electrodeposition techniques. In general, gravitational forces are expected to be more critical in EPD than in ELD. However, the gravitational effects are quite negligible when sub-100 nm particles are used [275].

Different EPD options are developed to suit raw NPs and final film characteristics (Table 4.3). Co-electrodeposition from baths containing both NPs and electrolytes or other additives will likely provide flexibility in composition and structure and also improve film characteristics [276]. Nanosized particles of Al_2O_3, Cr, diamond, SiC, Au, SiO_2, ZrO_2, TiO_2, PS, and Si_3N_4 with a size range of 4–800 nm have been co-eletrodeposited on conductive substrates [276]. Besides the electrolyte composition, the following parameters have been shown to have a direct effect on film characteristics [276]:

a. Applied current density;
b. Particle type and concentration;

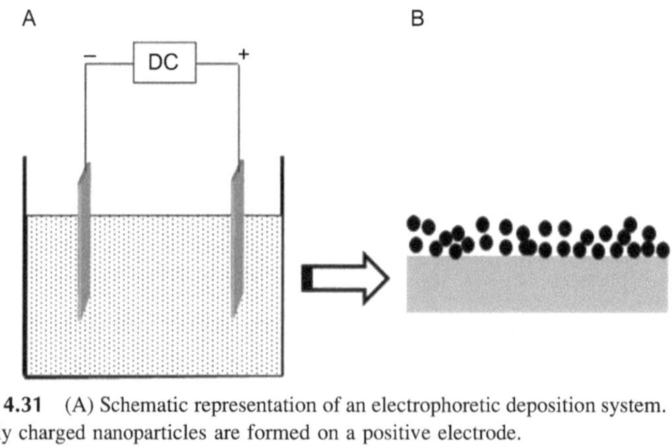

FIGURE 4.31 (A) Schematic representation of an electrophoretic deposition system. (B) Film of negatively charged nanoparticles are formed on a positive electrode.

TABLE 4.3 Different Types of Nanostructured Films and Materials Obtained by Electrodeposition from Nanoparticles and Ion Salts [276]

Methods of electrodeposition	Types of nanostructure materials			
	Nanoparticles in a metal deposit	*Nanomultilayers*	*Nanotubes/nanowires*	*Nanocrystalline materials*
Direct current (DC)	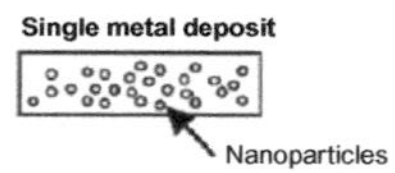	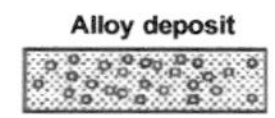	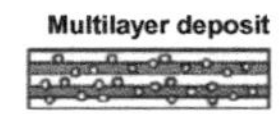	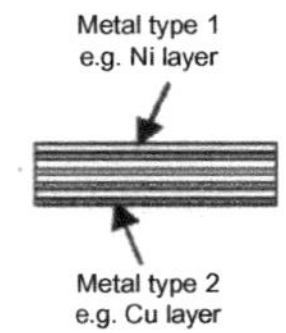
Pulsed direct current (PDC)				
Pulsed reverse current (PRC)			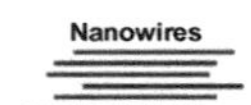	
Potentiostatic (P)				
Pulsed potentiostatic (PP)	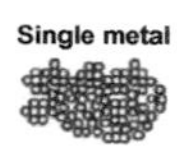		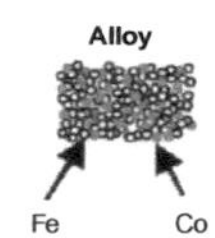	

c. Bath agitation;
d. Electrode movement;
e. pH and bath temperature.

The electrostatic surface potential (zeta potential), often referred to as ζ-potential, is a key parameter during EPD. The electrophoretic mobility is proportional to the ζ-potential for large particles. A slight deviation may occur for submicron particles. Depending on the medium, size, and shell composition, more complicated μ–ζ relationships are observed.

EPD is particularly suitable for the deposition of high-density ceramic films with thickness ranging from 1 nm to a few 100 μm [277]. The other alternative techniques such as tape casting and extrusion are used to fabricate much thicker layers. Film thicknesses up to several 100 μm and 1 mm are obtained using tape casting and extrusion, respectively [277]

High-density barium titanate ($BaTiO_3$) film was obtained by EPD from monodispersed $BaTiO_3$ NPs with an average size of 10 nm [278]. A $BaTiO_3$ concentration of 0.075 mol/l was used as the precursor material. The Pt/Ti/SiO_2/Si cathode and the stainless-steel anode plate are separated by a 2-cm gap. Control of the rheologic properties of TiO_2 NP-based suspension in acetylacetone is critical to obtaining defect-free EPD films [279]. Optimization of film quality was accomplished by varying NP concentration (0.3–2.5 wt%) and pH (1–9). It was found that good-quality TiO_2 deposition on stainless steel is achievable with a concentration of 1 wt.% and a pH of 4.

Using hybrid organic–inorganic sols including SiO_2 and ZrO_2, Castro et al. obtained a high-quality 12 μm-thick coating with excellent corrosion resistance [280]. Often, this thickness of film is required to obtain good anticorrosion properties. A high pH was used to create negative charge at the particle surface allowing film growth using EPD at constant current. Functionalized anionic and cationic polymeric NPs have been incorporated into electrolytic zinc solution to form nanocomposite coatings [281]. This electroplating solution improves the corrosion resistance of electrogalvanized steel.

Film quality seems to depend on the applied voltage and incubation time [278]. Indeed, denser $BaTiO_3$ thin films have been obtained using 5 V when compared to those obtained using 15 V. It has been shown that electrophoretically deposited films using NPs are of better quality than films obtained from microparticles [282]. Indeed, EPD films obtained from 60 nm $YBa_2Cu_3O_{7-\delta}$ particles showed much less cracking.

As shown in Fig. 4.32, the co-deposition process could be divided into five consecutive steps. First, NPs are surrounded by charge species (step #1), followed by convection toward the cathode (step #2) and diffusion through a hydrodynamic boundary layer (step #3) and finally through a concentration boundary layer (step #4). In the last step (step #5), particles are physisorbed on the substrate [276].

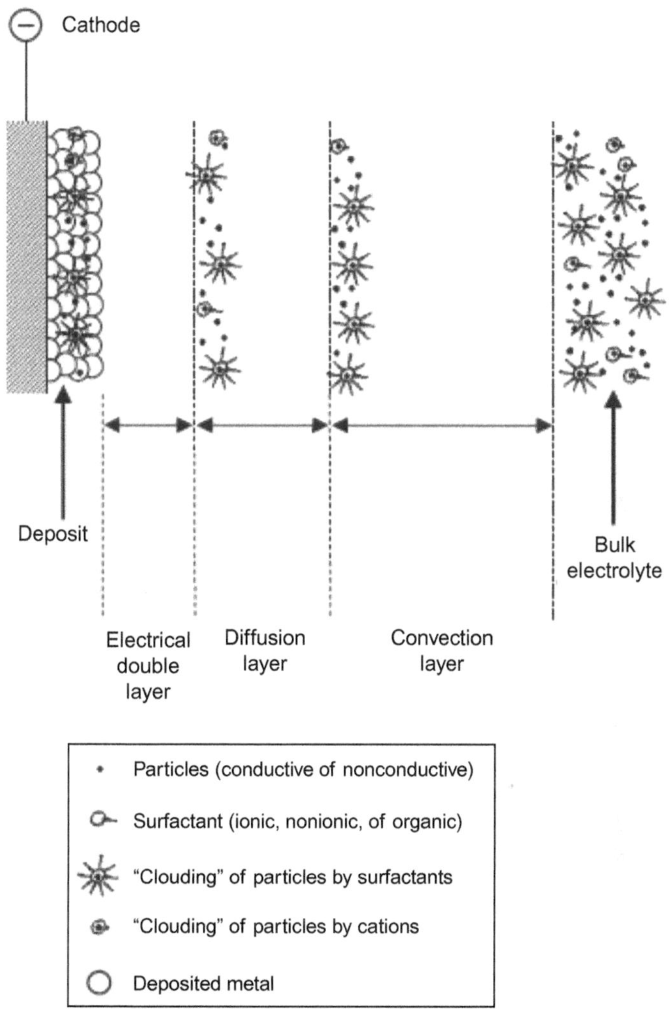

FIGURE 4.32 Schematic representation of the five processing steps used in neutral and conductive nanoparticle electrodeposition into a metal deposit: bulk electrolyte (cm long) consisting of particles and ions mixtures; a convective layer of particles (mm long); a diffusion layer (100 μm long); an electrical double layer (nm long); and a deposition layer [276].

The presence of microparticles and NPs dispersed in a metallic matrix improves corrosion and wear resistance properties. In the case of pure or mixed nickel coatings, the presence of nanocrystalline structures increases the overall hardness properties of the film. Nanocrystalline structures are controlled by adding NPs during electrodeposition. NP addition also increases the Vickers microhardness of nickel-based films [276].

EPD is also suitable for manufacturing laminated and functionally graded materials (FGM) from submicron particle suspensions [283]. FGM are obtained by judiciously adjusting the suspension composition as a function of time [283]. This setup was used to fabricate WC-Co hard metal from a mixture of WC-Co-5Ti(C,N) and WC-10Co powder suspensions. A reduction of the Ti(C,N) content toward the core of the plate is also achievable [283]. To maintain homogeneity, suspensions are stirred continuously.

Another EPD variant called occlusion electrosynthesis has been developed by Rajeshwar's group [284]. Metal-semiconductor nanocomposites were obtained by electrodeposition of TiO_2 NPs and metal hexacyanoferrate (MHCF) on gold substrate [284]. This technique may also provide the possibility of grading structure.

It is possible to produce micrometer and submicrometer patterns using EPD-like techniques [285]. Dispersed NPs using an ultrasonic bath are electrodeposited on a prepatterned surface [285]. NPs are polarized in a dielectric fluid once an electric field is applied [286]. This technique is currently evaluated for high-resolution xenography. Another procedure for surface patterning has been obtained using dielectrophoretic forces [190]. Using 100 nm Al NPs, an electrostatic image has been formed on a patterned insulating surface [190].

Under a nonuniform electric field, dielectrophoresis forces are reported even with noncharged particles. It has been estimated that the DEP forces are larger than the disturbing forces. This includes Brownian forces for particles above 1 μm. DEP provides several specific advantages [190]. Utilization of DEP for thin film deposition minimizes hydrogen-induced porosity. DEP can also be used for precise nanostructures when combined with nanofabrication techniques [287,288]. Before DEP film deposition, specific suspension formulation along with surface patterning using an electron beam and electrical charging are required [190]. After deposition, proper rinsing and drying are also needed.

NPs carrying different biomolecules are electrophoretically printed onto a polymeric substrate containing nanopatterns [289]. Referred to as electrophoretic printing lithography (EPL), this technique allows bimolecular nanopatterning at high spatial resolution. The spatial resolution is limited only by the electrode size and the NP diameter. Combining EPL with a microfluidic system could allow high-efficiency and low-cost nanoarray fabrication [289].

Higher NP content in nanocomposite film has been obtained when the substrate is aligned horizontally in the plating bath (Fig. 4.33). In this configuration, sediment co-deposition, improved microhardness, and wear resistance of the Ni-Al_2O_3 nanocomposite coating are obtained. Indeed, the mechanical properties compare well with those of Ni-Al_2O_3 obtained using the conventional electroplating technique.

EPD could be easily integrated to an R2R process. Continuous EPD processes giving rise to patterned films has also been demonstrated [268]. Using

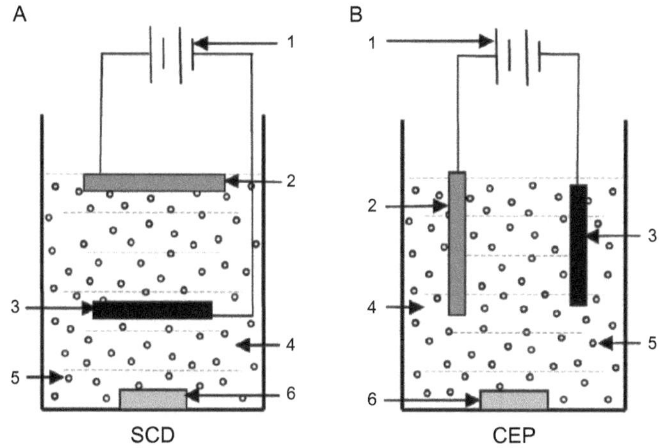

FIGURE 4.33 Schematic representation of sediment co-deposition (SCD) and conventional electroplating (CEP) using a nanoparticle-containing bath: (1) DC power; (2) Ni anode; (3) mild steel substrate; (4) electroplating solution; (5) Al_2O_3 nanoparticles; (6) magnetic stirring bar [290]. (For color version of this figure, the reader is referred to the online version of this chapter.)

a tape carrier with a multicomponent particulate allowed a patterned and uniform film thickness [268]. Using silver palladium (Ag/Pd) alloy powder (around 300 nm) dispersed in acetic acid, a uniform conductive layer is obtained. The film thickness is less than four times larger than the average particle diameter. There should not be any problem in applying this technique using sub-100 nm NPs.

Using high DC voltage, 3.2 nm diameter CdSe NPs are electrodeposited on both anode and cathode simultaneously. Both electrodes consist of a gold layer deposited on a silicon wafer precoated with Ti. The two electrodes are separated by a distance of 1.4 mm in a hexane solution containing NPs. A relatively uniform distribution over a large-area thick film (3.4 μm) has been achieved [288]. Using this technique, it is possible to obtain a robust film with micrometer spatial futures.

5.1.2 *Electrostatic Self-Assembly*

ESA bears many similarities to the LBL technique. It is quite simple and provides a lot of flexibility in terms of film composition. The ESA deposition system includes a plastic dish rotated by a stepper motor, two beakers containing polycation solutions, two beakers containing polyanion solutions, and two beakers containing water [266]. The substrate is fixed on a holder of the dipping mechanism. The stepper motor and the dipper are interfaced to the PC to allow precise control of film deposition [266]. A drying stage between each deposition step is also used. ESA provides precise thickness control within the 1 nm range. However, it may take several hours (if not days) depending on the targeted film thickness. Each layer deposition step may take around 20 min.

ESA takes advantage of opposite charges on the surface of NPs induced by cation and anion groups of electrolytes. Several polycations and polyanions have been used for ESA [266]. Furthermore, NPs having different sizes and compositions are easily incorporated following poly-ion modifications.

5.1.3 Others

Numerous other electrical techniques have been used to produce films from nanomaterials. They include electrospinning, electrodynamics, and electroless techniques. Electrospinning has been used in academia and industry to produce porous structures using mostly nanofibers [291–294]. However, a variety of polymer–NP nanocomposite solutions have also been successfully deposited by electrospinning [293,295]. PbS NPs have been successfully integrated into polymer–fiber matrices by electrospinning. First, PVP fibers containing lead ions are electrospun through a charging nozzle. Well-dispersed PbS NPs, with diameters of about 5 nm, are obtained after subjecting the deposited film to H_2S gas. Charged monodisperse aerosol NPs are deposited onto a surface patterned via electrodynamic focusing [296]. This technique does not require a conducting surface.

The electroless technique is probably the most promising technique of this group, as it does not require any external electric input. However, the presence of a chemical reducing agent in the solution is required to complete the deposition. Electroless depositions are often used to produce supported NPs on different substrates. Silver NPs with an average size of 55 nm printed on alumina substrates are used as a seed layer for subsequent electroless copper plating [297]. Combining these two techniques allows the formation of controlled conductive patterns at a relatively high rate. The coverage of deposited silver NPs on the surface of hydrogenated Si nitride (SiNx:H) was found to depend significantly on the $R = NH_3/SiH_4$ ratio [298]. The NH_3/SiH_4 ratio was also found to affect the average particle size. Three possible scenarios (Fig. 4.34) leading to NP deposition are illustrated for the different *R* values [298]:

- **(i)** Small dissolution rate and high surface density of Si precipitates (Fig. 4.34A): silver NPs are first deposited on an Si cluster surface (t_0), followed by a growth step (t_1), and then a coalescence step (t_2). Large interconnected silver nanostructures are thus formed on the Si-rich surface.
- **(ii)** Medium dissolution rate and medium surface density of Si precipitates (Fig. 4.34B): silver NPs are nucleated and formed on the Si structure initially formed at the SiNx:H surface (t_0). Additional silver NPs are developed on the Si structure, which promotes coalescence (t_1 and t_2). Aided by chemical dissolution, the coalescence effect of silver NPs is reduced with the promotion of additional seed sites.

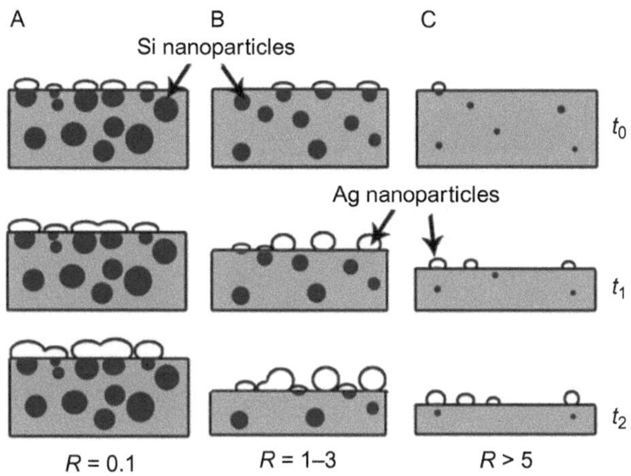

FIGURE 4.34 Schematic representation of the different kinetic scenarios of Ag nanoparticle deposition: (A) small dissolution rate, (B) medium dissolution rate, and (C) fast dissolution rate [298].

(iii) Fast dissolution rate and low surface density of Si precipitates (Fig. 4.34C): following the initial nucleation stage, only a small number of new Si seed sites appear (t_1 and t_2) due to the low volume fraction of Si nanoclusters.

Electroless deposition from a solution containing nano- and microparticles has been used to produce nanocomposite films [299]. For example, using hard inorganic particle inclusions improves the mechanical properties of a plastic-based coating. An electroless plating solution containing silver NPs was used to prepare a nanocomposite coating of Ni–P/Ag [300].

5.2 Magnetic

Precise assembling of nanostructured magnetic film is very important for several applications including data storage media [301]. Magnetic forces could be used in various setup designs to assemble NPs into different shapes. Densely packed films with long-range order are obtained from magnetic NPs. Magnetic traps with tunable size have been built and used to assemble both micron and submicron particles [185]. Magnetic-field-induced chains of cobalt NPs are also formed from a solution [301].

5.3 Mechanical—Langmuir–Blodgett

Highly dense 2D arrays of polymer-coated NPs could be obtained using LB techniques [302,303]. First, NPs are dispersed on a liquid surface in an LB trough. These NPs are adsorbed on a solid substrate upon application of

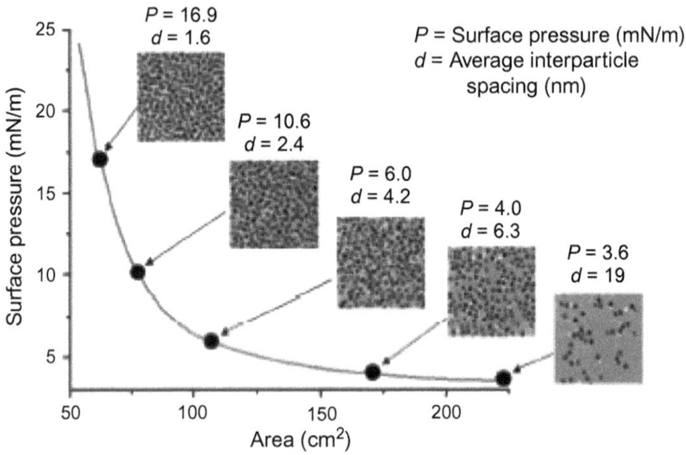

FIGURE 4.35 Evolution of the surface pressure of polymer-capped monodispersed platinum nanoparticles on an oxide surface, following a compression stage using the Langmuir–Blodgett (LB) technique [302].

pressure. The surface density (Fig. 4.35) is controlled by varying the applied surface pressure [302,303].

The LB technique allows precisely controlling the interparticle separation in a 2D plane. However, a transition toward a 3D structure is observed when stronger forces are applied [304]. Narrow NP size distribution, less volatile solvent, and shorter capping ligand length have been found to form ordered 2D phases [305]. Short-chain alkanethiolate-capped gold NPs form close-packed 2D films. Thiolated gold NPs have been assembled into high-density film over a hydrogen-terminated silicon surface using the LB technique [306]. Solvent evaporation is required before LB compression to form a domain size up to 10 μm.

A combination of microcontact printing and LB technique has been used to fabricate a micropatterned 2D monolayer film from a solution of cobalt NPs [307]. A well-organized Langmuir film of NPs is first transferred onto a patterned (PDMS) mold. Afterward, the pattern is transferred by microcontact printing onto another substrate. Guo et al. also reported patterned *micro*-dot arrays of γ-Fe_2O_3 NPs using a combination of LB and soft lithography [308]. Monodisperse γ-Fe_2O_3 NPs stabilized with oleic acid are first deposited using LB on PDMS stamps and then the patterns are transferred using microcontact printing.

5.4 Mechanical Consolidation Techniques

Powder consolidation to form ceramic films and objects is a well-established process within both academia and industry. Metal powder compaction permits achieving desired shapes and dimensions. It also allows control of the level of porosity and strength. Several compaction processes and techniques have been successfully used at the industrial scale even when submicron NPs were used. Indeed, densification of submicron particles to form high-density films and parts has been used for decades, and could even be traced back to antiquity [309]. High-temperature sintering including pressureless and pressure-assisted techniques is widely used. In the case of cold compaction, first a compaction is performed to form a green body. This is followed by high-temperature sintering. The timescale and temperature of sintering depend on the degree of interparticle physical contacts and the size of the pores in the green body.

Common consolidation methods used mostly in ceramic forming include dry and semi-dry pressing (die compaction and isostatic pressing), casting of slurry (slip and tap casting), and deformation of a plastic mass (extrusion, injection molding). Material formulation is critical [118]. Each forming method will require specific material formulation [118]. For example, in dry-pressing processes, powder or free-flowing granules are often used. Slip and tape casting require free-flowing slurry with low and high binder contents, respectively. Moist mixture of powder and binder solution precursors is used in extrusion processes. Numerous green body shapes (small, large, intricate, thin sheets, elongate) are achieved using these techniques. Depending on the application, monodisperse, biomodal, or continuous particle size distributions are used. Furthermore, various additives (solvents, dispersants, binders, and plasticizers) are used. Most of these techniques have found practical applications for microparticle-based ceramic materials. There are fewer applications based on NPs.

Nanocomposite-based consolidation techniques such as MIM extrusion have been covered elsewhere in this chapter. Consolidation includes compaction, hot pressing, and sintering and high-energy methods [310]. Other non-conventional consolidation methods that have been used include microwave sintering, field-assisted sintering, and shockwave consolidation [311]

Densification of powder using a two-step compaction process is a well-established process for ceramic materials. First, a high-density green body is formed using static pressure, and then a high-temperature sintering is applied to induce grain growth. This process does not often lead to good results when submicron particles are used. This is due to increasing interparticle repulsion and also to the relatively higher density of surface contamination. The main disadvantages of consolidation techniques include agglomeration and coarsening, which could lead to loss of nanostructure. Furthermore, increased interparticle reactivity could lead to bulk and surface contaminations. However, these disadvantages can be turned into advantages if high-density films are desired using low-temperature processing.

Consolidation of NPs into fully densified bulk components, with the retention of particle integrity or without, is of paramount technological importance. Numerous mechanical and thermomechanical processes have been used to fabricate bulk nanocomposite structures [312]. Sintering often leads to higher density although at the expense of increased grain size [313]. The onset of sintering temperature is often much lower when NPs are used [313]. In the case of WC-Co particles, this onset is reduced by more than 300 °C when the particle size is reduced from 1000 to 10 nm [313]. The variation of the relative density as a function of temperature changes significantly depending on the NP size range [313].

Another important parameter is the state of the green material before sintering. In particular, the state of aggregation and/or agglomeration is critical [313]. Temperature, pressure, size distribution, and surface contamination are the most critical parameters influencing the mechanical compaction process of NPs. This is particularly relevant in the case of NPs with high-melting points [310]. The compressibility of powder does not change significantly as a function of the particle size [310]. The difficulty in obtaining higher-density materials is ascribed to poor plasticity and large interparticle friction [310]. Higher-temperature compaction (hot pressing) seems to be able provide a possible avenue for higher-density samples. This has been demonstrated in the case of TiO_2 NPs with a grain size of 14 nm. Indeed, at 550 °C, fully dense film has been obtained [310].

Dynamic compaction involving high-pressure pulses is used to densify submicron particles [314,315]. Powder forging and high-velocity compaction provide relatively higher density when compared to conventional compaction [315]. Independent of the compaction method, three compaction stages are often distinguished. In the first stage, called transitional restacking, intimate interparticle contact is generated. The second stage is called plastic deformation. In this stage, a plastic deformation of the particles leads to a quasi-elimination of the void pores. In the third and final stage, called cold working, fixation of dislocations and/or particle attrition takes place, leading to material of even higher density. Explosive compaction and magnetic pulse compaction are two examples of dynamic compaction techniques [312–315].

Combining temperature and pressure control has led to higher density while minimizing grain size growth [316]. This could be done using hot pressing, hot isostatic pressing, or sinter forging [316]. The heating rate and the presence of gas are equally critical [317]. For example, the average grain size is reduced from around 700 to 100 nm when the heating rate is increased from 200 to 3000 °C/h [317].

Among the numerous mechanical consolidation approaches [312,318], three techniques have been used in the case of nanostructured materials:

a. Severe plastic deformation techniques that include equal-channel angular pressing;
b. High-pressure torsion;
c. Shock wave consolidation techniques.

TiO_2 NPs used in numerous industrial applications have two important phases: anatase and rutile. The anatase phase becomes rutile between 400 and 1200 °C at atmospheric pressure. It has been shown that high pressure has a drastic effect on the nucleation rate and phase transformation of TiO_2 NPs [319,320]. During sintering, grain growth of TiO_2 NPs (32 nm) has lower activation at low temperature (below 700 °C) and higher activation energy at higher temperature (above 700 °C) when compared to TiO_2 microparticles (3 μm) [321]. This could explain why nanomaterials show grain stability at high temperature before they start showing a dramatic increase in grain size.

Thermomechanical techniques are also used to overcome purely mechanical processes. Hot pressing has been successfully used to densify ceramic and metallic nanopowders [312]. The sensitivity and other characteristics of the hot-pressed materials compare quite well with the sintering process [312]. Hot isostatic pressure using high hydrostatic pressure and high temperature to consolidate nanopowders is used in numerous applications [312]. Another relevant thermomechanical technique is related to hot extrusion.

Karaman et al. have developed an equal-channel angular extrusion (ECAE) NP consolidation process. ECAE is a bottom-up technique allowing high compaction efficiency of solid NPs into bulk nanocrystalline materials (Fig. 4.36). Bimodal grain size distribution allowed higher material strength. This technique could be used at high or low temperature.

Conventional hot pressing has also been used to fabricate fully dense yttria-stabilized zirconia (Y-TZP) composites from 30 nm ZrO_2 NPs [264]. These composites contain up to 50 vol% of tungsten carbide (WC) nanosized

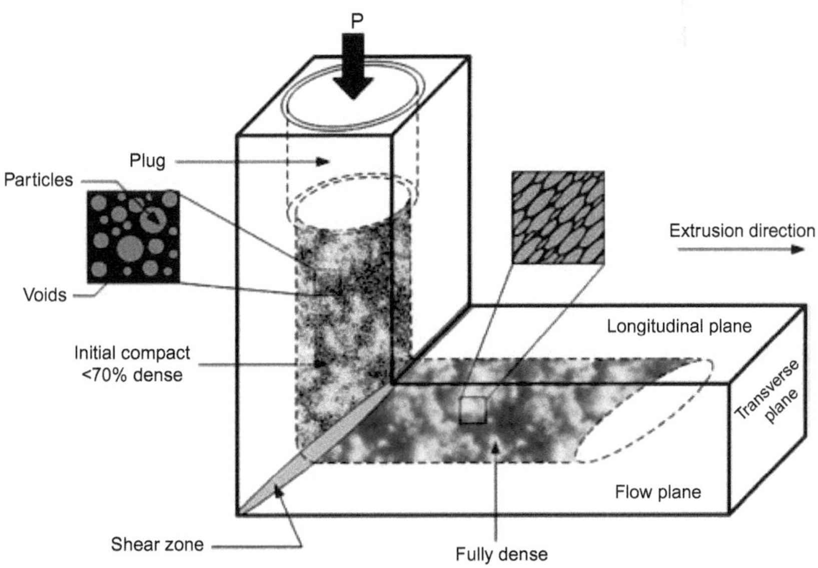

FIGURE 4.36 Schematic layout of the equal-channel angular extrusion (ECAE) nanoparticle consolidation process [322].

particles. The hardness, strength, and fracture toughness of the Y-TZP–WC prepared from nanosized WC particles are relatively higher than that of micron-sized particle-based ceramics [264].

Controlled film density is critical in newly developed dye solar cells (DSC). Indeed, high transfer electric current and large reaction interface require high-density and low-density materials, respectively. The effect of room temperature compression on photovoltaic properties of TiO_2 NP films obtained by spray deposition has been studied and optimized [323]. Low-temperature densification is required to avoid material degradation. A simple compression as the posttreatment step of a TiO_2 NP film obtained by spray deposition enhances the efficiency of DSC [323]. Indeed, a compression pressure of 15 kN/cm^2 gives rise to higher power efficiency than a pressure of 10 kN/cm^2.

Forming highly compact and dense films from dispersed NPs is highly desired in some applications. This is certainly the case with oxygen separation membranes and most thin film-based solar cells. Furthermore, as discussed elsewhere, reducing the grain size (*d*) is beneficial to the mechanical properties of polycrystalline bulk material [324]. However, it remains quite challenging to reach bulk material density using any consolidation process, given the metastable nature of the starting nanocrystalline materials. Indeed, most consolidation processes lead to extensive grain growth and/or agglomeration.

The densification mechanism of NPs depends on the size and composition. For example, densification of 11 nm MgO NPs proceeds via plastic deformation, whereas densification of 34 nm YAG NPs proceeds via nano-grain rotation aided by particle surface softening [325]. However, other parameters could have significant effects on the properties of the NP-based bulk materials. Agglomeration and aggregation of NPs is deleterious to the sintering process. For example, the sintering temperature and time usually increase with increasing agglomeration. Higher temperature leads to significant grain growth, which is not desired for some applications. Thus, nonagglomerated NPs are important for high-density films with small grains. Surfactants are often used to prevent NP agglomeration.

Higher-energy methods including shock compression have also been studied [326]. The sock compaction method leads to high-density diamond ceramic. Self-propagating high-temperature synthesis under a high pressure of 50 MPa has been used to produce dense (98%) NiAl crystallites [327].

Last but not least, lamination could provide simple and cost-effective NP-based films in functional and durable devices. It is often difficult to build such a structure using simple and cost-effective processes. Combining low-temperature and nonvacuum deposition techniques with lamination could provide a complete alternative manufacturing process for current silicon-based PV (Fig. 4.37). The process will consist of the following steps:

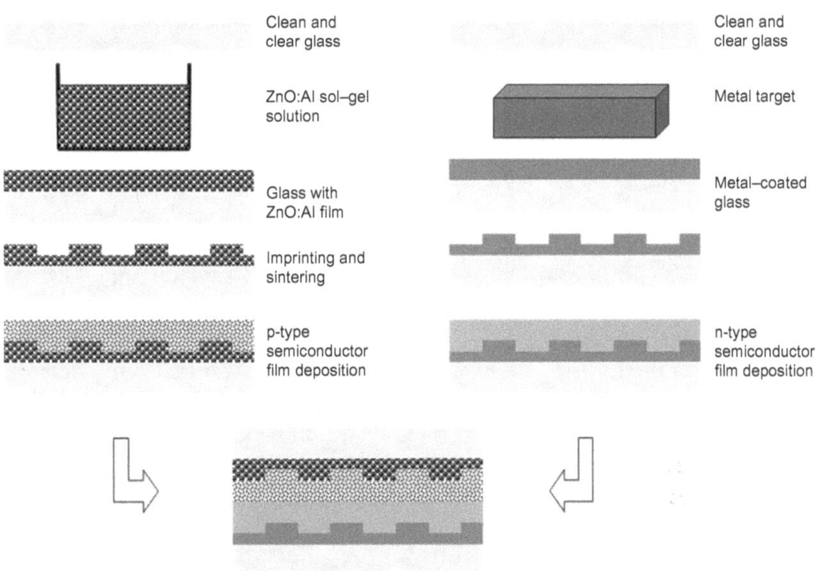

FIGURE 4.37 A nonvacuum process for photovoltaic solar cell integration assembly [328]. (For color version of this figure, the reader is referred to the online version of this chapter.)

a. Prepare two highly transparent glass sheet;
b. Prepare a sol–gel of Al-doped ZnO;
c. Apply a thin film of ZnO:Al (collector);
d. Create a patterned sol–gel film;
e. Use sol–gel film sintering;
f. A p-type semiconductor is formed on the ZnO:Al;
g. A thin metal (collector) is formed on the second glass sheet;
h. A p-type semiconductor is formed on the metal-coated glass sheet;
i. Lamination is done of the coated glass pans.

In this design [328], it is possible to fine-tune light transmission and electrical power generation by using patterned transparent and conductive sol–gel structures. Furthermore, relative alignment and distance between patterns on each electrode are critical toward optimizing light transmission and other optoelectronic properties.

5.5 Nonmechanical Consolidation Techniques

As discussed earlier, there are three challenges facing ceramic NP processing [329]:

a. Preparation of nonagglomerated NPs;
b. Preparation of crack-free green bodies following cold compaction;
c. Preparation of high-density material by sintering without grain growth.

To reach the full potential of ceramic-based NPs, often referred to as nanoceramics, alternative processes are needed in some cases [329–338]. Among these alternative processes, one could mention field-assisted sintering technique (FAST), also referred to as spark plasma sintering (SPS), and plasma-activated sintering (PAS). PAS is often referred to as pulse electric current sintering (PECS) in Japan or plasma pressure compaction (PPC) in the USA.

An increased number of interparticle contacts combined with an increased number of NPs involved leads to an increased frictional resistance to the applied pressure during compaction. A pressure in the range of about 1 GPa or more is required to obtain 51%, 62%, and 60% green densification from ZrO_2 particles (10 nm), TiO_2 (15 nm), and 3Y-TZP (10 nm) NPs, respectively [329]. In the case of SPS, uniaxial pressure and heating are applied [329].

PECS at 1600–1900 °C has been used to densify pure and 1 wt% VC-doped submicron WC particles without metallic binder addition [330]. Pure ZrO_2 nanopowder with crystallite size of 27 nm has been densified into (Nd,Y)-TZP (tetragonal ZrO_2 polycrystalline) ceramics by pressureless sintering in air at 1450 °C [331]. The resulting ceramics showed hardness and toughness properties similar to those of conventional CeO_2-stabilized TZP.

The heating rate of Ni NP sintering using FAST is critical for the process of interparticle gliding [332]. A moderate heating rate of about 210 °C was found to be beneficial for the densification of Ni NPs having a size of 50–80 nm. Silver microparticle-based paste is widely used in interconnect technology, particularly for device bonding and joining to substrate. High temperature (above 650 °C) and/or pressure (40 MPa) are required to densify deposited metal particulate paste [333]. Using NPs, a pressureless and low-temperature (250 °C) paste has been developed by Lu et al. [333]. Iron NPs are consolidated to 95% density using the PPC technique [334]. SPS has been used quite extensively to densify powders rapidly without substantial grain growth [335]. Sintering temperature during the SPS process increases the average grain above a critical temperature [335]. Grain size distribution of 20, 40, and 150 nm has been obtained on a milled Fe–C alloy process by SPS at 400, 500, and 600 °C, respectively [335].

Laser cladding has been used to densify a loose powder film on a solid substrate. A laser beam is used to obtain a crack-free coating from powder predeposited on the metallic substrate. Using laser cladding, it is possible to deposit a high-melting point alloy on a low-melting point substrate [336]. Laser-cladding approaches include two-step laser cladding and one-step laser cladding [337]. In the one-step cladding approach, one can distinguish three variants: paste laser cladding, powder injection laser cladding, and wire feeding laser cladding [337]. The laser-cladding process can be applied simultaneously with powder deposition.

Microwave sintering of a TZP composite based on submicron ZrO_2 particles showed finer and more uniform microstructures compared to the conventionally sintered samples [330]. Both sintering processes provide fully dense ZrO_2 ceramic grades co-doped with different amounts of CeO_2 and Y_2O_3. Wear resistance of low carbon steel has been improved after coating with a Co–CeO_2 composite obtained from a mixture of cobalt micropowder and CeO_2 NPs [338]. Furthermore, these films are smooth and defect-free.

6 SPRAY DEPOSITION

There are two groups of spray technologies commonly used for coating deposition: spray atomization and thermal spray [339]. Spray atomization deposition is by far the simplest and the easiest to scale in the case of NP deposition [339]. The heating plate option is used for decomposition of the precursors and/or heat treatment of the deposited materials (spray pyrolysis). Thus, spray pyrolysis consists of depositing droplets and aerosol onto a heated substrate. Aerosol-assisted chemical vapor deposition (AACVD), in which aerosols are suspensions of solid or liquid particles in a gas carrier obtained by atomization, provides an efficient process for formation of high-quality thin film.

In the case of spray atomization, one can also distinguish various subgroups. Conventional spray pyrolysis consisting of solution droplet deposition onto a heated substrate is widely used. Other aerosol deliveries of precursors onto a heated substrate have also been developed. Thermal spray methods including plasma spray and high-velocity oxygen fuel (HVOF) are receiving more attention. Plasma spray technology is now widely used to produce various coatings for SOFC [340], such as thermal barrier coatings [341]. Plasma spray can be applied on either dry powder or solution powder. Plasma spray of dry nanopowder is often problematic because of its low efficiency that often leads to clogging of the nozzle plasma gun and inconsistent flow [342].

Spray technology using a solution material precursor is the most widely used. Depending on the composition of the spray solution, there are two categories of spray processing. The spray solution could be either a suspension of powder particles in a solvent, or a solution of precursors.

Plasma spray is the most versatile thermal spray processing technique with numerous process variables. Indeed, parameters such as high energy density, high heating and cooling rate, high temperature, and high flux are used to optimize the coating properties. Depending on the type of spray deposition technique, dispersion characteristics are different. In some cases, highly concentrated dispersion (called slurry or slip) is required. Homogeneous dispersion is often required to reach a certain level of reproducibility in the quality of the films. The following methods could be used in conjunction with spray and other nanocomposite-based deposition techniques [343]:

a. High-speed stirrer obtained by rotor–stator disperser;
b. High shear mills;
c. Attritor.

In some cases, chemical aids are needed to improve the dispersibility of NPs. These NPs are either charged or modified chemically to increase interparticle repulsion.

6.1 Conventional Spray Deposition

Spray atomization and deposition is a widely used technological process [344]. Spray deposition is a relatively simple and inexpensive process [345]. It is also easy to scale up, although film uniformity may be an issue if very large surface areas are used. It is quite simple to form porous films. Dense films often require posttreatment of the green materials. High-quality CdTe [346] and $CuInSe_2$ [347] are obtained on heated SnO_2- and molybdenum-coated glass substrates, respectively.

Crack-free 7–14 μm-thick TiO_2 films were achieved by heating the transparent conductive oxide (TCO)-coated glass substrate during the spray deposition of the powder suspensions [323]. Crack-free films have been obtained after frying by using an intermittent deposition to provide enough time for the solvent to evaporate. Intermittent spray deposition has been effectively used in spray pyrolysis of precursor solutions to obtain high-quality films [348].

NP spraying could be used for the active and ohmic contact layer formation [349,350]. Solid-containing ink is atomized either pneumatically or ultrasonically to form an aerosol before deposition. A pneumatic atomizer is schematically shown in Fig. 4.38. Generated metal aerosol has been used as a direct-write technique for front contact metal printing of highly efficient solar cells [350].

Another alternative spray formation is based on electrospray. Pneumatic dispersion of preformed NPs has been electrosprayed. Subsequent and simultaneous plasma treatment of deposited NPs to remove ligand gives rise to highly adhesive particles onto the substrate [349].

Spraying techniques have been used to dry colloidal solutions leaving behind powder-like materials. This process has been also used to form precisely ordered NPs. Different scenarios for silica-particle integration on the surface of charged colloidal particles have been developed [351]. In one scenario, aggregate silica particles are formed in a micron-size droplet. In a second scenario, porous structures are formed following the attraction between the silica and PS particles. In a third scenario, hollow aggregates are produced by oppositely charged silica and PS particle attraction.

6.2 Flame Spray Deposition

High-velocity suspension flame spraying (HVSFS) has been used to produce high-density ceramic films [352,353]. The use of a piston pump or pressure vessel as suspension feeder to the HVOF torch allowed the formation of high-quality films (Fig. 4.39). Sub-20 nm alumina and titania powders are obtained using this technique.

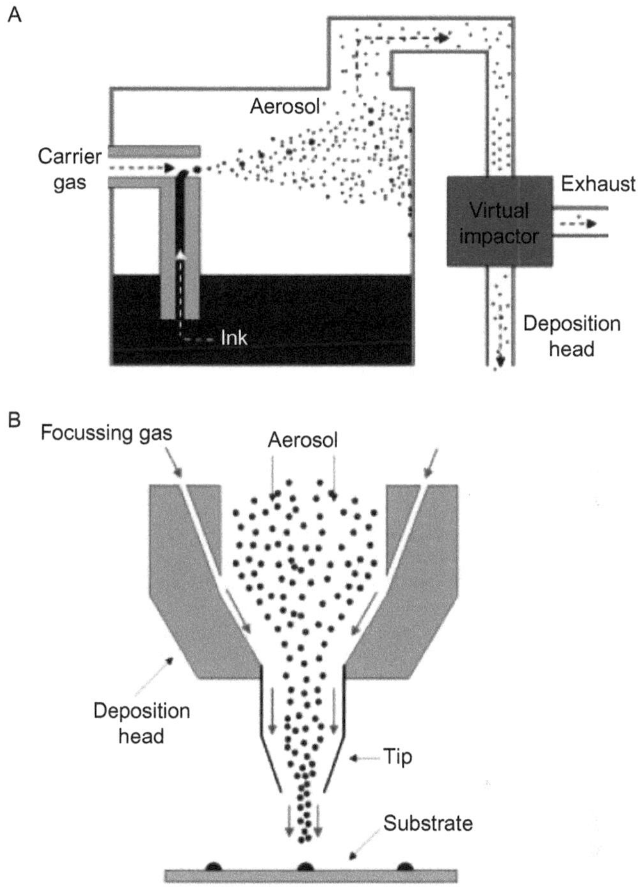

FIGURE 4.38 (A) Pneumatic atomizer and (B) deposition head [350].

Dense coatings with an improved microstructure have been achieved with several suspension compositions. These suspensions consist of an inorganic solid dispersed in an organic solvent. Characteristics of the particle suspension will have a direct effect on the coating properties. Suspensions containing titanium oxide nanopowders (n-TiO_2), chromium (III) oxide (n-Cr_2O_3), yttrium-stabilized zirconia (n-YSZ), n-hydroxyapatite (n-HAP), and glass powders have been evaluated (Table 4.4). The n-oxide suspensions give rise to dense and homogeneous oxide coatings with lower surface roughness when compared to conventional coatings. However, film adhesion is relatively lower.

6.3 Suspension Plasma Spray

It is important to distinguish suspension plasma spray (SuPS) from the solution precursor plasma spray (SuPPS) [342,354,355]. SuPS has been used

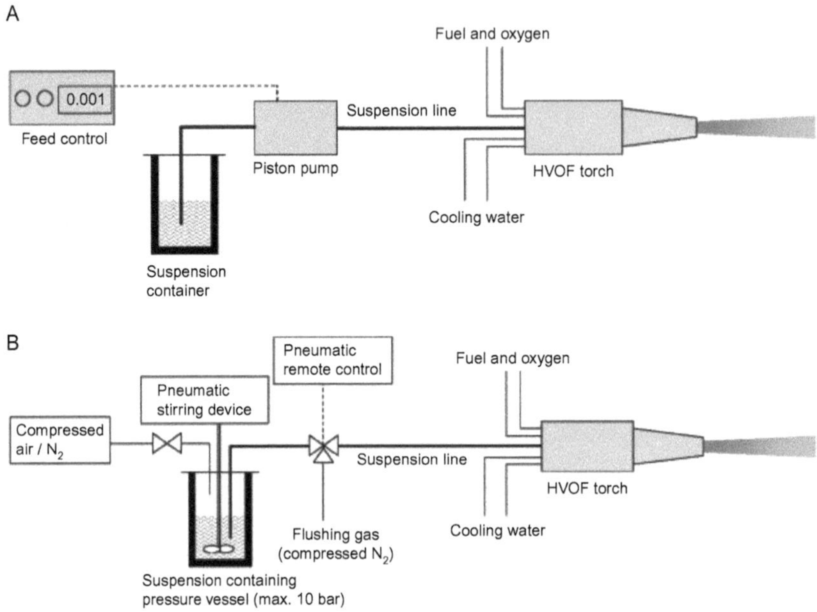

FIGURE 4.39 Simplified experimental setup for high-velocity suspension flame spraying (HVSFS) using a high-velocity oxyfuel (HVOF) torch with two different suspension feeders: (A) piston pump and (B) pressure vessel [352]. (For color version of this figure, the reader is referred to the online version of this chapter.)

with powder solutions with various particle sizes and shapes to fabricate coatings and films [354]. Except for the solution composition, SuPS and SuPPS are quite similar (Fig. 4.40). Superior thermal barrier coatings have been obtained using SuPS and SuPPS [357]. Dense and porous films are achievable with both techniques. Furthermore, SuPS and SuPPS are suitable for thicker film coatings (above 10 μm), which are difficult to achieve with other techniques.

The photocatalytic properties of TiO_2-coating films obtained by conventional plasma and liquid plasma spraying have been analyzed [355]. TiO_2 anatase slurry as feedstock instead of TiO_2 anatase powder seems to provide higher photocatalytic activity decomposition of nitrogen oxides. Integrity of the nanostructured precursor following plasma spray deposition is important for some applications [358]. Using XRD and SEM to record size distribution of the feedstock and coated film of zirconia particles, Liam et al. showed that nanostructure character is preserved when using a broad particle size distribution [358]. A combination of inkjet printing and plasma arc discharge has been used to manufacture ceramic coatings from zirconia with particle sizes of about 100 nm [359].

TABLE 4.4 Characteristics of Different Suspensions Used for HVSFS [353]

Suspension	Solid content *(wt.%)*	Primary grain size *(nm)*	Volume ratio isopropanol/water	Additives	Remarks
n-TiO_2	10	12	90/10	DOLAPIX CE64	Highest viscosity moderate processability
n-Cr_2O_3	10	80	90/10	TRUSAN 471	Strong agglomeration but good processability
n-YSZ	10	10	100/0	–	Good processability
n-$Ca_5(PO_4)_3OH$	13	100	Ethylene glycol	–	Good processability
FOA glass frit	20	2600	50/50	DOLAPIX CE64	Good processability
AW glass frit	20	2600	50/50	DOLAPIX CE64	Good processability

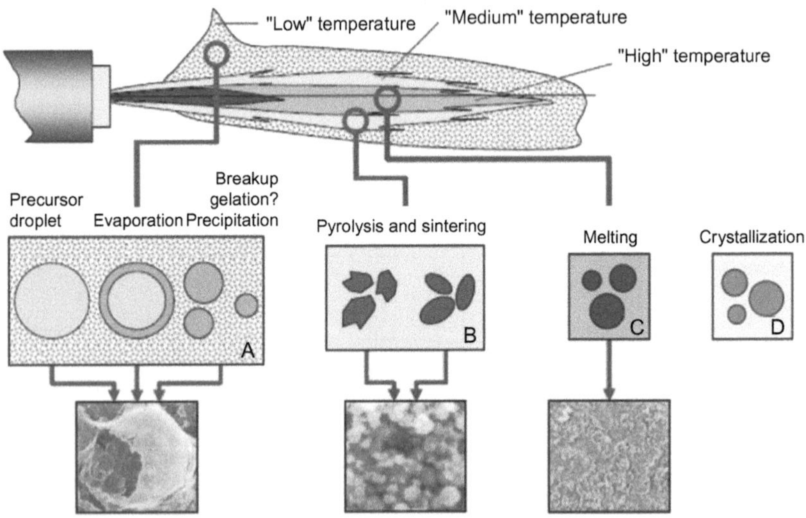

FIGURE 4.40 Schematic representation of solution precursor plasma spray (SuPPS) process. (A) The droplet evaporation and solution precipitation, (B) pyrolysis and sintering, (C) melting, and (D) crystallization steps are also indicated with the corresponding material structures [356]. (For color version of this figure, the reader is referred to the online version of this chapter.)

6.4 Electrostatic Spray-Assisted Vapor Deposition

Electrospraying has been long recognized for its ability to form films with precise control over thickness and other properties [349,360]. Various electrospray setups have been reported. Electrospray could be accomplished vertically or horizontally [345]. Electrospray deposition is a versatile technique allowing good control over deposited film characteristics [345]. In particular, dense nanostructured films are achievable.

Electrostatic spray-assisted vapor deposition (ESAVD) developed by Choy's group [361] has been used to deposit high-quality thin films from solution precursors and suspensions. As schematically shown in Fig. 4.41, the ESAVD process involves formation and deposition of atomized precursor droplets across an electric filed. These droplets undergo heterogeneous chemical reaction near a heated substrate giving rise to strongly adherent films with tunable properties. Photocatalytic activity of TiO_2 film obtained by ESAVD using titanium diisopropoxide bis(2,4-pentanedionate) and TiO_2 NPs has been reported [361]. High surface area with good photocatalytic activity is obtained using both methods.

Another approach has been developed using electrostatic and fluid forces to carry submicron-size aerosol droplets onto electrically conductive surfaces [362–365]. Liquid source misted chemical deposition (LSMCD) is a variant of the electrostatic-driven mist deposition technique [362]. Although only precursor-based solutions have been used, it is quite obvious that this system

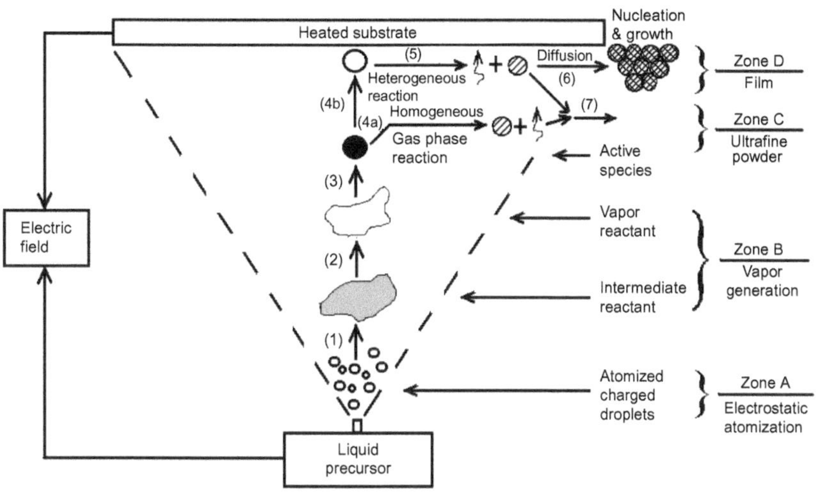

FIGURE 4.41 A schematic diagram of the ESAVD deposition process [361].

could be easily used to deposit high-quality film from NP suspensions. A mother solution consisting of NPs and/or ionic precursors can be used. This solution is first atomized and delivered through a shower head and a field screen to a wafer (Fig. 4.42). A modification of the first-generation LSMCD adding an aerosol generation has been reported [364,365]. This modification allows combining the versatility of the spin-on technique with the large-scale manufacturability of CVD [364].

6.5 Ultrasonic Pulverization

Aerosol deposition following pulverization provides numerous advantages for NP integration from solution. Typically, an aerosol is formed using an ultrasonic generator followed by film deposition [366]. Thickness and refractive index are precisely controlled using this process [366]. An excitation frequency variation has allowed controlling aerosol properties and optimizing the deposition process.

6.6 Supercritical Spray Deposition

Besides NP synthesis, supercritical fluid could also be used to spray deposit NPs (Fig. 4.43). Combined with a strong electric field, a film is obtained from a supercritical CO_2 solution containing charged NPs. Rapid expansion of supercritical solutions has been used to produce tunable NPs on solid substrates (Fig. 4.43). This is achieved by controlling the synthesized particles and electrically charging the substrates.

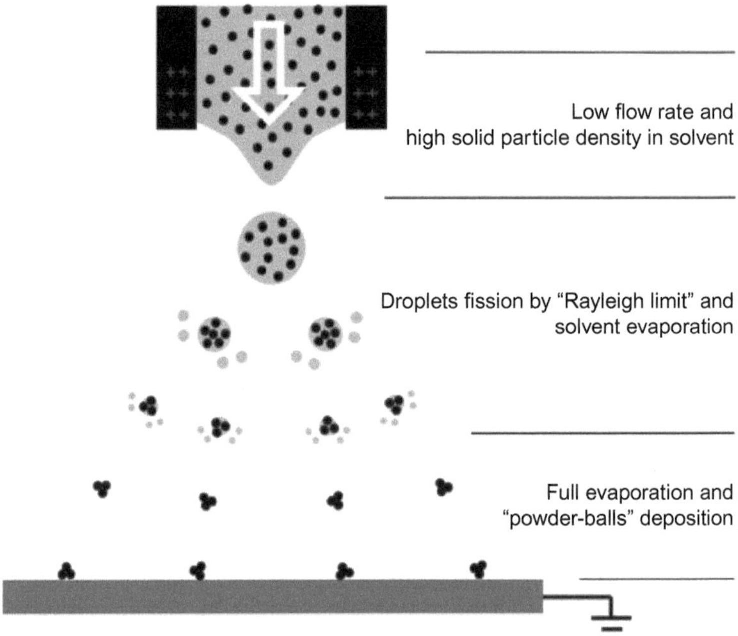

FIGURE 4.42 Schematic layout of an electrostatic slurry spray deposition (ESSD) system [362]. (For color version of this figure, the reader is referred to the online version of this chapter.)

6.7 Supersonic (Cold) Spray Deposition

To avoid high-temperature posttreatment, supersonic deposition of NPs has been developed [368]. A hypersonic plasma particle deposition technique was used to directly deposit SiC NPs onto a substrate [369]. Note that these SiC NPs are formed *in situ* in the plasma reactor. Monodispersed crystalline SiC NPs have been obtained using this technique as confirmed by TEM of SiC NPs of a scraped film deposited on the substrate [369]. This technique provides films of lower density when compared to other thermal plasma deposition techniques [369].

Hypersonic plasma is used to form the NPs directly from chemical precursors. Bulk structures are obtained following ballistic particle deposition [370]. Using this technique, films with superior mechanical properties are obtained. Needle-like tower structures have been also obtained [370]. Based on this, it is possible to manufacture MEMS-scale objects with nanometric resolution.

7 OTHERS

Numerous NP integration techniques have not been covered in the previous sections. A few additional deposition techniques are discussed in the following sections.

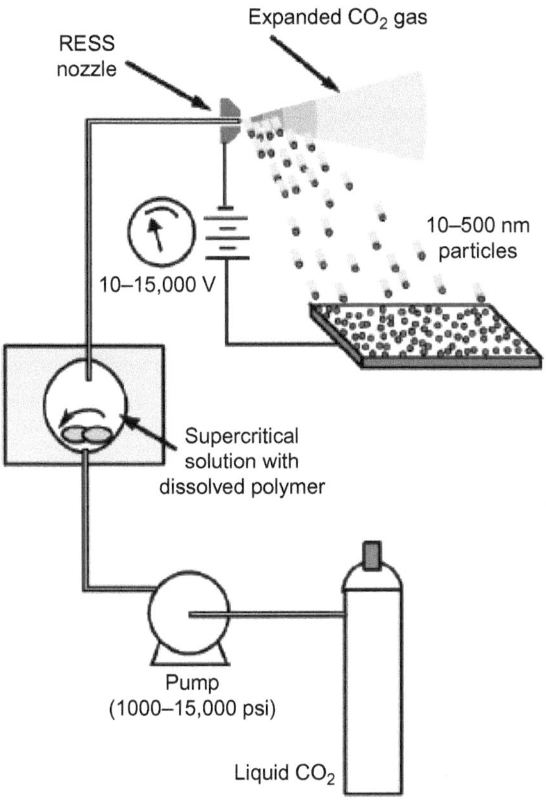

FIGURE 4.43 Experimental setup for the deposition of nanoparticles from the rapid expansion of supercritical solutions (RESS). Schematic representation of the device used for electrostatic charging particles in an RESS expansion followed by their deposition onto a substrate in a strong electric field [367].

7.1 Electrochemical Deposition

Some aspects of electrical deposition have been discussed elsewhere in this book. Although electrochemical deposition techniques have been widely used to fabricate nanostructured coatings in microelectronics and other industries [371], there are very few reports of the use of electrochemical processes to form NPs on a solid substrate. Two electrochemical approaches are used to fabricate these nanostructured films. In the first approach, called the direct approach, NPs are formed directly following a nucleation-growth process on the surface of an electrode. In the second approach, called the indirect approach, elements are first electrodeposited, and a posttreatment leads to NP formation.

Kinetically controlled growth process produces NPs with reduced size distribution when compared to diffusion control processes. The growth regime is set by the amplitude of the applied overpotential [372]. Various metallic NPs have been successfully deposited onto a highly oriented pyrolytic graphite. Low current density is used to allow slow metal particle growth [372]. Deposition time and the nature of the overpotential are shown to have direct effect on particle size distribution.

Electrochemical deposition under high overpotential has been used to fabricate large-area and high-density gold NP arrays on an Si(100) substrate [373]. SEM and XRD have been used to show relatively small gold NP size and gap distribution following a reaction time of 3 min [373]. Longer reaction time gives rise to different size and gap distributions [373]. Zinc ferrite NP deposition has been obtained using the indirect approach [374]. Following a deposition from a bath containing the salts of zinc and iron, deposited materials are converted electrochemically into oxide before undergoing crystallization to form the $ZnFe_2O_4$ structure [374].

7.2 Aerosol-Assisted Chemical Vapor Deposition

The aerosol technique is used for NP integration using two different approaches depending on whether the NPs are already formed or not. Aerosol deposition using NP precursors has been described elsewhere in this book. AACVD is often used when precursors of conventional atmospheric pressure chemical vapor deposition (APCVD) are volatile or thermally unstable [361].

Using preformed gold NPs in toluene, Parkin's group reported nanostructured gold films by AACVD [375]. They did not observe any significant change in NP size and shape following AACVD [375]. Highly durable gold-based nanocomposite films with tunable optical properties have also been obtained using AACVD [375].

7.3 Cluster Beam Deposition

Barborini et al. have described a cluster beam deposition (CBD) technique to produce and deposit NPs [376,377]. In this CBD, NPs produced by a pulsed microplasma source are introduced through an orifice into a vacuum with an aerodynamic focuser. The particle jet is introduced into a differentially pumped deposition chamber. Co-deposition of preformed NPs with atomic vapor allows wider control over the size, volume fraction, and density of the film [378].

7.4 Colloidal Structure

Assembling NPs into or onto NP or microparticle carriers is challenging in potential applications including those in the biomedical field. Availability of silica particles ranging from sizes of 20 nm or below to several microns, their biocompatibility, nontoxicity, and ease of functionalization make them the best candidates. Several architectures based on silica NPs have been developed including core silica matrix, hybrid, and silica matrix embedded with molecular and atomic species [379–383]. Such structures could provide avenues for NP integration [379–381] or even lab-on-a-particle architectures [382]. Numerous techniques are viable for the formation of such structures. These include LBL, electrostatic interactions, deposition–precipitation methods, and chemical attachment to functionalized silica NPs [383].

7.5 Solvothermal

Metals, alloys, and semiconductors are deposited over a large area with relatively high density under solvothermal conditions [384]. This approach allows *in situ* NP synthesis followed by deposition. Smooth and uniform films with tunable porosity are achievable on flat and complex surface topographies. Several NP compositions have been evaluated including Au, Ag, CdS, Fe, Co, and FePt [384]. Micropatterned gold NP films are obtained on glass substrates containing a copper grid [384].

7.6 Seeded Growth

Monodispersed NPs are obtained from much smaller seed particles. This seeded growth could also be used to produce micro- and macroscale materials [385,386]. Controlling morphology and structure is feasible by taking advantage of the faceted NP structures. Indeed, there is a correlation between the bulk-like structure and the detailed structure of the seed particles [385,386]. It is also possible to prepare hybrid structures from a mixture of different seed NPs [387].

7.7 Bioinspired Composites

Biocomposites obtained directly by mimicking naturally occurring materials may provide several advantages in terms of functional properties and sustainability [388]. Macromolecule scaffolds provide an ideal template for the growth of the silica or polysilicic acid, allowing control over the final structure. Macromolecules self-assemble to form spatially ordered domains in aqueous solutions [389]. Polysilicic acid could be added to this scaffold to provide a template synthesis. Dissolved polysilicic acid forms silica nanostructures when poly-L-lysine (PLL) is used as a scaffold [389]. The structure of the PLL macromolecules dictate the final geometrical structure of the mineral materials. External forces including electrical, magnetic, and mechanical forces are also used to control the final shape [389].

Hierarchically organized silica NPs have been obtained using diatom cells [388]. Gold NPs imbedded in biofilms have been obtained by the reaction of tetrachloroauric (III) acid, acetic acid, and chitosan under mild biomimicking conditions [390]. In this preparation, acetic acid is used as the reducing agent and chitosan as a catalyst and a stabilizing agent.

REFERENCES

[1] R. Shenhar, V.M. Rotello, Acc. Chem. Res. 36 (2003) 549–561.
[2] H. Lowen, J. Phys. Condens. Matter 13 (2001) R415–R432.
[3] A. Jaworek, J. Mater. Sci. 42 (2007) 266–297.
[4] B.J. Wiley, D. Qin, Y. Xia, ACS Nano 4 (2010) 3354–3359.
[5] R. Moreno-Atanasio, S.J. Antony, R.A. Williams, Particuology 7 (2009) 106–113.
[6] Q. Li, U. Jonas, X.S. Zhao, M. Kappl, Asia-Pac. J. Chem. Eng. 3 (2008) 255–268.
[7] D. Velegol, J. Nanophoton. 1 (2007) 1–25.
[8] J. Brant, J. Labille, J.Y. Bottero, M.R. Wiesner, In: M.R. Wiesner, J.Y. Bottero (Eds.), Environmental Technology, New York, McGraw-Hill, 2007, pp. 231–294.
[9] R. Messier, J. Nanophoton. 2 (2008) 1–21.
[10] A.F. Routh, W.B. Russel, AIChE J. 44 (1998) 2088–2098.
[11] M.A. Winnik, Curr. Opin. Colloid Interface Sci. 2 (1997) 192–199.
[12] M.A. Winnik, J. Feng, J. Coat. Technol. 68 (1996) 39–50.
[13] D.J. Lipomi, E.A. Weiss, George M. Whitesides, In: J. Garcia-Martinez (Ed.), Nanotechnology for the Energy Challenge, Wiley-VCH, Weinheim, 2009, pp. 231–279.
[14] G. Decher, In: G. Decher, J.B. Schlenoff (Eds.), Multilayer Thin Film, Wiley-VCH Weinheim, 2003, pp. 1–46.
[15] S. Maenosono, T. Okubo, Y. Yamaguchi, J. Nanopart. Res. 5 (2003) 5–15.
[16] W.J. Parak, L. Manna, F.C. Simmel, D. Gerion, P. Alivisatos, In: G. Schmid (Ed.), Nanoparticles: From Theory to Applications, Wiley-VCH, Weinheim, 2004, p. 4.
[17] S. Forster, M. Konrad, J. Mater. Chem. 13 (2003) 2671–2688.
[18] S.R. Forrest, Chem. Rev. 97 (1997) 1793–1896.
[19] N.A. Kotov, MRS Bull. 26 (2001) 992–997.
[20] V.T. Liveri, Controlled synthesis of nanoparticles in microheterogeneous systems, Springer, New York, 2006.
[21] J. Zhang, Z.-L. Wang, J. Liu, S. Chen, G. Liu, Self-Assembled Nanostructures, Kluwer Academic, New York, 2002.
[22] N.A. Kotov (Ed.), Nanoparticle Assemblies and Superstructures, Taylor & Francis, Boca Raton, FL, 2006.
[23] C. Bechinger, Curr. Opin. Colloid Interface Sci. 7 (2002) 204–209.
[24] J.D. Sherman, Proc. Natl. Acad. Sci. U.S.A. 96 (1999) 3471–3478.
[25] X.S. Zhao, X.Y. Bao, W. Guo, F.Y. Lee, Mater. Today 9 (2006) 32–39.
[26] F. Schth, Annu. Rev. Mater. Res. 35 (2005) 209–238.
[27] N.Z. Logar, V. Kaučič, Acta Chim. Slov. 53 (2006) 117–135.
[28] A. Tavolaro, E. Drioli, Adv. Mater. 11 (1999) 975–996.
[29] M.G. Warner, J.E. Hutchison, Nat. Mater. 2 (2003) 272–277.
[30] J.A. Moulijn, F. Kapteijn, A. Stankiewicz, In: A. Stankiewicz, J.A. Moulijn (Eds.), Re-engineering The Chemical Processing Plant, Marcel Dekker, New York, 2004, pp. 355–392.

[31] V. Tomasic, F. Jovic, Appl. Catal. A Gen. 311 (2006) 112–121.
[32] P.O. Thevenin, P. Goving, S.G. Järås, Cat. Tech. 7 (2002) 10–22.
[33] F. Iskandar, M.K. Okuyama, In: H.S. Nalwa (Ed.), Encyclopedia of Nanoscience and Nanotechnology, vol. 8, American Scientific Publishers, New York, 2004, pp. 259–270.
[34] J. Weitkamp, Solid State Ionics 131 (2000) 175–188.
[35] B.C. Tappan, S.A. Steiner, III E.P. Luther, Angew. Chem. Int. Ed. Engl. 49 (2010) 4544–4565.
[36] R.T.K. Baker, S.A. Stevenson, J.A. Dumesic, G.R. Raupp, S.T. Tauster, In: S.A. Stevenson, J.A. Dumesic, R.T.K. Baker, E. Ruckenstein (Eds.), Metal-Support Interactions in Catalysis, Sintering and Redispersion, Van Nostrand Reinhold, New York, 1987, p. 50.
[37] G.Q. Lu, X.S. Zhao, In: G.Q. Lu, X.S. Zhao (Eds.), Nanoporous Materials: Science and Engineering, World Scientific, Singapore, 2004, pp. 1–12.
[38] J.B. Uytterhoehoeven, Acta Phys. Chem. 24 (1978) 53–69.
[39] A. Corma, Chem. Rev. 97 (1997) 2373–2419.
[40] J.C. Van der Waal, H. Van Bekkum, J. Porous Mater. 5 (1998) 289–303.
[41] B.C. Gates, Chem. Rev. 95 (1995) 511–522.
[42] M.E. Davies, Nature 417 (2002) 813–821.
[43] A. Corma, M.J. Díaz-Cabañas, J.L. Jordá, C. Martínez, M. Moliner, Nature 443 (2006) 842–845.
[44] L. Tosheva, V.P. Valtchev, Chem. Mater. 17 (2005) 2494–2513.
[45] L. Valtchev, K.N. Bozhilov, J. Phys. Chem. B 108 (2004) 15587–15598.
[46] G.L. Drisko, A. Zelcer, V. Luca, R.A. Caruso, G. Soler-Illia, Chem. Mater. 22 (2010) 4379–4385.
[47] G. Schmidt, J. Mater. Chem. 12 (2002) 1231–1238.
[48] Y. Lei, W. Cai, G. Wilde, Prog. Mater. Sci. 52 (2007) 465–539.
[49] L.M. Bronstein, Top. Curr. Chem. 226 (2003) 55–89.
[50] F. Schuth, Annu. Rev. Mater. Res. 35 (2005) 209–238.
[51] V. Tomasic, F. Jovic, Appl. Catal. Gen. 311 (2006) 112–121.
[52] H.P. Kormanna, G. Schmida, K. Pelzerb, K. Philippot, B. Chaudret, Z. Anorg. Allg. Chem. 630 (2004) 1913–1918.
[53] S. Shingubara, J. Nanopart. Res. 5 (2003) 17–30.
[54] W. Cai, M. Tan, G. Wang, L. Zhang, Appl. Phys. Lett. 69 (1996) 2980–2982.
[55] X.M. Cai, K.W. Cheng, C.C. Oey, A.B. Djurixic, W.K. Chan, M.H. Xie, P.C. Chui, Thin Solid Films 491 (2005) 66–70.
[56] S. Altwasser, R. Glaser, A.S. Lo, P.-H. Liu, K.J. Chao, J. Weitkamp, Microporous Mesoporous Mater. 89 (2006) 109–122.
[57] J.F. Bengo, M.V. Cagnoli, N.G. Gallegos, A.M. Alvarez, L.V. Mogni, M.S. Moreno, S.G. Marchetti, Microporous Mesoporous Mater. 84 (2005) 153–160.
[58] S.K. Andersen, T. Johannessen, M. Mosleh, S. Wedel, J. Tranto, H. Livbjerg, J. Nanopart. Res. 4 (2002) 405–416.
[59] B. Sakintuna, Y. Yurum, In. Eng. Chem. Res. 44 (2005) 2893–2902.
[60] F. Beguin, E. Frackowiak, In: Y. Gogotsi (Ed.), Nanomaterials Handbook, CRC Press, Boca Raton, FL, 2006, pp. 713–737.
[61] H. Zhou, S. Zhu, M. Hibino, I. Honm, J. Power Sour. 122 (2003) 219–223.
[62] G.G. Wildgoose, C.E. Banks, R.G. Compton, Small 2 (2006) 182–193.
[63] B.M. Kim, S. Qian, H.H. Bau, Nano Lett. 5 (2005) 873–878.
[64] A. Stein, Z. Wang, M.A. Fierke, Adv. Mater. 21 (2009) 265–293.
[65] Z. Wang, N.S. Ergang, M.A. Al-Daous, A. Stein, Chem. Mater. 17 (2005) 6805–6813.

[66] R.A. Vaia, J.F. Maguire, Chem. Mater. 19 (2007) 2736–2751.
[67] M.R. Bockstaller, R.A. Mickiewicz, E.L. Thomas, Adv. Mater. 17 (2005) 1331–1349.
[68] V.I. Raman, In: Y. Gogotsi (Ed.), Nanomaterials Handbook, CRC Press, Boca Raton, FL, 2006, pp. 585–603.
[69] T. Ishida, M. Haruta, Angew. Chem. Int. Ed. Engl. 46 (2007) 7154–7156.
[70] S.A. Johnson, E.S. Brigham, P.J. Olivier, T.E. Mallouk, Chem. Mater. 9 (1997) 2448–2458.
[71] S. Rifai, C.A. Breen, D.J. Solis, T.M. Swager, Chem. Mater. 18 (2006) 21–25.
[72] C.L. Lu, B. Yang, J. Mater. Chem. 19 (2009) 2884–2901.
[73] J. Matsui, K. Akamatsu, N. Hara, D. Miyoshi, H. Nawafune, K. Tamaki, N. Sugimoto, Anal. Chem. 77 (2005) 4282–4285.
[74] R. Voicu, K. Faid, A. Farah, F. Bensebaa, R. Barjovanu, Ch. Py, Y. Tao, Langmuir 23 (2007) 5452–5458.
[75] A. Voronov, A. Kohut, W. Peukert, Langmuir 23 (2007) 360–363.
[76] P. Ulanski, J.M. Rosiak, In: H.S. Nalwa (Ed.), Polymeric Nano/Microgels, Encyclopedia of Nanoscience and Nanotechnology, vol. 8, American Scientific Publishers, New York, 2004, pp. 845–871.
[77] Q. Wang, L. Wang, M.S. Detamore, C. Berkland, Adv. Mater. 20 (2008) 236–239.
[78] N.T. Whilton, B. Berton, L. Bronstein, H. Henze, M. Antonetti, Adv. Mater. 11 (1999) 1014–1018.
[79] J.K. Oh, R. Drumright, D.J. Siegwartb, K. Matyjaszewski, Prog. Polym. Sci. 33 (2008) 448–477.
[80] S.S. Kistler, Nature 127 (1931) 741–741.
[81] D.W. Schaffer, Science 243 (1989) 1023–1027.
[82] M.L. Anderson, R.M. Stroud, C.A. Morris, C.I. Merzbacher, D.R. Rolison, Adv. Eng. Mater. 2 (2000) 481–488.
[83] C. Moreno-Castilla, F.J. Maldonado-Hodar, Carbon 43 (2005) 455–465.
[84] D.R. Rolison, Science 299 (2003) 1698–1701.
[85] O.D. Velev, S. Gupta, Adv. Mater. 21 (2009) 1897–1905.
[86] J.L. Mohanan, I.U. Arachchige, S.L. Brock, Science 307 (2005) 397–400.
[87] L.C. Cotet, V. Danciu, V. Coveanu, I.C. Popescu, A. Roig, E. Molins, Rev. Roumaine Chim. 52 (2007) 1077–1081.
[88] T. Gacoin, K. Lahlil, P. Larregaray, J.-P. Boilot, J. Phys. Chem. B 105 (2001) 10228–10235.
[89] R.M. Crooks, M. Zhao, L. Sun, V. Chechik, L.K. Yeung, Acc. Chem. Res. 34 (2001) 181–190.
[90] O.D. Velev, A.M. Lenhoff, Curr. Opin. Colloid Interface Sci. 5 (2000) 56–63.
[91] S.G. Romanov, A.S. Susha, C.M. Sotomayor Torres, Z. Liang, F. Caruso, J. Appl. Phys. 97 (2005) 086103–086105.
[92] H.M. Yates, et al., Adv. Funct. Mater. 15 (2005) 411–417.
[93] A. Yu, F. Meiser, T. Cassagneau, F. Caruso, Nano Lett. 4 (2004) 177–181.
[94] Eun Sik Kwak, Wonmok Lee, Nam-Gyu Park, Junkyung Kim, Hyunjung Lee, Adv. Funct. Mater. 19 (2009) 1093–1099.
[95] Y.W. Chung, I.C. Leu, J.H. Lee, M.H. Hon, J. Alloys Compd. 433 (2007) 345–351.
[96] S. Howorka, J. Mater. Chem. 17 (2007) 2049–2053.
[97] R.A. McMillan, Nat. Mater. 1 (2002) 247–253.
[98] D.V. Talapin, E.V. Shevchenko, A. Kornowski, N. Gaponik, M. Haase, A.L. Rogach, H. Weller, Adv. Mater. 13 (2001) 1868–1871.
[99] R.A. McMillan, J. Howard, N.J. Zaluzec, H.K. Kagawa, R. Mogul, Y.-F. Li, C.D. Paavola, J.D. Trent, J. Am. Chem. Soc. 127 (2005) 2800–2801.

[100] H. Yoshimura, Colloids Surf. A Physicochem. Eng. Asp. 282–283 (2006) 464–470.
[101] A. Navrotsky, Proc. Natl Acad. Sci. U.S.A. 101 (2004) 12096–12101.
[102] Y. Shin, G.J. Exarhos, Cellulose 14 (2007) 269–279.
[103] S.A. Davis, S.L. Burkett, N.H. Mendelson, S. Mann, Nature 385 (1997) 420–423.
[104] S. Förster, T. Plantenberg, Angew. Chem. Int. Ed. Engl. 41 (2002) 688–714.
[105] S. Kronholz, S. Rathgeber, S. Karthäuser, H. Kohlstedt, S. Clemens, T. Schneller, Adv. Funct. Mater. 16 (2006) 2346–2354.
[106] K. Liu, H. Fu, K. Shi, F. Xiao, L. Jing, B. Xin, J. Phys. Chem. B 109 (2005) 18719–18722.
[107] G.-R. Yi, J.H. Moon, V.N. Manoharan, D.J. Pine, S.M. Yang, J. Am. Chem. Soc. 124 (2002) 13354–13355.
[108] J. Zhang, Z. Sun, B. Yang, Curr. Opin. Colloid Interface Sci. 14 (2009) 103–114.
[109] C.B. Murray, C.R. Kagan, M.G. Bawendi, Annu. Rev. Mater. Sci. 30 (2000) 545–610.
[110] Y. Min, M. Akbulut, K. Kristiansen, Y. Golan, J. Israelachvili, Nat. Mater. 7 (2008) 527–538.
[111] Y.S. Lee, Self-Assembly and Nanotechnology: A Force Balance Approach, Wiley, New Jersey, 2008, pp. 149–170.
[112] L. Motte, A. Courty, A.-T. Ngo, I. Lisiecki, M.-P. Pileni, In: M.-P. Pileni (Ed.), Nanocrystals Forming Mesoscopic Structures, Wiley-VCH, Weinheim, 2005, pp. 1–47.
[113] M. Akbulut, A.R.G. Alig, Y. Min, N. Belman, M. Reynolds, Y. Golan, J. Israelachvili, Langmuir 23 (2007) 3961–3969.
[114] G.S. Lazarov, N.D. Denkov, O.D. Velev, P.A. Kralchevsky, K. Nagayama, J. Chem. Soc. Faraday Trans. 90 (1994) 2077–2083.
[115] A.R.G. Alig, M. Akbulut, Y. Golan, J. Israelachvili, Adv. Funct. Mater. 16 (2006) 2127–2134.
[116] B.G. Prevo, D.M. Kuncicky, O.D. Velev, Colloids Surf. A Physicochem. Eng. Asp. 311 (2007) 2–10.
[117] R.W. Carpick, N. Agrait, D.F. Ogletree, M. Salmeron, Langmuir 12 (1996) 3334–3340.
[118] M.N. Rahaman, Ceramic Processing and Sintering, second ed, Marcel Dekker, New York, 2003, pp. 181–247.
[119] J. Kolny, A. Kornowski, H. Weller, Nano Lett. 2 (2002) 361–364.
[120] T.H. Galow, A.K. Boal, V.N. Rotello, Adv. Mater. 12 (2000) 576–579.
[121] N. Hunt, R. Jardine, P. Bartlett, Phys. Rev. E 62 (2000) 900–913 and references therein.
[122] Z.H. Nie, W. Li, M. Seo, S.Q. Xu, E.J. Kumacheva, J. Am. Chem. Soc. 128 (2006) 9408–9412.
[123] S.J. Park, A.A. Lazarides, C.A. Mirkin, R.L. Letsinger, Angew. Chem. Int. Ed. Engl. 40 (2001) 2909–2912.
[124] Y. Lin, A. Boker, J.B. He, K. Sill, H.Q. Xiang, C. Abetz, X.F. Li, J. Wang, T. Emrick, S. Long, Q. Wang, A. Balazs, T.P. Russell, Nature 434 (2005) 55–59.
[125] S.C. Glotzer, M.A. Horsch, C.R. Iacovella, Z. Zhang, E.R. Chan, X. Zhang, Curr. Opin. Colloid Interface Sci. 10 (2005) 287–295.
[126] A. van Blaarderen, MRS Bull. (2004) 85–87.
[127] G. Schmid, U. Simon, Chem. Commun. (2005) 697–710.
[128] A.L. Rogach, D.V. Talapin, E.V. Shevchenko, A. Kornowski, M. Haase, H. Weller, Adv. Funct. Mater. 12 (2002) 653–664.
[129] H. Hofmeister, In: A. Vaseashta, D. Dimova-Malinovska, J.M. Marshall (Eds.), Nanostructured and Advanced Materials for Applications in Sensor, Optoelectronic and Photovoltaic Technology, Nato Science Series, vol. 240, Springer, Netherlands, 2004, pp. 131–134.
[130] Z. Tang, N.A. Kotov, Adv. Mater. 17 (2005) 951–962.

[131] K. Mertens, V. Putkaradze, D. Xia, S.R.J. Brueck, J. Appl. Phys. 98 (2005) 034309–034315.
[132] P.D. Cozzoli, T. Pellegrino, L. Manna, Chem. Soc. Rev. 35 (2006) 1195–1208.
[133] L. Li, S.I. Stupp, Angew. Chem. 117 (2005) 1867–1870.
[134] Y. Qin, S.M. Lee, A. Pan, U. Gösele, M. Knez, Nano Lett. 8 (2008) 114–118.
[135] B.A. Korgel, D. Fitzmaurice, Phys. Rev. Lett. 80 (1998) 3531–3536.
[136] X. Li, W. Xu, J. Zhang, H. Jia, B. Yang, B. Zhao, B. Li, Y. Ozaki, Langmuir 20 (2004) 1298–1304.
[137] P.M. Mendes, J.A. Preece, Curr. Opin. Colloid Interface Sci. 9 (2004) 236–248.
[138] S. Sun, C.B. Murray, D.R. Weller, L. Folks, A. Moser, Science 287 (2000) 1989–1992.
[139] A.N. Shipway, E. Katz, I. Willner, Chem. Phys. Chem. 1 (2000) 18–52.
[140] X.M. Lin, H.M. Jaeger, C.M. Sorensen, K.J. Klabunde, J. Phys. Chem. B 105 (2001) 3353–3357.
[141] T.P. Bigioni, X. Lin, T.T. Nguyen, E. Corwin, T.A. Witten, H.M. Jaeger, Nat. Mater. 5 (2006) 265–270.
[142] F. Zamborini, J. Hicks, R. Murray, J. Am. Chem. Soc. 122 (2000) 4514–4515.
[143] T. Nakanishi, B. Ohtani, K. Uosaki, J. Phys. Chem. B 102 (1998) 1571.
[144] J.J. Davis, P.D. Beer, In: C.I. Contescu, K. Putyera (Eds.), Dekker Encyclopedia of Nanoscience and Nanotechnology, Taylor & Francis Group, LLC, New York, 2009, pp. 2888–2911.
[145] S. Coe-sullivan, J.S. Steckel, W.-K. Woo, M.G. Bawendi, V. Bulovic, Adv. Funct. Mater. 15 (2005) 1117–1124.
[146] P.J. Thomas, G.U. Kulkarni, C.N.R. Rao, J. Phys. Chem. B 104 (2000) 8138–8144.
[147] M.P. Pileni, In: M.-P. Pileni (Ed.), Nanocrystals Forming Mesoscopic Structures, Wiley-VCH, Weinheim, 2005, pp. 307–316.
[148] M. Hu, Y. Yamaguchi, T. Okubo, J. Nanopart. Res. 7 (2005) 187–193.
[149] C.J. Kiely, J. Fink, M. Brust, D. Bethell, D.J. Schiffrin, Nature 396 (1998) 444–446.
[150] B. Kim, S.L. Tripp, A. Wei, J. Am. Chem. Soc. 123 (2001) 7955–7956.
[151] C.P. Collier, T. Vossmeyer, J.R. Heath, Annu. Rev. Phys. Chem. 49 (1998) 371–404.
[152] F.X. Redl, K.S. Cho, C.B. Murray, S. O'Brien, Nature 423 (2003) 968–971.
[153] S.A. Claridge, A.W. Castleman, S.N. Khanna, C.B. Murray, A. Sen, P.S. Weiss, ACS Nano 3 (2009) 244–255.
[154] J. Cheng, C.A. Ross, A. Mayes, Nat. Mater. 3 (2004) 823–828.
[155] M. Hu, S. Chujo, H. Nishikawa, Y. Yamaguchi, T. Okubo, J. Nanopart. Res. 6 (2004) 479–487.
[156] E.V. Shevchenko, D.V. Talapin, A.L. Rogach, A. Kornowski, M. Haase, H. Weller, J. Am. Chem. Soc. 124 (2002) 11480–11485.
[157] R. Klajn, K.J.M. Bishop, M. Fialkowski, M. Paszewski, C.J. Campbell, T.P. Gray, B.A. Grzybowski, Science 316 (2007) 261–265.
[158] O. Balmes, J.-O. Malm, G. Karlsson, J.-O. Bovin, J. Nanopart. Res. 6 (2004) 569–576.
[159] D. Vanmaekelbergh, P. Liljeroth, Chem. Soc. Rev. 34 (2005) 299–312.
[160] M.P. Pileni, J. Phys. Chem. B 105 (2001) 3358–3371.
[161] S. Matshushita, S. Onoue, In: M.-P. Pileni (Ed.), Nanocrystals Forming Mesoscopic Structures, Wiley-VCH, Weinheim, 2005, pp. 119–155.
[162] D. Xia, D. Li, Y. Luo, S.R.J. Brueck, Adv. Mater. 18 (2006) 930–933.
[163] Y. Gao, Z. Tang, Small 7 (2011) 2133–2146.
[164] R.R. Bhat, J. Genzer, B.N. Chaney, H.W. Sugg, A. Liebmann-Vinson, Nanotechnology 14 (2003) 1145–1152.

[165] P. Akcora, H. Liu, S.K. Kumar, J. Moll, Y. Li, B.N.C. Benicewicz, L.S. Schadler, D. Acehan, A.Z. Panagiotopoulos, V. Pryamitsyn, V. Ganesan, J. Ilavsky, P. Thiyagarajan, R.H. Colby, J.F. Douglas, Nat. Mater. 8 (2009) 354–359.
[166] R.K. Iler, J. Colloid Interface Sci. 21 (1966) 569–594.
[167] G. Decher, J.B. Schlenoff (Eds.), Multilayer Thin Films: Sequential Assembly of Nanocomposite Materials, Wiley-VCH GmbH & Co, Weinheim, 2003.
[168] G. Decher, Science 277 (1997) 1232–1237.
[169] K. Varahramyan, Y. Lvov, Proc. IMechE 220 Part N (2007) 29–37.
[170] A.I. Abdelrahman, A.M. Mohammad, T. Okajima, T. Ohsaka, J. Phys. Chem. B 110 (2006) 2798–2803.
[171] M. Gao, X. Zhang, B. Yang, F. Li, J. Shen, Thin Solid Films 284–285 (1996) 242–245.
[172] Y.K. Cho, R. Wartena, S.M. Tobias, Y.-M. Chiang, Adv. Funct. Mater. 17 (2007) 379–389.
[173] D. Lee, M.F. Rubner, R.E. Cohen, Nano Lett. 6 (2006) 2305–2312.
[174] A. Mamedov, N.A. Kotov, N.N. Mamedova, In: C.I. Contescu, K. Putyera (Eds.), Dekker Encyclopedia of Nanoscience and Nanotechnology, Taylor & Francis Group, LLC, New York, 2009, pp. 1062–1069.
[175] Y.S. Lee, Self-Assembly and Nanotechnology: A Force Balance Approach, New Jersey, Wiley, 2008, p. 151.
[176] PHYSorg, 27 Nov 2006. http://www.physorg.com/news83849970.html.
[177] J. Werfel, R. Nagpal, Int. J. Robotics Res. 27 (2008) 463–479.
[178] L. Berti, G.A. Burley, Nat. Nanotechnol. 3 (2008) 81–87.
[179] S. Mann, W. Shenton, M. Li, St. Connolly, D. Fitzmaurice, Adv. Mater. 12 (2000) 147–150.
[180] M. Motornov, J. Zhou, M. Pita, V. Gopishetty, I. Tokarev, E. Katz, S. Minko, Nano Lett. 8 (2008) 2993–2997.
[181] S.W. Lee, C. Mao, C.E. Flynn, A.M. Belcher, Science 296 (2002) 892–895.
[182] H. McNally, M. Pingle, S.W. Lee, D. Guo, D.E. Bergstrom, R. Bashir, Appl. Surf. Sci. 214 (2003) 109–119.
[183] M.M. Maye, D. Nykypanchuk, M. Cuisinier, D. van der Lelie, O. Gang, Nat. Mater. 8 (2009) 388–391.
[184] A.P. Philipse, D. Maas, Langmuir 18 (2002) 9977–9984.
[185] B.B. Yellen, O. Hovorka, G. Friedman, Proc. Natl. Acad. Sci. U.S.A. 102 (2005) 8860–8864.
[186] R. Sheparovych, Y. Sahoo, M. Motornov, S.M. Wang, H. Luo, P.N. Prasad, I. Sokolov, S. Minko, Chem. Mater. 18 (2006) 591–593.
[187] J. Jimenez, R. Sheparovych, M. Pita, A.N. Garcia, E. Dominguez, S. Minko, E. Katz, J. Phys. Chem. C 112 (2008) 7337–7344.
[188] S. Sun, S. Anders, H.F. Hamann, J.U. Thiele, J.E.E. Baglin, T. Thomson, E.E. Fullerton, C.B. Murray, B.D. Terris, J. Am. Chem. Soc. 124 (2002) 2884–2885.
[189] A.M. Kalsin, M. Fialkowski, M. Paszewski, S.K. Smoukov, K.J.M. Bishop, B.A. Grzybowski, Science 312 (2006) 420–424.
[190] H. Fudouzi, M. Kobayashi, N. Shinya, J. Nanopart. Res. 3 (2001) 193–200.
[191] X. Xiong, A. Busnaina, J. Nanopart. Res. 10 (2008) 947–954.
[192] D. Wang, H. Mohwald, J. Mater. Chem. 14 (2004) 459–468.
[193] Y. Yin, Yu Lu, B. Gates, Y. Xia, J. Am. Chem. Soc. 123 (2001) 8718–8729.
[194] J. Aizenberg, P.V. Braun, P. Wiltzius, Phys. Rev. Lett. 84 (2000) 2997–3000.
[195] D.J. Harris, J.C. Conrad, J.A. Lewis, Philos. Trans. R. Soc. A 367 (2009) 5157–5165.
[196] P.A. Kralchevsky, N.D. Denkov, Curr. Opin. Colloid Interface Sci. 6 (2001) 383–401.

[197] S.C. Tjong, Mater. Sci. Eng. R 53 (2006) 73–197.
[198] M.Q. Zhang, M.Z. Rong, K. Friedrich, In: H.S. Nalwa (Ed.), Encyclopedia of Nanoscience and Nanotechnology, vol. 7, American Scientific Publishers, New York, 2004, pp. 125–160.
[199] M. Adeogun, A Closer Look at Nanofillers and Polymer Nanocomposites in: SRI Business Insights, Menlo Park (CA), April 2006.
[200] S.S. Sternstein, A.-J. Zhu, Macromolecules 35 (2002) 7262–7273.
[201] S. Beun, T. Glorieux, J. Devaux, J. Vreven, G. Leloup, Dent. Mater. 23 (2007) 51–59.
[202] A. Bansal, Nat. Mater. 4 (2005) 693–698.
[203] P.W. de Oliveira, C. Becker-Willinger, M.H. Jilavi, Adv. Eng. Mater. 12 (2010) 349–361.
[204] F. Muller, W. Peukert, R. Polke, F. Stenger, J. Miner. Process. 748 (2004) S31–S41.
[205] M. Iijima, N. Sato, M. Tsukada, H. Kamiya, J. Am. Ceram. Soc. 90 (2007) 2741–2746.
[206] C. Hegedus, F. Pepe, D. Lindenmuth, D. Burgard, J. Coat. Technol. 5 (2008) 42–52.
[207] H.-W. Lin, W.-H. Hwu, M.-D. Ger, J. Mater. Process. Technol. 206 (2008) 56–61.
[208] M.E. Mackay, A. Tuteja, P.M. Duxbury, C.J. Hawker, B. Van Horn, Z. Guan, G. Chen, R.S. Krishnan, Science 311 (2006) 1740–1743.
[209] B. Vincent, Adv. Colloid Interface Sci. 4 (1974) 193–277.
[210] H. Althues, J. Henle, S. Kaskel, Chem. Soc. Rev. 36 (2007) 1454–1465.
[211] C. Li, N. Murase, Langmuir 20 (2001) 2229–2233.
[212] Y. Zhou, H. Itoh, T. Uemura, K. Naka, Y. Chujo, Langmuir 18 (2002) 277–283.
[213] S. Tian, J. Liu, T. Zhu, W. Knoll, Chem. Mater. 16 (2004) 4103.
[214] B.C. Sih, M.O. Wolf, Chem. Commun. (2005) 3375–3384.
[215] S. Taniguchi, M. Minamoto, M.M. Matsushita, T. Sugawara, Y. Kawada, D.J. Bethel, Mater. Chem. 16 (2006) 3459–3465.
[216] L.F. Francis, Mater. Manuf. Process. 12 (1997) 963–1015.
[217] K. Norrman, A. Ghanbari-Siahkali, N.B. Larsen, Annu. Rep. Prog. Chem. C 101 (2005) 174–201.
[218] H. Li, Y. Chen, C. Ruan, W. Gao, Y. Xie, J. Nanopart. Res. 3 (2001) 157–160.
[219] R.E. Mistler, E.R. Twiname, Theory and Practice, The American Ceramic Society, Westerville, OH, 2000.
[220] A.I.Y. Tok, F.Y.C. Boey, K.A. Khor, J. Mater. Process. Technol. 89–90 (1999) 508–512.
[221] L.P. Meier, L. Urech, L.J. Gauckler, J. Eur. Ceram. Soc. 24 (2004) 3753–3758.
[222] T. Chartier, A. Bruneau, J. Eur. Ceram. Soc. 12 (1993) 243–247.
[223] L.H. Luo, A.I.Y. Tok, F.Y.C. Boey, Mater. Sci. Eng. A 429 (2006) 266–271.
[224] T. Isobe, Y. Hotta, K. Watari, J. Am. Ceram. Soc. 90 (2007) 3720–3724.
[225] S.B. Reddy, P.P. Singh, N. Raghu, V. Kumar, J. Mater. Sci. 37 (2002) 929–934.
[226] H. Santa-Nokki, J. Kallioinen, T. Kololuoma, V. Tuboltsev, J. Korppi-Tommola, J. Photochem. Photobiol. A Chem. 182 (2006) 187–191.
[227] M.I. Bodnarchuk, M.V. Kovalenko, S. Pichler, G. Fritz-Popovski, G. Hesser, W. Heiss, ACS Nano 4 (2010) 423–431.
[228] F.C. Krebs, Sol. Energ. Mater. Sol. Cell 93 (2009) 394–412.
[229] G. Harsanyi (Ed.), Sensors in Biomedical applications, CRC Press, LLC, Boca Raton, 2000.
[230] P. Calvert, Chem. Mater. 13 (2001) 3299–3305.
[231] Alexander Kamyshny, Matti Ben-Moshe, Shai Aviezer, Shlomo Magdassi, Macromol. Rapid Commun. 26 (2005) 281–288.
[232] B. Ong, Semiconductor ink advances flexible displays, Laser Focus June 1st, 2004.

[233] R.A. Street, W.S. Wong, S.E. Ready, M.L. Chabinyc, A.C. Arias, S. Limb, A. Salleo, R. Lujan, Mater. Today 9 (2006) 32–37.
[234] A. Kamyshny, J. Steinke, S. Magdassi, Open Appl. Phys. J. 4 (2011) 19–36.
[235] B. Ryu, Y. Choi, H. Park, J. Byun, K. Kong, J. Lee, H. Chang, Colloids Surf. A 270–271 (2005) 345–351.
[236] J.U. Park, M. Hardy, S. Jun Kang, K. Barton, K. Adair, D. Kishore Mukhopadhyay, Ch. Young Lee, M.S. Strano, A.G. Alleyne, J.G. Georgiadis, P.M. Ferreira, J.A. Rogers, Nat. Mater. 5 (2007) 1–8.
[237] D.Z. Wang, S.N. Jayasinghe, M.J. Edirisinghe, J. Nanopart. Res. 7 (2005) 301–306.
[238] W.J. Tseng, C.N. Chen, J. Mater. Sci. 41 (2006) 1213–1219.
[239] H.H. Lee, K.S. Chou, K. Huang, Nanotechnology 16 (2005) 2436–2441.
[240] Y. Zhao, Q. Zhou, L. Liu, J. Xu, M. Yan, Z. Jiang, Electrochim. Acta 51 (2006) 2639–2645.
[241] S.B. Fuller, E.J. Wilhelm, J.M. Jacobson, J. Microelectrochem. Syst. 11 (2002) 54–60.
[242] A. Atkinson, J. Doorbar, A. Hudd, D.L. Segal, P.J. White, J. Sol-Gel Sci. Technol. 8 (1997) 1093–1097.
[243] E. Tekin, P.J. Smith, S. Hoeppener, A.M.J. van den Berg, A.S. Susha, A.L. Rogach, J. Feldmann, U.S. Schubert, Adv. Funct. Mater. 17 (2007) 23–28.
[244] D. Huang, F. Liao, S. Moles, D. Redinger, V. Subramania, J. Electrochem. Soc. 150 (2003) G412–G417.
[245] S.H. Ko, H. Pan, C.P. Grigoropoulos, Ch.K. Luscombe, J.M. Frechet, D. Poulikakos, Nanotechnology 18 (2007) 345202–345211.
[246] H.S. Kim, J.S. Kang, J.S. Park, H.T. Hahn, H.C. Jung, J.W. Joung, Comp. Sci. Technol. 69 (2009) 1256–1264.
[247] E.B. Guttoff, C.E. Kendrick, AIChE 33 (1987) 141–145.
[248] S. Ishizuka, Method of Simultaneous Multi-Layer Application, USPTO 4,976,999.
[249] M.S. Carvalho, H.S. Khesghi, AIChE J. 46 (2000) 1907–1917.
[250] E. Katoh, Y. Tsubaki, Ink jet recording sheet, US Patent 6,699,536, 2004.
[251] D. Cohen, E.B. Gutoff, Modern Coating and Drying Technology, VCH Publishers, New York, 1992.
[252] W.B. Chu, J.W. Yang, T.-J. Liu, C. Tiu, J. Guo, Colloids Surf. A Physicochem. Eng. Asp. 302 (2007) 1–10.
[253] B.D. Gates, Q. Xu, M. Stewart, D. Ryan, C.G. Willson, G.M. Whitesides, Chem. Rev. 105 (2005) 1171–1196.
[254] M.A. Auger, P.L. Schilardi, I. Caretti, O. Sanchez, G. Benitez, J.M. Albella, R. Gago, M. Fonticelli, L. Vazquez, R.C. Salvarezza, O. Azzaroni, Small 1 (2005) 300–309.
[255] Z. Liang, K. Li, Q. Wang, Langmuir 19 (2003) 5555–5558.
[256] T. Kraus, L. Malaquin, H. Schmid, W. Reiss, N.D. Spencer, H. Wolf, Nat. Nanotechnol. 570 (2007) 570–576.
[257] K. Salaita, Y. Wang, C.A. Mirkin, Nat. Nanotechnol. 2 (2007) 145–155.
[258] B. Basnar, I. Willner, Small 5 (2009) 28–44.
[259] B.A. Parviz, Declan Ryan, G.M. Whitesides, IEEE Trans. Adv. Packaging 26 (2003) 233–241.
[260] J.A. Lewis, J.E. Smay, J. Stuecker, J. Ceserano III, J. Am. Ceram. Soc. 89 (2006) 3599–3609.
[261] M.A. Randolph, Commercial Assessment of Roll-to-Roll Manufacturing of Electronic Displays, Master of Engineering in Materials Science and Engineering, Massachusetts Institute of Technology, Massachusetts, 2006.

[262] Printed Electronics, Observatory Nano: ICT Sector Focus Report, April 2010. www.observatorynano.eu/project/filesystem/files/ObservatoryNanoFocusReport_Printed Electronics.pdf.
[263] K. Nishiyabu, S. Tanaka, Metal Powder Rep. 63 (2008) 28–32.
[264] G. Anné, S. Put, K. Vanmeensel, D. Jiang, J. Vleugels, O. Van der Biest, J. Eur. Ceram. Soc. 25 (2005) 55–63.
[265] C. Capiglia, J. Yang, N. Imanishi, A. Hirano, Y. Takeda, O. Yamamoto, J. Power, Sources 119–121 (2003) 826–832.
[266] A. Nabok, Organic and Inorganic Nanostructures, Artech House, Boston, 2005 (Chapter 2).
[267] E.A. Olevsky, Mater. Sci. Eng. R23 (1998) 41–110.
[268] J. Van Tassel, C.A. Randall, J. Mater. Sci. 39 (2004) 867–879.
[269] D.-H. Tsai, S.H. Kim, T.D. Corrigan, R.J. Phaneuf, M.R. Zachariah, Nanotechnology 16 (2005) 1856–1862.
[270] C.R. Barry, H.O. Jacobs, Nano Lett. 6 (2008) 2790–2796.
[271] L. Besra, M. Liu, Prog. Mater. Sci. 52 (2007) 1–61.
[272] O.O. Van der Biest, L.J. Vandeperre, Annu. Rev. Mater. Sci. 29 (1999) 327–352.
[273] J.B. Talbot, E. Sluzky, S.K. Kurinec, J. Mater. Sci. 39 (2004) 771–778.
[274] A.R. Boccaccini, I. Zhitomirsky, Curr. Opin. Solid State Mater. Sci. 6 (2002) 251–260.
[275] I. Shao, P.M. Vereecken, R.C. Cammarata, P.C. Searson, J. Electrochem. Soc. 149 (2002) C610–C614.
[276] C.T.J. Low, R.G.A. Wills, F.C. Walsh, Surf. Coat.Technol. 201 (2006) 371–383.
[277] A.R. Boccaccini, J.A. Roether, B.J.C. Thomas, M.S.P. Shaffer, E. Chavez, E. Stoll, E.J. Minay, J. Ceram. Soc. Jpn. 114 (2006) 1–14.
[278] J. Li, Y.J. Wu, H. Tanaka, T. Yamamoto, M. Kuwabara, J. Am. Ceram. Soc. 87 (2004) 1578–1581.
[279] M.J. Santillane, F. Membrives, N. Quaranta, A.R. Boccaccini, J. Nanopart. Res. 10 (2008) 787–793.
[280] Y. Castro, B. Ferrari, R. Moreno, A. Duran, J. Sol-Gel Sci. Technol. 23 (2002) 187–189.
[281] O. Kammona, K. Kotti, C. Kiparissides, J.P. Celis, J. Fransaer, Electrochim. Acta 9 (30) (2009) 2450–2457.
[282] N. Sato, M. Kawachi, K. Noto, N. Yoshimoto, M. Yoshizawa, Physica C 357–360 (2001) 1019–1022.
[283] O. Van der Biest, L. Vandeperre, S. Put, G. Anné, J. Vleugels, Adv. Sci. Technol. 45 (2006) 1075–1084.
[284] N.R. de Tacconi, K. Rajeshwar, R.O. Lezna, Electrochim. Acta 45 (2000) 3403–3411.
[285] C.R. Barry, N.Z. Lwin, M.G. Steward, H.O. Jacobs, Proceedings of the 2004 NSF Design, Service and Manufacturing Grantees and Research Conference, pp. 1–7.
[286] D. Li (Ed.), Encyclopedia of Microfluidics and Nanofluidics, Springer, New York, 2008, pp. 350–357.
[287] K.D. Hermanson, et al., Science 294 (2001) 1082–1086.
[288] S.I. Khondaker, et al., Appl. Phys. Lett. 85 (2004) 645–647.
[289] Yu. Chang, Suxian Huang, Yong Chen, Small 5 (2009) 63–66.
[290] Q. Feng, T. Li, H. Yue, K. Qi, F. Bai, J. Jin, Appl. Surf. Sci. 254 (2008) 2262–2268.
[291] S. Ramakrishna, K. Fujihara, W.-E. Teo, T. Yong, Z. Ma, R. Ramaseshan, Mater. Today 9 (2006) 40–50.
[292] P.W. Gibson, H.L. Schreuder-Gibson, D. Rivin, AIChE J. 45 (1999) 190–195.
[293] R. Luoh, H.T. Hahn, Comp. Sci. Technol. 66 (2006) 2436–2441.
[294] O.V. Salata, Curr. Nanosci. 1 (2005) 25–33.

[295] X. Lu, Y. Zhao, C. Wang, Adv. Mater. 17 (2005) 2485–2488.
[296] H. Kim, J. Kim, H. Yang, J. Suh, T. Kim, B. Han, S. Kim, D. Seong Kim, P.V. Pikhitsa, M. Cho, Nat. Nanotechnol. 1 (2006) 117–121.
[297] C.-Y. Kao, K.-S. Chou, Electrochem. Solid-State Lett. 10 (2007) D32–D34.
[298] T. Nychyporuk, Z. Zhou, A. Fave, M. Lemiti, S. Bastide, Sol. Energ. Mater. Sol. Cell 94 (2010) 2314–2317.
[299] R.C. Agarwala, V. Agarwala, Sadhana 28 (2003) 475–493.
[300] H. Ma, F. Tian, D. Li, Q. Guo, J. Alloys Compd. 532 (2008) 1–4.
[301] G. Cheng, D. Romero, G.T. Fraser, A.R.H. Walker, Langmuir 21 (2005) 12054–12059.
[302] G.A. Somorjai, A.M. Contreras, M. Montano, R.M. Rioux, Proc. Natl. Acad. Sci. U.S.A. 103 (2006) 10577–10583.
[303] H. Song, F. Kim, S. Connor, G.A. Somorjai, Peidong Yang, J. Phys. Chem. B 109 (2005) 188–193.
[304] S. Chen, Anal. Chim. Acta 496 (2003) 29–37.
[305] J.R. Heath, C.M. Knobler, D.V. Leff, J. Phys. Chem. B 101 (101) (1997) 189.
[306] S. Huang, G. Tsutsui, H. Sakaue, S. Shingubara, T. Takahagi, J. Vac. Sci. Technol. B 19 (2001) 115–120.
[307] J. Park, W.-R. Lee, S.-S. Bae, Y.J. Kim, K.-H. Yoo, J. Cheon, S. Kim, J. Phys. Chem. B 109 (2005) 13119–13123.
[308] Q. Guo, X. Teng, S. Rahman, H. Yang, J. Am. Chem. Soc. 125 (2003) 630–631.
[309] J.R. Gorza, In: C.C. Koch (Ed.), Nanostructured Materials: Processing, Properties, and Applications, W. Andrew, Norwich, NY, 2007, p. 173.
[310] R.A. Andrievski, J. Mater. Sci. (1994) 614–631.
[311] G.S. Upadhyaya, Powder Metallurgy Technology, Cambridge International Science Publishing, England, 2002.
[312] V. Viswanathan, T. Laha, K. Balani, A. Agarwal, S. Seal, Mater. Sci. Eng. R 54 (2006) 121–285.
[313] Z.Z. Fang, H. Wang, Int. Mater. Sci. 53 (2008) 326–352.
[314] M.J.G. Jak, In: Ph. Knauth, J. Schoomann (Eds.), Nanostructured Materials, Kluwers Academic Publishers, Netherlands, 2002, pp. 55–71.
[315] G. Sethi, N.S. Myers, R.M. German, Int. Mater. Rev. 53 (2008) 219–234.
[316] K. Lu, Int. Mater. Rev. 53 (2008) 21–37.
[317] A. Polotai, K. Breece, E. Dickey, C. Randall, A. Ragulya, J. Am. Ceram. Soc. 88 (2005) 3008–3012.
[318] T.C. Lowe, R.Z. Valiev, JOM (2004) 64–68.
[319] B.H. Kear, J. Colaizzi, W.E. Mayo, S.-C. Liao, Scr. Mater. 44 (2001) 2065–2068.
[320] S.C. Liao, K.D. Pae, W.E. Mayo, Mater. Sci. Eng. A204 (1995) 152–159.
[321] B.B. Panigrahi, M.M. Godkhindi, J. Nanopart. Res. 8 (2006) 627–633.
[322] I. Karaman, M. Haouaoui, H.J. Maier, J. Mater. Sci. 42 (2007) 1561–1576.
[323] J. Halme, J. Saarinen, P. Lund, Sol. Energ. Mater. Sol. Cell 90 (2006) 887–899.
[324] E. Ma, Scr. Mater. 49 (2003) 663–668.
[325] R. Chaim, Mater. Sci. Eng. A 443 (2007) 25–32.
[326] K. Kondo, S. Sawai, J. Am. Ceram. Soc. 73 (1990) 1983–1991.
[327] S. Dong, P. Hou, H. Cheng, H. Yang, G. Zou, J. Phys, Condens. Matter 14 (2002) 11023–11030.
[328] F. Bensebaa, Transparent and/or Photovoltaic Solar Cell and Module, CIPO CA 2621665, 19 February 2008.
[329] A. Mukhopadhyay, B. Basu, Int. Mater. Rev. 52 (2007) 257–288.
[330] S.G. Huang, K. Vanmeensel, O. Van der Biest, J. Vleugels, Int. J. Refract. Metals Hard Mater. 26 (2008) 41–47.

[331] J. Vleugels, T. Xu, S. Huang, Y. Kan, P. Wang, L. Li, O. Van der Biest, J. Eur. Ceram. Soc. 27 (2007) 1339–1346.
[332] V.Y. Kodash, J.R. Groza, K.C. Cho, B.R. Klotz, R.J. Dowding, Mater. Sci. Eng. A 385 (2004) 367–371.
[333] J.G. Bai, Z.Z. Zhang, J.N. Calata, G.-Q. Lu, IEEE Trans. Components Pack. Technol. 29 (2006) 589–593.
[334] R. Kalyanaraman, S. Yoo, M.S. Krupahankara, T.S. Sudarshan, R.J. Dowding, Nanostruct. Mater. 10 (1998) 1379–1392.
[335] H.W. Zhang, R. Gopalan, T. Mukai, K. Hono, Scr. Mater. 53 (2005) 863–868.
[336] Q. Meng, L. Geng, D. Ni, Mater. Lett. 59 (2005) 2774–2777.
[337] E. Toyserkani, A. Khajepour, St. Corbin, Laser Cladding, CRC Press, Boca Raton, 2005 (Chapter 2).
[338] M. Li, S. Zhang, H. Li, Y. He, J.-H. Yoon, T.-Y. Cho, J. Mater. Process. Technol. 202 (2008) 107–111.
[339] D. Perednis, L.J. Gauckler, J. Electroceram. 14 (2005) 103–111.
[340] R. Hui, Zh. Wang, O. Kesler, L. Rose, J. Jankovic, S. Yick, R. Maric, D. Ghosh, J. Power Sour. 170 (2007) 308–323.
[341] Y.H. Sohn, B. Jayaraj, S. Laxman, B. Franke, J.W. Byeon, A.M. Karlsson, JOM (2004) 53–56.
[342] E. Brinley, K.S. Babu, S. Seal, JOM (2007) 54–59.
[343] I. Morrison, fourth ed, Kirk-Othmer Encyclopedia of Chemical Technology, R.E. Kirk and D.F. Ohmer (Eds.), vol. 8, John Wiley & Sons, New York, 2003, pp. 697–739.
[344] E.J. Lavernia, Y. Wu, Spray Atomization and Deposition, John Wiley & Sons, Chichester, 1995.
[345] J. Schoonman, Solid State Ionics 135 (2000) 5–19.
[346] M. Pehnt, D.L. Schulz, C.J. Curtis, K.M. Jones, D.S. Ginley, Appl. Phys. Lett. 67 (1995) 2176–2178.
[347] D.L. Schulz, C.J. Curtis, R.A. Fliton, H. Wiesner, J. Keane, R.J. Matson, K. Jones, P.A. Parilla, R. Noufi, D.S. Ginley, J. Electron. Mater. 27 (1998) 433–437.
[348] Y. Sawada, Ch. Kobayashia, Sh. Sekia, H. Funakubo, Thin Solid Films 409 (2002) 46–50.
[349] F. Schulz, S. Franzka, G. Schmid, Adv. Funct. Mater. 12 (2002) 532–536.
[350] A. Mette, P.L. Richter, M. Horteis, S.W. Glunz, Prog. Photovolt. Res. Appl. 15 (2007) 621–627.
[351] S.Y. Lee, L. Gradon, S. Janeczko, F. Iskandar, K. Okuyama, ACS Nano 4 (2010) 4717–4724.
[352] A. Killinger, M. Kuhn, R. Gadow, Surf. Coat.Technol. 201 (2006) 1922–1929.
[353] R. Gadow, A. Killinger, J. Rauch, Surf. Coat.Technol. 202 (2008) 4329–4336.
[354] P. Fauchais, Gh. Montavon, M. Vardelle, J. Cedelle, Surf. Coat.Technol. 201 (2006) 1908–1921.
[355] F.-L. Toma, G. Bertrand, S.O. Chwa, C. Meunier, D. Klein, C. Codd, Surf. Coat.Technol. 200 (2006) 5855–5862.
[356] P. Fauchais, V. Rat, J.-F. Coudert, R. Etchart-Salas, G. Montavon, Surf. Coat.Technol. 202 (2008) 4309–4317.
[357] E.H. Jordan, L. Xie, X. Ma, M. Gell, N.P. Padture, B. Cetegen, A. Ozturk, J. Roth, T.D. Xiao, P.E.C. Bryant, J. Therm. Spray Technol. 13 (2004) 57–66.
[358] R.S. Lima, A. Kucuk, C.C. Berndt, Mater. Sci. Eng. A313 (2001) 75–82.
[359] P. Blazdell, S. Kuroda, Surf. Coat.Technol. 123 (2000) 239–246.
[360] A. Jaworek, A.T. Sobczyk, J. Electrostat. 66 (2008) 197–219.
[361] X. Hiu, K.L. Choy, Chem. Vap. Depos. 12 (2006) 583–596.

[362] J. Lee, I. Park, H. Lee, D. Shin, J. Power Sour. 212 (2012) 35–42.
[363] J. Ruzyllo, D.-O. Lee, P. Roman, M. Horn, P. Mumbauer, M. Brubaker, R. Grant, IEEE Advanced Semiconductor Manufacturing Conference, 2001.
[364] N. Solayappan, L.D. Mcmillan, C. Paz de Araujo, B. Grant, Integr.Ferroelec. 18 (1997) 127–136.
[365] www.primaxxinc.com (accessed 29.09.07).
[366] P. Marage, M. Langlet, J.C. Joubert, Thin Solid Films 238 (1994) 218–227.
[367] J.L. Fulton, George S. Devermana, Clement R. Yonkera, Jay W. Grate, James De Young, James B. McClain, Polymer 44 (2003) 3627–3632.
[368] C. Huang, W.T. Nichols, D.T. O'Brien, M.F. Becker, D. Kovar, J.W. Keto, J. Appl. Phys. 101 (2007) 064902–064912.
[369] S.L. Girshick, J. Hafiz, J. Phys. D Appl. Phys. 40 (2007) 2354–2360.
[370] L. Steven, J.L. Girshick, J. Nanopart. Res. 10 (2008) 935–945.
[371] P.C. Andricacos, C. Uzoh, J.O. Dukovic, J. Horkans, H. Deligianni, IBM J. Res. Dev. 42 (1998) 567–574.
[372] H. Liu, F. Favier, K. Ng, M.P. Zach, R.M. Penner, Electrochim. Acta 47 (2001) 671–677.
[373] J. Fang, H. Yoy, B. Ding, X. Song, Electrochem. Commun. 9 (2007) 2423–2427.
[374] M.K. Roy, H.C. Verma, J. Magn. Magn. Mater. 306 (2006) 98–102.
[375] R.G. Palgrave, I.P. Parkin, J. Am. Chem. Soc. 128 (2006) 1587–1597.
[376] E. Barborini, P. Piseri, A. Bassi, A. Ferrari, C.E. Bottani, P. Milani, Chem. Phys. Lett. 300 (1999) 633–638.
[377] K. Wegner, E. Barborini, P. Piseri, P. Milani, Kona 24 (2006) 54–69.
[378] C. Binns, et al., J. Phys. D Appl. Phys. 38 (2005) R357–R379.
[379] W.S. Seo, J.H. Lee, X. Sun, Y. Suzuki, D. Mann, Z. Liu, M. Terashima, P.C. Yang, M.V. Mcconnell, D.G. Nishimura, H. Dai, Nat. Mater. 5 (2006) 971–976.
[380] V. Salgueirin-Maceira, M.A. Correa-Duarte, Adv. Mater. 19 (2007) 4131–4144.
[381] W. Cai, X. Chen, Small 3 (2007) 1840–1854.
[382] A. Burns, H. Ow, U. Wiesner, Chem. Soc. Rev. 35 (2006) 1028–1042.
[383] Y. Piao, A. Burns, J. Kim, U. Wiesner, T. Hyeon, Adv. Funct. Mater. 18 (2008) 3745–3758.
[384] B. Mukherjee, N. Ravishankar, Nanotechnology 18 (2007) 1–9.
[385] S.E. Skrabalak, Y. Xia, ACS Nano 3 (2009) 10–15.
[386] Y. Xia, Y. Xiong, B. Lim, S.E. Skrabalak, Angew. Chem. Int. Ed. Engl. 48 (2009) 60–103.
[387] Y. Chen, K. Munechika, I.J.L. Plante, A.M. Munro, S.E. Skrabalak, Y. Xia, D. Ginger, Appl. Phys. Lett. 93 (2008) 053106–053109.
[388] E. Ruiz-Hitzky, K. Ariga, Y.M. Lvov (Eds.), Bio-inorganic Hybrid Nanomaterials, Wiley-VCH Verlag GmbH & Co. KGaA, Weinheim, 2008.
[389] S.V. Patwardhan, N. Mukherjee, M. Steinitz-Kannan, S.J. Clarson, Chem. Commun. (2003) 1122–1123.
[390] D.S. dos Santos, P.J.G. Goulet, N.P.W. Pieczonka, O.N. Oliveira, R.F. Aroca, Langmuir 20 (2004) 10273–10277.

Chapter 5

Clean Energy

1 INTRODUCTION

Energy demand and environmental impact are intimately related issues. Fossil fuel provides most of the current energy demand. According to a recent International Energy Agency (IEA) report, coal/peat, oil, and natural gas cover around 27%, 32%, and 21% of the overall energy demand, respectively [1]. The rest is provided by nuclear and renewable sources. It is expected that fossil fuel will maintain its dominance for at least the next 20 years [1]. However, based on current rates of oil production and discoveries, global oil peak oil is projected to occur around 2020 [2]. Higher energy demands in the future will likely increase greenhouse gas (GHG) emissions under business-as-usual conditions. New nonconventional oil and gas resources will likely have a higher environmental footprint. Increase in GHG emissions with its potential for global warming and its negative effects on quality of life will remain a major challenge for decades to come. Under the business-as-usual scenario, this energy and environment crisis is expected to become even worse in the future. Currently, the energy sector accounts for almost 90% of the GHG emissions in the USA. Transportation, electricity, and heat represent nearly 60% of the energy demand [3]. These sectors should receive special attention from technology developers and policy makers. Fuel combustion, industry, and agriculture are the other important contributors to energy demand and GHG emissions.

Energy efficiency along with clean and renewable sources of energy provides sustainable solutions to reducing dependency on polluting and dwindling fossil sources. Power generation and transportation sectors are the largest energy users. Numerous clean power generation technologies are currently being used at the demonstration or commercial stage. This includes solar, wind, and bioenergy. High cost and/or the demand for rare raw materials make these technologies unsustainable in the long term. The transportation sector is a big energy consumer and large contributor to the overall GHG emissions.

Nanoparticles (NPs) are omnipresent in numerous energy-related material processes. A few industries are taking advantage of the potential offered by NPs, particularly those related to energy production, distribution, and

Interface Science and Technology, Vol. 19. http://dx.doi.org/10.1016/B978-0-12-369550-5.00005-7

utilization. Catalysis and electrical power generation are by far the most important areas, and therefore are covered in relatively greater detail. In this chapter, a broad view of nanoparticle (NP) utilization in various energy-related areas is provided. Five major areas are reviewed including energy conversion, energy storage, energy transport and distribution, energy saving, and alternative manufacturing processes. As summarized in Table 5.1, NP utilization in energy-related technologies is rich and diverse. Several NP-based technologies are (or will be) incorporated in different vehicle models [4]. Besides the well-established NP-based catalyst converter, alternative solar films, green coatings, high-efficiency fuel, sensing, and glazing are key areas where NPs could contribute. There are also numerous other potential applications that include fuel cell (FC) and batteries [5].

NPs have been used for nearly a century in catalysis. It is only in the last two decades that a more concerted effort has been directed specifically toward using NPs for energy-related applications. Catalyst activity of colloidal platinum was reported by Bredig in 1901 [6]. Sabatier reported hydrogenation of numerous substances on powder metals in the beginning of the last century [7]. Catalysis is probably the first large-scale commercial application of NPs. Indeed, platinum NPs stabilized in an alumina matrix have been used for reforming in the petrochemical industry since the late 1940s [8]. Today, catalysis plays a vital role in virtually all

TABLE 5.1 Current and Future Energy-Related Applications of Nanoparticle

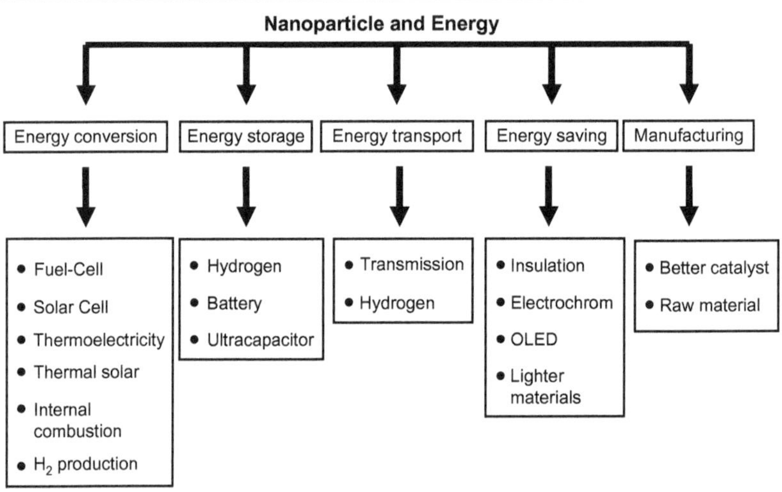

industries, taking advantage of the NPs' high surface area of NPs. Catalysis is particularly ubiquitous in the chemical industry and materials manufacturing.

Catalysts will continue to play a key role in developing clean and sustainable chemical processes through the optimization of raw materials and energy utilization. They are used in a variety of manufacturing processes. The value of manufactured catalysts is over US$8 billion, contributing to products with a value of over US$3000 billion, in goods and services [9]. The majority of these catalysts are used in petrochemistry, polymerization, environment, and fine chemicals.

Besides reducing energy and raw material consumption, there are numerous opportunities for NP-based catalysts to minimize by-products and waste. Today, waste control is becoming a critical issue in many industries. For example, optimized catalysts allow minimizing the waste by-product amount [10]. Oil refining and bulk chemicals produce about 10^7 and 10^5 tons of waste per year, respectively [10].

Electric power requirements for consumer electronics, transportation, residential, commercial, and industrial sectors are areas where NP integration will likely have the biggest impact in the future [11]. Distributed and renewable power generation receives significant attention in this chapter. As indicated in Fig. 5.1, the final utilization of primary energy worldwide consists of electricity (38.1%), heat and industrial use (44.3%), and transport (17.6%) [12]. A detailed understanding of energy source and energy consumption characteristics is important. This will allow the development of an

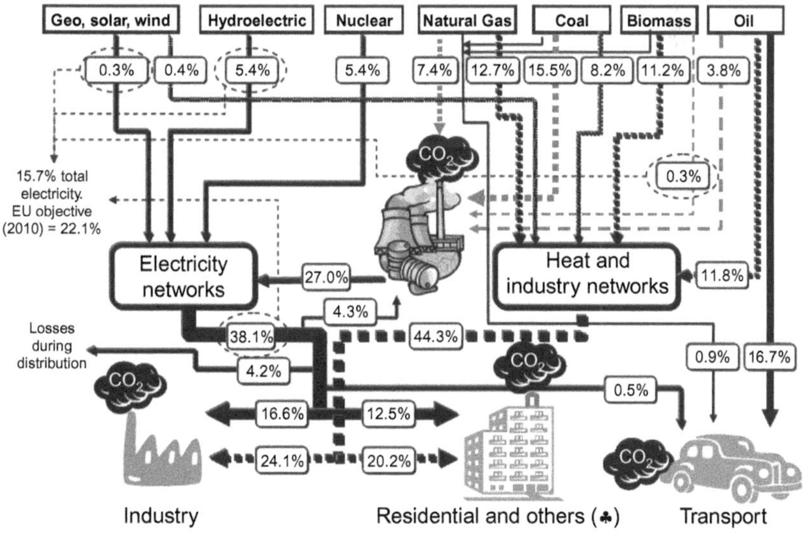

FIGURE 5.1 Sources and relative distribution of primary energy production and consumption [12]. (For color version of this figure, the reader is referred to the online version of this chapter.)

appropriate alternative, and clean and sustainable solutions. Direct electricity generation from renewable sources will help mitigate the increased needs for fossil fuel-based electrical power generation. With 4.2% of the primary energy lost during electricity distribution, corresponding to about 10% of the total electricity generated, increasing distributed power generation will eliminate long transmission lines and help reduce energy losses. Based on a worldwide primary energy consumption of about 125,000 TW h in 2004, this loss is quite significant.

Photovoltaic (PV) and FC can provide power output from about 1 watt (W) to several hundred kilowatts (kW) or more. FC and PV power generation can be used either in parallel with or independently of other large electrical utilities. In addition, distributed power generation has another advantage during high power demand and outages. It provides cost-effective and high-quality power. PV and FC power generation minimize reliance on fossil fuel and at the same time reduce GHG emissions. The growth of these technologies is hampered, however, by their current high cost due to expensive materials and processing.

The PV market is dominated today by the crystalline silicon solar cell with over 90% market share. At present, the contribution of PV power generation to the overall energy supply is limited. In spite of its exponential growth in the 2000s, solar energy including thermal and PV accounts for about 0.1% of the overall energy mix. Half of the PV module cost is related to the crystalline silicon-based wafer. The high cost of purifying silicon is considered the main hurdle in the development of PV. Various semiconductor NPs have been proposed as alternative PV materials to these silicon-based solar cells. NPs are easier to fabricate and process, although no successful commercial solution has yet been reported.

New NP-based materials and systems are also used in energy transport, storage, and saving. Applications of NP-based materials and systems will have direct and indirect impacts on the security and diversification of energy supply, and limit GHG emissions without affecting sustainable economical development. Several other important applications of NPs related to energy saving by using improved materials for thermal insulation, low-weight materials, and efficient manufacturing processes are also discussed.

2 CATALYSIS

Advanced catalysis materials are critical to addressing energy and global warming issues [13]. Catalysts could be described as materials helping to accelerate reactions without themselves undergoing any alteration or transformation. In practice, the properties of catalysts change to the point that they need to be discarded or recycled. Catalysts are often classified according to three main properties: activity, selectivity, and lifetime [14]. Catalyst activity refers to the conversion ratio from feedstock material to product. The catalyst activity is quantitatively defined by turnover number or turnover frequency (TOF) corresponding to the amount of product obtained per unit time. Selectivity is the ability of the catalyst to help synthesize one specific product out of a variety of different products. Lifetime is the time during which the catalyst is used without significant loss of its activity and selectivity. The overall cost of the catalyst during its lifetime is also important to define the performance-to-cost ratio.

Catalysis plays an important role in numerous industries. Around US$10 trillion worth of chemicals and materials are produced using various catalytic processes [15]. NPs already play a critical role in high reaction yield and also reduce high precursor cost. Indeed, catalysts based on NPs dispersed in high surface area support are widely used [16,17]. Various NP formulations are currently used in automotive catalytic converters. The basic microscopic and macroscopic structures of these catalytic converters have been established (Fig. 5.2). The 3D dispersion of NP catalysts on the catalyst substrate (see lower inset of Fig. 5.2) will improve the turnover frequency. Dispersed NPs can be obtained using different processes during or after synthesis [18].

Catalysts are often referred to as either homogeneous or heterogeneous, depending on whether they operate in a single- or multiphase reactor. Heterogeneous catalysis consists of active catalytic sites placed on the surface of a

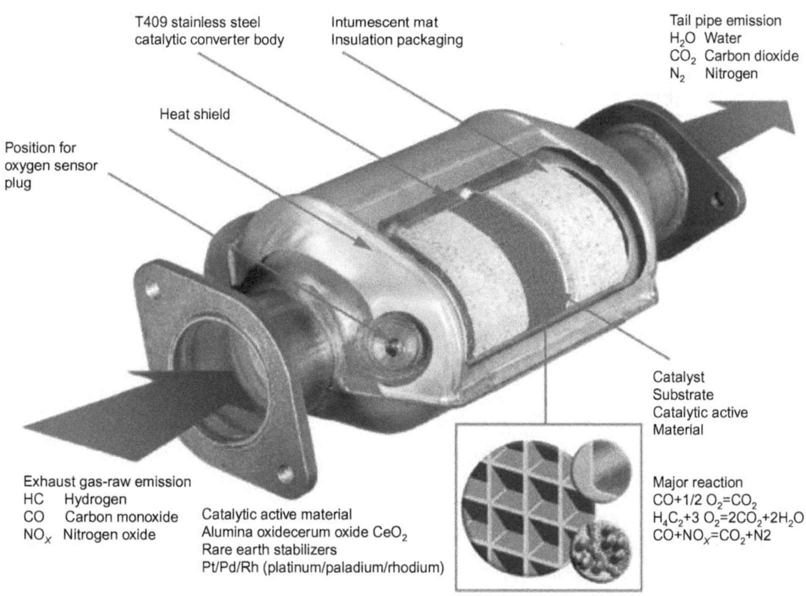

FIGURE 5.2 Macroscopic, microscopic, and nanoscopic structures of an automotive catalytic converter [17]. (For color version of this figure, the reader is referred to the online version of this chapter.)

solid support. As a consequence, it is easy to separate the product from the catalyst. There is no support in homogeneous catalysis. The ratio of solid waste to product is much higher when homogeneous catalysis is used. Homogeneous catalysis was developed about a century ago when the environment was not a major issue. In spite of its attributes, the homogeneous catalysis market share has diminished significantly. This is due to toxicity, handling issues, large waste, and high cost [19].

Classification of NP catalysts as homogeneous versus heterogeneous is difficult in the classical term. Although catalytic activity of free NPs in solution could be categorized as homogeneous, this is not quite the case when these NPs are deposited on a solid support. In this case, depending on the type of interaction between the NPs and the support, the catalysis process cannot be considered as purely heterogeneous. The catalytic characteristics of NPs can be loosely classified as intermediary between heterogeneous and homogeneous catalysis characteristics [20]. Often they are referred to as quasi-homogeneous catalysts.

Unsupported NP behavior is intermediary between that of supported and nonsupported catalysts (Fig. 5.3). On the macroscopic level, like any homogeneous catalyst, unsupported NP-based catalysts mix homogeneously with precursors and products. However, on the microscopic level, NPs on one side and the precursor/products on the other side are quite distinct phases. Often, the exact nature of the active catalytic site is still ill-defined [21,22]. The concept of quasi-homogeneous catalysis has been used to characterize the particularities of colloidal-based catalysts [22,23]. Quasi-homogeneous catalysis has been pursued to address the shortcomings of conventional homogeneous and heterogeneous catalysts [24]. Unsupported NP-based catalysts may give rise to catalytic properties that are intermediary between bulk-like heterogeneous catalysis and molecular-like homogeneous catalysis (Fig. 5.3). Separating NPs from catalysis products is potentially much easier in heterogeneous catalysis than in the case of conventional homogeneous catalysis.

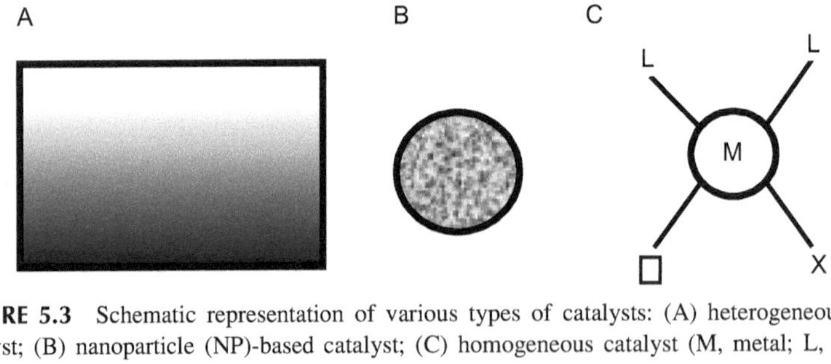

FIGURE 5.3 Schematic representation of various types of catalysts: (A) heterogeneous bulk catalyst; (B) nanoparticle (NP)-based catalyst; (C) homogeneous catalyst (M, metal; L, ligand; X, counterion; □, vacant site).

In the particular case of transition metal-based catalysts, the main difference between homogeneous and heterogeneous catalysis is related to the ratio of metal exposed to the reaction mixture [20]. Currently, there are several catalytic reactions where the nature of the active site, the reaction mechanism, and the exact role of the supported catalyst are still controversial [25,26]. Indeed, Finke et al. distinguished between a metal-particle heterogeneous catalyst and a metal-complex homogeneous catalyst [25,26]. In the case of very small NPs, the ratio of catalyst sites exposed to reactant could be as high as 50% or more. Furthermore, highly soluble monodisperse NPs form a homogeneous solution within the catalysis reactor.

Both homogeneous and heterogeneous catalysts may contribute to energy-related processes. Transition-metal catalysis is the main focus here given the large body of work in both the industrial and academic fields. Advantages of heterogeneous catalysts include wider range of processing parameters, easier catalyst separation and regeneration, and cost-effectiveness. However, when compared to their homogeneous counterparts, heterogeneous catalysts have also some limitations. For example, not all heterogeneous catalyst sites are accessible to reactants. Furthermore, they also lack selectivity.

2.1 Catalysis and NPs

There are numerous reviews covering NP applications in catalysis [27–30]. During the catalyst design stage for a particular process, the nature of the catalyst is important. The catalyst support often made of porous refractory materials must sustain the harsh conditions (pH, heat, and pressure) within the reactor. Understanding and controlling catalysis at the nanoscale level, including NP–support interface properties, is attracting a lot of interest [31]. This will certainly improve the efficiency of current catalysts. For example, particle size can be controlled using the following parameters: carrier surface area, nature of the precursor, chemical nature of the carrier surface, metal concentration, metal distribution, dispersion promoters, impregnation procedure, and thermal treatments [31]. These parameters and others provide the flexibility to control and fine-tune chemical reactions.

Gold, platinum, and palladium NPs have shown superior catalytic properties for a variety of applications [30]. The catalytic properties improve when the size of these NPs is reduced. However, NPs cannot be reduced below a critical size. Indeed, below a critical size, dispersed NPs may lose their solid-sate nature and electronic properties [32]. Interatomic distance increases when NP size is reduced to below a few nanometers [33]. Up to 13% interatomic increase in the case of copper has been reported [33]. Nonmetallic properties have been observed on catalytic active gold clusters, although bulk is catalytically not active [34].

Besides high yield, improving selectivity of current catalysts is highly desired. NP-based catalysts have the potential to yield 100% selectivity [35] with minimal waste [10]. NP-based catalyst properties may behave differently whether they are used in solid-state, liquid-phase, or gas-phase reactions. They offer the distinctive advantages of higher active surface area and selectivity. In the case of liquid phase, catalysis reactions are rate-limited. Indeed, diffusion coefficients in liquid phase are about a factor of 10^4 lower than in gas phase [36]. The diffusion rate of reactants to the surface of the catalyst bodies is proportional to $1/d^2$, where d is the diameter of the catalyst particle. Thus, it is advantageous to use small NPs to increase the rate of the reaction. Three areas of catalysis in which NPs could have significant potential for commercial applications are discussed:

a. Increased active surface area, leading to optimal usage of scarce and expensive noble elements;
b. Selectivity which still remain quite elusive, although NP preparations offer several parameters (size, shape, defect, shell coating) to optimize surface reactions. A high level of selectivity has been achieved using transition-metal complexes [37];
c. Recovery and/or recycling of catalysts and their role in green and sustainable development of the petrochemical industry and other sectors. Whether supported or unsupported, most industrial catalysts are produced using wet chemical processing including the precipitation and impregnation methods.

Among the drawbacks of NP-based catalysts is their stability and effectiveness under harsh conditions. Quick catalyst deactivation and attrition are considered the main aspects requiring more development. In this regard, it is no surprise that caged NPs within a zeolite structure are by far the most successful NP-based catalysts. These zeolite materials provide a nanosized channel path with a diameter between 0.4 and 100 nm or more; a chemical-, size-, and shape-selective matrix; and a high surface-to-bulk ratio while protecting the catalyst from the harsh environment. Furthermore, besides their function as a support, zeolites provide a combination of the intrinsic properties of acidity along with shape and size selectivity.

NPs are used in colloidal form or supported on various surfaces. They provide high surface area and energy that allow a high chemical reaction rate. The presence of low coordinate sites in small catalyst particles is also used to selectively carry out certain reactions. Dispersed nanocatalysts also offer the possibility of recovery and recycling of valuable spent raw material. Catalysts allow reduction in energy required for reaction by decreasing operation temperatures. Lowering catalyst diameter leads to high surface area and energy. Indeed, the percentage of surface atoms is reduced from around 90% to less than 20% when NP size is increased from 1 to 10 nm [38]. At the same time, the energy shift ΔE (difference between the core-level binding energy levels of the NP and bulk) of Pd is reduced from 1.2 to about 0.2 eV when the NP diameter is increased from 1 to 6 nm [39].

Two other parameters are expected to change significantly when the diameter of the particles is reduced to below a few nanometers. First, anisotropy results in a drastic change in the proportion of defects, edges, and crystallographic orientations. Another change that occurs is related to the electronic properties. These changes are related to the fact that the chemical reactivity of NPs (particularly of metal) depends on the lattice structure, bond length, bandgap, and Fermi level.

Catalysis involving NPs is rapidly evolving. Free-standing colloidal NPs or those supported on a substrate have been used. Inorganic NPs have been used mostly for catalyzing various chemical reactions including alcohol and hydrogen reductions. Reductions involving electrochemical, photochemical, and sonochemical processes have also been reported. These techniques have allowed flexibility in controlling NP parameters, which could affect the catalytic properties. For example, the shape of NPs was found to play a critical role in their catalytic activity. Catalytic activities of Pt NPs with tetrahedral, cubic, and spherical shapes are evaluated (Fig. 5.4). Tetrahedral shapes with the largest surface atoms at the edges and corners give rise to the highest catalytic activity.

Biocatalysts such as enzymes are used in biochemical industrial processes. Current biocatalysts face numerous challenges including shorter lifetime, low yield, and high lifetime cost [41]. Immobilization of enzyme on/in NPs

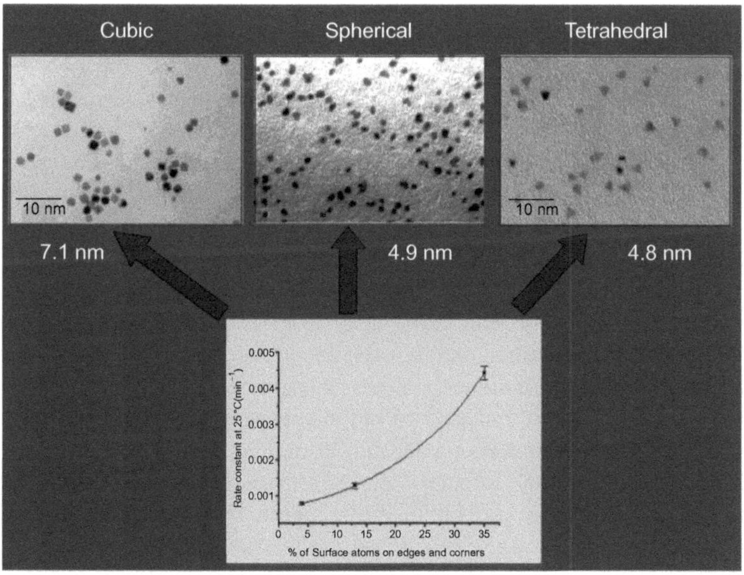

FIGURE 5.4 Variation of catalyst activity with the shape of Pt nanoparticles. The shape is represented by the % of surface atoms located on the corners and edges. *Reprinted with permission from Narayanan and El-Sayed [40]. Copyright 2005, American Chemical Society.* (For color version of this figure, the reader is referred to the online version of this chapter.)

has been suggested to address some of these challenges. Other NP-based biocatalysts are also pursued [41].

Results obtained on bulk catalysts do not translate directly to NP-based catalysts. One possible explanation is related to the significant change occurring to surface energy and structure. Most examples provided here are related to transition metal and transition-metal oxide NPs. Other compositions, such as carbon NPs, are currently being considered although with relatively limited success. Synthesis and application of metal clusters have attracted a lot of interest in academia [42]. Applications of semiconductor NPs in photocatalysis will be considered separately in this chapter.

2.1.1 Cluster-Based Catalysis

The fundamental aspects and potential applications of cluster-based catalysts have received some interest [34,43–55]. A difference between cluster-based catalysts and other more conventional NP catalysts has been proposed [26]. Clusters are often related to sub-1 nm-sized particles with well-defined chemical composition and structure at the atomic level. NP-based catalysts need to be stabilized first to prevent agglomeration. This could be done through either electrostatic or steric stabilization [43]. A third category, referred to as electrosteric, has been identified by Finke's group that allows preparation of well-defined Ir(0) nanoclusters [26,44]. Cluster size [32] and solvation level of the metal clusters [45] are critical. For example, a metal cluster with tunable size has been obtained via an electrochemical process [54].

As shown in Fig. 5.5, Ir(0) clusters are stabilized through both steric and electrostatic forces using $P_2W_{15}Nb_3O_{62}{}^{9-}$ polyoxoanion. This potentially enhances the catalytic properties as the reduced number of stabilizers per nanocluster allows more coordinatively unsaturated catalytic sites. Besides the nature of the stabilizer, size range, distribution, and the overall solubility of the NPs play a key role in defining catalyst behavior [46].

Steric stabilization is by far the most common stabilization approach. Numerous molecular systems have been used to provide steric repulsion including polymers, carbon black, siloxane polymers, dendrimers, microgels, surfactants, resorcinarenes, amines, phosphines, and oxides [26]. Electrostatic stabilization is induced by adding excess anions to the solution either prior to the formation or after the formation of the NPs. The presence of these anions near the surface of the NPs gives rise to Coulombic repulsions. As pointed out by Finke [47], purely cationic stabilization is counterintuitive, unless the presence of salts changes charge state at the NP surface.

Palladium cluster size effect on NO reduction by CO ($CO+NO \rightarrow 1/2N_2+CO_2$) has been evaluated [48]. Metal nanoclusters are suitable for FCs [49], as they provide optimized performance while using a minimal amount

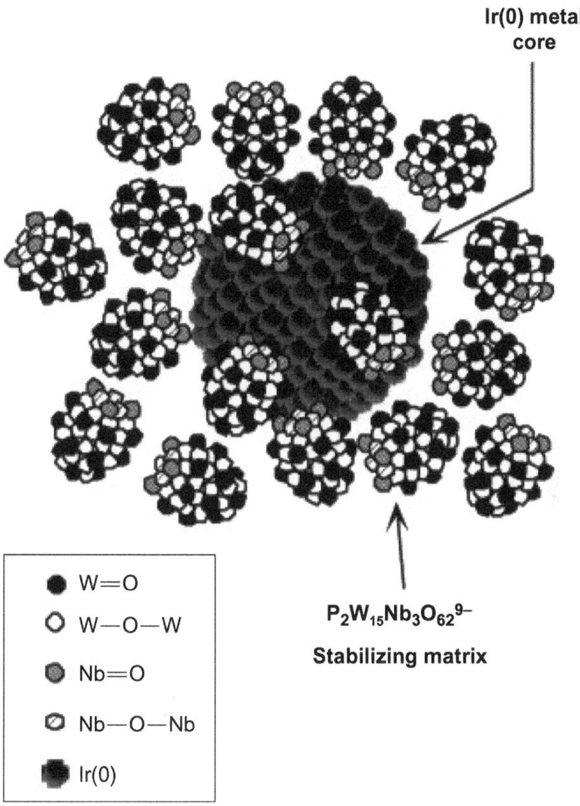

FIGURE 5.5 Idealized representation of Ir(0) metal core nanocluster stabilized by $P_2W_{15}Nb_3O_{62}^{9-}$ polyoxoanion and Bu_4N^+ (not shown) [26].

of expensive platinum. Trialkylaluminum is used as both reducing agent and protecting group for controlled seed growth [49]. Average nanocluster size is around 1.2 nm. Other metal nanoclusters have been synthesized using this approach [47–49]. Supported metal cluster catalysis has been claimed [34,52,53]. Stabilized Pd clusters have been shown to affect the Suzuki reaction [55].

Crystal planes play a significant role in catalysis in small NPs. Above 5 nm, exposed crystal planes of metal NPs are relatively independent of their size [50]. A few authors differentiate between ultrasmall clusters (about 1 nm or less in diameter), colloids (between 5 and 100 nm in diameter) on which catalysis is based [26,44], and clusters in the range of 1–5 nm [51]. Below 5 nm, NPs have a larger proportion of facets, edges, corners, microdefects, and unsaturated bonds. Furthermore, the electronic properties of ultrasmall particles could be significantly different. This area of research is quite active and promising, as it provides key parameters allowing the development of NP-based catalysts with tunable specificity and efficiency.

2.1.2 Homogeneous Catalysis

The relative market share of homogeneous catalysis has declined steadily in spite of its advantages [56]. High cost, low yield, and large waste are the main challenges. The concept of quasi-homogeneous catalysis has been introduced to address these issues [21,56–62]. As discussed already, NPs could address the shortcomings of homogeneous catalysis while retaining the advantages of heterogeneous catalysis [21,24,46]. For example, quasi-homogeneous NP-based catalysis could increase yield and reduce waste by-products. It is worth noting that small amounts of trace NPs or metals may lead to erroneous conclusions in bulk catalysis experiments. Synthetic chemists should pay particular attention to this issue, as it may lead to wrong conclusions. It has been suggested that true homogeneous catalysis is difficult to achieve experimentally [25]. Indeed, it is difficult to eliminate all trace metals that could have an effect on the overall catalytic activity.

Supported homogeneous catalysis is probably the best route to address current catalyst shortcomings [28,60,61]. Heterogeneous catalysis of gas over $Cu/ZnO/Al_2O_3$ is the dominant process for methanol production. Quasi-homogeneous hydrogenation catalysis using nickel and cobalt NPs has been reported [57]. Quasi-homogeneous synthesis of methanol using highly active copper NPs has also been reported [58]. Homogeneous hydrogenation catalysis has been observed using dendrimer-encapsulated Pd and Pt NPs as well [59]. Several groups have also claimed homogeneous catalysis based on Pd NPs [62].

2.1.3 Heterogeneous Catalysis

Heterogeneous catalysis is currently the most dominant catalysis solution. This subject has already been discussed in the previous section and is extensively covered in the literature [63]. Support materials in heterogeneous catalysis are critical for the stability and efficiency of the catalysts. Several structural and geometrical variations of silica- and alumina-based catalyst supports have been reported [64]. These design variations could be used for specific catalyst needs. Numerous processes have been developed for the fabrication of supported NP catalysts. The incipient-wetness approach, in which the catalyst support is first impregnated by catalyst precursors and then treated (thermally or chemically), gives rise to tunable and stable reduced metallic nanocatalysts. Stable silica–alumina substrates with tunable porous structures are widely used for this purpose [64,65].

On a nanoscale, heterogeneous NP-based catalysts are quite complicated to characterize and control. Electrons may need to cross at least two layers with different compositions following catalytic reaction. Often overlooked is the presence of an oxide layer between the metal-based catalyst and the stabilizing surfactant [49]. This oxide intermediate layer may reduce the catalyst activity although it could also provide long-term stability. Even this representation is simplified, as it does not account for other contributions such as the different facets and surface segregation.

Noble metal (gold, platinum, palladium, silver, rhodium, and ruthenium) and transition-metal (iron, nickel, copper) NPs are used in different catalytic reactions. Reviewing all potential catalytic reactions is beyond the scope of this book. Several recent review articles and books have been dedicated to this issue [21,30,43,65]. In early studies, support materials was mainly used to prevent NP agglomeration. NP supports are also used to provide thermal and chemical stability along with mechanical strength. Supports may or may not play a role during the process of catalysis. There are many parameters allowing the control of supported catalyst properties:

a. Nature and concentration of precursors;
b. Nature and surface area of the support;
c. Dispersion procedure;
d. Nature and distribution of the promoter;
e. Thermal treatments;
f. Nature and distribution of the metal catalysts.

Controlling these parameters is critical for the commercial success of supported catalysts. Two approaches are considered for preparing supported NPs. In the first approach, the support is impregnated with precursors. Then a reduction is conducted *in situ*. In the second approach, colloidal NPs are first formed and then deposited on a high surface area support. Both approaches have been successfully implemented in FC and other energy-related areas. However, the first approach has been more widely used.

Bulk gold is usually catalytically inactive. However, small gold NPs do show activity toward the oxidation of hydrogen and carbon monoxide [66–69]. Often catalytic activity is observed only with particles having a diameter of 50 nm or less [68]. In the case of Au–TiO_2, a maximum CO oxidation TOF is obtained with gold NPs of about 3 nm [34]. Above 4 nm, these supported gold NPs are not active for this CO oxidation. Supported gold NPs (10 nm or less) have been recently shown to provide specific catalytic properties [67].

In most cases, catalytic activity of gold NPs is observed only below 5 nm size. This result does not seem to depend on the material support composition (Al_2O_3, Fe_2O_3, $MgAl_2O_4$, SiO_2, and TiO_2) [69]. It has been postulated that the relative concentration of atoms located at the cornerstone of the bulk of NPs as a function of particle diameter is the key factor determining the TOF [69]. However, it is worth noting that smaller particles are more reactive but sinter much faster and often at lower temperatures [70]. Indeed, ultrasmall NPs are not stable, which leads to increased size after a few cycles. Furthermore, reducing metal NP size to below a critical value often leads to loss of the metallic properties.

2.1.4 Laboratory-Scale Catalysis

NP-based catalysts are currently being developed to provide cleaner fuels, and energy-efficient processes with a reduced footprint [71]. Hydrogenation, hydrosylation, and hydration reactions (mostly of unsaturated organic molecules) have been reported [72]. Although catalytic hydrogenation takes only a small market share of the industrial organic syntheses, it represents a significant contribution to the hydrotreatment in the crude oil industry. This step in the process allows the elimination of aromatics and unsaturated hydrocarbons (HCs) in transportation fuel [73]. A relatively high catalytic conversion efficiency is obtained only when using high catalyst concentration and longer reaction time [73]. High TOF is achievable using short reaction time and low NP catalyst concentration. For example, Pd NPs dispersed in silica showed high catalytic activity for the hydrogenation of 1-hexene (TOF = 38,247), cyclohexene (TOF = 33,000), and benzene (TOF = 10,000) [73]. Pd NP-based catalysts give rise to the lowest catalytic activity in the case of 2-hexanone, cyclohexanone, and benzonitrile hydrogenation [73].

Even if bulk gold is not active, its NP counterpart has shown promising catalytic activity in a chemical reaction. In particular, active CO combustion on gold NPs has received much attention [69,74–76]. Supported gold NPs catalyze numerous chemical reactions [75]. Gold NPs are catalytically active when their size is reduced below 10 nm even at low temperature [75]. Three factors contribute to the gold NP-based catalysis. They include strong bonding between gold NPs and the support, catalytic support type, and the NP size distribution.

Major concerns related to small catalyst NPs include chemical stability and sintering behavior particularly during high-temperature treatments [77]. Except for gold, other noble metals tend to form oxides on their surface even at low temperature. Agglomeration and aggregation may also occur. Effects of time length on the size distribution are summarized in Fig. 5.6. Data are presented in terms of frequency (%) as a function of particle size. The smooth curve represents a fit to the log-normal distribution function. The average particle size increases from 20 up to 500 nm when the sintering time is increased. The Ostwald ripening is likely the main sintering mechanism. Small NPs are reduced in size as atoms are transferred to larger NPs [77].

Bifunctional alloyed metal NPs are used to catalyze several chemical reactions [78]. Pt–Ru NPs dispersed in various polymers and oxide supports are found to efficiently catalyze methanol oxidation and other reactions [78]. High surface area, stability, and other electrical and mechanical properties make carbon nanotubes (CNTs) an ideal catalyst support for some applications. Several

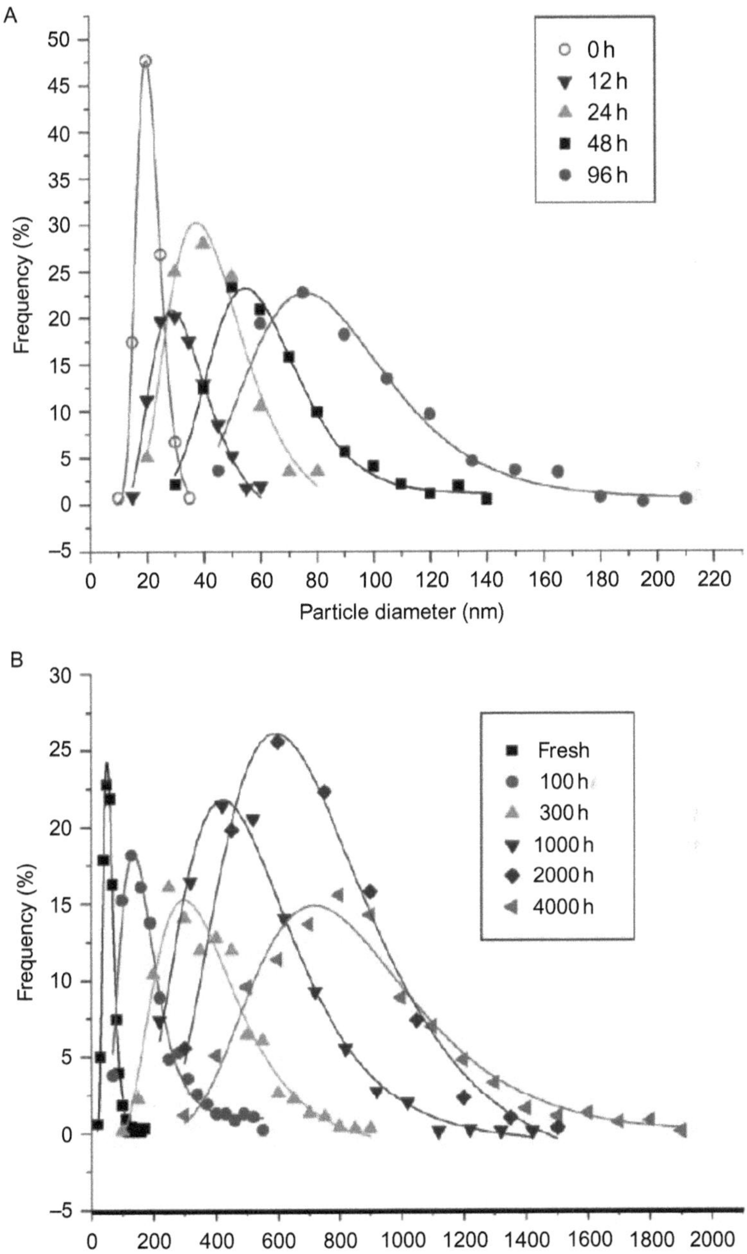

FIGURE 5.6 (A) Particle size distributions after sintering 7 wt% Pd/alumina at 900 °C for differing time lengths and (B) 15 wt% Pd/silica–alumina up to 4000 h. Data fittings are obtained using a log-normal distribution [77]. (For color version of this figure, the reader is referred to the online version of this chapter.)

manufacturers are currently scaling up their production, which may reduce their high cost. Commercialization is currently hampered by high cost. Low production yield is the main challenge. Metallic NPs have been used to catalyze single-walled carbon nanotube synthesis [79]. Cobalt NPs are widely used for CNT synthesis [80].

Fluidized bed reactors (FBR) are quite suitable as catalyst supports for large-scale applications. However, integrating NPs in FBR is quite challenging [81–83]. NP fluidization can achieve homogeneous (bubble-less) fluidization using high superficial gas velocity. Degussa aerosil R974 powder, with a primary particle size of 12 nm, was fluidized using nitrogen in a cylindrical vessel [81].

2.2 Catalysis in Bulk and Fine Chemical Manufacturing

There are significant differences between bulk and fine chemical manufacturing industries [20]. Contrarily to fine chemicals, bulk chemicals are produced in massive quantity (usually more than 10 ktons per year) with low variety, while using very few processing steps. The fine chemical industry is characterized by a wide variety of products and high added value. Furthermore, the fine chemical industry is labor-intensive with a limited number of producers.

In contrast to the bulk chemical industry, most fine chemicals are manufactured with minimal catalyst use. On the other hand, solvent use is much more common in fine chemical production. Solvents are believed to have significant cost and negative environment impacts. One attribute of NP-based catalysts is their potential for recycling [84]. Indeed, expensive noble catalyst NPs can be separated relatively easily from the matrix.

2.2.1 Bulk Chemical

NPs are used in catalyzing bulk chemical production reactions. Reaction products involving olefins, ethylene, propene and carbon monoxide, alcohols, vinyl-halogens, polyamides, and other basic chemicals have been obtained using NP catalysis.

It has been demonstrated that there is a direct relationship between particle size and the performance of rhodium-based catalysts used for CO hydrogenation [76,85,86]. Ojeda et al. showed that $Rh/\gamma\text{-}Al_2O_3$ catalyst TOF increases by a factor of 4 when the Rh particle size is increased from less than 5 to about 30 nm [85]. Interaction between the metal NPs and the oxide support also increases when the particle size is decreased. Increased metal–support interaction leads to surface Rh atom oxidation. Higher-density $Rh^{\delta+}$ species in smaller Rh particles leads to the formation of oxygenate species [85].

The catalytic effect of ZSM-5 zeolites on the oligomerization of ethylene has been reported [87]. This reaction produces higher HCs of the gasoline range. Zeolites with small crystallites (30–50 nm) showed higher ethylene conversion and high gasoline range selectivity. The longer catalytic lifetime of these small crystallites has been directly correlated to the change in specific area.

Vinyl acetate monomer (VAM) is a key raw material for the production of polyvinyl acetate and polyvinyl alcohol (PVOH or PVA) used in paints, adhesives, and surface coatings. VAM world production was about 4.3 million tons in 2003 [88]. VAM is prepared by reacting ethylene, acetic acid, and oxygen over a fixed-bed catalyst containing nanosized Pd particles. Particle size of Pd catalysts supported on a high-surface-area SiO_2 has been shown to have a direct effect on VAM synthesis [89]. Indeed, the rate of VAM formation is significantly enhanced along with the selectivity and apparent activation energy when the Pd particle size is reduced.

Methanol is considered as one the largest bulk chemicals in terms of volume. Today, natural gas remains the most economical source for methanol production using a two-step process. Synthetic gas (syngas, CO/H_2) is first produced using a reforming reaction. Syngas is obtained using steam methane reforming. CO and H_2 constituting the syngas are reacted over copper-based catalysts to form methanol. Methanol as an energy vector may have some advantages over hydrogen. Methanol is relatively easier to store and distribute. Methanol economy [90] proponents believe that methanol has better prospects than hydrogen to challenge electricity as the future energy vector. For example, methanol is considered a more suitable fuel for FC. Easier to transport than natural gas and hydrogen, methanol also provides an alternative for on-site hydrogen production [91]. Methanol synthesis from syngas has been shown to increase on supported nanosized ZnO catalysts [92].

Alternative non-oil-based sources for gasoline are currently sought. Natural gas and coal are more abundant and relatively cheaper than oil. Methanol-to-olefins (MTO) and methanol-to-gasoline (MTG) are often part of an integrated process that includes natural gas reforming for ethanol [93]. NP-based catalysts are used to improve the overall efficiency of MTO and MTG processes [93].

Direct conversion of methane to olefin and gasoline has attracted even more attention, instead of going though intermediary chemicals. A two-step reaction process based on Pt dispersed in zeolites is used for higher alkane production [94]. Zeolite matrix type may affect the catalyst activity for methane conversion [95]. Pt NPs stabilized with PVP showed good activity in the synthesis of higher HCs at low temperature [95]. The distribution and the molar amounts of C2+ per unit mass of Pt increase when the size of platinum NPs decreases. Larger metal particles are favorable for the selectivity of higher HC molecules [95].

2.2.2 *Fine Chemical*

Contrarily to the bulk chemical industry, fine chemicals are often produced via more conventional stoichiometric reactions including oxidations and reductions. Furthermore, due to their thermal instability, fine chemicals are often produced in the liquid phase at moderate temperatures. Nonsupported catalysts play a more significant role in the manufacturing process of fine chemicals.

In light of the dwindling fossil fuel resources, production of fine chemicals from renewable energy is attracting a lot of interest [96]. As a feedstock, the potential reserve of biomass including wood, sugarcane, corn, algae, and oil palm is unlimited and renewable. However, their conversion to fuel and chemicals is challenging. High capital cost and low conversion yield have limited their development so far. There are several examples of natural products based on fine chemicals extracted from various biomass sources. These include the anticancer drug paclitaxel and food colorant β-carotene. Solid catalyst recycling may improve production yield when compared to biocatalysts and chemo-catalysts [96]. As discussed earlier, NP-based catalysts combining homogenous and heterogeneous attributes could address the current shortcomings of these biocatalysts and chemo-catalysts.

Homogeneous catalysis is widely used in fine chemical synthesis. One drawback of homogeneous catalysis is the requirements for extra steps to separate catalysts from the product. It is conceivable that fixing these homogeneous catalysts to NP supports will aid easy recovery. The development of large-pore-size zeolite has allowed the immobilization of numerous homogeneous catalysts [60]. In some cases, enzyme immobilization has shown enhanced catalytic activity as well as improved stability and recycling capability [60]. It is expected that zeolite NPs will provide even higher surface activity and smaller diffusion path length particularly for support with lower pore size.

Cluster-based catalysts provide the potential for very precise chemical synthesis. Stabilized gold NPs have been pursued as potential catalyst alternatives for a sustainable development of the chemical industry [30]. These catalyst systems are used for different oxidation, hydrogenation, C–C coupling, and water–gas shift reactions. Polymer-stabilized gold NPs with a core size of 10 nm or less provide higher TOF [97]. Furthermore, these NPs allow precise tuning reaction products while using benign precursors and solvents. For example, a 98% yield glucose to gluconic acid reaction was reported when using 2.6 nm gold NPs supported by anion-exchange resin [97]. NP-based gold catalysis could be key to sustainable process chemistry. Other transition-metal NPs are also pursued in different industrial processes [30]. For example, Pd, Pt, Ag, Fe, and Ni NPs have also been successfully used.

Carbon-encapsulated FeNi NPs are developed for fine chemical manufacturing [98]. These FeNi NPs are relatively easier to separate after use. The carbon shell isolates the core magnetic NPs and prevents destructive reactions before and during the operation. Furthermore, the graphitic carbon surface is functionalized with catalytic bearing sites such as metal clusters, homogeneous catalysts, and enzymes [98]. Hydrogenation reactions involving partial reduction of functional groups have been achieved using the so-called Mazda process [99]. In this process, precious metal NPs are embedded in the catalyst support, preventing catalyst degradation and high efficiency. This process could be used in the fine chemical and pharmaceutical industries as demonstrated by BASF [99].

2.2.3 Catalyst Additives for Combustion

Combustion is a very inefficient conversion process. When suitable catalysts are used, heterogeneous combustion proceeds more efficiently than the homogeneous gas-phase combustion. Given their attributes, NPs are pursued to overcome the limitations of the combustion process. Indeed, lower processing temperature, higher efficiency, lower emission, and limited industrial waste are achievable using NP-based additives. One advantage of carrying out combustion at lower temperatures and lower fuel concentration is the reduction in NO_x emission. Metal oxide NPs have been shown to promote combustion reactions, leading to higher combustion efficiency and lower emissions.

Optimized NP catalysts accelerate the fuel combustion rate, allowing higher energy release and lower exhaust emission. A fabrication method of fuel additives containing ceria NPs doped with a divalent or trivalent metal has been recently disclosed [100]. CeO_2/BHA NPs prepared by reverse microemulsion allowed light-off with 1 vol% CH_4 stream in air at 400 °C [101]. This temperature is much lower than the temperature of 1300 °C required for maintaining a stable methane combustion flame.

Using cerium oxide NPs as fuel additives, Oxonical LLC claims that it boosts fuel efficiency by 5% and cuts soot emissions by about 15% [102]. Recently, H2OIL [103] announced that using nanosized water droplets of about 10 nm as additives allowed a 20% reduction in fuel consumption and a reduction of up to 50% in the most harmful emissions. Ceria NP size and concentration are directly linked to the combustion characteristics of different fuel mixtures [104]. The effect of composition (metallic and oxide), size (15 and 50 nm), and volume fraction (0%, 0.1%, and 0.5%) of aluminum NP addition on the ignition properties of diesel fuel has been recently investigated [105]. It was concluded that NP addition improves significantly the fuel ignition probability. Furthermore, the size and composition of the NPs does not seem to have any effect on the ignition properties. The effect of NP addition is probably due to an improvement in the radiative heat transfer.

2.2.4 Catalyst for Automobile Exhaust Emission Control

HC particulates and NO_x gases (NO and NO_2) are the main emission components from internal combustion engines requiring rigorous control. Indeed, they have harmful effects on human health and the environment. Control of emission from automobile exhaust using different NP-based catalytic converter formulations is considered one of the greatest technological successes of the twentieth century [106]. More stringent emission regulations on gasoline- and diesel-powered cars require the development of higher-performance catalysts. Three-way catalysts mitigate the three major pollutant emissions from car exhaust (CO, HCs, NO_x). These three pollutants are converted into CO_2, N_2, and H_2O. This three-way catalyst based on Pt–Pd–Rh ultrafine particles on Al_2O_3 is capable of simultaneously removing NO_x, HC, and CO.

Catalyst requirements for emission control are quite severe. Indeed, these catalysts should react instantaneously with a response time of less than 1 s. This corresponds to the contact time of gas with a solid catalyst. They should also be effective from room temperature (below 0 °C in cold regions) up to 60 °C and above. Noble metal (Pt, Pd) NPs dispersed on monoliths with narrow channels are effective in oxidizing CO and HCs. Alloying using additional elements is used for the reduction of other emission gases [16]. Instead of Pt, Rh is often used to reduce and/or eliminate carbon components and NO_x from diesel engine. Sulfur poisoning of noble metal catalysts is the main challenge. Given the high cost of noble metal-based catalysts, alternative materials are currently being considered. Well-dispersed CuO and CuO/Al_2O_3 NPs are being evaluated for NO_x emission control [29].

Monolith or honeycomb structure is the dominant catalyst support for automobile exhaust conversion. Cordierite containing two MgO, five SiO_2, and two Al_2O_3 parts is the preferred material composition. Ceramic honeycombs supporting Pt/Rh/Pd NPs are effective in the case of gasoline engines operating at a stoichiometric air/fuel ratio of 14.6:1. This is not the case of diesel engines operating under lean conditions with air/fuel ratios greater than 17:1 [106].

2.3 Petrochemistry—Hydrocarbon Refining and Petrochemical

The HC-refining sector consumes the highest amount of catalysts. Transportation fuel (gasoline, diesel, and kerosene) is the main product of oil refineries. Other primary petrochemicals consist of olefins (ethylene, propene, and butadiene) and BTX (benzene, toluene, and xylene) aromatics. By-products including lubricating oils, paraffin wax, and coke are also obtained. These chemicals are used to produce other bulk chemicals, including methanol, ethanol, glycol, and styrene. Although natural gas is often used as a fuel source, it is also used to produce syngas, which could be used later for chemicals and liquid fuel manufacturing via the Fischer–Tropsch process. Direct methane to chemicals is also attracting some interest, although there is no commercial process for this as yet. Besides steam cracking (hydrocracking) and thermocracking, catalytic cracking and catalytic reforming are two very important petrochemical processes. The majority of these processes involve heterogeneous catalysis, which includes acids and acidic mixed oxides, semiconductors and transition-metal oxides, transition metals and insulator oxides.

Optimizing catalyst size and shape is critical for cost reduction [14,106,107]. The catalyst particle should be as small as possible to increase surface area and optimize expensive catalyst utilization. At the same time, it should be stable in the reactor. There are three main heterogeneous catalyst compounds: metal oxide used for selective HC oxidation, metal sulfide used for hydrodesulfurization of petroleum, and transition metals used for hydrogenation and dehydrogenation. Nanostructured catalysts are believed to have an even greater impact on the petroleum and chemical processing industries in the near future.

Introduction of the fluid-bed catalytic cracking processes facilitated the utilization of supported NP-based catalysts [108]. Fluidization and collapsing beds have been obtained using 10 nm fumed silica [109]. Indeed, high bed expansion ratio without large bubbles is achieved [109]. NP-based catalysts have been used in an FBR to carry out catalytic cracking, vinylation, and oxidehydrochlorination [110]. An aerosol nanocatalyst-based process, referred to as an aerosol nanocatalyst fluidized bed reactor, has been shown to provide up to 10^5 times improvement over the standard industrial catalysts [110].

Size and shape of zeolite crystals are known to have a direct effect on the overall catalyst performance [111]. Pinnavaia's group has recently reported an aluminosilicate NP that provides high catalytic cracking of high-molecular-weight HCs [112]. Protozeolitic aluminosilicate nanoclusters and starch have been used to obtain catalysts with a combination of mesoporosity, hydrothermal stability, acidity, and longevity.

2.3.1 Hydrocarbon Refining

Improved catalysts for the petrochemical industry will help reduce carbon footprint and other negative environmental impacts [113]. Increased fuel (gasoline, diesel, and jet fuel) and lubricant requirements demand specific and efficient nanocatalysts. Bifunctional zeolite catalysts consisting of ultrasmall particles dispersed in acidic zeolites are widely used in commercial processes such as hydrocracking. These nanocatalysts are often obtained by introducing metallic ions (e.g., Pt, Ru, Mo, and Ni) in zeolites by ion exchange followed by hydrogen reduction.

Geometrical factors of catalyst supports including size also matter. Aggregated H-ZSM-5 zeolites consisting of small crystallites showed long catalytic lifetime and high activity. Sugimoto et al. showed that aggregated H-ZSM-5 consisting of small crystallites (<200 nm) have longer catalytic life and higher selectivity [114]. In the case of conversion reaction of MTG (methanol to HCs), these nanocatalysts give rise to higher selectivity for C5+ HC, aromatics, and C9+ aromatics when compared to large crystallites (about 4 μm) ZSM-5 [114]. Sotelo et al. showed that ZSM-5 NPs with an average size of 50 nm provide superior catalytic cracking of polyolefin (HDPE, LDPE, and PP) and lube oil when compared to larger ZSM-5 particles with an average size of 5 μm [115].

Utilization of biomass presents several challenges to refining. Differences in composition and structure may require the development of specific catalysts. NP-based catalysts may provide the best solution for catalytic upgrading of biorefinery oil from microalgae [116]. Indeed, oxide NPs provide superior catalytic properties during transesterification [116].

2.3.2 Desulfurization

Regulations for low sulfur transportation fuel ($S < 10$ ppm) has spurred the development of nanosized hydrodesulfurization catalysts. Furthermore, new measures have been enacted to reduce the release of sulfur dioxides in the atmosphere. Sulfide NPs of cobalt, nickel, and molybdenum embedded in porous alumina support have been used. Bimetallic Pt–Pd NPs dispersed in different titanium dioxide supports are used to convert SO_2 to SO_3 [117]. Temperature and the type of titania have been found to affect the conversion rate. Metal NPs embedded in alumina supports have been developed for hydrodesulfurization of difficult-to-remove refractory sulfur species in deep desulfurization processes [118].

2.3.3 Synthetic Gas (Syngas) and Hydrogen Production

Syngas consisting of CO and H_2 is used in many large-scale industrial processes for the production of methanol, ammonia, aldehyde, and gas-to-liquid (see later). Syngas generation accounts for more than half of the overall cost of these chemicals and fuels. Natural gas and light HCs are the major feedstocks for syngas manufacturing [119,120]. Syngas is obtained by an endothermic reaction of methane and water gas around 750 °C ($CH_4 + H_2O \rightarrow CO + 3H_2$), followed by an exothermic reaction at around 350 °C ($CO + H_2O \rightarrow CO_2 + H_2$). Syngas production requires significant capital investment with a large thermal budget. The overall production cost is estimated to be around $4/kg depending o the price of natural gas. Thus, improving efficiency to reduce cost is the main challenge of the syngas process.

Steam reforming (STR) and partial oxidation of methane (POM) and higher aliphatic HC remain the most economical processes. STR involves an endothermic reaction providing an H_2/CO ratio of 3 or more. POM is exothermic and provides an H_2/CO ratio of around 2 or less. Autothermal reforming (ATR) that combines STR and POM processes is more energy-efficient and economical for very-large-scale syngas production. ATR provides more flexibility in the value of the H_2/CO ratio. Various reactor designs have been reported. The catalyst often consists of highly dispersed NPs in a ceramic matrix [121].

A major drawback of POM and ATR processes is the requirement for high-purity oxygen along with an energy- and capital-intensive process. Ion-conductive ceramic-based membranes have been shown to provide a cost-effective and energy-efficient process for hydrogen generation [122]. Combining hydrogen membrane and ATR will produce purified hydrogen. Particle catalysts of about 5 mm diameter with high packing density are used [123]. A variety of catalysts have been used for partial oxidation of methane to form syngas. Active catalysts often consist of few nanometer NPs when noble metal or metal oxide is used as a support [124]. More than 95% of syngas production uses STM, POM, or a combination of the two. In most cases, catalysts are used to reduce operating temperature and improve efficiency. Spent catalysts could be regenerated after deactivation [125].

Reforming catalysts are often produced by either the coprecipitation or impregnation method, followed by washing, drying, calcining, and pelletizing or extrusion [126–128]. Size and surface properties of nickel nanocrystals have been reported to play a vital role [129,130]. Natural gas reforming uses nickel-based catalysts with a typical dispersion of 1% corresponding to particle size of 20–200 nm [119]. Nickel NPs can also be used to synthesize carbon nanofibers [129].

Well-dispersed metal NPs in a porous network are used for syngas and hydrogen production [130]. In the case of Ni nanocrystals supported on $MgAl_2O_4$, high-resolution transmission electron microscopy (TEM) showed well-defined nickel NPs about 1 nm large. Sintering occurs at high temperature giving rise to an increase in particle size. STM uses nickel supported on alumina or other thermoresistant carriers as the catalyst system. Precious metal-based catalysts have also been used [131]. The impregnation method followed by a reduction step at high temperature is often used to obtain high surface area catalysts. This gives rise to dispersed metallic nickel nanocatalysts. It has been shown that reduction of Ni catalysts at 500 °C gives rise to nanosized crystallites with a diameter between 6 and 15 nm. They are more stable under cooking conditions than the Ni catalysts reduced at 700°C, with catalyst size ranging from 10 to 36 nm [132].

An alternative and cleaner approach for syngas and hydrogen manufacturing is achieved by using mixed electronic–ionic membrane, allowing preferential diffusion of proton or oxygen ions. As summarized in Fig. 5.7, there are various concepts for oxygen permselectivity using both dense and porous ceramic membranes. Electric field, differential pressure, and porosity gradient are used to drive oxygen separation. These membranes could not only be used for oxygen production, but also be integrated into solid oxide fuel cell (SOFC) and reforming reactors.

A ceramic membrane has been recently developed by the Air Product group, allowing syngas manufacturing directly on an oxygen ion ceramic

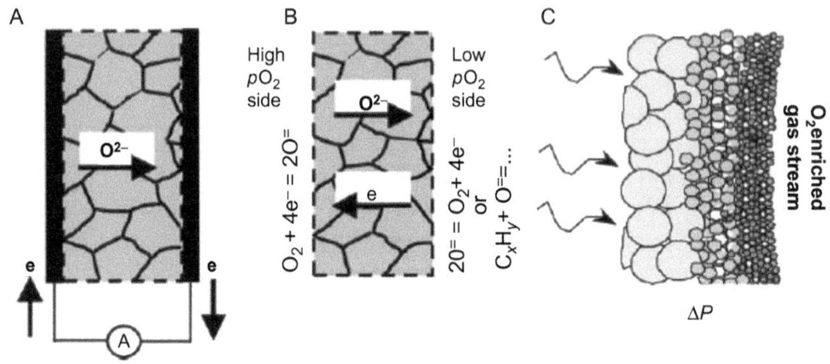

FIGURE 5.7 Different dense membrane concepts used for oxygen separation and transport in a reactor: (A) electric-driven membrane cell, (B) pressure-driven mixed ionic–electronic conductor (MIEC), (C) asymmetric porous membrane with a graded porosity [133].

membrane [122,134]. Ceramic materials for such membranes remain a major challenge facing commercialization of this new technology. One potential solution is to use Pr-doped $La_2Mo_2O_9$ NPs as reported recently by Subramania et al. [135]. Numerous other nanocrystalline-based ion conductor membranes have also been reported [136,137]. It has been established that depositing dispersed NPs on a dense membrane surface enhances the yield of partial oxidation reactions [133]. Indeed, the presence of 3D interface boundaries involving gas phase, catalyst, and membrane will facilitate high oxygen desorption and alkane adsorption around the active catalyst sites. High-resolution TEM analysis confirmed Pd nanoclusters obtained by laser vaporization deposition and by VMgO sol–gel film [133].

Existing materials for oxygen-permeable ceramic membrane reactors are either unstable under reducing atmosphere conditions (methane or syngas) or have low O^{2-} conductivity. Nanocomposites containing equal volume fractions of CeO_2 and $La_{0.2}Sr_{0.8}O_{3-\delta}$ are deposited on a thick porous substrate. The resulting asymmetric tubular reactor improves the overall physical and catalytic properties [138]. CeO_2 powder with an average diameter of 2 μm provides excellent support-buffer properties during and after sintering of the deposited submicron $La_{0.2}Sr_{0.8}O_{3-\delta}$ particles (Fig. 5.8).

Hydrogen production from methanol is economical [139]. Catalytic activity of Cu/ZnO NPs supported on alumina has been optimized for low-temperature methanol reforming [140]. Temperature, pH, and addition of boehmite are used as processing parameters. In particular, low-temperature precipitation promotes the formation of smaller copper crystallites providing high surface area. Hydrogen production from methanol steam reforming is efficiently catalyzed using nanoparticulate ternary oxides. Copper-based nanosized catalysts ($CuMn_2O_4$ and $CuCr_2O_4$ spinels or CuO/CeO_2 and CuO/ZnO) have been prepared using silica xerogel as templates [141]. These

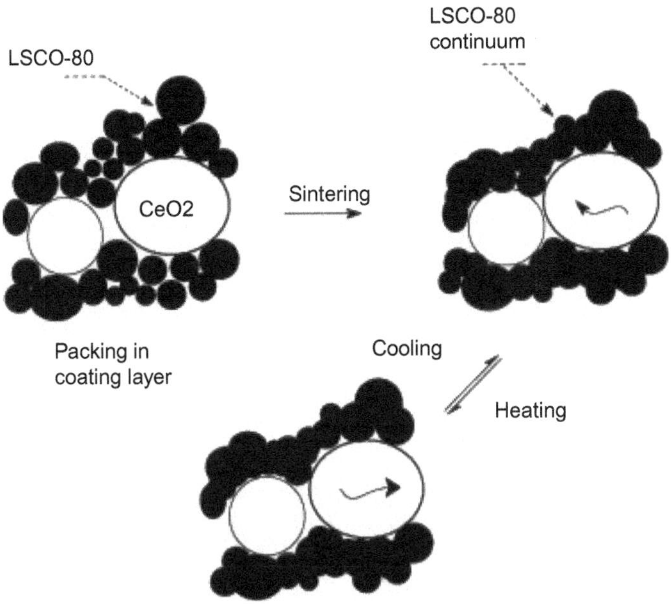

FIGURE 5.8 Schematic illustration of CeO_2 powder role during the sintering process of $La_{0.2}Sr_{0.8}O_{3-\delta}$ nanoparticles [138].

NPs showed high catalytic activity with selectivity toward CO_2 formation above 90%.

Very small PdZn NPs obtained by coprecipitation do not promote high catalytic activity [142]. These PdZn catalysts subjected to increased heat treatment allowed size control. It was found that catalysts with a mean diameter of 9 nm provide activity similar to that of catalysts with a mean particle diameter of 34 nm.

2.3.4 *Fisher–Tropsch Synthesis*

Ultraclean diesel and gasoline fuels are obtained from synthetic gas using the Fischer–Tropsch synthesis (FTS) process. Several commercial-scale plants have been built around the world. Although the first commercial diesel plant using the FTS process was built in the 1950s, it is only recently that interest in the FTS process has gained significant momentum. When using syngas as a raw material, two reactions help define the overall chemistry of the FTS process:

$$CO + 2H_2 \rightarrow (-CH_2-) + H_2O \qquad \text{(reaction I)}$$

$$2CO + H_2 \rightarrow (-CH_2-) + CO_2 \qquad \text{(reaction II)}$$

Two types of supported NP catalysts have been used. These catalysts are based on either cobalt or iron. Cobalt-based catalysts promote reaction I, whereas iron-based catalysts promote reaction II. The water–gas shift

$$CO + H_2O \rightarrow H_2 + CO_2$$

is catalyzed in the presence of iron. Fe_xO_y NPs embedded in a carbonaceous matrix have been used as catalysts [143]. These NPs are stable and provide selectivity for different HC chain lengths. Catalyst supports are mostly based on refractory oxides such as alumina, silica, aluminosilicate, and titania.

2.4 Photocatalysis

Photocatalysis is a light-driven redox reaction that often requires solid-state catalysts [144]. Efficient and cost-effective photocatalysis is obtained using nanosized semiconductors wide bandgaps [145,146]. The photochemistry-based energy conversion process has been hailed as the best alternative to the current fossil energy-based conversion [147]. Several catalysts have been developed to improve the efficiency of natural photosynthesis. Solar-to-electricity, light-to-mechanical work, and light-to-hydrogen are possible photoconversion processes that are currently under development or utilization.

Anatase titania NPs provide excellent photocatalytic properties. With a bandgap of about 3.1 eV, they absorb the ultraviolet (UV) region of the sunlight that leads to electron–hole creation and separation. These charges provide strong oxidizing power allowing organic molecule decomposition. It has been reported that the market of TiO_2-based photocatalysis may reach US$5 billion in the near future [148]. UV and visible radiation photocatalysis have been studied for decades because of their potential applications in automobiles (hydrophilic coating for mirror cleaning), air purification (photodecomposition of odors and volatile organic compounds (VOCs)), and building and structure (photodecomposition of surface contaminations) [149]. Often these photocatalytic reactions involve four steps [150]:

a. Light absorption leading to electron–hole pairs;
b. Migration of formed electron–hole pairs to the TiO_2 surface;
c. Formation of hydroxyl radicals and superoxide ions via redox reactions;
d. Photodecomposition of surface pollutants via reaction with active species on the TiO_2 surface.

In spite of all their advantages, titania-based photocatalysts suffer from low efficiency due to their high-energy bandgap. Indeed, the titania energy bandgap does not fit the solar spectrum distribution. Several material solutions have been developed to address this issue.

Photocatalysis may also provide a sustainable solution for next-generation fuels such as hydrogen. This subject is described in more detail elsewhere in this chapter. Photocatalysis is also a potential solution for CO_2 mitigation. Indeed, CO_2 photoreduction forming methanol at the semiconductor particle surface had been described several decades ago [150]. Low efficiency and high capital cost are major challenges. There are numerous requirements for any semiconductor photocatalyst to provide a cost-effective and an efficient solution:

a. Stability over a wide range of conditions;
b. Low bandgap (E_g) and its threshold wavelength: λ (μm) = 1.24/E_g (eV);
c. High absorption coefficient α defining the light extinction using $I = I_0$ exp ($-\alpha l$) equation. For example, a material with $\alpha = 10^5$ cm^{-1} at a given wavelength λ means that 90% of incident light is extinguished after crossing 100 nm of the semiconducting material at this wavelength;
d. Alignment of conduction and valence bands with oxidation and reduction states of the catalyzed reactants;
e. Rates of oxidation and reduction should be similar; note that organic oxidation by valence-band holes occurs at a much faster rate than oxygen reduction by conduction band electrons.

To optimize light absorption, and for bandgap, and band-edge optimization, TiO_2 doping was used [149,151]. NP size and composition structure are also critical parameters in photocatalysis [152]. The large TiO_2 bandgap (3.1 eV) does not match the solar spectrum. As shown in Fig. 5.9, MoSe NPs with an average diameter of 4.5 nm are more suitable. Nanosized

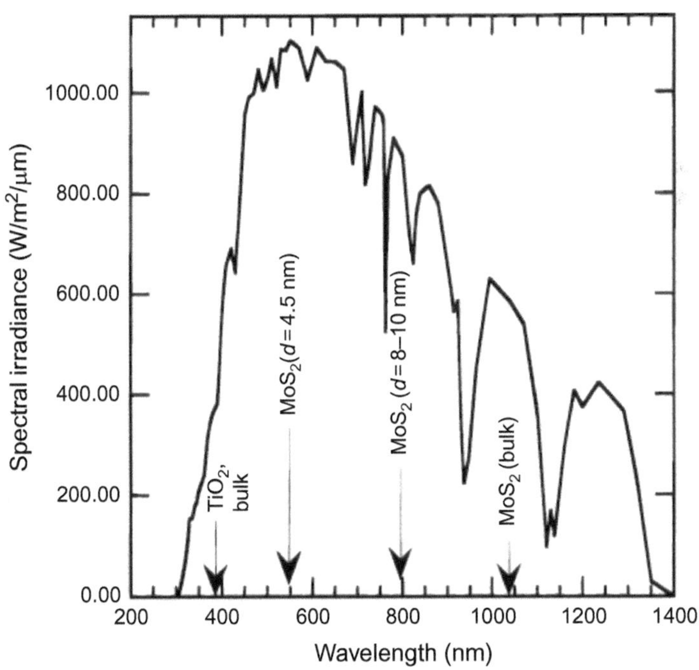

FIGURE 5.9 Spectral irradiance of the AM1.5D solar spectrum with the absorbance edge of various semiconductors. *Reprinted with permission from Thurston and Wilcoxon [153]. Copyright 1999, American Chemical Society.*

MoS_2 materials have been successfully used to catalyze the photooxydation of phenol [153].

There is a direct relationship between the photocatalytic properties on the one hand and the size and phase compositions of the NPs on the other hand [154]. Water splitting in a photoelectrochemical (PEC) cell using a titania photoanode with a wide bandgap was first reported by Fujishima and Honda in 1975 [155]. Although stable under harsh conditions, this cell provides very low conversion efficiency. Colloidal TiO_2 NPs are often used to build the photoanode. The use of other oxides with lower-energy bandgaps did not provide better results. Titania doping has been used successfully to improve quantum efficiencies for oxidation and reduction photoreactivity [156]. Nitrogen doping of titania has also been used to improve efficiency [157].

A multi-bandgap photoconversion concept has been developed to optimize solar radiation absorption and conversion [158]. A two-photoelectrode design circumvents the need for additional voltage bias [159]. A monolithically integrated PEC cell based on a tandem design with an efficiency of 12.4% has been reported [160]. GaAs and GaInP semiconductors films are used in this case. Besides its high cost, the stability of these materials is a major issue.

2.4.1 Organic Decomposition

Photocatalysts used for organic decomposition provide practical applications for sterilization, water purification, air purification, and building exterior self-cleaning, particularly in areas where there is no electricity [161,162]. Photocatalytic organic degradation in water using ZnO NPs has been recently reported [163]. Photocatalysts not only kill bacteria, viruses, fungi, and cancer cells but also decompose the cells. This may provide avenues for sustainable product extraction from algae and other single cells. In some cases, TiO_2 is more effective than any other antibacterial agent. For example, disinfection using titanium oxide is three times stronger than with chlorine, and 1.5 times stronger than with ozone. Water is also purified by pollutant oxidation and bacteria destruction using photocatalysis. Hydroxyl radicals help accelerate the breakdown of molecular bonds of VOCs (tobacco odor, formaldehyde, chlorofluorocarbon, nitrogen dioxide, urine, and gasoline), giving rise to less harmful gases. Architectural glass coated with a TiO_2 layer allows self-cleaning. Under sunlight, the coating promotes photooxidation of carbonaceous dirt on the surface, which is washed away by rainfall or water spray.

2.4.2 Water Splitting and Hydrogen Production

In spite of a large number of studies, direct water splitting using visible light is still inefficient and expensive [155,164,165]. Two approaches could be considered for water splitting: photoelectrochemistry and high-temperature electrolysis. In a typical PEC setup, a wide-bandgap oxide semiconductor is used as an anode to split the water molecule to form oxygen and protons.

These protons are carried to the cathode (here platinum) to complete the reaction, allowing hydrogen formation. The energy required per electron is 1.23 eV corresponding to the energy gap between the oxidation and reduction potentials. To overcome overvoltage losses and to increase production rate, applied voltage of up to 2 V is required. Often, a photo energy of 1.6 eV (0.77 μm), corresponding to the visible regime of the solar spectrum, is sufficient to drive the water-splitting reaction.

Among all the potential photocatalytic materials, TiO_2 remains the best choice because TiO_2 is very stable in aqueous electrolyte under solar radiation. Furthermore, the properties of TiO_2 are tunable as functions of size, doping, and surface modification. Finally, TiO_2 is widely available and inexpensive to produce and refine. However, the high bandgap limits the solar radiation absorption. Other semiconductor materials with smaller bandgaps including oxide, oxysulfides, oxynitrides, and composites are currently being developed [166]. The conduction band, valence band, and bandgap energy values of some of these alternative materials (e.g., CdS and GaP) in aqueous electrolyte are suitable for PEC [167]. However, their overall performance is still low. Modified TiO_2 photocatalysts used for water splitting provide higher efficiency. Indeed, high solar-to-chemical conversion efficiency in the range of 10% is achieved [168]. Doped ceria NP-based electrodes give rise to a better response in the visible region of the solar spectrum when compared to titania [169]. However, overall efficiency of PEC based on titania is higher.

A 3D nanoarray of interconnected indium phosphide (InP) NPs modified by $Fe_2S_2(CO)_6$ is used as a photocathode for hydrogen production [170]. First, an InP film is obtained using layer-by-layer (LBL) assembly by alternating exposure to the dithiol linker and InP solutions. The resulting film is soaked in toluene solution containing $Fe_2S_2(CO)_6$. Using this platform, a PEC efficiency of more than 60% is achieved with a single-wavelength light-emitting diode (LED) [170]. The PEC process does not rely on the excited states of organic molecules. The limited lifetime of photosensitizers has so far hindered the use of organic dye cells for hydrogen production.

Cu and/or Pt NPs deposited on nitrogen-doped titania nanotube arrays are used to convert efficiently CO_2 and/or H_2O gases to useful fuel and chemicals [171]. The nanotube wall thickness is critical for obtaining high-rate photocatalytic conversion.

An alternative PEC design is to use a suspension of catalyst NPs in aqueous solution. Suspension of photocatalyst NPs in water under irradiation leads to oxidation/reduction reactions on the semiconductor surface [172]. Surface modification should be used in this case.

2.4.3 Direct Methane Conversion

There is a significant amount of stranded natural gas reserve in remote areas. Given the low specific energy content in natural gas, there is a need of *in situ* transformation to increase energy density or added value. Currently, there is no direct commercial process to convert natural gas into denser chemical or fuel. Low HC yield is the main reason for the lack of a competitive direct methane conversion process.

Photocatalytic conversion of natural gas will improve the conversion efficiency of the gas-to-chemical process. In particular, nanostructure-based photocatalysis has been investigated [171]. A wide range of HC yields have been obtained by different material compositions for photocatalytic nonoxidative coupling methane (NOCM). In this experiment, 200 mmol methane at a pressure of 7 kPa with 0.2 g catalyst loading was photoirradiated by a 300 W Xe lamp for 3 h at room temperature [171]. Ga_2O_3 and TiO_2–SiO_2 photocatalytic materials showed the highest yield [171]. Nanostructured TiO_2–SiO_2 materials with a yield of 0.2 are potentially a cost-effective and sustainable solution as they consist of low cost and widely available elements.

2.5 Biocatalysis

Biomimicking has played a significant role in the development of NP-based catalysts [172]. High reaction rates and up to 100% selectivity are reported. Fundamental studies of natural catalysts based on inorganic nanoclusters modified by enzymes containing proteins have been made. Inorganic nanoclusters containing iron are the active sites of enzyme catalysts. These nanostructured biocatalysts allow a high rate of oxidation and reduction reactions.

Enzymes are widely used to catalyze biochemical reactions. There are several examples showing clear advantages when functionalized NPs are used to immobilize enzymes [41]. For example, the Brownian motion induced by the unique NP–enzyme dynamics provides some advantages for enzymatic reaction kinetics. Furthermore, attached enzymes could be more stable and more active. NPs have been used as biocatalysts and nanoenzymes [174].

The single-enzyme NP (SEN) concept has been developed to combine enzyme stabilization and high surface area of nanoporous structures [175]. As shown in Fig. 5.10, a protease (α-chymotrypsin) is stabilized as an SEN. It provides desirable characteristics such as stabilized activity, reduced mass transfer limitation, and enhanced processability [175]. A thicker and denser SEN network is confirmed by TEM analysis.

Enzymes are highly specific and efficient biocatalysts although relatively very fragile. Often, they require protection against environmental degradation. Magnetite-containing spherical silica NPs with an average size of 53 nm with entrapped enzymes showed high biological activities [176]. These biocatalysts are easily removed from the reaction mixture using an external magnetic field. They are endowed with long-term stability in wide temperature and pH

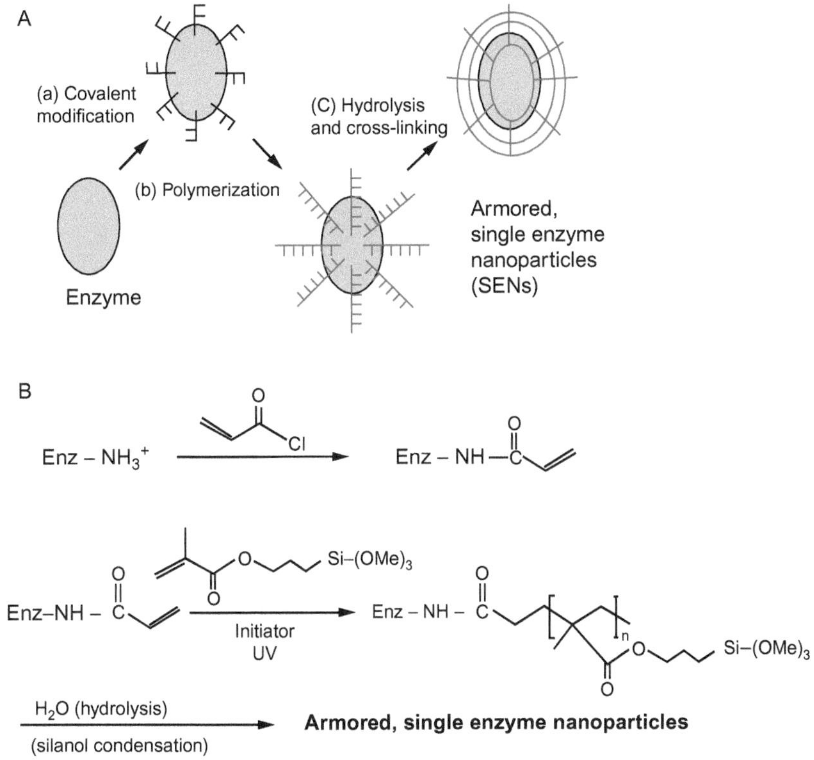

FIGURE 5.10 (A) Schematic representation of the three-step formation process of armored single-enzyme nanoparticles and (B) the detailed chemistry reaction occurring at each step. *Reprinted with permission from Kim and Grate [175]. Copyright 2003, American Chemical Society.* (For color version of this figure, the reader is referred to the online version of this chapter.)

ranges, as compared to conventional enzyme-based catalysts. Gold NP scaffolds are used to stabilize chymotrypsin (ChT) against denaturation at the air–water interface. This is accomplished via multivalent surface binding [177]. Indeed, the relatively strong gold NP segregation at the air–water interface eliminates the possibility of ChT interfacial denaturation.

3 FUEL CELLS AND ELECTROCATALYSTS

Although first reported in the early 1800s by Grove [178], it is only since the early 1960s that FCs have begun to be seriously considered for power generation. Basically, an FC is an electrochemical system consisting of an electrolyte sandwiched between two electrodes. It converts a fuel (e.g., H_2) to electrical power with minimum GHG emissions. Water is often the only major by-product. When the hydrogen fuel is produced from a renewable energy source, life cycle analysis shows that FC technology is far less polluting than the best current internal combustion engine. Several successful prototypes showed that FCs could be used for power generation in mobile (vehicles), stationary (power generation), and portable (device) applications. Although stationary applications remain the largest market, the capital cost disadvantages of FC are a major drawback. Power needs for portable (computer, cell phone) and mobile (cars) applications will likely see larger growth in the future.

The basic FC system consists of four main components: FC pack, fuel and oxidant feeding, exhaust gas and water management, and electrical output. Like batteries, FCs are electrochemical-based energy storage units. However, they show some differences, particularly in the way the energy is stored. FCs use chemicals (such as hydrogen or methanol) stored outside the electrochemical cell system, requiring a feeding system to the reactor. Typically, an FC contains two electrodes (anode and cathode) integrated onto two charge collectors including fuel/oxidant flow delivery. These two electrodes sandwich the electrolyte matrix.

When compared to conventional sources of electricity, FCs are cleaner, more efficient (no Carnot cycle limitation), quiet, and modular. Their high cost is probably the single most important hurdle. The cost of today's FCs in the range of \$10–\$100 per watt is prohibitive. FCs are at least an order of magnitude more expensive than the conventional power generation sources. Noble metal-based electrocatalysts are another major issue hindering further development of some FC technologies. Based on 50 million cars produced per year and 20 g of platinum per car, we need around 1000 tons of Pt per year, about eight times the current annual Pt production. Besides the high cost of platinum (about \$40,000/kg and still increasing) and other platinum-group metal catalysts, their availability may prevent mass marketing of FC cars in the future. Indeed, with the low platinum abundance in the earth's crust, this will contribute to further increases in Pt cost, and thus the overall FC cost [179]. The high cost of hydrogen fuel is another significant hurdle. As discussed already, hydrogen is currently produced from natural gas using the reforming process, and from water using different splitting processes. As high purity levels are required to avoid catalyst deactivation, hydrogen purification is mandatory.

Based on the type of fuel, working temperature, structure, and composition of the electrodes and electrolytes, at least six types of FCs are deemed to be of interest (Fig. 5.11). SOFC operates at 800–1000 °C and molten carbonate fuel cell (MCFC) operates at 550–650 °C. Other FC types operating at lower temperatures have also been developed including alkaline fuel cell (AFC) (60–90 °C), polymer exchange membrane fuel cell (PEMFC) (60–90 °C), direct methanol fuel cell (DMFC) (60–90 °C), and phosphoric acid fuel cell (PAFC) (180–220 °C). Figure 5.11 summarizes the chemical processes occurring during normal operation of these FCs. Maximum efficiencies of 60%, 45%, 60%, 34%, 50%, and 70% have been estimated for industrial AFC, PAFC, MCFC, DMFC, PEMFC, and SOFC, respectively.

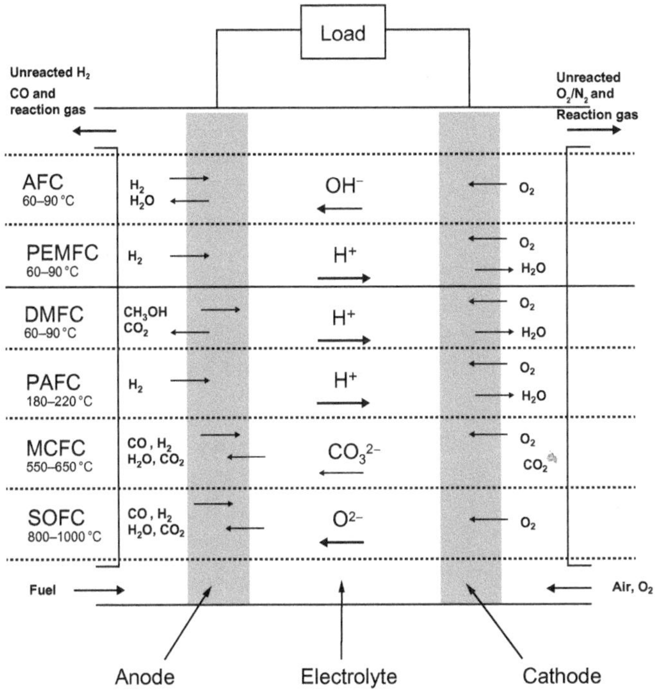

FIGURE 5.11 Main reactions and processes occurring in different types of fuel cells. *Reprinted with permission from Winter and Brodd [173]. Copyright 2004, American Chemical Society.*

DMFC and SOFC are probably the two most promising FCs that could reach wider use in the near future. NP incorporation in FC will help in lowering the cost and improving performance. Materials used for the catalyst, membrane, and bipolar plate are the main cost items. Platinum-based NP catalysts are currently used in commercial PEMFC, DMFC, and PAFC. Silica-based nanocomposites are also used to improve the performance of the electrolyte membrane. Increased water retention was achieved at high temperature using silica nanocomposites. Ceramic nanopowders are used in efficient solid-state electrolytes for SOFC. Alloyed catalyst NPs improve performance of DMFC, providing great potential for portable devices and mobile cars using the existing infrastructure [180]. Newly developed direct borohydride fuel cell (DBFC) could be more suitable for applications requiring small operation space [181].

Scientific, technological, and commercial developments of FCs have been reviewed in numerous publications [182–185]. Although incremental improvements are still possible through engineering optimization, electrocatalyst cost remains the single most important issue hindering the development of most FCs [186,187]. In the following section, a few examples are provided of NP utilization in different types of FCs.

3.1 Polymer Exchange Membrane Fuel Cell

PEMFC has attracted a lot of interest due to its simplicity, high efficiency, fuel flexibility, and low operation temperature [188–191]. The PEMFC operation is based on a four-phase boundary system [192] involving metallic NP catalyst (Pt), support (Vulcan carbon), ion conductor (acid electrolyte or a proton-exchange polymer), and fuel (H_2). These components contribute to:

a. Transport of the fuel (H_2) to catalyst sites;
b. Electro-oxidation of fuel to protons;
c. Transport of protons from catalyst site via a proton-conducting medium;
d. Transport of electrons from catalyst sites via the carbon support.

Non-Pt catalysts and other carbon supports have been reported and evaluated [193]. Pd, Au, Pd–Pt, CoN/C, and FeN/C catalyst compositions have been suggested [194]. Pt–Ru is discussed in the next section. The oxygen reduction reaction activity of Pd–Pt is about two times that of a single-component Pt-based catalyst [194]. Sulfonated silica NPs have been used for PEMFC [195]. These hydrophilic silica NPs are incorporated into a poly(styrene–NaSS–UAN) random copolymer [195].

3.2 Direct Methanol Fuel Cell

With an efficiency of about 35%, DMFC has the lowest efficiency in FC technology. However, DMFC is a cost-effective technology when compared to other FC technologies. Furthermore, for transportation applications, it provides the unique advantage of being able to use existing distribution infrastructure.

Alloyed Pt–Ru NPs with a 1:1 atomic ratio have been recognized as the best electrocatalyst system for DMFC in terms of activity and stability [192]. The vast majority of successful DMFC prototypes are based on supported catalysts. However, there are a few reports showing that unsupported NPs could provide excellent catalytic properties [196]. Liu et al. reviewed the various methods used for manufacturing Pt–Ru NPs for DMFC catalysts [197]. These methods are summarized in Fig. 5.12. The different approaches seem to provide relatively high activity. However, it is difficult to make a fair comparison as numerous factors may play a critical role. Specific activity (activity by unit catalyst mass) is an important figure of merit, although it is not often reported in the literature. Choice of catalyst supports also provides various avenues to optimizing anode properties. Carbon black, nanostructured carbon such as CNT, and mesoporous carbon have been used.

Like PEMFC, DMFC deployment is severely hindered by the scarcity of Pt. Utilization of expensive and rare Ru to produce allowed metal catalysts

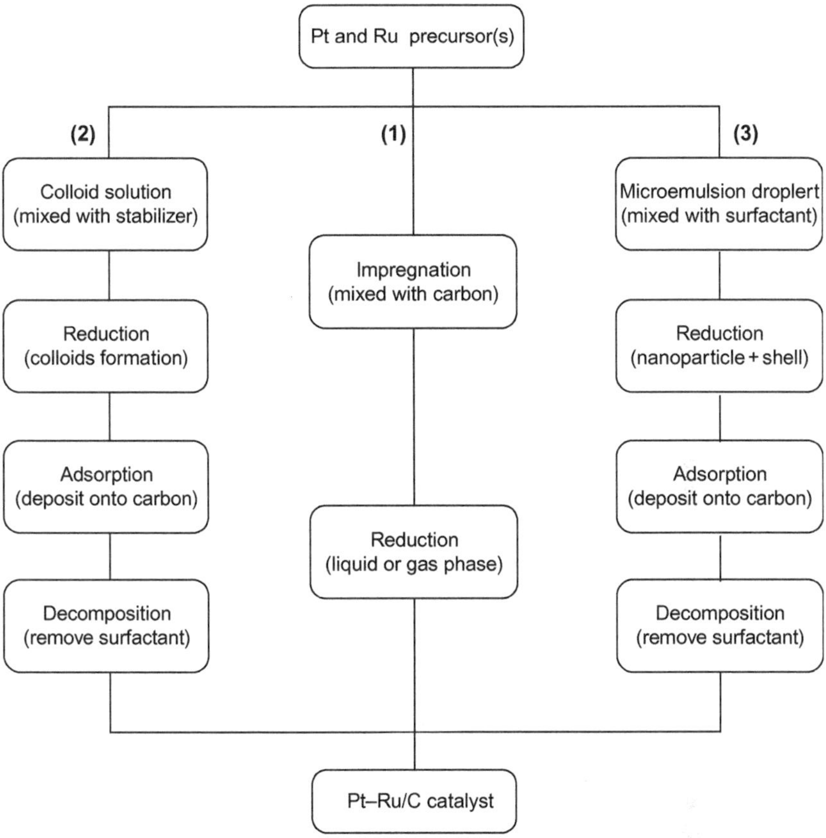

FIGURE 5.12 Flowchart of three synthetic approaches used in Pt–Ru catalyst fabrication and integration: (1) the impregnation method, (2) the colloidal method, and (3) the microemulsion method [197].

makes the overall cost even higher. However, improved catalyst and membrane design will likely improve energy density [198,199]. DMFC and DBFC are probably the best FCs for portable application [181]. As illustrated in Fig. 5.13, homogeneous NP catalyst distribution in conductive polymers will reduce the amount of expensive catalysts while providing an optimum 3D access to gaseous fuel. Pt–Ru NPs are prepared using the polyol approach and a conductive polymer as surfactant stabilizer. NPs obtained using this approach have been shown to provide very high catalytic activity per unit mass [198].

The concept of multiscale dispersion of Pt–Ru NP catalysts has been recently reported [200]. This concept describes the importance of alloyed particle dispersion and intraparticle dispersion among the constituent atoms [200]. In this work, the authors showed that the presence of TiO_2 can decrease the grain size from 3–4 nm in the case of Pt–Ru/C catalysts to

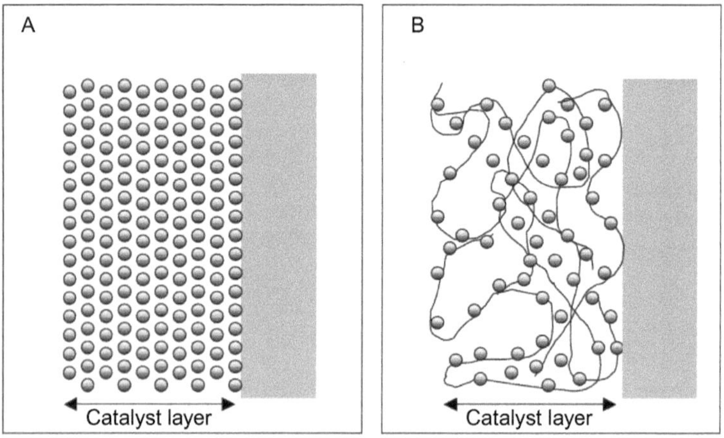

FIGURE 5.13 Pictorial representation of (A) aconventional anode structure without polymer and (B) a proposed anode structure of Pt–Ru nanocatalyst dispersed in a conductive polymer. *Reprinted with permission from Bensebaa et al. [198]. Copyright 2005, American Chemical Society.*

1–2 nm in the case of ternary Pt–Ru–Ti/C catalysts. Detailed characterizations indicated that the core consists of a Pt–Ru alloy-rich phase whereas the shell is Pt-rich. Multiscale dispersed Pt–Ru–Ti/C catalysts showed better activity toward methanol oxidation than commercial E-TEK 30 Pt–Ru/C.

Incorporating titania NPs 20 nm in diameter in polymer electrolytes has been shown to enhance the cell voltage and maximum power density of DMFC [201]. TiO_2 containing a Nafion membrane is crack-free and has higher protonic conductivity and lower methanol permeability. Titania NPs have also been used to increase the operation temperature up to 130 and 145 °C for PEMFC and DMFC, respectively [202]. A comparison between different preparation methods has been reported [203].

To address the cost issue of Pt–Ru, alternative electrocatalysts have been also considered. They include Pt–Sn, Pd–Ti, and Pd–Co–Au. Combinatorial techniques have been used to study systematically all possible multi-alloyed systems [204]. Pt–Sn catalysts showed promising results [205], although other compositions have also been considered [206]. So far, electrocatalyst performance of these alternative compositions is lower than that of Pt-based catalysts.

3.3 Solid Oxide Fuel Cell

Typically, an SOFC contains an oxygen-ion-conducting ceramic electrolyte membrane, a perovskite cathode, and a nickel cermet anode electrode. Y_2O_3-stabilized ZrO_2 (YSZ) films must be heated above 1000 °C to be effective in an SOFC. $BaCeO_3$ is a good alternative material requiring lower annealing temperature. However, it is difficult to obtain homogeneous composition and tubular shape, which adversely affects the overall performance [207]. Lee et al. showed that 100 nm $BaCeO_3$ NPs with uniform shapes are obtained using the citrate process followed by a calcination of the precursors at 900 °C [207]. A sol–gel method has been also used to prepare YSZ monodisperse NPs [208]. Electrical conductivity depends on the grain size used to for scandia-stabilized zirconia electrodes [209].

Tubular SOFC configuration has several advantages over planar configuration [210]. NP-based layers can be incorporated in tubular SOFC. Wet impregnation has been used to fabricate NP-based electrodes for SOFC [211]. NiO and YSZ NPs are heterocoagulated in a pH-controlled aqueous medium to form a nanostructured SOFC anode [212].

4 SOLAR CELL

Solar cells produce power by converting solar energy into electricity through the PV process. Although the PV effect was discovered nearly two centuries ago, the modern era of the solar cell industry started only in the early 1950s [213]. More than half a century ago, a group from Bell Telephone laboratories reported a power efficiency of 6% on a crystalline silicon-based p–n junction [213]. Power efficiency is defined as the ratio of the electrical power output to the impinging solar power under standard conditions. It took several decades of research and development (R&D) for the silicon solar cells to reach an efficiency of 24% in the laboratory and an average efficiency of 16% at the commercial scale [214].

Solar radiation is believed to be the only realistic alternative to fossil fuel. This prediction is based on the abundance, clean, and renewable attributes of solar energy. However, the high cost of crystalline silicon solar cells has prevented solar energy from taking a significant share of the overall energy mix. It is believed that thin film-based technologies [215] will significantly reduce the production cost of PV modules. Besides the different crystalline silicon technology options, three thin film-based PV modules are currently used for power generation. Amorphous silicon (a-Si), cadmium telluride (CdTe), and $CuIn_{1-x}Ga_xSe_2$ (also referred to as CIS or CIGS) have shown promising reductions in cost [216,217]. Within each PV technology group, there are different subgroup technologies often related to variation in the manufacturing process. In the case of CIGS solar cells, sputtering, evaporation, plating, and NP integration are used. Even in the case of NP-based CIGS (nano), one can distinguish different subgroups. CIGS and CdTe thin-film technologies are promising as they provide high laboratory-scale efficiencies at 20% and 16%, respectively.

Comparison of different PV power generation technologies is very difficult, even between well-established commercial modules. Raw material cost, semiconductor junction and metal-contacting fabrication processes, overall efficiency, lifetime, balance of system (BOS), and system integration are factors that could affect the unit electricity cost. Often, a comparison is made based on just the unit area cost of the PV film fabrication. In the case of polymer-based materials, polymer film deposition is potentially inexpensive [218] when compared to vacuum-based deposition techniques currently used for commercial PV modules. However, the high production cost of these polymers, low module efficiency, lifetime of the module, and the overall performance ratio should be considered to establish a complete ranking of these technologies [219,220].

To sustain PV industry growth, PV module and system prices must fall significantly. More than 90% of today's solar cells are based on crystalline (mono- and polycrystalline) silicon. Silicon is the second most abundant element in the earth's crust, mostly found as silicon dioxide. However, silicon dioxide reduction and silicon purification (at least 99.999% pure) require a significant amount of energy, capital investment, and know-how. Although the cost of high-purity silicon has been a major hurdle, recently its price has been reduced to less than $25/kg. This has translated into less than 1$/W for crystalline silicon modules.

Although not yet commercialized, III–V organic and dye-sensitized solar cell (DSC) PV technologies have been investigated for decades [221]. High efficiency but relatively expensive III–V semiconductor-based solar cells have been already used in spatial applications [222]. Major issues are preventing the deployment of organic solar cells and DSC. Low efficiency and stability of laboratory solar cells are the main disadvantages. Furthermore, low efficiency increases the BOS capital cost. Another issue is related to the environmental and financial costs involved in active PV organic material production. Currently, poly(3-hexylthiophene) (P3HT) [223] and phenyl C_{61}-butyric acid methyl ester (PCBM) [224] are used in plastic solar cells. The polythiophene synthesis process is not straightforward, particularly for the regioregular P3HT [225]. It is generally assumed, quite wrongly, that energy and environmental impact of P3HT is similar to that of PEDOT/PSS [226]. Furthermore, to achieve reported higher power efficiency, commercially available P3HT polymer requires further purification steps [227]. Even before purification, P3HT materials cost several hundred dollars a gram [228]. In comparison, the market spot price of purified silicon is four orders of magnitude cheaper, with a potentially lower environmental impact.

The cost gap between crystalline silicon-based and thin film-based PV technology has significantly decreased in the last few years [229,230]. Thin-film technology based on a monolithic integration over a large area involves reduced fabrication steps. Except in the case of the CIGS module, the overall power efficiency of crystalline silicon solar cells remains much higher than that of their thin-film counterparts. This is particularly true for commercial solar panels. Due to its maturity and synergy with the microelectronic industry, the crystalline silicon-based PV industry will remain the dominant technology in the near future.

Production cost per unit area and efficiency are two key parameters allowing a fair comparison of various PV technologies. High power efficiency and low manufacturing cost are the two most important characteristics. Availability of raw materials and environmental impact should also be carefully analyzed. CdTe, DSC, and CIGS PV technologies are hindered by limited resources, in spite of their favorable efficiency and cost values. A recent study by NREL classified existing commercial module technologies [219]. Based on their potential technology improvement, this study estimated the ratio of the relative cost to relative performance. The following value of the ratio (lowest number is related to the best commercial technologies) of six technologies has been estimated:

a. Silicon (standard): 1.00 (baseline);
b. Silicon (nonstandard): 0.85;
c. CIGS: 0.65;
d. CdTe: 0.77;
e. a-Si (single junction): 1.28;
f. a-Si (three junction): 1.05.

In any solar cell design, several geometrical and structural parameters are believed to play critical roles in the overall PV power efficiency (η). The overall power efficiency is provided by

$$\eta = \eta_{abs} \times \eta_{gen} \times \eta_{col}$$

where η_{abs}, η_{gen}, and η_{col} deal with solar spectrum absorption, charge generation, and charge collection through an external circuit, respectively. In the case of excitonic systems, the second contribution (η_{gen}) is divided into two separate contributions. One contribution quantifies exciton production efficiency, and another quantifies electron–hole separation efficiency. The typical energy losses of standard solar cells are grouped into four categories [231]:

a. Solar radiation with energy lower than the bandgap is not absorbed by the PV layer;
b. Solar radiation with energy above the bandgap is converted into heat;
c. Voltage loss factor is due to the fact that the open-circuit voltage is smaller than the bandgap;
d. Fill factor is lower than unity.

To address these loss factors, several concepts and designs have been developed. For this, the absorption coefficient needs to be optimized throughout the solar radiation spectrum. The power distribution of the solar spectrum consists of the following four components: UV (<390 mm), visible (290–750 nm), near-infrared (NIR) (750–1400 nm), and infrared (>14,000 nm). These four components contain 3.4%, 45%, 38%, and 14% of the overall solar spectrum power, respectively. An ideal PV material should absorb solar radiation efficiently across a broad spectral range. This is not the case with the conventional solar cell design.

Two limiting factors in conventional single-junction solar cells are identified. First, absorbed photon with energy above the energy bandgap is mostly lost as heat. This heat will contribute to increased temperature detrimental to the overall efficiency. Second, solar radiation with energy lower than the bandgap does not contribute to the overall PV process. Various third-generation concepts based on NPs have been recently pursued to improve efficiency and/or lower the fabrication cost [232–236]. Different concepts have been investigated toward increasing power efficiency. Tandem cells [237,238], concentrator [239], multiple exciton generation [240], and luminescence converter [241] designs have been reported. These concepts are (or could be) facing some challenges during their implementation in commercial production. To compete with crystalline technology, these third-generation concepts must consider nonvacuum large deposition techniques. Third-generation PV technologies provide the following advantages:

a. Tunable bandgap to match the solar spectrum;
b. Carrier multiplication (or multiple exciton generation) to use nonabsorbed solar radiation instead of contributing to the increase in the solar cell temperature, thereby even lowering efficiency. Carrier multiplication is usually weak in bulk materials;
c. A 3D heterojunction instead of the 2D junction;
d. Amenability to low-cost integration platforms.

One or more of these factors have been integrated in different NP-based PV films. Several examples are provided here. But none of them has so far been included in a commercial module. The efficiency of current nanostructured solar cells is still relatively low, although it has been increasing at a fast pace in the last decade (Fig. 5.14). In spite of the plethora of NP-based solar cells, none has so far reached commercialization. Besides low efficiency, numerous engineering obstacles should be addressed [230].

NP-based solar cells can be classified according to their geometrical design, composition, or a combination of both. In this chapter, the focus is mainly on four categories of PV materials based on semiconductor NPs. They include organic nanostructured solar cells, dye-sensitized cells (DSSC), hybrid NP–polymer cells (NPC), and tandem and third-generation solar cells.

4.1 Organic Nanostructured Solar Cell

Since the first all-organic active PV solar cell was reported by Tang in 1986 [243], there has been tremendous activity in the area of vacuum- [244] and solution-processed [245–248] organic PVs. Nanostructured 3D heterojunction morphology plays a critical role in the solar cell's performance [249]. A 3D heterojunction solar cell consists of a nanostructured polymer–fullerene PV film sandwiched between two electrodes including indium-tin-oxide (ITO) coated with a hole-transporting layer PEDOT/PSS and an aluminum top electrode.

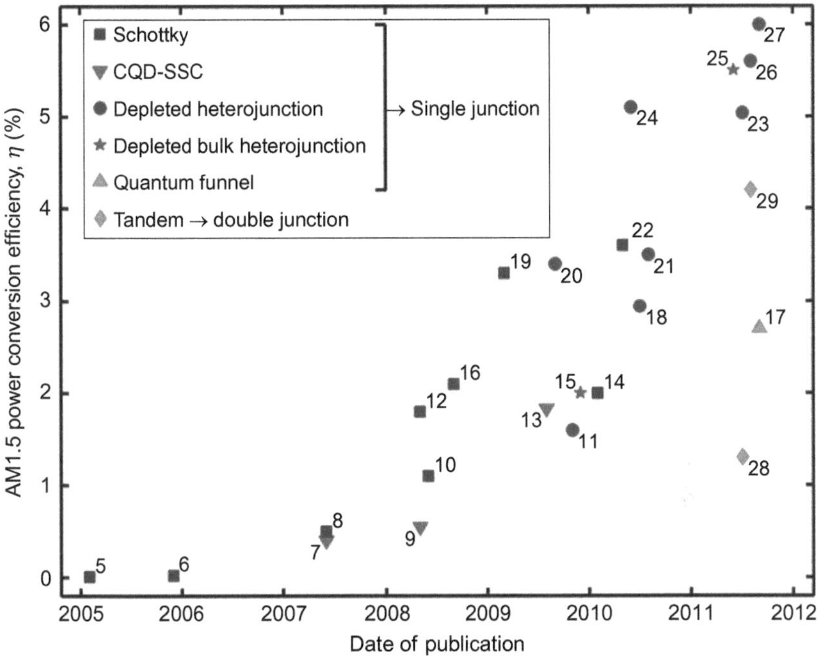

FIGURE 5.14 Evolution of nanostructured laboratory solar cell performance since 2005. *Reprinted with permission from Kramer and Sargent [242]. Copyright 2011, American Chemical Society.* (For color version of this figure, the reader is referred to the online version of this chapter.)

The efficiency of the photoinduced charge transfer process between the p-type PPV polymer backbone and the n-type fullerene forming a 3D polymer (donor)–PCBM (acceptor) heterojunction depends on two main factors [246]. First, there should be an overlap in the energy levels of the PPV polymer and modified C_{60} fullerene (PCBM). The contour plot of power efficiency has been calculated as a function of the bandgap and the lowest unoccupied molecular orbital (LUMO) level of the donor polymer [250]. Second, the physical mixture should be uniform throughout the PV film. PCBM–polymer nano- and micromorphology structures are critical to achieving high power efficiency. Both PCBM-rich and PCBM-poor domains are found to contain PCBM nanocrystals [251].

Nanocomposite films using modified C_{60} and conductive polymers give rise to a bulk heterojunction with relatively high power efficiency for plastic PV. Within the heterojunction, the C_{60} agglomerate gives rise to different morphological structures. It has been well established that microscopic morphology of the films directly affects the device characteristics [252]. Morphology could be modified during film preparation and posttreatment. In particular, annealing has been shown to improve solar cell performance [251]. With an outer diameter of about 10.18 Å, it is possible that a PCBM nanocrystal is formed. Annealing gives rise to a two orders of magnitude increase in photocurrent. This is directly related to the formation of finer morphological crystalline structures at the submicron level.

All-organic nanocomposite solar cells have been obtained by blending polymeric NPs [253]. This approach facilitates the control of the phase separation dimensions within the nanocomposite PV film. First, a miniemulsion technique is used to prepare dispersions of 50 nm NPs. Pure polymer components or a mixture of high-mobility hole-transporting polymer PFB [poly(9,9-dioctylfluorene-2,7-diyl-*co*-bis-*N*,*N*′-(4-butylphenyl)-bis-*N*,*N*′-phenyl-1,4-phenylenediamine)] and a high-electron-mobility F8BT [poly-(9,9-dioctylfluorene-2,7-diyl-*co*-benzothiadiazole)] is used in each particle. A mixture of these miniemulsion NPs dispersed in water is prepared. Then the mixture solution is dropped on a substrate followed by spin coating to produce the PV film. The highest external quantum efficiency (=4%) was obtained with polymer blend NPs containing PFB/F8BT using a weight ratio of 1:2 in each individual NP. Conjugated polymer MEH–PPV NPs have also been integrated into nanocomposite PV film, yielding relatively high solar efficiency [254]. The power efficiency of polymer-based solar cells is often limited by their high bandgap [255]. Unfortunately, it is very challenging to synthesize stable, low bandgap polymers. This approach could be used with oligomers or even with inorganic materials.

4.2 All-Inorganic NP Cell

The various all-inorganic solar cells remain the only commercial technologies as of today and for the near future. Small-bandgap inorganic materials (around 1 eV or less) are more suitable for solar radiation absorption optimization. Bulk CdS (2.4 eV) and, to a lesser extent, CdSe (1.7 eV) and CdTe (1.4 eV) are already outside the optimal bandgap region. Corresponding NP bandgaps are even higher. All thin film-based commercial solar panels are made of inorganic active PV materials ($CuIn_{1-x}Ga_xSe_{2-y}S_y$, amorphous silicon, and CdTe). High deposition costs due to slow deposition, high equipment cost, and low yield are the main drawbacks of thin-film technologies.

Efficient electronic transport within the NP array film remains the biggest challenge of PV devices [256]. The charge collection factor within a nanostructured design is increased via a bi-continuous nanostructure. Five factors have been identified to affect charge transport involving the different interfaces within the solar cell [256,257]:

a. Polymer–NP interface: poor wavefunction overlap at this interface overlap will prevent exciton dissociation;

b. NP–NP interface: good wavefunction overlap between neighboring NPs will reduce recombination of electrons and holes;

c. Polymer–hole transport layer interface: poor electrical contact across this interface layer will increase resistance;
d. NP–photocathode interface: poor electrical contact across this interface layer will increase resistance;
e. Photoanode–hole transport layer interface: poor electrical contact across this interface layer will increase resistance.

It is often difficult to build such a cell using simple and cost-effective processes. Figure 5.15 provides one possible solar cell design that may satisfy the interfacial requirements of the different layers. Combining low-temperature and non-vacuum deposition techniques with lamination could provide a competitive manufacturing process as an alternative to current silicon-based PV [257].

Silicon nanocrystals embedded in SiO_2 are used as a luminescence down-converter. This works as follows: low-wavelength photons, with low conversion efficiency in standard silicon solar cells, are absorbed in the silicon nanocrystal layer (converter layer). Afterward, these radiations are converted and reemitted with high efficiency in the red region. These red-emitted wavelengths are converted much more efficiently in the standard silicon layer [258,259]. Si nanocrystals obtained by pulverization of porous silicon are embedded into spin-on-glass antireflecting SiO_2. Solution-based processes are used for coating nanocrystals onto standard silicon-based solar cells [241].

Tunable QDs prepared by electrochemical etching of bulk silicon [260] showed that both 1 nm Si NPs (blue luminescent) and 2.85 nm Si NPs (red luminescent) provide enhanced power efficiency (around 60%) in the UV–blue range [261]. In the visible region, the enhancements are $\sim$10% for the

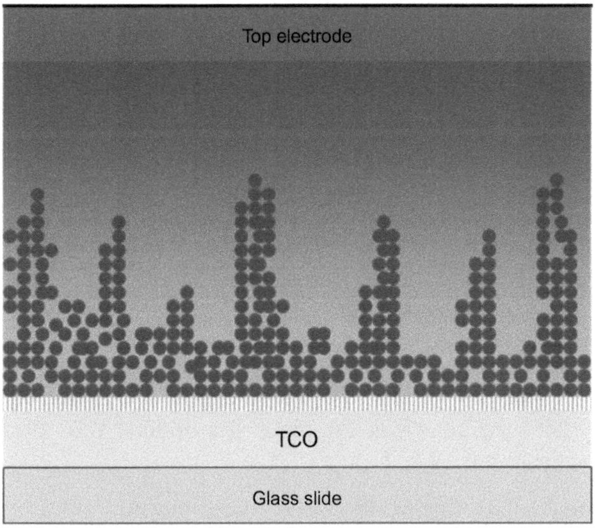

FIGURE 5.15 Gradient assembly of a bi-continuous nanostructure. (For color version of this figure, the reader is referred to the online version of this chapter.)

red NPs and ~3% for the blue NPs. This enhancement is due to large increases in the voltage of the solar cells [262].

A novel solar cell consisting of In_2S_3 buffer (0–200 nm) sandwiched between a $CuInS_2$ (p-type) layer and TiO_2 NPs deposited on a dense TiO_2 layer has been reported [262]. Several structural parameters are used to control the efficiency of this 3D $CuInS_2/TiO_2$ nanocomposite solar cell [262]. In particular, film thickness and structure, buffer layer thickness, and TiO_2 particle size have a direct effect on the solar cell performance. Thicker TiO_2 films and smaller TiO_2 particles sizes give rise to lower performance due to increased electric resistivity. As expected, the 3D heterojunction structure leads to better performance when compared to a planar structure. Optimal performance has been achieved when adding a 60 nm In_2S_3 buffer layer thickness. Thicker buffer film decreases leakage conductance but lowers the solar radiation transmission and increases the series resistance of the cell.

An all-inorganic nanocomposite film consisting of CdSe NPs and TiO_2 having promising PV properties has been reported [263]. Bifunctional molecules are used to link the two NPs. The photoresponse and photoconversion are optimized by varying the CdSe particle size and the TiO_2 structure. Smaller-sized CdSe NPs show higher incident photon to charge carrier efficiency (IPCE), also referred as quantum efficiency (QE). Larger particles have better absorption in the visible region but showed lower electron injection into TiO_2 when compared to smaller-sized CdSe NPs. Slightly higher IPCE has been achieved using tubular TiO_2 when compared to spherical morphology [263]. Numerous other nanostructured architectures have been developed to optimize the solar-to-electricity conversion efficiency [264].

Nanocrystalline silicon-based thin-film PVs are of interest because they can be obtained on a large scale using plasma-enhanced chemical vapor deposition (PECVD) [265]. TEM, XRD, and Raman characterizations showed that high-efficiency solar cells are obtained with crystalline nc-Si films. These data are consistent with molecular dynamics simulations (Fig. 5.16). Indeed, crystalline nanosized grain dispersion in a thin matrix of a-Si is achieved by using proper temperature and silane dilution. This work is quite relevant to current efforts to increase efficiency of amorphous silicon modules.

4.3 Dye-Sensitized Cell

In 1991, the group led by Gratzel developed a DSC by mimicking the photosynthesis process. This group reported a solar-to-electricity conversion of about 10% [266]. As shown in Fig. 5.17, the original DSC design is quite simple and consists of:

a. Transparent conductive oxide (TCO)-coated glass;
b. 10 μm-thin film of TiO_2 obtained using 10 nm TiO_2 NPs followed by a sintering step;

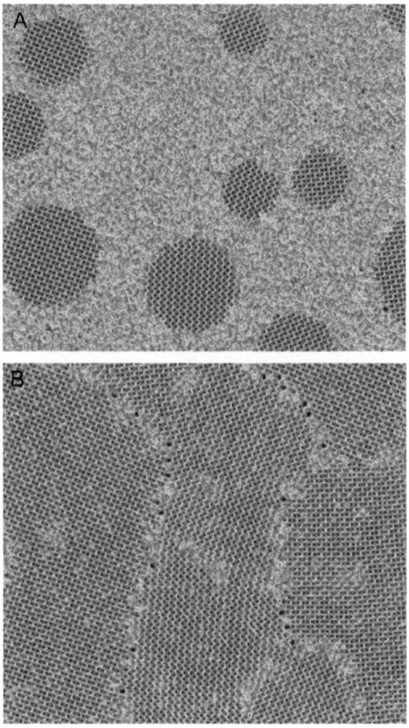

FIGURE 5.16 Snapshot (TEM-like) images obtained by molecular dynamic (MD) simulations (A) before and (B) after annealing the nc-Si/a-Si model system obtained by inserting crystalline Si nanoparticles in an amorphous matrix [265]. (For color version of this figure, the reader is referred to the online version of this chapter.)

c. Monolayer of organic dye molecule sensitizer coated on the porous TiO_2 surface;
d. Electrolyte liquid;
e. Second Pt-coated electrode.

Following light excitation of the sensitizer (S), a photoelectron is injected into the conduction band of the TiO_2 film. After passing the load and reaching the counter-electrode, the dye molecule and the redox system are regenerated successively. The open-circuit voltage of the solar cell corresponds to the difference between the redox potential of the mediator and the Fermi level of the nanocrystalline film. The different energy levels are shown in Fig. 5.17. In particular, the TiO_2 Fermi level is indicated with a dashed line. The best efficiency obtained on small-area DSC is around 11% [267]. Large-area modules are about 50% less efficient. DSSC is now at a precommercial development stage [268–273]. A precommercial manufacturing process has been reported [270]. The simplicity and potential of DSC manufacturing [271,272] have attracted significant interest from academia and industry [273].

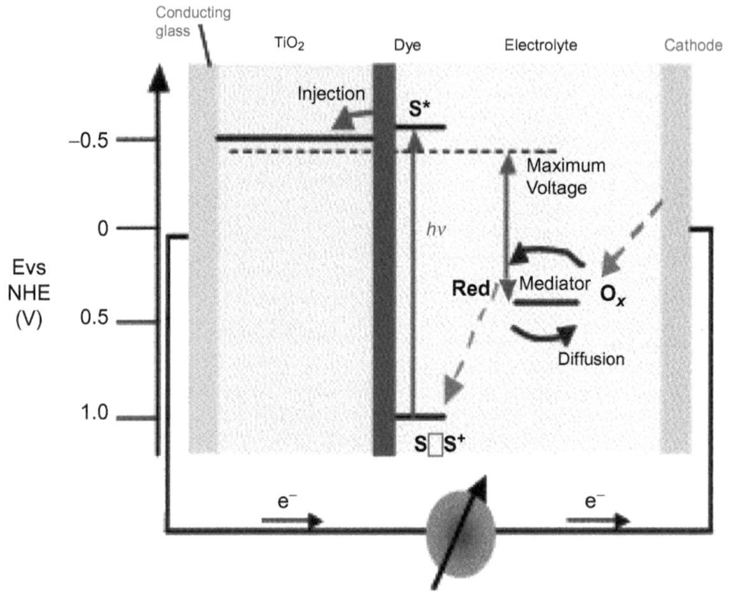

FIGURE 5.17 Schematic representation and operation mode of the dye-sensitized solar cell. Potentials are referred to the normal hydrogen electrode (NHE) [267].

Various synthetic routes have been used to prepare nanocrystalline TiO_2 including sol–gel, mechanochemical, and combustion [274]. The key to high-efficiency DSC is the porosity of the TiO_2 layer. This porosity depends not only on the TiO_2 particle size distribution but also on the sintering process [275]. Both conventional and microwave sintering have been performed without any significant difference on the porosity [275]. Chemical sintering has also been proposed as a more cost-effective method for fabricating DSC on plastic solar cells using lower-temperature processing [276]. A low-temperature binder-free sintering process of TiO_2 NPs has been recently reported [276]. Addition of ammonia into acidic colloidal solution reduces surface charges. A creamy paste with a viscosity nearly 500 times higher is obtained. Typically, resulting binder-free paste is deposited onto a TCO-coated substrate using a doctor-blade technique. Network geometry of the TiO_2 nanoparticulates does affect the electrical properties of the solar cell [277]. Current collector efficiency has been shown to be affected by the surface area of DSCs [278].

DSCs are limited by relatively weaker radiation absorption efficiency in the solar spectrum region. One way to improve absorption efficiency is by the deposition of metallic NP layers. Indeed, the surface plasmon effect will enhance the local electric field strength. An absorption enhancement factor of 149 has been reported when Ru-dye is adsorbed on nanostructured silver [279].

Utilization of liquid electrolyte in DSC is a major issue. Alternative solid-state electrolytes are being considered [280]. Different polymers have been used to replace liquid electrolytes [281]. Solid-state DSC is easily amenable to flexible substrates [282]. A new variant of the solid-state DSC, referred as extremely thin absorber solar cell, has been developed to address the stability issue. In this new concept, the electrolyte is replaced by a transparent p-type inorganic semiconductor [283].

4.4 Hybrid NP–Polymer Cell

Inorganic-polymer solar cells, or hybrid NPC, have received a lot of interest [284–287]. Optimization of NP–polymer interfaces is key to improving charge separation and collection. Different configurations of the NP–polymer interface have been identified. Strongly adsorbed polymers and weakly bound polymers in nanocomposites are of particular interest. The first geometry has been identified with NPs smaller than the chain. The second one involves an end-functionalized polymer chain that adsorbs onto the NPs. A simplified representation of the active PV layer (Fig. 5.18A) and solar cell (Fig. 5.18B) are highly key NPC parameters. In particular, up to six interfaces have been identified. Controlling and optimizing these interfaces is critical to minimize series resistance across the cell.

A maximum power efficiency of 10% has been predicted for NPC [287]. Interest in NPC could be traced back to two reports. The first one, published by Yu et al., showed enhanced PV efficiency using a 3D donor–acceptor heterojunction [288]. The second report showed a power conversion efficiency of 1.7% using a CdSe nanorod–polymer nanocomposite [289,290]. One of the advantages of the NPC design is the possibility for a band alignment through the choice of a wide variety of NP compositions and sizes. The energy levels of the semiconducting NPs could be adjusted by changing the size and the composition [287]. For example, ZnO, CdSe, or PCBM could be used as NPs. This tunability allows optimized solar absorption, charge separation, and charge collection. With the advent of new small-bandgap polymers [255] used in combination with a wide range of energy levels from different QDs, higher efficiency is achievable. CdTe NPs embedded in MEH–PPV forming a PV film have been reported although with relatively low power efficiency [291]. Recently, an efficiency of 5.4% has been reported using nanocomposite systems of CdTe NPs (donor) and water-soluble C_{60} derivatives (acceptor) [292].

4.5 Multijunction and Third-Generation Solar Cells

Non-concentrated first-generation (crystalline silicon) and second-generation (thin film) solar cells are currently limited by the theoretical conversion efficiency. This maximum efficiency is often referred by the Shockley–Queisser

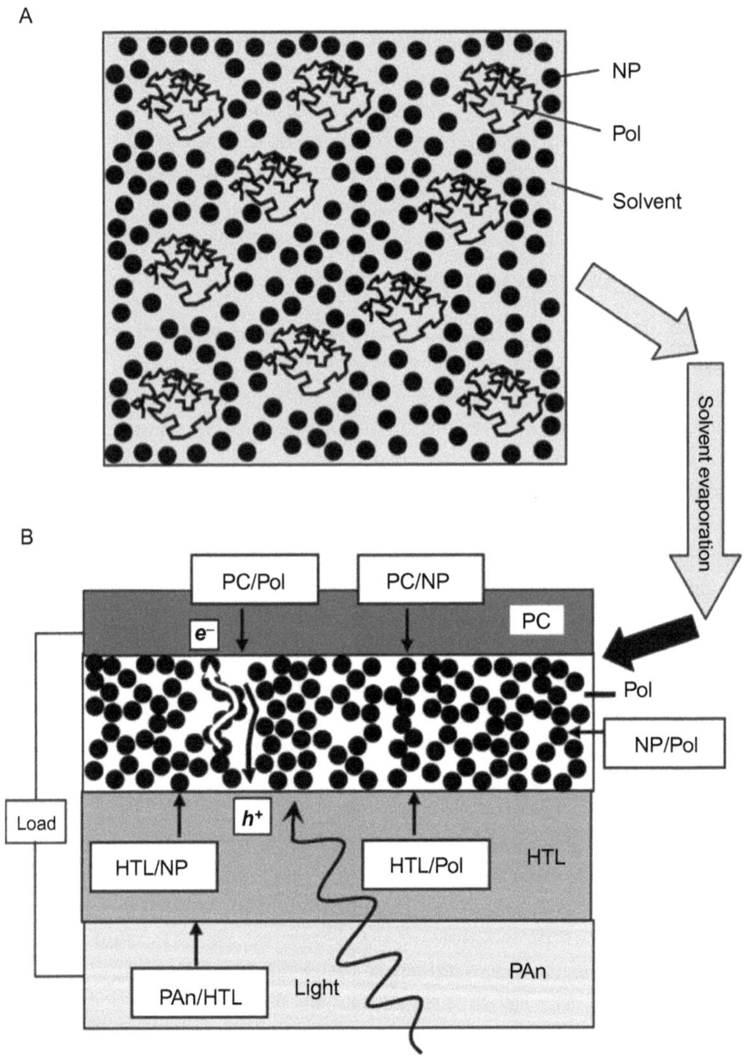

FIGURE 5.18 Schematic representation of the NP–Pol interface in (A) the NPC design and (B) the NPC solar cell. NP, Pol, PC, Pan, and HTL correspond to nanoparticle, polymer, photocathode, photoanode, and hole-transporting layer, respectively [287].

limit estimated at 33.7%. Various concepts of third-generation solar cells based on tandem structure, "multiple exciton generation" or "carrier multiplication," and other concepts have been developed [293,256] to overcome this limit. Higher efficiency at constant or lower production cost will provide a lower unit solar panel and a BOS prices. Another potential advantage of using tunable semiconductor NPs for solar cells is their potential utilization in fabrication of more cost-effective and efficient tandem solar cells [294].

Various tandem cell concepts have been recently developed without [236,238,294,295] and with [239] concentrators. A record 40.7% efficiency at 240 suns has been reported using metamorphic $Ga_{0.44}In_{0.56}P/Ga_{0.92}In_{0.08}As/$ Ge triple junction solar cells [296]. Even higher efficiency, approaching 60%, is within reach using solar concentrators. Incorporation of InAs QD and arrays in GaAs could significantly enhance light absorption and, thus, efficiency.

Multijunction quantum dot solar cells allow building a high-efficiency solar cell using tunable QD solutions [297]. Three junction tandem cells consisting of three QD (PV1, PV2, and PV3) layers with different bandgaps are shown in Fig. 5.19. The PV1 layer with the smallest NPs (QDs) will absorb mostly the high energy radiation (blue). The PV2 layer with the intermediary-sized NPs (QDs) will absorb mostly the green radiation. The PV3 layer with the largest NPs (QDs) will absorb mostly the low energy radiation (red). These PV layers are separated by transparent intermediary layers, consisting of a transparent (or semitransparent) counter-electrode, a recombination layer, and a TCO layer. The TCO layer is used as a conductive substrate for the following cell. Optimal band alignment is required to obtain high efficiency. The different bandgaps of the QDs and the conduction band

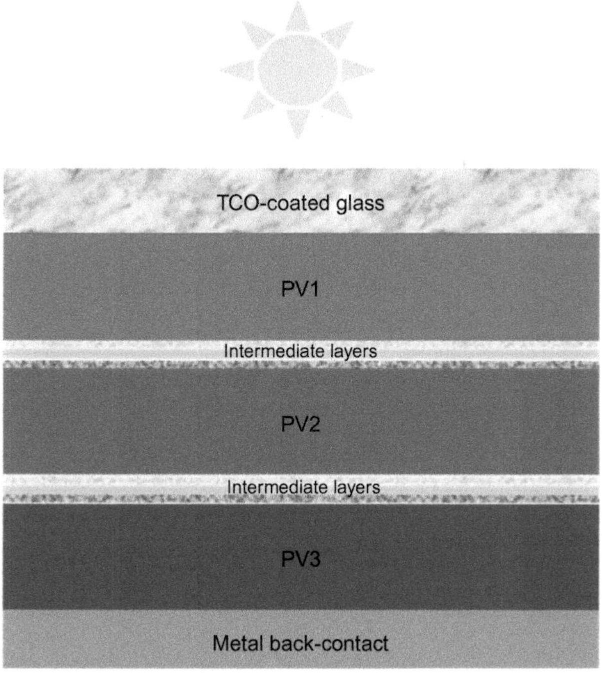

FIGURE 5.19 Schematic representation of a nanoparticle-based triple junction solar cell. PV1, PV2, and PV3 correspond to three photovoltaic layers separated by an intermediary film consisting of three layers. (For color version of this figure, the reader is referred to the online version of this chapter.)

levels of the different wide-bandgap semiconductors are also shown. Tunability of the QD surface provides flexibility in energy level alignment [297]. Molecular beam epitaxy is used to prepare these multi-stacked InAsGaAs self-assembled quantum dots. The formation of intermediate bandgaps has been exploited for optimum photon photocurrent generation [298]. Intermediate bandgaps are predicted to allow efficiency of above 60% [299].

It is often overlooked that commercially available tandem amorphous-microcrystalline silicon solar cells involve a submicron-crystalline structure (Fig. 5.20). Currently, a conical structure about 200 nm high with a circular base of about 9 nm provides the highest efficiency. A case could be made that higher efficiency is achievable using sub-100 nm nanocrystalline silicon embedded in amorphous silicon. However, the optoelectronic properties of this proposed design seem to degrade at a much faster rate when 10 nm crystallite silicon is used [301].

Various silicon nanostructured materials have been developed for third-generation solar cells [302]. Inclusion of 3–5 nm core–shell nanocrystalline silicon embedded in a hydrogenated multilayered silicon structure leads to high stabilized efficiency [303].

Recently, there have been reports of tandem cells using solution-based processing of QDs [304,305]. Taking advantage of colloidal QD bandgap tunability, multijunction solar cells could be built to optimize solar absorption. A tandem solar cell based on PbS NPs with a power efficiency of 4.2% has been recently reported [305]. This tandem cell consists of a bottom cell with a bandgap of 1.6 eV PbS and a top cell with a bandgap of 1 eV PbS. A graded recombination layer was used to tune the work function at the interface of the two cells. The current fabrication process of solution-based tandem cells is still too expensive compared to that of amorphous silicon tandem cells.

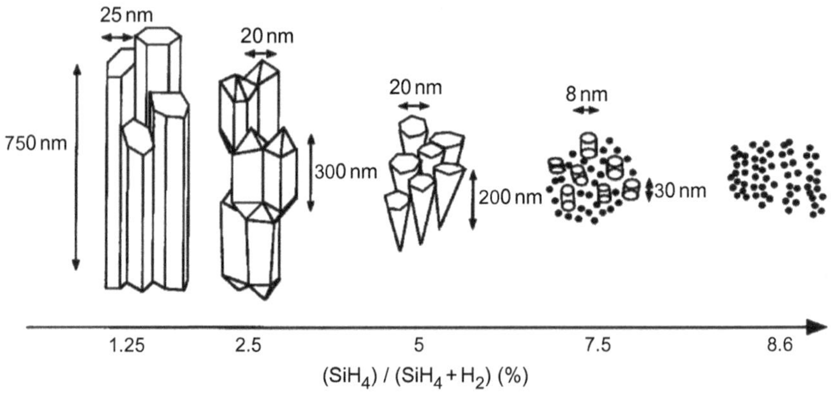

FIGURE 5.20 Geometrical dimension of micro- and nanocrystalline silicon materials obtained using VHF-PECVD using different silane/hydrogen ratios. The small black dots obtained using hydrogen content represent amorphous material [300].

4.6 Other Applications of NPs in Solar Cells

NP-based solar cell R&D is mostly focused on the development of the PV layer. Other components of the solar cell and modules could also benefit from the optoelectronic properties of NPs. For example, TiO_2 NPs could be used as a dielectric film in numerous layers of silicon-based solar cells [306]. NPs could be used effectively in antireflective coating, masks for electroless coating, improved and compatible passivating layers, and phosphorus diffusion barriers.

Current solar cell designs based on a single junction are limited to a maximum theoretical power efficiency of about 33%. Several solutions have been developed to overcome this limit [307,308]. Improving low conversion efficiency in the blue and UV solar spectrum of current commercial modules is the simplest approach to improve power efficiency. The goal is to shift the lower wavelength component of solar radiation to a higher wavelength before it reaches the photoconductive layer. Luminescent down-shifting (LDS) using luminescent material improves power conversion efficiency on crystalline silicon, CdTe, a-si, and CIGS systems [309]. Different LDS systems applied to crystalline silicon have been reported [309]. CdSe and CdS NPs have also been integrated onto the LDS films of microcrystalline silicon and CIGS-based solar cells, respectively [309]. However, currently, organic dye is inexpensive in comparison with NPs. For example, Lumogen-F dyes are commercially available at the cost of less than $14/kg [309].

Different approaches have been developed to increase PV efficiency based on the concept of LDS [310]. Eu^{3+} embedded in polymeric, silica, and hybrid sol–gel are currently being considered. They provide high stability with improved overall efficiency. The simplest design for wavelength down-shifting is to integrate absorbers in the encapsulation layer. Eu^{3+} organic complexes have been shown to improve efficiency of crystalline silicon-based solar cells [310]. Indeed, an increase of the total power delivered by the encapsulated silicon cells has been observed. In this approach, minimal fabrication process changes are introduced. Indeed, the addition of the luminescent materials (organic dyes or NPs) to the PVA before the fabrication of the PVA sheet will add only incremental cost besides the raw material cost.

A plasmonic system based on metallic NPs placed within or near the PV-active layer is being actively pursued as a way to enhance the overall power efficiency [311]. A planar converter consisting of QDs is used to optimize the solar spectrum in silicon. A plastic layer containing QDs is applied on top of a solar cell. Both blue and green radiations are absorbed by the QDs and reemitted in the red. This QD layer is used to red-shift low wavelength, allowing optimization of the overall solar cell efficiency [312].

Antireflective coating sol–gel based on SiO_2 NPs allows an increase in solar transmission from 90% to 95% [99]. The fabrication process consists of sol–gel preparation, a formulation that includes additives, dip coating, drying, and thermal treatment. A high-porosity film (up to 50% porosity) is obtained. The resulting refractive index is between 1.25 and 1.3, corresponding to a maximum transmission above 95%. After weighting with the total solar range, an overall peak power increase of 3.5% is achieved. This coating solution could be used to improve the conversion efficiency of both PV and solar thermal panels.

The addition of a 12-monolayer coating of 20 nm gold NPs on top of (n-type) a-Si:H–(p-type)c-Si heterojunctions improves PV efficiency from 7.5% to 9.5% [313]. This improvement is due to increased absorption of light due to the surface plasmon resonance of metal NPs. However, when 18 monolayers of gold NPs are deposited, the efficiency suddenly drops to less than 2%. Optimal gold NP coverage is used to improve the efficiency [314]. A similar behavior was observed in thin-film silicon PV. Light trapping is often used to increase absorption (hence cell efficiency) in thin-film solar cells [315]. Surface texturing with about 10 μm feature size has been used to increase light trapping. However, higher surface recombination losses may reduce the overall conversion efficiency [316]. The effect of submicron texturing is discussed later.

It has been reported that gold NPs absorbed on a silicon surface increase the p–n junction photocurrent response [317]. Surface plasmon emission from metallic NPs has been successfully used to shift the lower wavelengths of impinging solar light to higher wavelengths. Given the sensitivity of the surface plasmon resonance (SPR) to NP size and shape, one could optimize the solar radiation absorption. Devices consisting of silicon p–n junction photodiodes and a top layer containing gold NPs of variable-size (50, 80, and 100 nm) were fabricated and characterized [318]. Photocurrent response spectra for a reference Si p–n junction photodiode without surface NP modification and for photodiodes with layers of gold NPs of 50, 80, or 100 nm diameter are measured. The ratio of the photocurrent response of modified devices to that of the reference device allowed comparing photon–electricity conversion efficiency as a function of the wavelength. The normalized photocurrent behaves differently when gold NPs are added. The 100 nm gold NPs increase the response at longer wavelengths. Smaller NPs increase response at the shorter wavelength range [318].

The dispersing NP medium is also important. The scattering properties of silver NPs depend on the medium in which they are dispersed. Indeed, the scattering cross section is given by [319]

$$C_{\mathrm{scat}}(\lambda) = 3\mathrm{V}/2\pi(2\pi/\lambda)^4 |(\varepsilon_\mathrm{p}/\varepsilon_\mathrm{m} - 1)/(\varepsilon_\mathrm{p}/\varepsilon_\mathrm{m} + 2)|^2$$

where V, ε_p, and ε_m are the particle volume, the dielectric function of the particle, and the dielectric function of the embedding medium, respectively. When $\varepsilon_\mathrm{p} = -2\varepsilon_\mathrm{m}$, C_{scat} will become very large, corresponding to the surface plasmon resonance state [319]. Computed values of the C_{scat} indicated that the scattering cross section of silver is several orders of magnitude higher than gold for a given NP size. Scattering distribution is limited in space and also by the substrate and interface properties. Indeed, differences in the cross sections of the different layers contribute to higher interfacial scattering [319].

A practical example of potential efficiency enhancement using down-conversion has been reported by Inns et al. [320]. A solution of silica particles (150–200 nm diameter) is deposited on crystalline silicon before adding an aluminum layer to form a high-quality back surface reflector. Internal quantum efficiency at long wavelengths (>900 nm) is improved by a factor of 2. This improvement is attributed to reduced parasitic absorption in the aluminum layer [320].

Luminescent solar concentrator (LSC) based on down-conversion is another simple way to improve efficiency [321]. LCS consists of luminescent materials embedded in a highly transparent plastic sheet. Following radiation absorption, luminescent particles emit isotropically at longer wave lengths. Taking advantage of the internal reflection in the plastic sheet, high radiation transfer of the emitted wavelength to the solar cell is obtained. The simplicity of this approach allows solar cells with higher efficiency with a marginal increase in cost. Furthermore, contrarily to conventional PV concentrator, both direct and diffuse light are concentrated. LSC is of particular interest for building-integrated PV applications. Indeed, high power efficiency can be achieved when these cells are installed on both the roof and side wall of a building.

Early work focused on high luminescence quantum efficiency (LQE) organic dyes. More stable inorganic QDs have received more attention in spite of their relatively lower LQE. Lanthanide ion-doped fluoride NPs are used in up-conversion applications from NIR to visible range [322]. In particular, this property could be used to improve solar cell efficiency in the NIR region while reducing heat load. Combined use of PbS QDs and fluorescent up-converter materials embedded in an oxide matrix improves the photocurrent in bifacial solar cells by about 60% [323].

5 ENERGY STORAGE

Storage is critical to efficiently manage electricity demand–response and improve service quality. It is important to store energy for usage in remote and portable applications; electrical energy can also be stored during low demand for use when there is high demand [324]. Energy storage is used in electrical system regulation and spinning reserve (power outage and random fluctuations) to stabilize frequency and eliminate voltage sags and surges. It is used in peak shaving and load leveling to accommodate higher power demand during the day. It is also used in renewable energy systems to match the fluctuating supply to the changing demand.

Storage capacity and discharge rates dictate the technology choice. There are numerous physical forms of energy storage including thermal, potential, kinetic, electromagnetic, and electrostatic [325]. Batteries and ultracapacitors based on electrochemical reactions are the most common, although not the most economical, solutions for portable and backup applications. For large capacity storage, pumped hydro, compressed air, and flywheel have been used in the past, but they are not considered economical and/or sustainable solutions. Hydrogen and other chemical materials such as methanol, ethanol, and ammonia are considered a good alternative form of energy storage. They present different technological barriers that NP-based technologies could address. For example, high yield and specific NP-based catalysts will lower the energy cost of gas-to-chemical and gas-to-liquid fuel processes.

Electric power storage is a key component in vital sectors including vehicle transportation, electronic devices, and renewable power generation. Matching demand and supply of electric power is critical for the large-scale deployment of renewable energy sources in the future [326]. This is particularly the case of wind and solar energies, two fast-growing power generation sources hindered by their intermittence. PV power generation has been used in the past for water pumping, lighting, telecommunication, portable power, and cathodic protection in off-grid applications. Storage is critical when off-grid renewable power generation is used, allowing continuous accessibility to clean and renewable power.

For electric vehicles (EV), driving range (Fig. 5.21) is critical for the choice of battery technology [327]. Existing battery technologies provide up to an order of magnitude shorter driving distances than gasoline-powered internal combustion engines. Early EV models used lead–acid batteries or NiMH batteries. Improvement in energy density of these (and other) battery technologies may help close this gap in the future. In particular, high surface area nanomaterials will certainly contribute toward this goal [328].

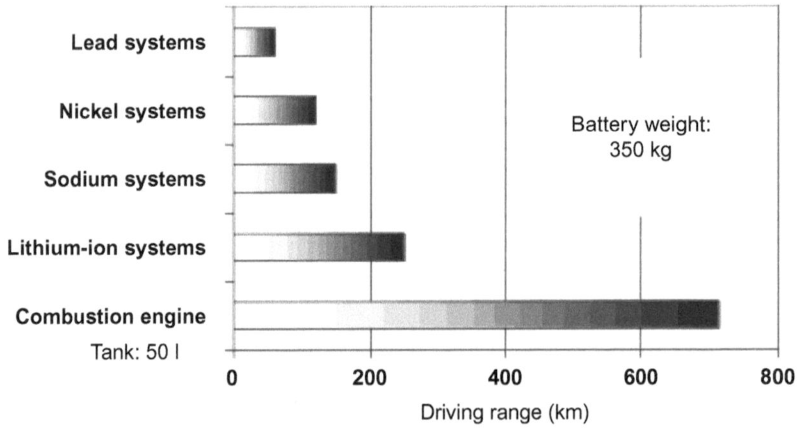

FIGURE 5.21 Driving ranges of different 350 kg battery systems and 50 l gasoline-powered engines. *Reprinted with permission from Winter and Brodd [173]. Copyright 2004, American Chemical Society.*

Lithium-ion batteries (LIBs or Li-ion batteries) widely used in portable devices seem the logical technology choice for EV applications as they have the highest energy and power density [173,329].

Four categories of electrical power storage have been developed. They include conventional capacitors, supercapacitors, primary (nonrechargeable) batteries, and secondary (rechargeable) batteries. Conventional batteries allow high-level energy storage, but their discharge rate is relatively slow. Capacitors provide high-level power discharge but their energy is relatively low. Supercapacitors may provide both high energy storage and high power discharge. Supercapacitors have a clear advantage in terms of power density (or the rate of charge/discharge) and lifetime. On the other hand, rechargeable batteries have higher energy density and low-temperature dependence [330–332]. Only secondary batteries are considered in this evaluation, excluding potentially inexpensive primary battery technologies [333]. Hydrogen is not cost-effective under current production and delivery schemes [334].

For a proper evaluation of the different battery technologies, specific criteria should be considered. The importance of each criterion may depend on the application. For example, specific energy density is of utmost importance for electrical cars. This criterion is not the most important for stationary applications, where cost/kWh is the most important parameter. In the case of batteries used in intermittent power generation (e.g., PV), the following criteria must be considered [331]:

a. Simplicity, longevity, reliability, public acceptance, sustainability;
b. High energy capacity at different discharge rates;
c. Low equalizing and boost charging requirement;
d. Efficiency at different solid oxide cell (SOC) levels;
e. Low self-discharge and fast response time;
f. Wide operating temperature range;
g. Robust design and low maintenance requirement;
h. Cost-effectiveness;
i. Manufacturing under stringent quality control.

In spite of the plethora of storage technologies, none has so far addressed all the technical and economic criteria. Bulk electric energy storage using batteries appears to be the most mature and acceptable technology [330]. For smaller capacity (<1 MW), battery storage is a proven technology providing numerous advantages. Lead–acid batteries and, to some extent, sodium–sulfur batteries are often used although their unit energy cost is relatively high. Lifetime prediction of battery under specific operating conditions is critical to identify the most suitable technology [332]. In the case of renewable power generation, where storage capacity and unit cost are critical, gravimetric energy density is not critical.

Energy storage needs depend on the applications. It is about 0.5 and 5 W h for a watch and cell phone, respectively. In the case of electric cars, around

5 kW h or more is needed. EVs and personal digital assistants are two large and continuously expanding markets requiring significant improvement in the characteristics of energy storage. Numerous NP-based batteries have been recently developed to improve specific and volume energy density. Although anode and cathode materials attract large interest, other components of battery systems may benefit from NP characteristics. In particular, electrolytes could use nanocrystalline materials to enhance ionic conductivity and mechanical properties [335]. The separator, consisting of electrically insulating materials containing pores to allow lithium-ion diffusion back and forth between the two electrodes, could gain from NP attributes. One key issue with the separator is related to temperature cycling during charge and discharge. Cycling often leads to materials breakdown causing electrical shorts which may lead to battery explosion.

Besides LIBs, several technologies are currently under consideration for vehicle applications. Energy density is the most critical parameter. Often, energy and power density values based just on electrode material weight are considered in the scientific literature. Practical energy density that also includes the packaging materials should be used as a figure of merit. For example, Na–S and Na–$NiCl_2$ have a theoretical gravimetric energy density about two times higher than that of Li-ions, although the practical energy densities are relatively the same [336]. This difference stems from the additional thermal management system in the case of Na–S and Na–$NiCl_2$ batteries. Given the fact that gravity and volumetric energy densities are not critical in the case of utility storage, Na–S and Na–$NiCl_2$ are expected to compete with lead–acid batteries for this application in the near future. Ni–Cd and Pb–acid batteries have relatively low energy densities.

Any new technological development needs to consider raw material availability and the environmental impact of uncontrolled release of toxic materials during manufacturing and utilization, and after disposal. In particular, elements such as Li and Ni are relatively scarce. Therefore, their use should be optimized through minimization of metal load and by recycling. Utilization of nanomaterials in general and NPs in particular will help achieve these goals.

5.1 Conventional Batteries

A basic battery consists of an electrolyte (liquid, polymer, or inorganic oxide) sandwiched between two electrodes. Battery systems are divided into two categories. Primary (single, throw-away batteries or nonrechargeable) and secondary (rechargeable) batteries are often used for different applications. In this section, the focus is on secondary batteries where NPs will have some impact including improving battery performance. Complex formulations including electro-active materials, binders, and electron conductive fillers could benefit from the recent developments in NP technology [337]. Formation of the solid electrolyte interface (SEI) is particularly critical to the operation of LIBs.

The choice of battery technology is difficult given the numerous options existing within each type of battery. There are also avenues to improve the electrolyte and the overall system packaging. However, the cathode and the anode are receiving more attention. In the case of LIBs, a plethora of cathode and anode material compositions are available to potentially address energy, power, and lifetime requirements [338]. To obtain high-energy batteries, $LiNi_xCoyM_{1-x-y}O_2$ (layered) or $LiMn_{2-x}M_xO_4$ (spinel) for the cathode and Si (Sn or Sb) or MO_x for the anode are used, respectively [338]. In the case of high power, $LiMn_{2-x}Al_xO_4$ (spinel), $LiNi_xCo_{1-2x}Mn_xO_2$, or $LiFePO_4$ (olivine) for the cathode are used in combination with hard carbon, graphite, and $Li_4Ti_5O1_2$ for the anode, respectively [338]. Finally, in the case of long lifetime batteries, $LiFePO_4$ (olivine) and $LiMn_{2-x}Al_xO_{4+\delta}$ are used as the cathode in conjunction with $Li_4Ti_5O_{12}$ and graphite for the anode [338].

Performance/cost ratio and reliability are hindering current battery technology utilization in electrical vehicles and intermittent distributed power generation. Electrochemical storage technologies have different degrees of maturity in different market niches. Lead–acid, NiCd, and NiMH are quite mature technologies. Advanced batteries include Li, Li/CF_x, Li/polymer, and Li-ion. Technical progress has been somewhat incremental in most cases [338–340]. Thin-film batteries can provide much higher energy and power densities. Indeed, volumetric and gravimetric energy densities of thin-film LIBs are about three times higher than those of conventional LIBs [339]. These thin-film-based batteries are currently produced on nonflexible substrates using mostly the physical deposition technique [339]. A thin-film battery consisting of a polymer-based solid electrolyte is attractive, but requires alternative production techniques. A flexible thin-film battery is easier to produce and assemble in high volume.

Lithium-based batteries are considered the best technology option for portable applications. However, current LIB performance and cost are still insufficient for mobile and stationary applications. Research on improving battery characteristics should address the following issues [341]:

a. High energy and power density;
b. Long lifetime using stable materials;
c. Low, manageable environmental impact due to toxicity;
d. Cost-effectiveness using efficient fabrication processes;
e. Safe operation.

To enhance interfacial properties, 2D and 3D architectures have been proposed to replace 1D architectures of conventional layered films. Strictly speaking, existing commercial batteries are often a combination of at least two architectures. Batteries made of aerogel- and sol–gel-based electrodes provide the best approach to obtain a 3D architecture. Rod and plate electrodes are not of 3D structure, although they provide a cost-effective and efficient approach to electric charge collection. Beside charge transport, other parameters should be considered when comparing the different architectures [328].

Power density is related to the mean diffusion time given by the following equation [338]:

$$\tau_{eq} = L^2/2D$$

where L and D correspond to the diffusion length and diffusion coefficient, respectively. Higher τ values are obtained by either decreasing L or increasing D. Electronic doping allows higher D values. Using nanostructured materials in the anode and/or cathode allows reduction in diffusion length (L). For example, by reducing the value of L from 10 mm (the typical size of commercial electrode materials) to 100 nm, the t_{eq} is reduced from 5000 to 0.5 s [338]. In this example, a typical value of $D = 10^{-10}$ cm^2/s is used. Several approaches have been developed to obtain nanostructured electrodes (Fig. 5.22). Both interdigitated and spongy geometries have advantages and drawbacks.

NPs could play a significant role in improving the performance of batteries by reducing the Li^+ diffusion path. This enhances the absorption and storage kinetics of Li-ions without possibly affecting the overall structural stability [338]. NPs will increase not only the rate capability (high power) by lowering the diffusion length, but also electroactivity toward Li insertion. Indeed, NPs provide high surface area and high surface energy allowing higher energy density. Furthermore, NP-based electrodes could also improve the reversibility

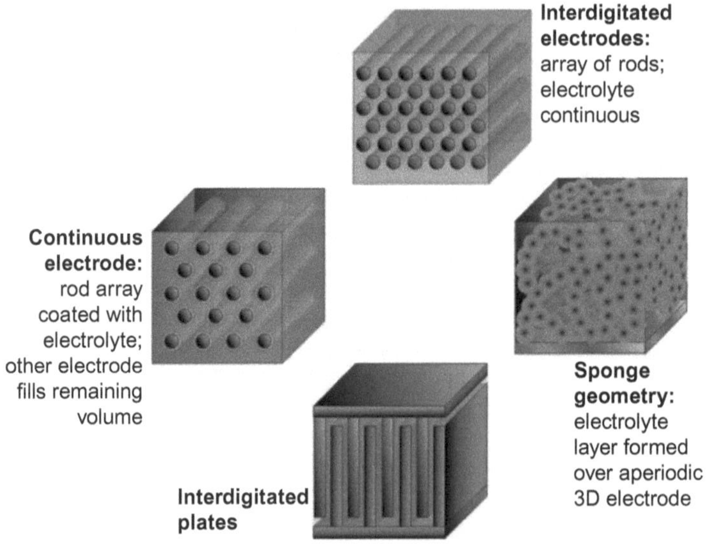

FIGURE 5.22 Different design options to enhance interfacial reactions in batteries. *Reprinted with permission from Long et al. [328]. Copyright 2004, American Chemical Society.* (For color version of this figure, the reader is referred to the online version of this chapter.)

during charge and discharge [342], which will help improve lifetime and, thus, unit energy cost.

NP-based batteries may suffer pronounced agglomeration during cycling. To enhance the structural stability of the lithium batteries, the particle radius should be smaller than the critical nucleation radius [338]. Indeed, small particles would easily accommodate the structural changes occurring during the cycling process involving Li insertion and removal. This is the case of $LiMnO_2$ batteries exhibiting significant capacity fading during cycling attributed to structural instability. Nanocrystalline structures can accommodate much higher Li-intercalation capacity than their bulk crystalline counterparts. When small NPs are used instead, higher Li accommodation is allowed near the surface. Overall, NPs can accommodate higher Li insertion at a higher rate.

NPs could improve battery characteristics in different ways. First, they can increase energy storage capacity:

$$C_{\text{bat}} = (1/3.6)(nF/M_{\text{m}})$$

where n, F, and M_{m} are the number of electrons transferred during the oxidation–reduction of the material, F is Faraday's constant, and M_{m} is the molecular mass of the compound (g/mol) [343]. Second, they can reduce the mean diffusion time. This is achieved by reducing the charge diffusion length and/or the diffusion coefficient [343]. Third, they can increase surface energy by reducing particle size, which improves electroactivity. Fourth, smaller NPs afford high interfacial contact between electrode and electrolyte, increasing Li^+ flux at the interface [344].

In spite of these advantages, nanomaterials including NP-based electrodes [345,346] are certainly not the only panacea for the limitations of the current storage battery. NP integration in batteries could be limited due to high production cost. Furthermore, high volumetric energy/power density and higher surface energy could lead to side reactions. Altairnano recently reported a safer precommercial LIB that can address the requirement of EV design [347,348]. They used nanosized lithium titanate oxide ($Li_4Ti_5O_{12}$) to replace the graphite used in the conventional LIBs. Energy density above 100 W/kg and a specific power above 1 kW/kg are achievable. Anode materials with the $Li_4Ti_5O_{12}$ composition have been obtained using a wet process [349]. A123 system used nanophosphate materials in the cathode. It claims to have the highest power density in W/kg of all commercially available LIBs. The cell can be continuously discharged to 100% depth-of-discharge at 35 C rate and could sustain discharge pulses as high as 100 C. The phosphate-based battery has a nominal voltage of about 3.3 V/cell with a peak charge voltage at 3.60 V. This is slightly lower than the cobalt-based LIBs. Valance Technology was the first to commercialize the phosphate-based LIB [350].

NP-based batteries have shortcomings as well [338]. High surface energy materials could lead to low thermodynamic stability. Indeed, small particles have a loose structure and a tendency to agglomerate particularly following electrochemical cycling. High surface energy may also lead to unwanted surface reaction raising safety issues. To prevent these adverse effects, several approaches have been proposed [338]. A mesoporous and multifunctional polymer matrix has been used in the past. Another approach is to add active NPs (e.g., Sn, RuO_2, V_2O_5) into inactive hollow carbon spheres (HCs). The pores are filled with electrolyte. Resulting hierarchical 3D mixed conducting networks on both nanoscale and microscale levels allow higher stability and high conductivity. Improved charge/discharge rates and storage capacity have also been achieved [338]. Up to 74 wt% of Sn is achievable allowing a high theoretical specific capacity of up to 831 m Ah/g.

5.1.1 Lead–Acid Battery

Applications using NPs have been reported in the case of lead–acid batteries. A positive-grid alloy formulation based on Pb–Ca–Sn and Ba has been developed to improve the lifetime of lead–acid batteries [351]. Enhanced mechanical strength and corrosion resistance will improve the durability of these batteries. The presence of NPs gives rise to precipitates when Ba is added. This will likely improve lead–acid characteristics.

5.1.2 NiMH

NiMH batteries have been extensively used in portable devices, and remain the most commercially successful battery chemistry for electrical vehicles. Higher power density and increased rate of discharge are two characteristics of NiMH batteries that need further improvement. Current NiMH design includes insertion of 5–7-nm-diameter metallic nickel particles throughout the oxide layer. The insertion of these NPs increases catalytic activity, enabling dramatic increases in NiMH power [352,353].

As shown in Fig. 5.23, insertion of nickel NPs combined with engineered voids within the metal hybrid alloy allows high surface area. As a consequence, improved surface activities with rapid ionic and water diffusion are obtained. These properties lead to high specific power in the range of 1–900 W/kg [352].

High capacity, good kinetics, and long life cycle of NiMH batteries are enabled by the relative disorder within the metal–electrolyte interface [354]. Indeed, metallic nickel alloy inclusions of less than 7 nm provide high surface area and large catalytic activity to the metal hydride [354]. Other NiMH compositions have been developed [355].

5.1.3 Lithium-Ion Battery

Improved safety, higher specific energy, and lower cost are three areas requiring improvement in LIB. NP integration into electrodes, electrolytes,

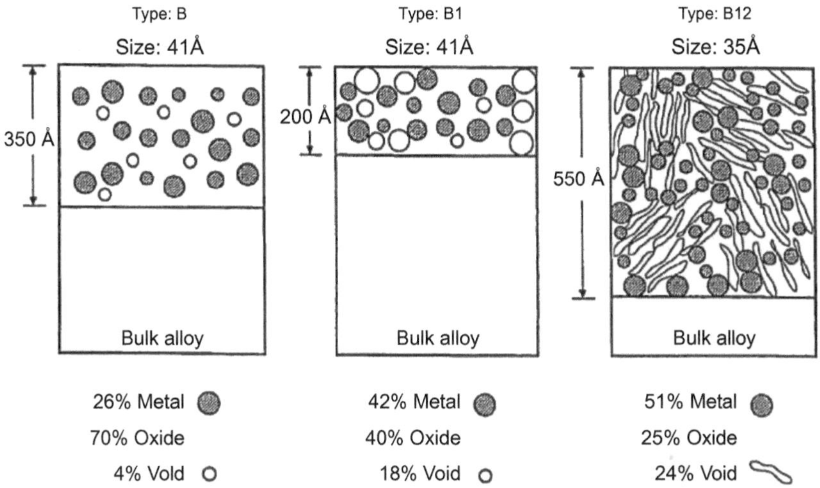

FIGURE 5.23 Schematic representation of the oxide layer consisting of engineered void volumes to improve reactants and metal nanoparticle transfer [352].

and separator materials compositions are discussed in the following sections.

5.1.3.1 Separator

Al_2O_3 particles dispersed in the PVdF matrix have been evaluated as ceramic powder-based separators (CPS) for LIB [356]. The CPS film based on 10 nm Al_2O_3 NPs (Fig. 5.24) showed superior characteristics. SEM characterization of the separator obtained with 10 and 300 nm Al_2O_3 particles showed clear microstructural differences (Fig. 5.24).

5.1.3.2 Electrolyte

Battery safety remains the biggest technology challenge of LIB used in EVs. A nonflammable electrolyte with larger energy window (E_g) between the LUMO and highest occupied molecular orbital is needed. Liquid, amorphous polymer, and crystalline polymer electrolytes have been used in the past [344]. Solid-state ion conductors offer numerous advantages over liquid [357,358]. Additives that accelerate the formation of an SEI layer are needed. This will prevent lithium plating on the anode during battery charging [357].

Solid NPs are potentially good candidates as solid additives for the development of nonflammable electrolytes. Conductivity and electrochemical properties of ionic conductors are size-dependent [335,359]. Polyethylene oxide (PEO)-based electrolyte has been used in LIB instead of liquid electrolyte [360]. To lower operational temperatures of PEO-based polyelectrolyte without degrading its mechanical and electrical properties, ceramic NPs have been added [361]. A nearly 2 orders of magnitude enhancement in the conductivity

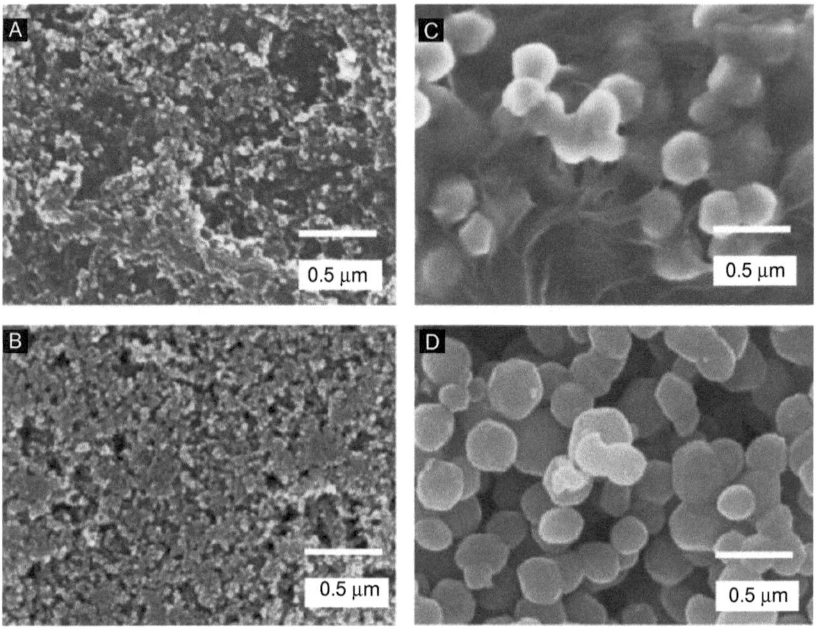

FIGURE 5.24 SEM of ceramic powder-based separator (CPS) using different sizes and powder loads: (A) 0.01 μm particle size with low powder load; (B) 0.01 μm particle size with high powder load; (C) 0.3 μm particle size with low powder load; (D) 0.3 μm particle size with high powder load [356].

has been reported after adding 13 nm TiO_2 and 5.8 nm Al_2O_3 to the PEO matrix [361]. Solid electrolyte nanocomposites based on poly-(ethylene glycol) and 15-nm-diameter silicate NPs provide high conductivity (10^{-4} S/cm) and high cation transference number [362].

Other LIB electrolytes have been prepared by mixing PEO–LiX ($X = CF_3SO_3$, ClO_4, and PF_6) with inorganic NPs (TiO_2, SiO_2, and Al_2O_3) [363]. Smaller particles seem to provide higher ionic conductivity. The highest ionic conductivity has been measured using $LiPF_6$ and Al_2O_3. Single-ion-conducting silicate-based polymer nanocomposites have also been reported elsewhere [364]. Additional mechanical stability is inferred by the presence of these ceramic fillers through the interaction between the NP surface group on the one hand and the polymer chain and salt anion on the other hand [365].

5.1.3.3 Cathode

Widespread use of LIB beyond the consumer gadget is hampered by the poor performance/cost value of cathode materials. In particular, the high cost of precursor materials, low surface area, and unstable mechanical and electrochemical properties have been identified. The importance of nanostructured cathodes is now well established [366]. The demand for high discharge in some portable applications has led to the development of new geometrical designs for battery electrodes. The main limitation of current lithium batteries is caused by the slow solid-state diffusion of Li^+ within the electrode layers [367]. Two designs have been proposed to increase the Li^+ diffusion rate (see Fig. 5.25). These designs may give rise to specific nanoscale-related effects [367].

Rate capability tests using two different electrode densities (3.4 and 2.8 g/cm^3) and four $LiCoO_2$ particle sizes (50, 100, 300, and 1000 nm) showed that particles with a diameter of 300 nm provide the best results [368]. Numerous NP compositions have been developed for use in LIB. NPs allow high surface area, higher electrochemical reactivity, and a reduced lithium-ion diffusion pathway. Recent work has shown that these novel materials improve power density, lifetime, and charge/discharge rate. Several reports have shown that these advanced materials enhance the number of cycles at high C-rates. Electrodes made of transition-metal oxide NPs showed a capacity of at least 700 mAh/g over 100 cycles. Several companies including XOLIOX (NTERA), Altair Nanotechnologies, and NanoGram have announced the development of NP-based LIBs with high discharge rates.

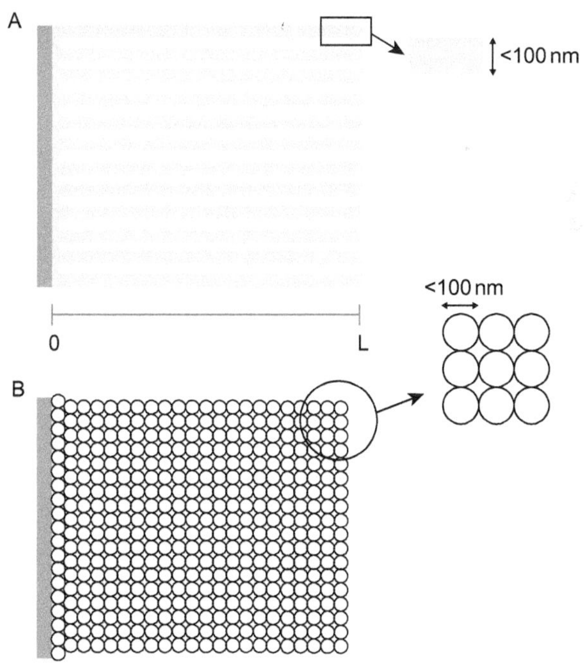

FIGURE 5.25 Schematic representations of highly porous electrodes using (A) 2D nanofibers and (B) 3D nanoparticles. Note that the porous electrode phase and electrolyte are continuous along the length *L*. *Reprinted with permission from Bueno and Leite [367]. Copyright 2003, American Chemical Society.*

Today's commercially available LIBs are based mostly on microcrystalline transition-metal oxides such as $LiCoO_2$, $LiMn_2O_4$, $LiMnO_2$, LiV_2O_5, and $LiFePO_4$. High surface area nanomaterials used as cathodes have been pursued to improve battery characteristics in the past [369,370]. Indeed, NP-based anode and cathode lead to improved electrochemical response. This is due to higher electrode surface area, leading to higher charge/discharge rates. Furthermore, shorter path lengths for Li^+ ion transport result in an increase in power capabilities. Another important factor is the ability to accommodate the strain of lithium insertion/removal. Often, NP-based active electrodes could be subjected to large volume expansion–contraction. This change occurs during the lithium alloying–dealloying process following charge and discharge cycles. The volume changes lead to severe mechanical strains limiting the lifetime of the battery.

The drawbacks of NP-based electrodes have been described [371]. High surface energy metallic NPs are a safety hazard as they are flammable and may cause explosions. The tendency of high-area NPs to aggregate upon cycling leads to degradation of the battery characteristics.

The comparison of $LiNi_{0.5}Mn_{1.5}O_4$ intercalation electrodes comprised of micro- or nanosized particles showed that NPs enable faster kinetics and a more reversible electrochemical behavior [372]. $LiMn_2O_4$ NPs obtained from commercially cheap MnO_2 precursors have showed promising electrochemical properties of interest to LIB [373]. After calcination, highly pure $LiMn_2O_4$ spinel consisting of sub-micrometer cubic crystal agglomerates is obtained. Nanosized agglomerates are also observed. A stable capacity of 115 mAh/g has been obtained using this material [373]. Particle sizes and the morphology of $LiMn_2O_4$ powders have been shown to have a direct effect on power density and rate of discharge [374]. $LiMn_2O_4$ NPs with an average diameter of 15 nm prepared via a sacrificial hard template have been shown to provide improved performance [375]. Polymer-coated $LiMn_2O_4$ NPs have been used to minimize surface degradation of the cathode material [376]. Surface modification of the $LiMn_2O_4$ particles extended the lifetime of the LIB by preventing Mn^+ dissolution [376]. Capacities in the 180–200 mAh/g range have been reported during the cycling of cathodes prepared using $LiNi_{0.5}Mn_{0.5}O_2$ NPs. These NPs are obtained by combustion a mixture of sucrose and metal nitrates [377].

Recently, a direct relationship between $LiMnPO_4$ sol–gel characteristics and LIB performance has been reported [378]. As shown in Fig. 5.26, smaller particles provide a higher reversible capacity. Indeed, calcinations at 520 °C give rise to particle sizes of about 140 nm and a reversible capacity of 130 mAh/g. Dry ball milling of calcined powder reduces the primary particle by about 30%. Furthermore, at faster charging rates, electrochemical performance improved when smaller $LiMnPO_4$ particles were used [378].

The discharge rate of the Li-ion cell using $LiCoO_2$ NPs showed improvement over microparticles [379]. These $LiCoO_2$ NPs are obtained from

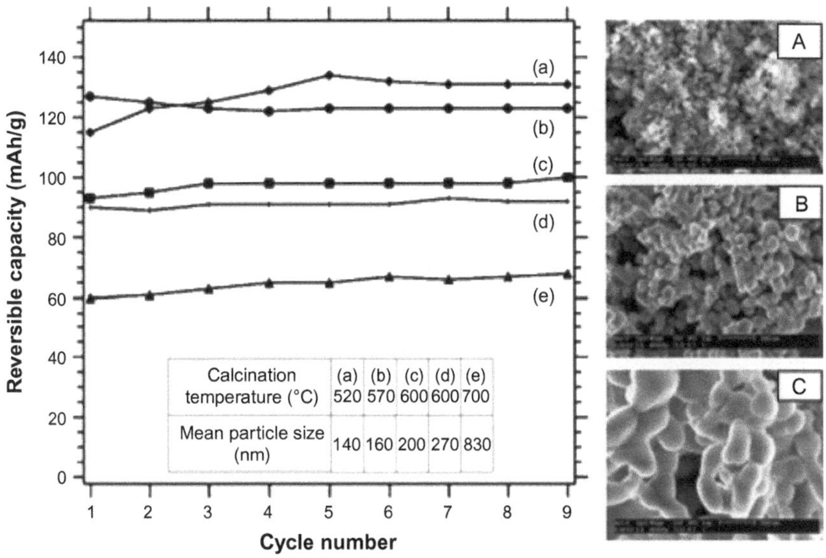

FIGURE 5.26 Reversible capacity of LiMnPo4 with different particle sizes obtained using different calcination temperatures [378].

calcination of cobalt acetate and lithium acetate at 600 °C. The discharge capacity of these $LiCoO_2$ NPs is about 100 mAh/g, which is two times higher than the capacity of the ones obtained from 5 μm $LiCoO_2$ particles. Uniform $LiCoO_2$ NPs with tunable diameter have been synthesized using sol–gel methodology. An initial discharge capacity of 149 mAh/g with good cycling stability was measured on these sol–gel materials [380]. An ultrathin-film electrode has been prepared by the inkjet printing method, indicating the potential of using NPs for thin-film batteries, to replace the vacuum deposition methodology.

To prevent metal-ion dissolution from $LiNi_{0.8}Co_{0.16}Al_{0.04}O_2$-based cathodes, the use of an additional coating layer consisting of $Co_3(PO_4)_2$ NPs showed promising results [381]. Anomalous size effects have been observed in oxide NP-based LIB. This has been ascribed to excess surface free energy of oxide NPs [382]. Si–Ni NPs (20–40 nm) were produced by arc-melting bulk Si–Ni ingots in a 50%Ar and 50%H_2 mixture [383]. These NPs have been used to produce a high performance lithium storage system with a capacity of 1304 mAh/g. Using the combustion flame method, Amatucci et al. showed higher Li^+ intercalation capacity in nanocrystalline compared to microsized V_2O_5 particle host materials [384]. Furthermore, they showed for the first time that reversible intercalation of Y^{3+} is feasible in nanocrystalline V_2O_5. Intercalation of divalent and trivalent cations can result in significantly improved gravimetric capacity.

Lithium iron phosphate ($LiFePO_4$) NP-based batteries are believed to be able to address the cost issue of $LiCoO_2$ batteries [385]. In particular, iron precursors are cheaper than cobalt precursors. $LiFePO_4$ NPs with an average size of 50 nm are prepared by coprecipitation in aqueous solution and then tested as anode and cathode materials for LIBs. When using cathode $LiFePO_4$, a capacity of 485 mAh/g has been achieved [386].

Several parameters are expected to affect the performance of $LiFePO_4$ NPs. The size-dependent miscibility gap of $Li_{1-x}FePO_4$ particles, the coexistence of ferrous and ferric species in the solid solution, the lattice parameters, and the bond lengths of the phosphate phases could affect the electronic properties [387]. These factors are expected to alter the electrochemical properties. Coating of these NPs with an ultrathin and highly conductive layer is often done to control these electronic properties. The core–shell structure of $LiFePO_4$ plays a critical role in achieving high energy and long lifetime [388]. This structure accommodates the juxtaposition of the movement of the phase boundary and the shrinking of the $Li_{1-x}FePO_4$ NP core during the charge/discharge process [388].

Adding sugar to precursors before heating leads to a uniform carbon layer on $LiFePO_4$ NPs [389]. This method increases the rate capability significantly. The combination of an $LiFePO_4$/C cathode and a gel–polymer electrolyte gives rise to high charge and discharge rates [390]. A relatively low-cost synthesis method of a $LiFePO_4$ NP-based composite has been obtained by a two-step milling process [391]. A mixture containing CH_3—$COOLi \cdot 2H_2O$, Fe_2O_3, Fe, and $NH_4H_2PO_4$ has been used in this case [391]. After sintering, an initial discharge capacity of 162.4 mAh/g was measured on these $LiFePO_4$–C nanocomposites. After the 30th cycle, a discharge capacity of 166.6 mAh/g was measured.

Starting from aniline and inexpensive precursors, Wang et al. prepared highly crystalline $LiFePO_4$ NPs with a core diameter of 20–40 nm [392]. The addition of Fe^{3+} ions to a solution containing $PO_4{}^{3-}$ ions and aniline gives rise to a precipitate consisting of $FePo_4$ core particles coated by a polyaniline shell. The shell consisting of 1–2-nm-thick semigraphitic carbon is obtained following an *in situ* polymerization [392]. Posttreatment leads to the formation of $LiFePO_4$/carbon nanocomposites. High and stable capacity after 1000 cycles has been achieved using this olivine structure.

Combining $LiFePO_4$ cathode materials and solid-state electrolytes provides low cost and safety attributes. Furthermore, enhanced capacity at high power has been recently demonstrated by replacing partially or totally the inactive carbon and PTFE binder by conductive polymers [393]. In particular, it was found that electrodeposited carbon-coated $LiFePO_4$/PPy enhances the capacity and rate capability of LIB.

5.1.3.4 Anode

Anode materials for lithium batteries have been reviewed [394]. Several elements of the group IV (C, Si, Ge, Sn) and group V (P, Sb) and their composites including Ag and Mg have been investigated [395]. Agglomerated graphite NPs are currently used in newly developed LIBs. Replacement of conventional graphite microparticles by NPs will certainly increase charge/discharge (insertion/removal of Li^+) rates. However, when graphite NPs are used, catastrophic reactions may occur following lithium deposition and build-up [344]. Graphite materials have been used in anode batteries without any safety issues [396]. Numerous alternative materials [371,394,397–406] including silicon, tin, carbon, and alloyed materials have been widely used for improving anode performance.

New anode materials should allow higher redox potential along with high charge rate and low strain. Numerous lithium titanate and Li-based (Li_xM_y) alloys have been developed. Anode material composition of $Li_{4.4}Sn$ with an energy capacity of more than 1000 mAh/g has been reported. The drawback of these alloys is the relatively large expansion/contraction following the formation of the alloys that could lead to mechanical failure [344]. Using small Li_xM_y NPs may alleviate this issue. SnO_2 NPs having 3, 4, and 8 nm diameters are obtained using hydrothermal synthesis at 110, 150, and 200 °C, respectively [400]. NPs with 3 nm diameter provide the highest initial capacity at 740 mAh/g without any capacity loss. The only drawback of these SnO_2-based Li batteries is their relatively large irreversible capacity. This has been attributed to the severe reaction with the electrolyte within the porous structure [400]. The irreversibility issue has been addressed by reducing the volume expansion of the Li_xSn alloy upon integration of the tin oxide in a porous template [401]. A new anode material for LIB using NPs of tin confined in a microporous carbon matrix has been reported. Reversible capacity with a nominal value of 305 mAh/g was achieved using a carbon matrix with a 2-nm-diameter pore [401].

Tin NPs with an average particle size of 3.5 and 10.0 nm have been prepared and dispersed in graphite and used as anode material for LIB [398]. The small particle size provides higher and reversible storage capacity [398]. Tin-based NPs prepared by infiltration into gel structures followed by high-temperature calcinations are used as an anode [371]. Specific capacity of 200 mAh/g has been measured at 5 C rate. Insertion of carbon-coated $Sn_{92}Co_8$ NPs in the anode of LIB showed improvement in retention capacity of the carbon-coated nanoalloys [403].

Silicon insertion has been demonstrated to improve anode material performance. However, its capacity for fading and degradation after cycling has hindered its commercialization [397]. Promising nanoscale silicon anode intercalation has been reported [397]. Morita and Takami used Si NPs embedded into a silicon dioxide and carbon matrix for the development of a high-capacity anode [403]. These new nanocomposite anodes allowed high capacity (700 mAh/g) and long life cycle (above 200 cycles). Silicon–carbon core–shell NPs prepared via resorcinol–formaldehyde microemulsion polymerization showed good long-term stability [397]. However, their capacity is lower than that of bare silicon [404].

Carbon-based NPs with sizes ranging from 40 to 80 nm are prepared by electrochemical reduction and used as anode materials for LIB [405]. Lithium titanate $Li_{4/3}Ti_{5/3}O_4$ NPs with an average diameter of 12 nm prepared by electrospray pyrolysis are shown to allow fast cycling [406]. With a constant redox potential at 1.56 V versus Li/Li^+, the fast Li-ion diffusion was not possible with bulk lithium titanate. NP-based lithium titanate batteries have been reported to provide more than 30% improvement in energy density and more than 10 times faster charge and discharge times.

5.2 Thin-Film Batteries

All-solid-state thin-film batteries are attracting significant attention, particularly in applications for microbatteries. Like any LIB, thin-film LIB consists of a solid-state electrolyte layer sandwiched between two current collectors coated by a lithium anode film and lithium cathode film (e.g., $LiCoO_2$). LBL manufacturing using thin-film-based deposition techniques (often gas phase) will result in an overall stack thickness of less than 20 μm [407]. Conventional LIB is about two orders of magnitude thicker. In these thin-film batteries, a solid-state electrolyte is used instead of the liquid electrolyte used in conventional batteries. In the case of Li-based thin-film batteries, operation principles are similar to those of conventional batteries. Thin-film batteries provide improved life cycle and high-temperature processes and operations. Furthermore, these thin-film batteries can be fabricated in any shape or size with minimal additional cost.

Currently, most thin-film batteries are obtained by gas-phase vacuum deposition techniques. Magnetron sputtering and, to a lesser extent, electron beam evaporation are used. Alternative techniques are currently being considered. They include spin coating, inkjet printing, sol–gel, and spray deposition methods. Techniques such as CVD and laser deposition are also worth considering. Numerous factors should be considered before deciding on the right solution. Capital and operation costs, yield, and product quality are the main characteristics.

An all-solid-state cell with $Li/LiPON/CuWO_4$ battery containing $CuWO_4$ NPs has been obtained using radio-frequency sputtering deposition [408]. Alternative thin-film deposition should be considered for reducing the cost. Spin coating of solid polymer electrolytes has been used for all-solid-state rechargeable $Li/polymer/LiMn_2O_4$ thin-film batteries [409]. Inkjet printing was successfully used to fabricate anodes for thin-film batteries from SnO_2 NPs [410]. Based on a film thickness of 2.3 μm, an initial discharge capacity of 812.7 mAh/g has been measured. Lithium phosphorous oxynitride polymer is often used for the electrolyte layer.

5.3 Supercapacitor

Three different electrochemical capacitors are currently used [341]. Besides the conventional capacitor, two other designs have been developed to improve performance over cost characteristics. They include double layers using high surface area carbon as electrodes and different electrolytes (aqueous sulfuric acid or organic liquids), and asymmetric supercapacitors. Nanostructured carbon, transition-metal oxide (e.g., RuO_x), and electronic conducting polymers are a few materials that are under consideration for high-performance capacitors.

The supercapacitor market is estimated at about $200 million [411], and it is expected to increase significantly in the future given its potential application in electric transportation and renewable energy storage. Supercapacitors, also referred to as electrochemical capacitors, ultracapacitors, or electrical double-layer capacitors, are capacitors that are able to hold significantly more charge [412]. Supercapacitors are thus different from electrostatic capacitors and electrolytic capacitors [411].

Like any capacitor, but unlike batteries, the supercapacitor's storage mechanism is based on a physical process. Contrary to batteries, there is no chemical reaction when supercapacitors are charged. The energy is stored by increasing static charge electron density on the surface. Supercapacitors allow higher power density by up to 10 times when compared to traditional batteries. Supercapacitors have much higher specific power than batteries. However, today's commercial supercapacitors carry at least an order of magnitude less energy per unit volume than equivalent-weight batteries. Thus, weight reduction is probably the most critical issue of supercapacitors. This could be achieved by introducing high/low density materials with high energy density.

To increase the energy density of supercapacitors, an asymmetric structure has been developed. A traditional design of a supercapacitor consists of an electrolyte (basic or acid aqueous solution) sandwiched between two high surface area electrodes, forming two complete capacitors connected in series. On each side, between the electrolyte and the electrode, a double charge layer is formed. The overall capacitance, C, is estimated from the classical law: $C^{-1} = C_1{}^{-1} + C_2{}^{-1}$ where C_1 and C_2 are the capacitance of the two electrodes. Stored energy is given by the following relation:

$$E = \mathrm{CV}^2/2$$

where C and V correspond to the capacitance and voltage, respectively.

The capacitance is related to the dielectric constant (ε), thickness of the dielectric material (d), and its geometric area (A) by the following relation: $C = \varepsilon S/d$. Since charge is stored only at the interface and not within the entire electrode, supercapacitors usually have lower charge density than batteries. But their discharge rate (about 1 s^{-1}) is usually much faster because of the limited contribution from the electrolyte layer.

Supercapacitors are attracting considerable interest as an intermediate power source not addressed by conventional capacitors and rechargeable batteries. Indeed, there is a mismatch between the fast growth in power required by portable devices and the low discharge rate of commercial batteries. Today's batteries cannot respond to large bursts of power. LIBs have been developed to increase power density of traditional batteries, although they are limited by their low charge and discharge rates. Supercapacitors are believed to meet or exceed most requirements of hybrid vehicles [413]. With a discharge time of a few milliseconds, they could produce large electrical currents. However, supercapacitors still suffer from lower energy density and their relatively bulkier and heavier structure. The power and energy capacity of supercapacitors are limited by separator ionic conductivity and the electrode surface area [341]. Physical stability and the chemical reversibility are the limiting factors in batteries. These factors do not play a role in the case of supercapacitors. Thus, the lifetime of a supercapacitor is almost unlimited. However, site reactions may lead to reduction in the lifetime.

Ideally, a supercapacitor may consist of micro- and nanostructures [414–416]. High surface area and conductivity are two important criteria. Given that positive and negative charges in the electrolyte are of unequal sizes, it is important to precisely engineer the porous structure of the two electrodes. Various porous carbon materials have been developed for capacitors and supercapacitors including activated carbons, aerogels, xerogels, and CNTs [415]. Carbon with surface area of up to 2000 m^2/g is required. Other materials (oxides, conductive polymers) have also been investigated [416]. Various oxide materials are currently used in supercapacitors showing higher performance although at a higher cost [411]. These oxide-based supercapacitors are used in missiles and aerospace applications, where cost is not the main criterion.

In spite of the recent progress, the lack of substantial development in carbon material design is limiting the energy density of supercapacitors [417]. Indeed, the relatively small energy density carried by supercapacitors is preventing their large-scale utilization. The effect of extremely narrow pores on capacitance has been demonstrated [418]. However, there is still some controversy regarding the relative size of these versus the solvated electrolyte ions. For high rate discharge and lower energy density, such as in the case of pulse power applications, it is more beneficial to use relatively larger microporous materials.

Several transition-metal oxides can provide higher density and better stability [419]. Gravimetric capacitance has been suggested to decrease with increase in RuO_2 particle size [420]. Indeed, the capacitance is reduced by two orders of magnitude when the RuO_2 particle diameter is increased from about 1 to 10 nm. Given the variation of the specific capacitance and the specific surface area as a function of the RuO_2 particle size [420], a strong case for using smaller nanoparticulates to achieve higher capacitance could be made. Machida et al. showed that the electrochemical cycle lifetime is enhanced by five orders of magnitude when RuO_2 NPs are coated with a conducting polymer [421]. Gravimetric capacitance of 400 F/g has been reported on $Ru_{0.35}V_{0.65}O_2$ oxide having an average particle size of 5 nm [422].

Xerogel SnO_2-based supercapacitors have been shown to provide high specific and volumetric capacitance [416]. Surface grafting treatment has been performed prior to calcination to avoid particle coarsening. Manganese dioxide NPs prepared by sonochemistry showed high specific capacitance (above 200 F/g), which compares quite well with templated MnO_2[423]. $MnFe_2O_4$/carbon black nanocomposite has been used as a supercapacitor with good electrochemical properties [424].

5.4 Hydrogen Production and Storage

Although hydrogen is better known by the general public for its usage in hydrogen bombs and more recently in FCs, it is mostly used in refining, reforming, and manufacturing various chemicals. It is also used as a reducing agent in metallurgy. More than 40 million tons of hydrogen is consumed worldwide per year, worth about $120 billions/year, with a nearly double-digit yearly growth [334]. More than 96% of this hydrogen is produced from nonrenewable sources. With all its potential advantages as an energy vector, hydrogen has numerous issues related to its production, storage, and distribution [12]. Although production aspects have already been discussed, a broad description of the different production techniques is provided in the following section. Potential utilization of NPs for hydrogen is also discussed and compared to state-of-the-art commercial storage technologies.

5.4.1 Hydrogen Production

There are numerous avenues for hydrogen production, but only a few of them are economically viable and practical [425,426]. Fossil fuels (oil, gas, and coal) remain the dominant sources of hydrogen with about 96% market share. Following are the main practical processes for hydrogen production:

a. Steam reforming of natural gas using nickel catalysts;
b. Room temperature water electrolysis: electrochemical dissociation of water into H_2 and O_2 using electrical power;
c. High-temperature water electrolysis: electrochemical dissociation of heated water into H_2 and O_2 using less electrical energy;
d. PEC: photons from the sun are used to generate charge and induce an electrochemical dissociation of water at a semiconducting surface;
e. Thermal water splitting: thermal heat is used to heat and split water molecules at around 1000 °C;
f. Thermochemical splitting: a combination of heat and chemical catalysts is used to split water.

With a current target of \$2–3/kg of H_2, including production, delivery, and dispensing, most renewable processes are not economical [427]. Future technology developments in electrolysis using wind energy, in thermochemical water splitting using concentrated solar energy, in PEC, and in biomass-to-hydrogen processes are promising. For example, electrolysis of water using renewable sources requires an electricity cost of about 0.05/kWh to reach this target. Furthermore, these distributed sources of hydrogen will reduce the need for capital-intensive transport, distribution, and storage infrastructures.

HC direct decomposition, partial oxidation, and steam reforming remain the dominant processes [121,428]. The first two processes require high temperatures (around 1000 °C or more) and produce significantly more CO as a product or by-product. Furthermore, all the HC processes are energy- and technology-intensive. Another promising avenue is based on photoelectrochemistry and water splitting using an electrolyzer. Every kilogram of hydrogen necessitates around 40 kWh of electricity for a typical electrolysis setup. At a cost of 0.10/kWh, this makes hydrogen quite expensive. Nickel NPs with an average size of 10 nm have been shown to significantly enhance hydrogen output and efficiency when using water electrolysis [429]. Thermolysis uses thermal heating to split chemical-containing hydrogen such as methane and water. A two-step water-splitting thermochemical cycle has been recently developed using a Zn/ZnO redox system at low temperature.

While steam reforming retains the lion's share of the large-scale hydrogen production market, electrolyzers are competing in small-scale industries. Depending on natural gas cost and production process, H_2 production cost from natural gas could be lowered below \$1/kg [430]. Availability of an economical and clean platform for hydrogen production and storage will help FC commercialization and realization of the hydrogen economy scenario [12]. Hydrogen production using renewable and clean energy sources are the best long-term route. For example, electrolysis could be used to produce clean hydrogen at off-peak power at or near the site of utilization. Hydrogen using electrolysis is about two to three times more expensive than natural gas reforming [430]. This difference could be offset by the high capital cost of infrastructure.

Centralized production suits petroleum quite well given its high energy density. A decentralized production of hydrogen is more suitable given its lower energy density. With decentralized production of hydrogen, at least in the early deployment stages of hydrogen economy, there is no need for an expensive infrastructure for packaging and transportation of hydrogen. There are already numerous renewable technologies developed for a decentralized hydrogen production [334]. With plenty of water and solar resources, water splitting is the best route to produce hydrogen [431].

Irrespective of the production and storage mode, NPs and nanomaterials could play a key role in improving the economic viability of hydrogen as an energy vector [337]. Indeed, reforming nanocatalysts play a key role in hydrogen production [338]. A methanol reformer for on-board hydrogen production is a potentially effective solution: the current gasoline distribution infrastructure can be used while circumventing the hydrogen storage requirements. Methanol-to-hydrogen production efficiency of 5% is achieved using concentrated solar and $CuO/ZnO/Al_2O_3$ catalyst [432]. Higher hydrogen production rate could be obtained using catalyst NP when compared to bulk catalyst.

Hydrogen obtained using solar fuel provides the highest energy and lowest CO_2 emission per unit mass [159]. The overall efficiency of the solar-to-hydrogen process could be improved. SOC is used either to produce electricity from hydrogen (SOFC), or to produce hydrogen from renewable electricity (solid oxide electrolyzer cell, SOEC) [433]. Depending on the chemical process and hydrogen source, three high-temperature thermochemical routes using concentrated solar radiation have been developed for hydrogen production [434]. Water is used in the thermolysis and thermochemical cycle. Fossil fuel (e.g., natural gas) is used in solar reforming. In solar cracking and gasification, water and fossil fuel are used. Process and reactor design are critical for the viability of these technologies. For the thermochemical process, two concepts are considered [435]. One concept uses reacting particles forming a cloud in a stainless-steel cavity. The second concept uses reacting particles in a fluidized bed in a stainless-steel reactor. In both cases, concentrated solar radiation is focused through a quartz window to heat the reactor.

A two-step thermochemical water-splitting method based on the ZnO/Zn redox reaction has been developed [436–439]. A first endothermic step using solar thermal energy produces oxygen and Zn(g) vapor following solid ZnO(s) dissociation at above 2000 K. A second exothermic step produces hydrogen and ZnO(s) following hydrolysis of Zn(l). Ultrafine ZnO NPs are used in this process [439]. Oxygen and sub-100 nm Zn/ZnO core–shell NPs are obtained. High surface area Zn/ZnO NPs allow enhanced heat and mass transfer along with faster chemical reaction. An iron-based oxide process has also been developed [435].

Hydrogen production using high temperature electrolysis of water steam could reduce electricity consumption by up to 30%. SOFC/SOEC design could lead to cost reduction and overall GHG emission reduction. In SOFC, H_2 is fed through the anode and oxygen through the cathode. In SOEC, H_2O is fed through a porous cathode [433]. Typically, the SOEC consists of a 300 μm Ni/YSZ porous support layer on which 10 μm of Ni/YSZ and 10 μm of YSZ are spray-painted successively to form the anode and electrolyte layers, respectively. The YSZ material corresponds to yttrium-stabilized zirconia. After sintering, the cathode (or hydrogen electrode) consisting of $(La_{1-x}Sr_x)_aMnO_3$–$YSZ_{(1-y)}$ or LSM/YSZ is also spray-painted to obtain a 10-μm-thick layer [433]. As described elsewhere, Ni/YSZ and LSM/YSZ properties are improved by using submicron particles.

Hydrogen separation and purification can be accomplished through a membrane using Ni NP layers [440]. These layers consisting of 3–5 nm nickel

NPs dispersed in an amorphous silica matrix are formed on a mesoporous anodic alumina capillary. These films showed highly reversible hydrogen adsorption properties. Furthermore, differences in gas permeance (at 773 K) have been reported with gas molecules having different kinetic diameters [440]. These difference are obtained with nanocomposite membranes using different Ni/(Si + Ni) atomic ratios [440]. Hydrogen permeability in this membrane is about five times higher when compared to molecular sieve material [440].

As described elsewhere in this chapter, PEC cells [160] could become competitive in the long term [158]. Photocatalysts have been developed for both electrochemical and thermal water splitting [159,164]. Small- and large-scale membrane reactors to produce hydrogen are currently under development at the precommercial stage. Supported Cu_2O and CuO NPs have been shown to catalyze hydrogen production when H_2O/CH_3OH solution was irradiated by UV or visible light [441]. This study also showed that UV light is an order of magnitude more efficient than visible light. Furthermore, adding CuO will improve the activity by a factor of 3. Numerous thermochemical processes using different nanomaterials have been reported [442].

5.4.2 Hydrogen Storage

Centralized and distributed hydrogen production and utilization are currently used [427]. In both cases, storage is required although with different capacity requirements. Several hydrogen storage issues must be addressed. First, safety is important given the high volatility and high energy of hydrogen gas. Second, storage cost is currently too high with durability issues. Third, the weight and volume are presently too high. Fourth, the overall energy efficiency should be improved. Fifth, the refueling time is too long.

Numerous hydrogen-absorbing materials and concepts have been considered to address these issues [13,443–460], although most are not considered economical in the near term. Tank volume and safety issues have been addressed with low-pressure liquefied hydrogen, although more than 30% of stored energy is wasted during liquefaction [448]. A compressed hydrogen tank of 700 bars with hydrogen content in weight of about 11% has been demonstrated [448]. These two storage solutions are below the industry gravimetric and volumetric capacity goals. Indeed, it is suggested that a storage capacity of 6.5 wt% and 62 kg H_2/m^3 are required for the automotive and other applications. For example, hydrogen storage in steel and composite material cylinders allows a maximum of about 1 wt% and 3%, respectively.

Nanomaterials in general and NPs in particular may provide viable alternatives [455]. Very high surface area with controlled surface chemistry will allow not only high hydrogen adsorption but also controlled H_2 binding strength with adsorbing sites [456]. Numerous nanostructured materials with large surface area based on carbon structures, metals, and metal alloys have been considered [445,451,452]. Aerosol submicron ZnO particles have been used as redox systems [447]. Increased surface area using smaller particles will improve hydrogen storage capacity [450]. Percentage of surface atoms is estimated to increase from less than 4% for particles larger than 100 nm to about 70% for NPs of about 1 nm. Furthermore, kinetics and chemical potentials are also dependent on the particle size.

Nickel (Ni) NPs dispersed in amorphous silica (Si–O) are shown to absorb hydrogen at high temperature [440]. Nickel NPs embedded in mutiwalled carbon nanotubes are prepared using an incipient-wetness impregnation procedure [457]. Up to 2.8 wt% hydrogen absorption–desorption capacity has been achieved under a moderate temperature and pressure conditions. Nickel-catalyzed hydrogen dissociation has been suggested as an effective reversible hydrogen storage method. Stoichiometric Mg_2Ni intermetallic NPs are produced by arc-melting Mg and Ni. These NPs showed excellent hydrogen absorption behavior at low temperature [458]. Hydrogen storage contents of 2.77, 2.93, and 3.03 wt% at 523, 573, and 623 °C, respectively, have been reported within minutes. This suggests very low activation energy. $NaAlH_4$ microparticles doped with 2% mol of TiN NP catalysts have been shown to provide 4.5 wt/% hydrogen storage capacity [459]. High hydrogen solubility in Pd clusters with different sizes has been reported [460]. Relatively slower absorption and desorption kinetics have been observed with Pd clusters imbedded in polymeric matrix when compared to surfactant-stabilized Pd clusters [460].

6 ENERGY SAVING

Energy saving in domestic, commercial, and industrial sectors is probably the best avenue to reduce fossil fuel consumption. NP integration into devices and processes could help reduce energy consumption in different ways. First, it could reduce the current rate of raw material utilization. Indeed, most raw materials have high energy content embedded within their life cycle, starting from mining and purification to the transformation stages. Second, utilization of higher surface ratio nanomaterials in vehicles and airplanes will reduce transportation-related fuel consumption. Third, high-performance functional NP-based materials will reduce energy consumption by providing, for example, more efficient insulation and lighting.

6.1 Surface Finishing

A significant amount of energy is embedded within manufactured products. Increasing the energy payback of these products is critical. In the case of plastic-based coatings used in automotive, aerospace, and exterior wood dispersing, photodegradation due to UV exposure is a common failure mechanism that limits the lifetime of the product. UV-absorbing TiO_2 and ZnO

NPs provide significant protection not only to the matrix itself but also to the underlying substrate.

Surface finishing such as hard chrome coating generates significant amounts of toxic and hazardous by-products [461,462]. Development of alternative coating processes and raw materials without losing current performance and cost advantages is challenging. NPs are very effective in improving anticorrosion properties of thin films [463]. NPs embedded in metal matrix films are obtained easily using eletrophoretic deposition [464]. These electrodeposited nanocomposite films can withstand very high temperatures. Oxide NP additives preventing UV degradation have been used in cosmetics, wood finish, and windshield coatings [465]. Visible transparency of polyefin films with a coating incorporating 20 nm ZnO NPs is much better than when using larger ZnO particles [465]. Furthermore, incorporation of about 1% ZnO NPs improves outdoor UV degradation. This will improve product lifetime.

6.2 Insulation and Glazing

With a yearly market over US$ 20 billion, thermal insulation has attracted intense interest from material scientists and engineers. In the case of home insulation, numerous materials are used to reduce the different heat transfers between warm and cold sides. Conduction of heat flow, convection of heat flow through air movement, and radiation through reemission of absorbed radiation are the main heat exchange processes.

6.2.1 General

Thermal insulation will prevent significant heat flow from one side of the insulation to another. Fiberglass, rock wool, and slag wool are currently used.

Although they are expensive to fabricate, aerogels are also of interest for thermal insulation. Aerogels have low conductivity, high porosity, and high dielectric constant. Aerogel products have very low thermal conductivity compared to established materials [466]. Carbon and silica aerogels are currently pursued [467].

Aerogels based on SiO_2 NPs with a typical size range of 10–100 nm have been used to limit thermal losses in translucent windows (Fig. 5.27). Outstanding insulation and optical performance has been reported with a double-glazed translucent window. Indeed, thermal losses of the aerogel window are even lower than those of triple-E pane window and an opaque insulation [468]. Aerogel can be obtained through various processes [469]. Often, NPs are used as starting material to produce these aerogels [470].

6.2.2 Glazing Technology for Insulation

Window coating to control visible and infrared transmission is widely used. Window glazing includes solar heat control, spectral control, and smart or chromogenic window (electrochromic and photochromic). Often the

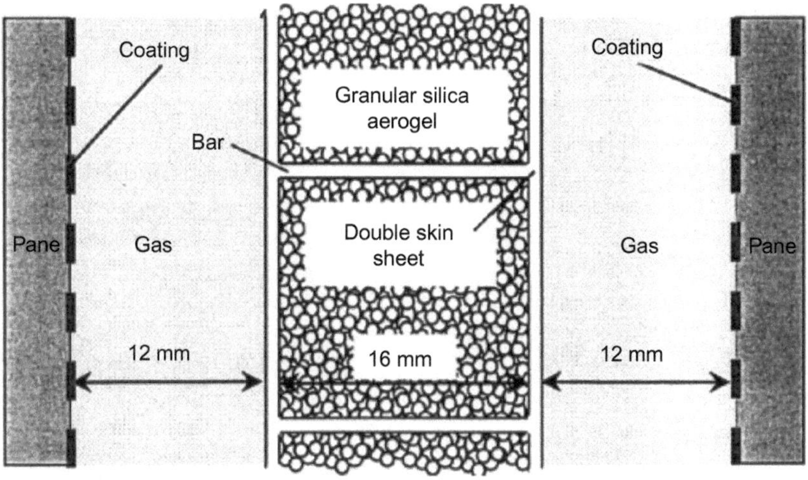

FIGURE 5.27 Schematic view of a translucent double-glazing window filled with granular aerogel [468].

fabrication cost of these coating solutions is the main factor hindering their commercialization.

6.2.2.1 Solar Heat Control

Developing highly transparent windows to visible radiation while blocking the infrared portion of the solar radiation remains an active research area. SnO_2 coating endowed with a high energy bandgap provides high transparency to visible and infrared solar radiation. Upon doping, it forms various TCO layers such as antimony (ATO), fluorine (FTO), and indium (ITO)-based materials. The main role of the doping is to increase conductivity and control optical properties of the film. The onset of the plasma absorption of the free carriers being located in the NIR leads to a glass-coated substrate with TCO to block the infrared wavelength [471]. Several NP-based TCO coatings have been described in Chapter 7. Silica aerogel has also been proposed as insulation for flat solar collectors [472].

6.2.2.2 Spectrally Controlled Coating

Selective (or spectrally controlled) absorber coating for solar-to-heat conversion [473–475] absorbs solar radiation and transparent (low emittance) radiation in the infrared region. Furthermore, these coatings should be durable for several years at very high temperatures. While solar radiation extends from about 0.2 to 2 μm, the largest part of the thermal radiation is found above 2 μm. The coating should have high absorptance in the solar spectral region and a low emittance in the range of thermal radiation. A step-like reflectance spectrum with $R = 0$ in the solar region and $R = 1$ in the infrared range is ideal, although practically not achievable.

Four approaches have been developed to obtain highly selective and durable surfaces [474]. They include intrinsic selectivity materials (W, Cu_2S, HfC), absorber–reflector tandems, wavefront discriminators, and interference stacks. Absorber–reflector tandems provide the most viable commercial solutions. Two possible multilayer layouts are shown in Fig. 5.28. One layout consists of two layers. A first layer is used as infrared reflector and a second layer is used as solar absorber material. In a second layout, a nanocomposite layer of metallic NPs embedded in a ceramic matrix is used for low infrared emittance and as a solar absorber [476].

Metallic NPs embedded in polymer foils have also been studied as solar control glazing [477]. A layout of three or more layers with at least one nanocomposite-based layer, showed enhanced solar absorption properties [235,478–481]. However, the higher cost for manufacturing a multilayer system may hinder its commercial viability.

Nanocomposites of metallic particles embedded in a ceramic matrix are often labeled as cermet (ceramic metal). Optimum absorptance film thickness is estimated at about 80 nm for Ni- and Co-based cermet [480]. These materials provide strong absorption in the solar region due to interband transitions, enhanced absorption in the visible region due to plasmon resonance phenomenon, and transparency to thermal infrared wavelength.

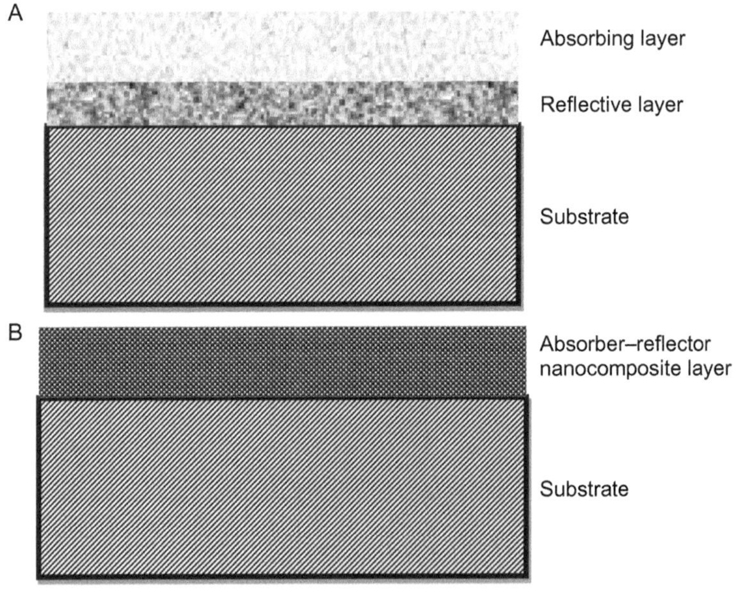

FIGURE 5.28 The two absorber–reflector tandem designs: (A) two separate materials for the absorber and reflector layers, (B) nanocomposite layer.

Dispersion of small metal NPs in a dielectric matrix offers a high degree of flexibility in composition choice of solar selective absorber. The most important requirement is the size of the particle, which must be much smaller than the shortest wavelength of the solar radiation. Therefore, typically, these NPs should be about 100 nm or less. Typically, metallic chromium is dispersed in dielectric Cr_2O_3. Electrodeposition is used to form Cr/Cr_2O_3 surface selection with a high absorption coefficient in the UV–visible–infrared range and low emittance in the far-infrared region. A low-emissivity coating has been obtained using Ag NPs with a size range of 18–48 nm [481].

NPs embedded in polymer foils could provide a lower cost solution. Different structural configurations have been reported [477]. This includes NPs sandwiched between two glass layers, NP film deposited on glass substrates, and NPs dispersed in bulk polymer structures [477]. Small and well-dispersed NPs in polyvinyl butyral, polyethylene terephthalate, polymethyl methacrylate, or PC can absorb infrared and NIR, and transparent radiation to the visible part of the solar irradiation. ITO NPs (30–40 nm) were found to provide better NIR radiation blocking than ATO NPs [477]. When used as laminates, these nanocomposites could reduce interior heat load.

6.2.2.3 Smart Windows

Electrochromic windows are energy-efficient glazing solutions [482]. They are relatively simple to build and operate (Fig. 5.29). Chromogenic materials allow reversibly controlling the optical properties (refractive index and extinction coefficient) of windows upon application of an electrical potential [483,484]. Large-area electrochromic films are achievable using sol–gel

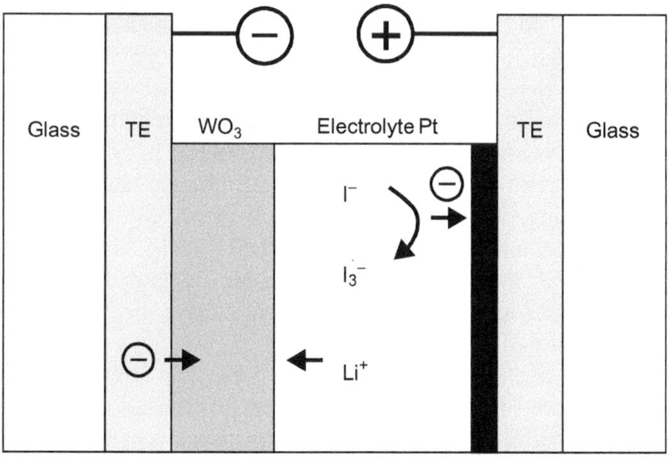

FIGURE 5.29 Schematic representation of an electrochromic device structure and operation principle [483].

processing. Photochromic windows powered by solar radiation without any external current are more efficient. Structure and operation principle of a photoelectrochromic device is similar to that of electrochromic devices. This device consists of two TCO glasses coated by a Pt and chromic material. An I_3^-/I^- redox couple and Li^+ ions provide solid ion conductivity. TiO_2 and WO_3 NPs have often been used in photoelectrochromic and electrochromic devices [485]. The high porosity is required to achieve high charge transfer and transport.

A colloidal solution based on ATO prepared by coprecipitation has been shown to provide remarkable electrochromic properties [486]. Thin films are obtained from concentrated colloidal solutions by repetitive spin coating. These films can be used for display and window applications [487]. Reversible transition transparency and opacity have been obtained with silver NPs dispersed in porous silica [488]. Photochromic WO_3 NPs with different sizes have been obtained using HWCVD [489]. Multicolor photochromism has been obtained using Ag–TiO_2 nanocomposites [490].

6.3 Efficient Lighting

With an efficiency of 2–5%, incandescent lamps are a significant source of waste energy. About 20% of electricity is used for lighting in North America. However, more efficient (~20%) compact fluorescence lamps (CFLs) use mercury (a toxic material) and emit significant amounts of potentially harmful radiation. The lifetime of CFLs is also limited. Potentially more efficient solid-state lighting (SSL)-based white lightings are poised to replace incandescent and fluorescent lighting [491]. Although fluorescent and SSL lightings currently have similar efficiencies, improved SSL using QDs will certainly increase their overall efficacy to 50% or more. LED efficacy surpassing 100 lm/W is currently commercialized. In addition, with more than 50,000 h, SSL lifetime is exceeding all other lighting technologies [492].

The first nonvacuum processed QD-LED has been reported by the Alivisatos group [493]. Nanocrystals Technology (Australia) has developed a quantum confined system that allows near-100% efficiency. Nanophosphor-based SSLs are promising. To avoid issues related to organic materials, Caruge et al. developed an all-inorganic QD-based LED [494]. Monolayers of QDs emitting in the red, green, and blue regions are integrated into a patterned LED (Fig. 5.30). These patterned monolayers have been obtained using microcontact printing.

Wavelength-tunable hybrid QD/organic LED has been obtained using CdSe/CdS/ZnS core–shell structure as an electron-transport layer (ETL) and a thermally polymerized polymer as a hole-transport layer (HTL) [496]. These devices give rise to narrow and saturated emission colors. Narrow bandwidth in the red, orange, yellow, and green regions with maximum luminance values

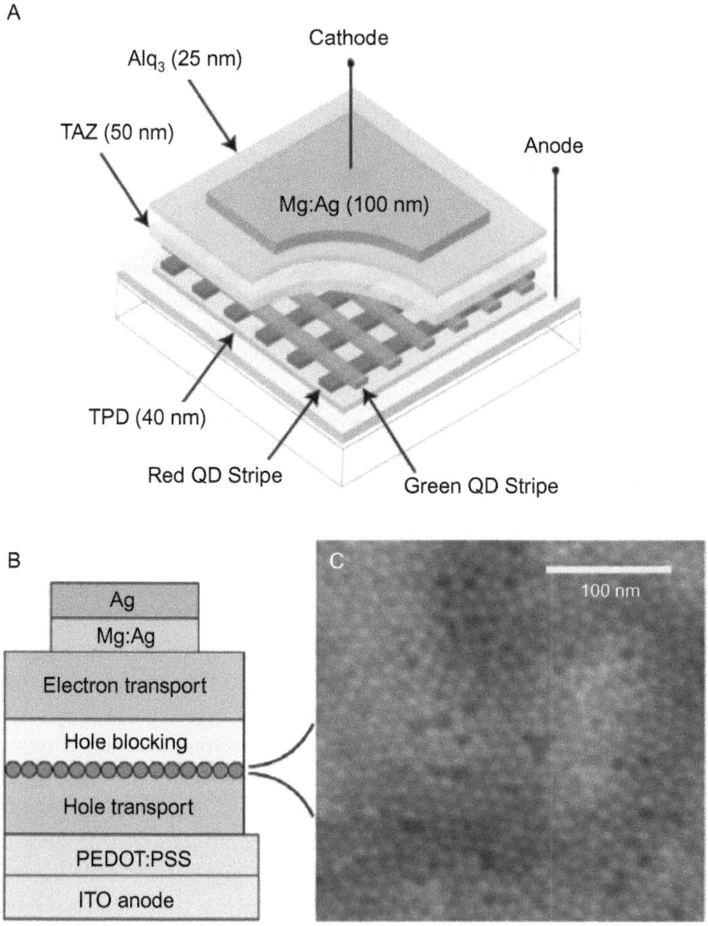

FIGURE 5.30 Schematic representation of (A) LED pixel patterned with 25 μm-wide stamp features and (B) cross section of the different layers forming the LED. (C) High-resolution AFM micrograph of close-packed QD monolayer. *Reprinted with permission from Kim et al. [495]. Copyright 2009, American Chemical Society.* (For color version of this figure, the reader is referred to the online version of this chapter.)

of 9064, 3200, 4470, and 3700 cd/m^{2,} respectively, has been obtained. An electroluminescence efficiency of almost 3 cd/A lasting several hundred hours has also been achieved.

Other approaches producing nanocrystal-based white light LEDs have been reported [497]. In one design, colloidal NPs were used as phosphors in a blue InGaN LED. In another design, an organic–inorganic hybrid white LED was built containing a NP layer, an HTL, and/or an ETL simultaneously emitting light to form a white emission. In a third design, NPs with different sizes simultaneously emit light at different wavelengths to achieve white emission. One of the advantages of LEDs based on semiconductor NPs is provided by the bandgap tunability, leading to changes in the emission color. Indeed, the electroluminescence and photoluminescence peaks of a CdSe film have been shown to shift by about 50 nm when the size of the NPs is changed from 3 to 5 nm [498].

A multicolor LED has been reported using CdSe/ZnS on a GaN substrate [499]. Transition-metal-ion doping of a wide-bandgap NP-based phosphor is currently being investigated [500]. NPs embedded within the luminophor systems could improve the overall energy efficiency by increasing excitation and host-to-impurity energy transfer efficiencies. This is achieved by optimizing the radiative recombination efficiency and minimizing the nonradiative recombination [500]. Metal NPs have been used to improve the light extraction efficiency of LEDs. Indeed, an electroluminescence enhancement factor of 8 has been obtained via excitation of surface plasmon resonance obtained with silver NPs embedded in thin silicon film [501].

6.4 Lighter and Stronger Materials

Lighter and/or stronger materials will allow a more efficient use of material and energy and therefore reduced environmental impact. Lighter and stronger materials will enable energy saving by allowing lighter vehicles, which will result in more fuel economy. Furthermore, lighter materials will result in energy saving through transport and handling.

With its high specific strength and good manufacturability, magnesium provides a good alternative to steel and even aluminum-based materials for lightweight vehicles. Currently, on average, about 16 kg of magnesium (mostly for the interior components) is used compared to 975 and 100 kg for steel and aluminum, respectively. Finding appropriate coatings to prevent magnesium corrosion is critical to increasing their share. Given the low melting point of magnesium-based materials, most coating processes are not suitable. Feil et al. have reported a NP-based coating process to take advantage of the low deposition and post-process temperatures [502]. Sol–gel is a simple and versatile coating process. Monodispersed particles and bimodal particles are applied to the surfaces of magnesium alloys AZ31 and AZ91 using dip and brush coatings. EPD has also been evaluated. The low sintering temperature (200 °C) of the inorganic NPs allows dense coatings.

Smaller grain size leads to lower internal strain and lower dislocation density [503]. This may result in a stronger and more resistant material, which will increase lifetime, thus reducing energy consumption and lowering environmental impact. High-temperature corrosion leads to an estimated 137 quadrillion joules of energy loss [504]. *In situ* and *ex situ* NP-reinforced metal matrix composite preparations have been shown to provide enhanced mechanical properties [505]. Corrosion degradation of materials is a major issue in the petrochemical industry. In the USA alone, the corrosion costs of the oil industries are around $200 billion a year [506].

Nanostructured materials including NPs used for base materials and for coating solutions could reduce the overall energy and capital costs [463]. Silica NPs are used as reinforcing materials and are the main components of coating films. Silica NPs could replace toxic hexavalent chromium in surface pretreatment [507].

The inclusion of NPs in a polymeric matrix improves the mechanical properties of nanocomposites [508,509]. It was found that Young's modulus increases as the size of particles is reduced below 1000 nm [508]. Clays and polymer monomers are used as precursors to produce nanocomposites. Nanocomposite-based parts used in cars and trucks could improve not only their esthetics and safety but also their fuel economy [510]. Although nanoclays are currently used in some new truck models as body moldings, silica NPs can also be used.

A variety of NP compositions have been incorporated in the construction sector [511]. NPs have been shown to improve the strength, lightness, and durability of the construction materials. For example, silica NPs have been used for mechanical reinforcing of concrete and flame-proofing [511]. Fe_2O_3 NPs increase the compressive strength and abrasive resistance of concrete. TiO_2 NPs affect the degree and kinetics of cement hydration [511].

Nanoporous metal foams (NMFs) combine the thermomechanical and catalytic properties of metal and high surface area materials [512]. Ultralow density and high strength-to-weight ratio of alumina provide excellent alternatives to reduce weight. Gold NPs have been used to produce Au aerogel [512].

NP dispersion enhances the mechanical properties of steel-based components while reducing their density [513,514]. Using ceramic NPs may lead to superplastic deformation due to grain boundary sliding. This property will help prevent brittleness in conventional ceramic products. Improved ductile properties have been obtained using submicron TiO_2 particles [515].

Aerogel consisting of interconnected NPs with tunable density from around 0.003 to 0.35 g/cm^3 is considered the lightest solid material [516]. Furthermore, aerogels are also endowed with exceptional thermal conductivity (e.g., insulation), and acoustic (e.g., low sound speed), optical (e.g., low refractive index) and electrical (e.g., low dielectric constant) attributes, allowing applications in building, transducers, sensors, and integrated circuits, respectively. Given these attributes, aerogels could certainly become disruptive materials in the future. Their economical production and integration are the main challenges that prevent their wider commercialization.

Inorganic nanomaterial utilization as lubricant components has been studied extensively [517]. Commercial lubricants consist of synthetic HCs and additive compounds. Lubricants consisting of alkaline metal core diameters in the range of 2.5–3.5 nm stabilized by a surfactant are already used commercially [517]. Numerous inorganic NPs are also used directly as additives to improve the lubricant properties. MoS_x NPs remain the most widely studied additives although with very little commercial success [517].

WS_2 and MoS_2 NPs with 10 or more shell layers could sustain heavy loads as solid lubricants [518]. These closed-cage structures are referred to by a prefixed term IF, indicating their fullerene-like structure. Friction coefficients of ball-on-flat pairs rubbed with paraffin oil and paraffin oil mixed with 0.5% and 1% of IF-WS_2 NPs have been measured [518]. At high loads, addition of IF-WS_2 NPs significantly lowers the friction coefficient. The advantage of these fullerene-like NPs over platelet additives as lubricants is not yet well understood. However, the chemical reactivity of the platelet additives may lead to their degradation over time.

Corrosion and wear resistance coatings extend the lifetime of metallic parts, thus lowering energy and raw material utilization. Numerous examples of nanomaterial applications in corrosion and wear control have been reported [463]. They include NP-based sol–gel coatings, high-temperature nanocrystalline alloys, and NP-based ceramic films.

NPs are also expected to have a significant impact on the aviation industry. Improved multifunctional coatings in the aerospace industry will not only reduce the weight of the airplane but also improve its thermomechanical properties [519].

7 OTHER APPLICATIONS OF NPS

Besides the applications described so far, there are numerous other sectors in which NPs could have some impact in the near and long term. A few examples are provided in different sectors.

7.1 Biomass and Bioenergy

Processes involving solar energy conversion to chemical energy are important. Applications of NPs in photosynthesis and photocatalysis have been discussed elsewhere in this chapter. Bioenergy conversions often require highly efficient NP catalysts. Ceria NPs have been shown to improve the biodiesel performance and emission characteristics [520]. The utilization of renewable biomass in NP fabrication has also been discussed elsewhere in the book. There are other NP-related applications involving biomass.

Polysaccharides (cellulose, starch, and chitin) can be extracted from biomass as NPs [521]. The extraction process of value-added products from biomass is often challenging. Biomass components such as microfibrillated cellulose are mechanically strong, resisting most conventional treatments.

Appropriate chemical modification is often required. Polysaccharide NPs could be formulated in aqueous latex dispersions.

Microalgae are currently being investigated as a sustainable source for high-quality protein and biofuel [522]. Microalgae do not compete with other food-related biomass for agriculture land. Scalable algae oil and conversion processes to produce fatty acid methyl esters (or biodiesel) have been demonstrated. However, low yield and high energy input in some key process steps are major hurdles. Aminopropyl functionalized 2D-hexagonally packed mesoporous silica NP materials with pore diameters of 10 nm (AP-MSN-10) provide high capacity and selectivity of free fatty acid (FFA) [523]. The AP-MSN-10 materials are obtained from the surfactant micelle-templated mesoporous silica NPs with an average crystallite size of 10 nm. Following FFA adsorption from microalgae, a methyl esterification is carried out. Fine-tuning the properties of these mesoporous materials may also help separate and convert other high-value biomolecules produced by microalgae and other microorganisms.

Production and harvesting of microalgae are energy- and capital-intensive. When compared to ponds, photobioreactors are believed to provide a more sustainable solution. However, light-harvesting efficiency of natural and artificial light is still low. Recently, silver NPs were shown to increase growth by about 30% [524]. Taking advantage of the plasmonic properties of metallic NPs, a faster growth is achieved by enhancing the blue-region irradiance of the impinging light. The possibility of tuning light scattering of metallic particles through shape and size control will certainly improve efficiency beyond the current level.

7.2 Solar Thermal and Thermophotovoltaic

Solar thermal has been used in two main application cases. In one case, hot water was produced for domestic and industrial use. In the second case, direct and indirect water vapor was produced from the concentration of solar power to drive a steam turbine in electrical power production. Using ultrafine Zn/ZnO particles, efficient biomass-to-fuel conversion is achievable [525]. Concentrated solar radiation is used to drive this high-temperature thermochemical process. These ZnO NPs serve simultaneously the functions of radiant absorber, thermal insulator, and chemical reactant [525].

Thermophotovoltaic (TPV) energy conversion systems consist of four components: absorber layer, emitter layer, PV cell, and mirror [526]. Solar cell efficiency is increased through the use of an emitter layer as a wavelength coupler between the absorber and the PV cell. The mirror layer reinjects the nonabsorbed wavelength into the device. With their ability to modify the spectral and angular distribution of thermal radiation, photonic bandgap materials could be used in TPV systems for frequency- and angular-selective radiating elements [526]. NP-based coatings could be used in all four components to improve the overall efficiency of TPV systems.

7.3 Thermoelectricity

The thermoelectric effect allows converting waste heat into electric power. There are already commercial wristwatches powered by thermoelectric power taking advantage of the temperature difference between the body and the surrounding environment. Combining thermoelectric and PV effects, higher solar-to-electricity conversion efficiency is achievable. PV absorbs around 58% of solar energy between 200 and 800 nm, while the rest of the solar energy in the range of 800 nm up to about 2500 nm can be converted by the thermoelectric effect [527].

The performance of thermoelectric materials is often defined by the following unitless figure of merit [528]:

$$\mathrm{ZT} = \sigma S2T/\kappa$$

where σ, S, κ, and T are electrical conductivity, the thermoelectric power, the thermal conductivity of the material, and the absolute temperature, respectively.

Today, bulk thermoelectric materials have a value of around a unit or less. Materials with ZT above a unit are being researched. Using low-dimensional systems, including NPs, ZT value of higher than a unit is achievable [528]. Structural details of thermoelectric materials have different effects on the mid- and long-wavelength phonon scattering [528]. Furthermore, electronic transport of hot and cold electrons is also affected. It has been established that atomic defects are effective at scattering short-wavelength phonons.

Concepts to improve ZT using NPs have been studied [528]. NPs embedded in the matrix materials are effective in mid- and long-wavelength phonon scattering. Grain boundaries have also been shown to play an effective role in long-wavelength phonon scattering. Using NPs, enhanced thermoelectric performance could be obtained by lattice thermal conductivity reduction and/or an increase in the electrical power factor. Thermal conductivity reduction is achieved by adding NPs (Fig. 5.31).

Several bulk thermoelectric materials have been developed [529,530]. The majority of these materials gives rise to low efficiency and/or larger dimensionality. Nanocomposites have been considered to address these two shortcomings. Thermal conductivity has been shown experimentally to decrease by an order of magnitude with the increase in interface area per unit volume [530,531]. PbTe–PbSeTe NPs have been shown to have overall improved thermoelectric characteristics when compared to bulk materials [532]. Bi_2Te_3 and PbTe nanomaterials including NPs obtained via various wet syntheses are considered as thermoelectric materials [533].

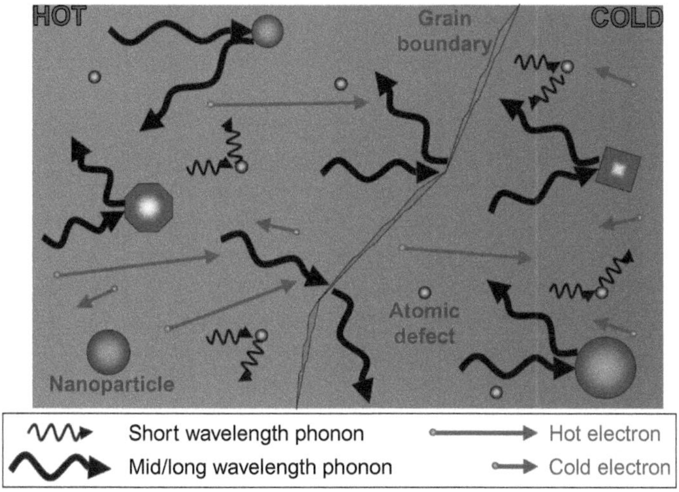

FIGURE 5.31 Schematic diagram illustrating phonon scattering mechanisms via atomic defects, nanoparticles, and grain boundaries [528]. (For color version of this figure, the reader is referred to the online version of this chapter.)

7.4 Propellants

Thermite powder composites depend on high contact area at the interface between NPs, fuel, and the oxygen to increase the amount and rate of the released energy [534]. Aluminum and MoO_3 NPs with diameters in the range of 20–50 nm resulted in a faster reaction. When compared to conventional thermite powders, three orders of magnitude higher reaction velocities are achieved with aluminum NPs [534]. Aluminum NPs are currently developed and optimized for use as propellants. The large surface area of aluminum NPs gives rise to a faster reaction rate with oxygen. This provides faster energy release to propel projects [535,536].

7.5 Nanofluid for Heat Transfer

Heat transfer is omnipresent in industrial processes and consumer products. Requirements for improved efficiency and increased power density per unit surface and volume have led to the development of new materials for heat transfer systems. Nanofluid consisting of NPs dispersed in a base fluid provides improved heat transfer capability [537–539]. Aqueous and nonaqueous solvents (oil, ethylene glycol) are often used as base solvents. Sub-100 nm NPs consisting of Cu and Al_2O_3 have been used as colloids [539]. Heat exchange plays a significant role in cooling systems and also in solar thermal power generation. Nanofluid consisting of NPs suspended in a fluid improves the overall system efficiency [540].

7.6 Nuclear Materials

Applications of NPs in various aspects of the nuclear power generation value chain are claimed. However, there are very few peer-reviewed publications on this subject. Secrecy and complication in testing potential NPs in this field are probably the main reasons. Magnetic iron/iron oxide core-shell NPs have been studied to extract radioactive actinides from spent nuclear fuel [541]. Currently, nitric acid is used in large-scale processes to dissolve spent fuels, followed by solvent extraction of uranium and plutonium. A NP-based breeding pellet has been proposed to minimize nuclear secondary waste [542]. The proposed pellet will contain particles with gradient grain size distribution. Absorption by fertile material is optimized and the structural and shielding material absorption is minimized.

7.7 Heavy Oil Extraction

Potential reserves of heavy oil deposits in Canada and Venezuela are comparable to the conventional oil reserves in the Middle East [543]. The oil (tar) sands in Alberta contain about 2×10^{11} bbl of recoverable oil. However, the cost and environmental impact of extracting the heavy oil are relatively high. Several novel techniques are under development for enhanced oil recovery techniques [543].

A larger part of the remaining oil sand deposit in north Alberta consists of underground oil. *In situ* recovery of underground deposits of this heavy oil provides several advantages [543,544]. Sub-100 nm inverse micelles with water inside (water in oil) are added to the injected steam (Fig. 5.32). This helps in lowering the viscosity, reducing H_2S and heavy metals, and *in situ* upgrading. Propagation of multimetallic NP catalysts (34 nm) in oil sand media has been recently reported [545]. This could be used during the steam-assisted gravity drainage (SAGD) recovery process, allowing *in situ* heavy oil upgrade.

Oil sand produces hundreds of millions of cubic meters of fluid fine tailings per year. Tailing ponds are used to store these waste materials. These tailings consist of particles with a wide range of sizes and compositions. Organic solids consisting of particle sizes of 1–110 μm contain large amounts of organic matter contributing to low recovery and potentially long-term contamination [546,547]. Ultrafine particles including NPs may take more time to settle unless appropriate treatment is applied. Furthermore, these ultrafine particles contain significant amounts of HC, contributing to low recovery and contamination [548]. The surface modification described may facilitate settling and utilization as reinforcing agents in nanocomposites.

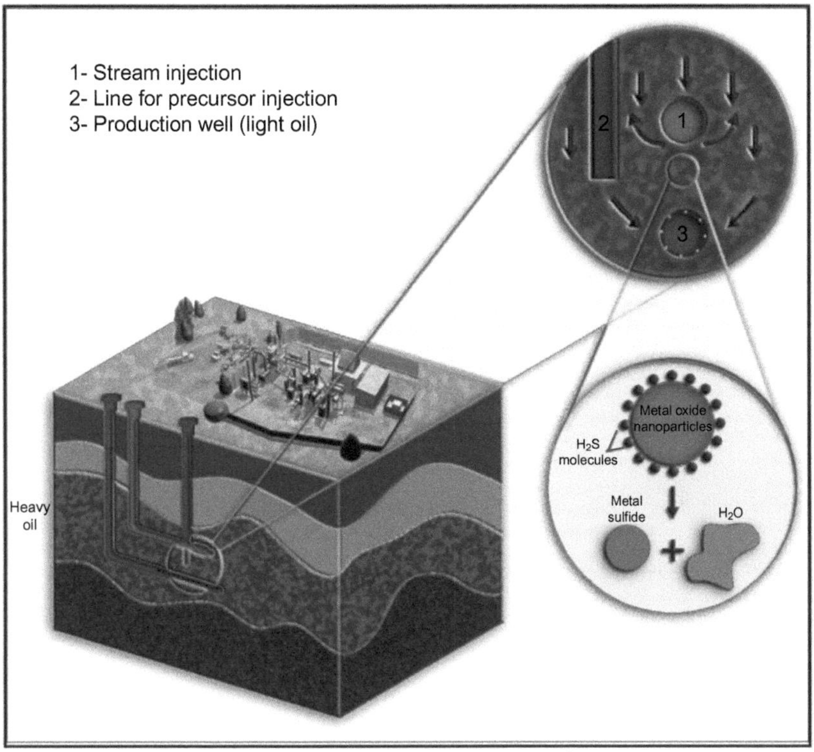

FIGURE 5.32 Pictorial representation of metal oxides sorbent during heavy oil recovery and upgrading [544]. (For color version of this figure, the reader is referred to the online version of this chapter.)

REFERENCES

[1] Key World Energy Statistics, OECD/IEA. www.iea.org/textbase/nppdf/free/2010/key_stats_2010.pdf, 2010.

[2] C.J. Campbell, Oil Crisis, Multi-Science Publishing, Essex, UK, 2005.

[3] K.A. Baumert, T. Herzog, J. Pershing, Navigating the Numbers: Greenhouse Gas Data and International Climate Policy (2005), World Resources Institute, www.pdf.wri.org/us_greenhouse_gas_emissions_flowchart.pdf (accessed 22.10.10).

[4] V. Rieke, G. Bachmann, J. Nanopart. Res. 6 (2004) 435–446.

[5] H. Presting, U. Konig, Mater. Sci. Eng. C 23 (2003) 737–741.

[6] H.J.S. Sand, Proc. R. Soc. Lond. 74 (1904/1905) 356–369.

[7] K. Rideal, Obituary Notices Fellows R. Soc. 4 (1942) 63–66.

[8] V. Haensel, In: B.T. Brooks, C.E. Boord, S.S. Kurtz, L. Schmerling (Eds.), The Chemistry of Petroleum Hydrocarbons, vol. II, Reinhold, New York, 1955, pp. 189–219.

[9] H.A. Wittcoff, B.G. Reuben, J.S. Plotkin, Industrial Organic Chemicals, Willey-Intersciences, New York, 2004, pp. 459–526.

[10] R.A. Sheldon, J. Chem. Technol. Biotechnol. 68 (1997) 381–388.

[11] Nanoscience Research for energy Needs, Report of the National Nanotechnology Initiative, Grand Challenge Workshop, 2004.
[12] G. Marban, T. Valdes-Solis, Int. J. Hydrogen Energy 32 (2007) 1625–1637.
[13] D.G. Vlachos, S. Caratzoulas, Chem. Eng. Sci. 65 (2010) 18–29.
[14] M.V. Twigg, Catalyst Handbook, second ed, Wolfe Publishing Ltd, London, 1989.
[15] P.K. Sharma, W. Miao, A. Giri, S. Raghunathan, In: C.I. Contescu, K. Putyera (Eds.), Taylor & Francis Group, LLC, New York, 2009, pp. 2799–2814.
[16] A.T. Bell, Science 299 (2003) 1688–1691; F. Raimondi, G.G. Scherer, R. Kutz, A. Wokaun, Angew. Chem. Int. Ed. Engl. 44 (2005) 2190–2209.
[17] http://www.cateran.com.au/cateran/data.aspx.
[18] A. Brumby, M. Verhelst, D. Cheret, Catal. Today 106 (2005) 166–169.
[19] X.S. Zhao, X.Y. Bao, W. Guo, F.Y. Lee, Mater. Today 9 (2006) 32–39.
[20] J.A. Moulin, M. Makkee, A. Van Diepen, Chemical Process Technology, John Wiley & Sons, Chichester, 2001.
[21] D. Astruc, F. Lu, J.R. Aranzaes, Angew. Chem. Int. Ed. Engl. 44 (2005) 7852–7872.
[22] S. Vukojevic, O. Trapp, J.D. Grunwaldt, Ch. Kiener, F. Schth, Angew. Chem. 117 (2005) 8192–8195.
[23] R. Xu, T. Xie, Y. Zhao, Y. Li, Nanotechnology 18 (2007) 55602–55606.
[24] A. Corma, Catal. Rev. 46 (2004) 369–417.
[25] J.A. Widegren, R.G. Finke, J. Mol. Catal. A Chem. 198 (2003) 317–341.
[26] L.S. Ott, R.G. Finke, Coord. Chem. Rev. 251 (2007) 1075–1100.
[27] J.M. Thomas, R. Raja, Chem. Rec. 1 (2001) 448–466.
[28] B.F.G. Johnson, Top. Catal. 24 (2003) 147–159.
[29] N. Coowanitwong, C.Y. Wu, M. Cai, S. Watano, Environ. Eng. Sci. 24 (2007) 138–148.
[30] J.M. Campelo, D. Luna, R. Luque, J.M. Marinas, A.A. Romero, Chemphyschem 2 (2009) 18–45.
[31] N. Pernicone, CatTeche 7 (2003) 196–205.
[32] T. Stace, Nature 331 (1988) 116–117.
[33] P.A. Montano, Phys. Rev. Lett. 56 (1986) 2076–2079.
[34] M. Valden, X. Lai, D.W. Goodman, Science 281 (1998) 1647–1651.
[35] G.A. Somorjai, R.M. Rioux, Catal. Today 100 (2005) 201–215.
[36] W. Teunissen, A.A. Bol, J.W. Geus, Catal. Today 48 (1999) 329–336.
[37] A. de Mejere, F. Diederich (Eds.), Metal-Catalyzed Cross-Coupling Reactions, Wiley-VCH, Weinheim, 2005.
[38] K.J. Klabunde, J. Stark, O. Koper, C. Mohs, D.G. Park, S. Decker, Y. Jiang, I. Lagadic, D. Zhang, Phys. Chem. 100 (1996) 12142–12153.
[39] C.N.R. Rao, G.U. Kulkarni, P.J. Thomas, P.P. Edwards, Chem. Eur. J. 8 (2002) 29–36.
[40] R. Narayanan, M.A. El-Sayed, J. Phys. Chem. B 109 (2005) 12663–12676.
[41] J. Kim, H. Jia, P. Wang, Biotechnol. Adv. 24 (2006) 296–308.
[42] G. Schmid, Chem. Rev. 92 (1992) 1709–1727.
[43] D. Astruc, In: D. Astruc (Ed.), Transition-metal Nanoparticles, in Catalysis, Nanoparticles and Catalysis, Wiley-VC, Weinheim, 2008, pp. 1–48.
[44] J.D. Aiken III, R.G. Finke, J. Mol. Catal. A Chem. 145 (1999) 1–44.
[45] K.J. Klabunde, Y. Li, B. Tan, Chem. Mater. 3 (1991) 30–39.
[46] Y. Lin, R.G. Finke, Inorg. Chem. 33 (1994) 4891–4910.
[47] R.G. Finke, In: D.L. Feldheim, C.A. Foss Jr. (Eds.), Metal Nanoparticles: Synthesis, Characterization, and Applications, Marcel Dekker, New York, 2002, pp. 17–54.

[48] K. Judai, S. Abbet, A.S. Worz, U. Heiz, C.R. Henry, J. Am. Chem. Soc. 126 (2004) 2732–2737.
[49] H. Bonneman, K.S. Nagabhushana, In: C.I. Contescu, K. Putyera (Eds.), Dekker Encyclopedia of Nanoscience and Nanotechnology, Taylor & Francis Group, LLC, New York, 2009, pp. 863–881.
[50] B.C. Gates, In: R.D. Adams, F.A. Cotton (Eds.), Catalysis by Di- and Polynuclear Metal Cluster Complexes, Wiley-VCH, Weinheim, 1998, pp. 509–538.
[51] B. Coq, F. Figueras, In: A. Wieckowski, E.R. Savinova, C.G. Vayenas (Eds.), Catalysis and Electrocatalysis at Nanoparticle Surfaces, Marcel Dekker, New York, 2003, pp. 847–876.
[52] A. Beck, A. Horvath, A. Sarkany, L. Guczi, Curr. Appl. Phys. 6 (2006) 200–204.
[53] H. Bonnemann, W. Brijoux, A. Schulze Tilling, K. Siepen, Top. Catal. 4 (1997) 217–227.
[54] M.T. Reetz', Wolfgang Helbig, J. Am. Chem. Soc. 116 (1994) 1401–1402.
[55] Y. Li, M.A. El-Sayed, J. Phys. Chem. B 105 (2001) 8938–8943.
[56] D.J. Cole-Hamilton, Science 299 (2003) 1702–1706.
[57] R. Xu, T. Xie, Y. Zhao, Y. Li, Nanotechnology 18 (2007) 1–5.
[58] S. Vukojevic, O. Trapp, J.-D. Grunwaldt, C. Kiener, F. Schuth, Angew. Chem. 117 (2005) 8192–8195.
[59] M. Zhao, R.M. Crooks, Angew. Chem. Int. Ed. Engl. 38 (1999) 364–366.
[60] X.S. Zhao, X.Y. Bao, W. Guo, F. Yin Lee, Mater. Today 9 (2006) 32–39.
[61] Z. Ma, S. Dai, ACS Catal. 1 (2011) 805–818.
[62] Y. Tsuji, T. Fujihara, Inorg. Chem. 46 (2007) 1895–1902.
[63] B.C. Gates, Chem. Rev. 95 (1995) 511–522.
[64] J. Weitkamp, Solid State Ion. 131 (2000) 175–188.
[65] F.M. Daytzenberg, Catal. Rev. 46 (2004) 335–368.
[66] M. Haruta, N. Yamada, T. Kobayashi, S. Iijima, J. Catal. 115 (1989) 301–309.
[67] M. Haruta, CATECH 6 (2002) 102–115.
[68] Q. Wang, A.E. Ostafin, In: H.S. Nalwa (Ed.), Encyclopedia of Nanoscience and Nanotechnology, vol. 5, American Scientific Publishers, Stevenson Ranch, 2004, pp. 475–503.
[69] B. Hvolbæk, T.V.W. Janssens, B.S. Clausen, H. Falsig, C.H. Christensen, J.K. Nørskov, Nano Today 2 (2007) 14–18.
[70] C.T. Campbell, S.C. Parker, D.E. Starr, Science 298 (2002) 811–814.
[71] J.M. Thomas, R. Raja, G. Sankar, R.G. Bell, D.W. Lewis, Pure Appl. Chem. 73 (2001) (2001) 1087–1101.
[72] B. Viswanathan, Bindhu Jacob, Catal. Rev. 47 (2005) 1–82.
[73] O. Domınguez-Quintero, S. Martınez, Y. Henrıquez, L. D'Ornelas, H. Krentzien, J. Osuna, J. Mol. Catal. A Chem. 197 (2003) 185–191.
[74] M. Arenz, U. Landman, U. Heiz, Chemphyschem 7 (2006) 1871–1879.
[75] M. Haruta, Chem. Rec. 3 (2003) 75–87.
[76] R.P. Underwood, A.T. Bell, Appl. Catal. 34 (1987) 289–310.
[77] A.K. Datye, Q. Xu, K.C. Kharas, J.M. McCarty, Catal. Today 111 (2006) 59–67.
[78] N. Toshima, T. Yonezawa, New J. Chem. 22 (1998) 1179–1201.
[79] A. Moisala, A.G. Nasibulin, E.I. Kauppinen, J. Phys. Condens. Matter 15 (2003) S3011–S3035.
[80] D.S. Bethume, C.H. Kiang, M.S. De Vries, G. Gorman, R. Savoy, J. Vazquez, Nature 363 (1993) 605–607.
[81] X.S. Wang, F. Rahman, M.J. Rhodes, Chem. Eng. Sci. 62 (2007) 3455–3461.
[82] L.F. Hakim, J.L. Portman, M.D. Casper, A.W. Weimer, Powder Technol. 160 (2005) 149–160.

[83] C. Zhu, Q. Yu, R.N. Dave, R. Pfeffer, AIChE J. 51 (2005) 426–439.
[84] A. Roucoux, J. Schulz, H. Patin, Chem. Rev. 102 (2002) 3757–3778.
[85] M. Ojeda, S. Rojas, M. Boutonnet, F.J. Pérez-Alonso, F.J. García-García, J.G. Fierro, Appl. Catal. Gen. 274 (2004) 33–41.
[86] L.E.Y. Nonnemann, A.G.T.M. Bastein, V. Ponec, Appl. Catal. 62 (1990) L23–L28.
[87] M. Yamamura, K. Chaki, T. Wakatsuki, H. Okado, K. Fujimoto, Zeolites 14 (1994) 643–649.
[88] R. Renneke, S. McIntosh, V. Arunajatesan, M. Cruz, B. Chen, Th. Tacke, H. Lansink-Rotgerink, A. Geisselmann, R. Mayer, R. Hausmann, P. Schinke, U. Rodemerck, M. Stoyano, Top. Catal. 38 (2006) 279–287.
[89] Y.-F. Han, D. Kumar, D.W. Goodman, J. Catal. 230 (2005) 353–358.
[90] A. Goeppert, G.A. Olah, G.K. Surya Prakash, Beyond Oil and Gas: The Methanol Economy, Wiley-VCH, Weinheim, 2006.
[91] D.R. Palo, R.A. Dagle, J.D. Holladay, Chem. Rev. (2007) 3992–4021.
[92] K.J. Klabunde, R.S. Mulukitla, In: K.J. Klabunde (Ed.), Nanoscale Materials in Chemistry, John Wiley & Sons, New York, 2001, p. 224.
[93] M. Stocker, Microporous Mesoporous Mater. 29 (1999) 3–48.
[94] E. Mielczarski, S. Monteverdi, A. Amariglio, H. Amariglio, Appl. Catal. Gen. 164 (1993) 215–228.
[95] J. Ma, S. Reng, D. Pan, R. Li, K. Xie, React. Funct. Polym. 62 (2005) 31–39.
[96] H. van Bekkum, L. Maat, In: G. Centi, R.A. van Senten (Eds.), Catalysis for Renewables: From Feedstock to energy Production, Wiley-VCH Verlag, Weinheim, 2007, pp. 101–118.
[97] T. Ishida, M. Haruta, Angew. Chem. Int. Ed. Engl. 46 (2007) 7154–7156.
[98] S.C. Tsang, V. Caps, I. Paraskevas, D. Chadwick, D. Thompsett, Angew. Chem. Int. Ed. Engl. 43 (2004) 5645–5649.
[99] M. Pagliaro, Nano-Age: How Nanotechnology Changes our Future, WILEY-VCH Verlag GmbH & Co. KGaA, Weinheim, 2010, pp. 82–83.
[100] G. Wakefield, USPTO 7,169,196.
[101] A.J. Zarur, J.Y. Ying, Nature 403 (2000) 65–67.
[102] http://www.oxonica.com/ (accessed December 2006).
[103] http://www.h2oil.com/ (accessed March 2007).
[104] H.B. Rotavera, A. Kumar, S. Seal, E.L. Petersen, Proc. Combus. Inst. 32 (2009) 811–819.
[105] H. Tyagi, P.E. Phelan, R. Prasher, R. Peck, T. Lee, J.R. Pacheco, P. Arentzen, Nano Lett. 8 (2008) 1410–1416.
[106] R.M. Heck, R.J. Farrauto, Appl. Catal. Gen. 221 (2001) 443–457.
[107] M. Halman, In: M. Gratzel (Ed.), Academic Press, New York, 1983, p. 507.
[108] B.H. Davis, In: J. Garcia-Martinez (Ed.), Nanotechnology for the Energy Challenge, Wiley-VCH, Weinheim, 2009, pp. 282–317.
[109] J. Jung, D. Gidaspow, J. Nanopart. Res. 4 (2002) 483–497.
[110] M.A. Glikin, I.M. Glikina, E. Kauffeldt, Adsorption Sci. Technol. 23 (2005) 135–143.
[111] G. Majano, S. Mintova, O. Ovsitser, B. Mihailova, T. Bein, Microporous Mesoporous Mater. 80 (2005) 227–235.
[112] Y. Liu, J. Pinnavaia, JACS 125 (2003) 2376–2377.
[113] T.G. Kaufmann, A. Kaldor, G.F. Stuntz, M.C. Kerby, L.L. Ansel, Catal. Today 62 (2000) 77–90.
[114] M. Sugimoto, H. Katsuno, K. Takatsu, N. Kawata, Zeolites 7 (1987) 503–507.
[115] J.L. Sotelo, R. Van Grieken, J. Aguado, D.P. Serrano, J.M. Excola, J.M. Menendez, In: M. Treacy, B. Markus, M. Bisher, J. Higgins (Eds.), 12th International Zeolite conference, Materials Research Society, Baltimore (Maryland), 1999, pp. 1441–1446.

[116] N.H. Tran, J.R. Bartlett, G.S.K. Kannangara, A.S. Milev, H. Volk, M.A. Wilson, Fuel 89 (2010) 265–274.
[117] S. Koutsopoulos, T. Johannessen, K.M. Eriksen, R. Fehrmann, J. Catal. 238 (2006) 206–213.
[118] A.C. Pulikottil, M. Santra, H. Biswas, M. Karthikeyan, D. Kumar Yadav, R. Prakash Verma, Process for preparing hydro-desulfurization catalyst, USPTO 6855653.
[119] J.R. Rostrup-Nielsen, Catal. Today 18 (1993) 305–324.
[120] M.A. Pefia, J.P. Gomez, J.L.G. Fierro, Appl. Catal. Gen. 144 (1996) 7–57.
[121] S. Freni, G. Calogero, S. Cavallaro, J. Power Sour. 87 (2000) 28–38.
[122] J.W. Phair, S.P.S. Badwal, Sci. Technol. Adv. Mater. 7 (2006) 792–805.
[123] W. Feng, T. Tan, P. Ji, D. Zheng, AIChE (2006) 2260–2270.
[124] S. Albertazzi, P. Arpentinier, F. Basile, P. Del Gallo, G. Fornasari, D. Gary, A. Vaccari, Appl. Catal. Gen. 247 (2003) 1–7.
[125] J.R. Rostrup-Nielsen, J. Sehested, J.K. Nørskov, Adv. Catal. 47 (2002) 65–138.
[126] L.E. Alzamora, J.R.H. Ross, E.C. Kruissink, L.L. van Reijen, J. Chem. Soc. Faraday Trans. 1 77 (1981) 665–681.
[127] A.B. Stiles, T.A. Koch, Catalyst Manufacture, second ed, Marcel Dekker Inc, New York, 1995.
[128] H.F. Rase, Handbook of Commercial Catalysts, CRC Press LLC, Boca Raton, 2000, pp. 403–466.
[129] S. Helveg, C. Lopez-Cartes, J. Sehested, P.L. Hansen, B.S. Clausen, J.R. Rostrup-Nielsen, F. Abild-Pedersen, J.K. Nørskov, Nature 427 (2004) 426–429.
[130] J.H. Sinfelt, Surf. Sci. 500 (2002) 923–946.
[131] R.J. Farruto, Y. Liu, W. Ruettinger, O. Ilinich, L. Shore, T. Giroux, Catal. Rev. 49 (2007) 141–196.
[132] K. Kochloefl, In: G. Ertl, H. Knözinger, J. Weitkamp (Eds.), Handbook of Heterogeneous Catalysis, vol. 4, Wiley-VCH, Weinheim, 1997, pp. 1819–1831.
[133] A. Julbe, D. Farrusseng, C. Guizard, Catal. Today 104 (2005) 102–113.
[134] P.N. Dyera, R.E. Richards, S.L. Russeka, D.M. Taylor, Solid State Ion. 134 (2000) 21–33.
[135] A. Subramania, T. Saradha, S. Muzhumathi, J. Power Sour. 167 (2007) 319–324.
[136] I. Kagomiya, T. Iijima, H. Takamura, J. Membr. Sci. 286 (2006) 180–184.
[137] Y.V. Frolova-Borchert, V.A. Sadykov, G.M. Alikina, A.I. Lukashevich, E.M. Moroz, D.I. Kochubey, V.V. Kriventsov, V.I. Zaikovskii, V.V. Zyryanov, N.F. Uvarov, Solid State Ion. 177 (2006) 2533–2538.
[138] X. Yin, L. Hong, Z.-L. Liu, J. Membr. Sci. 268 (2006) 2–12.
[139] D.R. Palo, R.A. Dagle, J.D. Holladay, Chem. Rev. 107 (2007) 3992–4021.
[140] Y. Kawamura, K. Yamamoto, N. Ogura, T. Katsumata, A. Igarashi, J. Power Sour. 150 (2005) 20–26.
[141] T. Valdes-Solıs, G. Marban, A. Fuertes, Catal. Today 116 (2006) 354–360.
[142] A. Karim, T. Conant, A. Datye, J. Catal. 243 (2006) 420–427.
[143] G. Yu, B. Sun, Y. Pei, S. Xie, S. Yan, M. Qiao, K. Fan, X. Zhang, B. Zong, J. Am. Chem. Soc. 132 (2010) 935–937.
[144] M.A. Fox, M. Dulay, Chem. Rev. 83 (1995) 341–357.
[145] B.L. Abrams, J.P. Wilcoxon, Crit. Rev. Solid State Mater. Sci. 30 (2005) 153–182.
[146] A. Hagfeldt, M. Gratzel, Chem. Rev. 95 (1995) 49–68.
[147] V. Balzani, A. Credi, M. Venturi, Chem. Sus. Chem. 1 (2008) 26–58.
[148] G.Q. Lu, X.S. Zhao, In: G.Q. Lu, X.S. Zhao (Eds.), Nanoporous Materials: Science and Engineering, World Scientific, Singapore, 2004, p. 1.

[149] Y. Taga, Thin Solid Films 517 (2009) 3167–3172.
[150] M. Halman, In: M. Gratzel (Ed.), Energy Resources Through Photochemistry and Catalysis, Academic Press, New York, 1983, p. 507.
[151] R. Asahi, T. Morikawa, T. Ohwaki, K. Aoki, Y. Taga, Science 293 (2001) 269–271.
[152] D. Beydoun, R. Amal, G. Low, S. McEvoy, J. Nanopart. Res. 1 (1999) 439–458.
[153] T.R. Thurston, J.P. Wilcoxon, J. Phys. Chem. B 103 (1999) 11–17.
[154] H.D. Jang, S.K. Kim, S.-J. Kim, J. Nanopart. Res. 3 (2001) 141–147.
[155] A. Fujishima, K. Honda, Nature 238 (1972) 37–38.
[156] M.R. Hoffmann, S.T. Martin, W. Choi, D.W. Bahnemannt, Chem. Rev. 95 (1995) 69–96.
[157] A.V. Emeline, V.N. Kuznetsov, V.K. Rybchuk, N. Serpone, Int. J. Photoenergy 2008 (2008) 1–18.
[158] S. Licht, J. Phys. Chem. B 105 (2001) 6281–6294.
[159] L.R. Sheppard, J. Nowotny, Adv. Appl. Ceram. 106 (2007) 9–20.
[160] Oscar Khaselev, John A. Turner, Science 280 (1998) 425–427.
[161] D.M. Blake, P.-C.h. Maness, Z. Huang, E.J. Wolfrum, J. Huang, W. Jacoby, Sep. Purif. Methods 28 (1999) 1–50.
[162] L. Frazer, Environ. Health Perspect. 109 (2001) A174–A177.
[163] C. Hariharan, Appl. Catal. Gen. 304 (2006) 55–61.
[164] A. Kudo, Int. J. Hydrogen Energy 31 (2006) 197–202.
[165] J. Nowotny, T. Bak, M.K. Nowotny, L.R. Sheppard, Int. J. Hydrogen Energy 32 (2007) 2609–2629.
[166] J. Zhu, M. Zäch, Curr. Opin. Colloid Interface Sci. 14 (2009) 260–269.
[167] M. Gratzel, Nature 414 (2001) 338–344.
[168] S. Takabayashi, R. Nakamuraa, Y. Nakato, J. Photochem. Photobiol. A Chem. 166 (2004) 107–113.
[169] A. Corma, P. Atienzar, H. Garcia, J. Chane-Ching, Nat. Mater. 3 (2004) 394–397.
[170] T. Nann, S.K. Ibrahim, P.-M. Woi, S. Xu, J. Ziegler, C.J. Pickett, Angew. Chem. Int. Ed. Engl. 49 (2010) 1574–1577.
[171] L. Yuliati, H. Yoshida, Chem. Soc. Rev. 37 (2008) 1592–1602.
[172] J. Grunes, et al., Chem. Commun. (2003) 2257–2260.
[173] M. Winter, R.J. Brodd, Chem. Rev. 104 (2004) 4245–4269.
[174] L. Pasquato, P. Pengo, P. Scrimin, Supramol. Chem. 17 (2005) 163–171.
[175] J. Kim, J.W. Grate, Nano Lett. 3 (2003) 1219–1222.
[176] H.-H. Yang, S.Q. Zhang, X.-L. Chen, Z.-X. Zhuang, J.-G. Xu, X.-R. Wang, Anal. Chem. 76 (2004) 1316–1321.
[177] B.J. Jordan, R. Hong, B. Gider, J. Hill, T. Emrick, V.M. Rotello, Soft Matter 2 (2006) 558–560.
[178] W.R. Grove, London Edinburgh Philos. Mag. J. Sci. Ser. 3 (14) (1839) 127–130.
[179] R.G. Cawthorn, S. Afr. J. Sci. 95 (1999) 481–489.
[180] S.K. Kamarudin, W.R.W. Daud, S.L. Ho, U.A. Hasran, J. Power Sour. 163 (2007) 743–754.
[181] Jung-Ho Wee, J. Power Sour. 161 (2006) 1–10.
[182] A. Hammett, Catal. Today 38 (1997) 445–457.
[183] S. Srinivasan, R. Mosdale, P. Stevens, C. Yang, Annu. Rev. Energy Environ. 24 (1999) 281–328.
[184] L. Carrette, K.A. Friederich, U. Stimming, Chem. Phys. Chem. 1 (2000) 162–193.
[185] S.M. Haile, Acta Mater. 51 (2003) 5981–6000.
[186] N.M. Markovic, P.N. Ross, CATTECH. 4 (2000) 110–126.
[187] D.R. Rolison, Science 299 (2003) 1698–1708.

[188] R. Dillon, S. Srinivasan, A.S. Aricò, V. Antonucci, J. Power Sour. 127 (2004) 112–126.
[189] S. Wasmus, A. Kuver, J. Electroanal. Chem. 461 (1999) 14–31.
[190] C. Stone, A.E. Morrison, Solid State Ion. 152–153 (2002) 1–13.
[191] P.M. Wee, Renew. Sustain. Energy Rev. 11 (2007) 1720–1738.
[192] D.R. Rolison, J.W. Long, J.C. Lytle, A.E. Fischer, C.P. Rhodes, T.M. McEvoy, M.E. Bourg, A.M. Lubers, Chem. Soc. Rev. 38 (2009) 226–252.
[193] K. Wikander, H. Ekstrom, A.E.C. Plamqvist, A. Lunblad, K. Holmberg, G. Lindbergh, Fuel Cell 6 (2006) 21–25.
[194] V. Mazumder, Y. Lee, S. Sun, Adv. Funct. Mater. 20 (2010) 1224–1231.
[195] Y.H. Sua, T.Y. Wei, C.H. Hsu, Y.L. Liu, Y.M. Sun, J.Y. Lai, Desalination 200 (2006) 656–657.
[196] A.S. Aricò, P. Cretì, E. Modica, G. Monforte, V. Baglio, V. Antonucci, Electrochim. Acta 45 (2000) 4319–4328.
[197] H. Liu, C. Song, L. Zhang, J. Zhang, H. Wang, D.P. Wilkinson, J. Power Sour. 155 (2006) 95–110.
[198] F. Bensebaa, A. Farah, D. Wang, C. Bock, X. Du, J. Kung, Y. Le Page, J. Phys. Chem. B 109 (2005) 15339–15344.
[199] F. Bensebaa, N. Patrito, Y. Le Page, P. L'Ecuyer, D. Wang, J. Mater. Chem. 14 (2004) 3378–3384.
[200] J.-M. Chen, L.S. Sarma, C.-H. Chen, M.-Y. ao Chenga, S.-C. Shih, G.-R. Wang, D.-G. Liu, J.-F. Lee, M.-T. Tang, B.-J. Hwang, J. Power Sour. 159 (2006) 29–33.
[201] Z. Liu, Bing Guo, J. Huang, L. Hong, M. Han, L.M. Gan, J. Power Sour. 157 (2006) 207–211.
[202] S. Licoccia, E. Traversa, J. Power Sour. 159 (2006) 12–20.
[203] T.C. Deivaraja, Jim Yang Lee, J. Power Sour. 142 (2005) 43–49.
[204] T. Kobayashi, A. Ueda, Y. Yamada, H. Shioyama, Appl. Surf. Sci. 223 (2004) 102–108.
[205] L. Jiang, G. Sun, Z. Zhou, W. Zhou, Q. Xin, Catal. Today 93–95 (2004) 665–670.
[206] J.W. Long, R.M. Stroud, K.E. Swider-Lyons, D.R. Rolison, J. Phys. Chem. B 104 (2000) 9772–9776.
[207] D.W. Lee, J.H. Won, K.B. Shim, Mater. Lett. 57 (2003) 3346–3351.
[208] C. Viazzi, A. Deboni, J. Zoppas Ferreira, J.P. Bonino, F. Ansart, Solid State Sci. 8 (2006) 1023–1028.
[209] G. Xu, Y.-W. Zhang, C.-S. Liao, Chun-Hua Yan, Solid State Ion. 166 (2004) 391–396.
[210] P. Singh, N.Q. Minh, Int. J. Appl. Ceram. Technol. 1 (2004) 5–15.
[211] S.P. Jiang, Mater. Sci. Eng. A 418 (2006) 199–210.
[212] Y. Sunagawa, K. Yamamoto, A. Muramatsu, J. Phys. Chem. B 110 (2006) 6224–6228.
[213] D.M. Chapin, C.S. Fuller, G.L. Pearson, J. Appl. Phys. 25 (1954) 676–677.
[214] A. Luque, S. Hegedu (Eds.), Handbook of Photovoltaic Science and Engineering, Wiley, England, 2003.
[215] R.W. Miles, G. Zoppi, I. Forbes, Mater. Today 10 (2007) 20–27.
[216] L.L. Kazmerski, Renew. Sustain. Energy Rev. 1 (1997) 71–170.
[217] M. Green, Prog. Photovolt. Res. Appl. 14 (2006) 383–392.
[218] F. Krebs, Solar Energy Mater. Solar Cells 93 (2009) 394–412.
[219] B. Von Roedern, ReFocus 7 (2006) 34–39.
[220] F. Bensebaa, Prog. Photovolt. Res. Appl. 19 (2010) 240–246.
[221] F. Dimroth, Phys. Stat. Sol. 3 (2006) 373–379.
[222] T.V. Torchynska, Semiconductor Phys. Quantum Electron. Optoelectron. 5 (2002) 63–70.
[223] G. Li, V. Shrotriya, J. Huang, Y. Yao, T. Moriarty, K. Emery, Y. Yang, Nat. Mater. 4 (2005) 864–868.

[224] J.C. Hummelen, B.W. Knight, F. LePeq, F. Wudl, J. Yao, C.L. Wilkins, J. Org. Chem. 60 (1995) 532–538.

[225] J. Roncali, Chem. Rev. 97 (1997) 173–205.

[226] A.L. Roes, E.A. Alsema, K. Blok, M.K. Patel, Prog. Photovolt. Res. Appl. 17 (2009) 372–393.

[227] R. Cugola, U. Giovanella, P. Di Gianvincenzo, F. Bertini, M. Catellani, S. Luzzati, Thin Solid Film 511–512 (2006) 489–493.

[228] Poly(3-hexylthiophene-2,5-diyl), Sigma-Aldrich Catalogue, item 445703.

[229] K.L. Chopra, P.D. Paulson, V. Dutta, Prog. Photovolt. Res. Appl. 12 (2004) 69–92.

[230] A. Slaoui, J.F. Guillemoles, l'actualité chimique 331 (2009) 41–49.

[231] T. Soga, In: T. Soga (Ed.), Nanostructured Materials for Solar Energy Conversion, Elsevier, Amsterdam, 2006, pp. 3–44.

[232] K.R. Catchpole, Philos. Trans. R Soc. A 364 (2006) 3493–3503.

[233] V.I. Klimov, J. Phys. Chem. B 110 (2006) 16827–16845.

[234] S. Kumar, G.D. Scholes, Microchim. Acta 160 (2007) 315–325.

[235] P. Oelhafen, A. Schuler, Solar Energy 79 (2005) 110–121.

[236] A.J. Nozik, Physica E 14 (2002) 115–120.

[237] J. Young Kim, K. Lee, N.E. Coates, D. Moses, T.Q. Nguyen, M. Dante, A.J. Heeger, Science 317 (2007) 22–225.

[238] P. Mahawela, G. Sivaraman, S. Jeedigunta, J. Gaduputi, M. Ramalingam, S. Subramanian, S. Vakkalanka, C.S. Ferekides, D.L. Morel, Mater. Sci. Eng. B 116 (2005) 283–291.

[239] M. Yamaguchi, T. Takamoto, K. Araki, Solar Energy Mater. Solar Cells 90 (2006) 3068–3077.

[240] R.J. Ellingson, M.C. Beard, J.C. Johnson, P. Yu, O.I. Micic, A.J. Nozik, A. Shabaev, A.L. Efros, Nano Lett. 5 (2005) 865–871.

[241] V. Svrcek, A. Slaoui, J.-C. Muller, Thin Solid Films 451–452 (2004) 384–388.

[242] I.J. Kramer, E.H. Sargent, ACS Nano 5 (2011) 8506–8514.

[243] C.W. Tang, Appl. Phys. Lett. 48 (1986) 183–185.

[244] P. Peumans, A. Yakimov, S.R. Forrest, J. Appl. Phys. 93 (2003) 3693–3723.

[245] H. Hoppe, N.S. Sariciftci, J. Mater. Chem. 16 (2006) 45–61.

[246] S. Günes, H. Neugebauer, N.S. Sariciftci, Chem. Rev. 107 (2007) 1324–1338.

[247] M.T. Lloyd, J.E. Anthony, G.G. Malliaras, Mater. Today 10 (2007) 34–41.

[248] S.E. Shaheen, D.S. Ginley, G.E. Jabbour, MRS Bull. 30 (2005) 10–19.

[249] J.K.J. van duren, X. Yang, J. Loos, C.W.T. Bulle-Lieuwma, A.B. Sieval, J.C. Hummelen, R.A.J. Jansen, Adv. Funct. Mater. 14 (2004) 425–434.

[250] M.C. Scharber, D. Mühlbacher, M. Koppe, P. Denk, C. Waldauf, A.J. Heeger, C.J. Brabec, Adv. Mater. 18 (2006) 789–794.

[251] X. Yang, J.K.J. van Duren, R.A.J. Janssen, M.A.J. Michels, J. Loos, Macromolecules 37 (2004) 2151–2158.

[252] Y. Kim, S. Cook, S.M. Tuladhar, S.A. Choulis, J. Nelson, J.R. Durrant, D.D.C. Bradley, M. Giles, I. McCulloh, C.-S. Ha, M. Ree, Nat. Mater. 5 (2006) 197–203.

[253] T. Kietzke, D. Neher, M. Kumke, R. Montenegro, K. Landfester, U. Scherf, Macromolecules 37 (2004) 4882–4890.

[254] C. Szymanski, C. Wu, J. Hooper, M.A. Salazar, A. Perdomo, A. Dukes, J. McNeill, J. Phys. Chem. B 109 (2005) 8543–8546.

[255] E. Bundgaard, F.C. Krebs, Solar Energy Mater. Solar Cells 91 (2007) 954–985.

[256] H.W. Hillhouse, M.C. Beard, Curr. Opin. Colloid Interface Sci. 14 (2009) 245–259.

[257] F. Bensebaa, Transparent and/or Photovoltaic solar cell and module, cipo ca 2621665, 19 February 2008.

[258] T. Trupke, M.A. Green, P. Wurfel, J. Appl. Phys. 92 (2002) 1668–1674.

[259] C. Strümpel, M. McCann, G. Beaucarne, V. Arkhipov, A. Slaoui, V. Švrček, C. del Cañizo, I. Tobias, Solar energy Mater. Solar Cells 91 (2007) 238–249.

[260] G. Belomoin, J. Therrien, M. Nayfeh, Appl. Phys. Lett. 77 (2000) 779–781.

[261] M. Stupca, M. Alsalhi, T. Al Saud, A. Almuhanna, M.H. Nayfeh, Appl. Phys. Lett. 91 (2007) 63107–63110.

[262] R. O'Hayre, M. Nanu, J. Schoonman, A. Goossens, Nanotechnology 18 (2007) 055702–055708.

[263] A. Kongkanand, K. Tvrdy, K. Takechi, M. Kuno, P.V. Kamat, J. Am. Chem. Soc. 130 (2008) 4007–4015.

[264] P.V. Kamat, J. Phys. Chem. C 111 (2007) 2834–2860.

[265] S. Pizzini, M. Acciarri, S. Binetti, D. Cavalcoli, A. Cavallini, D. Chrastina, L. Colombo, E. Grilli, G. Isella, M. Lancin, A. Le Donne, A. Mattoni, K. Peter, B. Pichaud, E. Poliani, M. Rossi, S. Sanguinetti, M. Texier, H. von Kanel, Mater. Sci. Eng. B 134 (2006) 118–124.

[266] M.K. Nazeeruddin, A. Kay, L. Rodicio, R. Humpbry-Baker, E. Miiller, P. Liska, N. Vlachopoulos, M. Gratzel, J. Am. Chem. Soc. 115 (1993) 6382–6390.

[267] M. Gratzel, J. Photochem. Photobiol. C 4 (2003) 145–153.

[268] J.-H. Yum, Peter Chen, Michael Gratzel, M.K. Nazeeruddin, Chem. Sus. Chem. 1 (2008) 699–707.

[269] H.J. Snaith, Adv. Funct. Mater. 20 (2010) 13–19.

[270] A. Hinsch, S. Behrens, M. Berginc, H. Bonnemann, H. Brandt, A. Drewitz, F. Einsele, D. Faßler, D. Gerhard, H. Gores, R. Haag, T. Herzig, S. Himmler, G. Khelashvili, D. Koch, G. Nazmutdinova, U. Opara-Krasovec, P. Putyra, U. Rau, R. Sastrawan, T. Schauer, C. Schreiner, S. Sensfuss, C. Siegers, K. Skupien, P. Wachter, J. Walter, P. Wasserscheid, U. Wurfel, M. Zistler, Prog. Photovolt. Res. Appl. 16 (2008) 489–501.

[271] B. O'Regan, M. Gratzel, Nature 353 (1991) 737–740.

[272] M. Spath, P.M. Sommeling, J.A.M. van Roosmalen, H.J.P. Smit, N.P.G. van der Burg, D.R. Mahieu, N.J. Bakker, J.M. Kroon, Prog. Photovolt. Res. Appl. 11 (2003) 207–220.

[273] J.M. Kroon, J. Bakker, H.J.P. Smit, P. Liska, K.R. Thampi, P. Wang, S.M. Zakeeruddin, M. Gratzel, A. Hinsch, S. Hore, U. Wurfel, R. Sastrawan, J.R. Durrant, E. Palomares, H. Pettersson, T. Gruszecki, J. Walter, K. Skupien, G.E. Tulloch, Prog. Photovolt. Res. Appl. 15 (2007) 1–18.

[274] T.V. Anuradha, S. Ranganathan, Bull. Mater. Sci. 30 (2007) 263–269.

[275] J.N. Harta, D. Menziesa, Y.-B. Cheng, G.P. Simona, L. Spiccia, Solar Energy Mater. Solar Cells 91 (2007) 6–16.

[276] N.G. Park, K.M. Kim, M.G. Kang, K.S. Ryu, S.H. Chang, Y.T. Shin, Adv. Mater. 17 (2005) 2349–2353.

[277] A.J. Frank, K.D. Benkstein, N. Kopidakis, J. van de Lagemaat, Does Network Geometry Influence the Electron Transport Dynamics in Mesoporous Nanoparticle Solar Cells? NREL/CP-520-33616, 2003.

[278] S. Chappel, L. Grinis, A. Ofir, A. Zaban, J. Phys. Chem. 109 (2005) 1643–1647.

[279] M. Ihara, K. Tanaka, K. Sakaki, I. Honma, K. Yamada, J. Phys. Chem. B101 (1997) 5153–5157.

[280] B. Li, L. Wang, B. Kang, P. Wang, Y. Qiu, Solar Energy Mater. Solar Cells 90 (2006) 549–573.

[281] L.B. Roberson, M.A. Poggi, J. Kowalik, G.P. Smestad, L.A. Bottomley, L.M. Tolbert, Coord. Chem. Rev. 248 (2004) 1491–1499.

[282] H. Lindstrom, A. Holmberg, E. Magnusson, S. Lindquist, L. Malmqvist, A. Hagfeldt, Nano Lett. 1 (2001) 97–100.

[283] A. Belaidi, R. Bayon, L. Dloczik, K. Ernst, M.Ch. Lux-Steiner, R. Konenkamp, Thin Solid Films 431–432 (2003) 488–491.
[284] S. Kumar, G.D. Scholes, Microchim. Acta 3 (2007) 315–325.
[285] E. Arici, H. Hoppea, F. Schafflerb, D. Meissnerc, M.A. Malikd, N.S. Sariciftci, Thin Solid Films 451–452 (2004) 612–618.
[286] M. Pientka, J. Wisch, S. Boger, J. Parisia, V. Dyakonova, A. Rogach, D. Talapinc, H. Weller, Thin Solid Films 451–452 (2004) 48–53.
[287] B.R. Saunders, M.L. Turner, Adv. Colloid Interface Sci. 138 (2008) 1–23.
[288] G. Yu, J. Gao, J.C. Hummelen, F. Wudl, A.J. Heeger, Science 270 (1995) 1789–1791.
[289] W.U. Huynh, J.J. Dittmer, A. Paul Alivisatos, Science 295 (2002) 2425–2427.
[290] W.U. Huynh, J.J. Dittmer, W.C. Libby, G.L. Whitting, A.P. Alivisatos, Adv. Funct. Mater. 13 (2003) 73–79.
[291] S. Kumar, T. Nann, J. Mater. Res. 19 (2004) 1990–1995.
[292] D.M. Guldi, I. Zilbermann, G. Anderson, N.A. Kotov, N. Tagmatarchis, M. Prato, J. Mater. Chem. 15 (2005) 114–118.
[293] M.A. Green, Physica E 14 (2002) 65–70.
[294] T. Takamoto, E. Ikeda, H. Kurita, M. Ohmori, Appl. Phys. Lett. 70 (1997) 381–383.
[295] J. Young Kim, K. Lee, N.E. Coates, D. Moses, T.-Q. Nguyen, M. Dante, A.J. Heeger, Science 317 (2007) 222–225.
[296] R.R. King, D.C. Law, K.M. Edmondson, C.M. Fetzer, G.S. Kinsey, H. Yoon, D.D. Krut, J. Ermer, R.A. Sherif, N.H. Karam, Adv. Optoelectron. (2007) 1–8.
[297] S. Ruhle, A. Zaban, M. Shalom, Chemphyschem 11 (2010) 2290–2304.
[298] S. Kamprachum, S. Thainoi, S. Kanjanachuchai, S. Panyakeow, 3rd World Conference on Photovoltaic Energy Conversion May 11–18, Osokn, Japan, 2003, pp. 259–261.
[299] A. Luque, A. Marti, Phys. Rev. Lett. 78 (1997) 5014.
[300] E. Vallat-Sauvain, U. Kroll, J. Meier, A. Shah, J. Pohl, J. Appl. Phys. 87 (2000) 3137–3142.
[301] S. Hazra, S. Ray, Solid State Commun. 109 (1999) 125–128.
[302] G. Conibeer, M. Green, R. Corkish, Y. Cho, E. Cho, C.W. Jiang, T. Fangsuwannarak, E. Pink, Y. Huang, T. Puzzer, T. Trupke, B. Richards, A. Shalav, K. Lin, Thin Solid Films 511–512 (2006) 654–662.
[303] S.Y. Myong, S.W. Kwon, K.S. Lim, M. Kondo, M. Konagai, Appl. Phys. Lett. 88 (2006) 083118–083120.
[304] X. Wang, G. Koleilat, J. Tang, H. Liu, I.J. Kramer, R. Debnath, L. Brzozowski, D. Aaron, R. Barkhouse, L. Levina, S. Hoogland, E.H. Sargent, Nat. Photon. 5 (2011) 480–484.
[305] A. Maria, P.W. Cyr, E.J.D. Klem, L. Levina, E.H. Sargent, Appl. Phys. Lett. 87 (2005) 213112–213114.
[306] B.S. Richards, J.E. Cotter, B. Honsberg, S.R. Wenham, 28th IEEE PVSC, 15–22 September 2000, Anchorage, Alaska, 2000.
[307] W. Shockley, H.J. Queisser, J. Appl. Phys. 32 (1961) 510–519.
[308] V.I. Klimov, Appl. Phys. Lett. 89 (2006) 123118–123120.
[309] E. Klampaftis, D. Ross, K.R. McIntosh, B.S. Richards, Solar Energy Mater. Solar Cells 93 (2009) 1182–1194.
[310] A. Le Donne, M. Acciarri, D. Narducci, Prog. Photovolt. Res. Appl. 17 (2009) 519–525.
[311] C. Hägglund, B. Kasemo, Opt. Express 17 (2009) 11944–11957.
[312] W.G.J.H.M. van Sark, A. Meijerink, R.E.I. Schropp, J.A.M. van Roosmalen, E.H. Lysen, Symposium on the Efficient Use of Solar Radiation in Photovoltaic Power Engineering, St. Petersburg, November 3–4, 2003, pp. 962–969.

[313] M. Losurdo, M. Giangregorio, G.V. Bianco, A. Sacchetti, P. Capezzuto, G. Bruno, Solar Energy Mater. Solar Cells 93 (2009) 1749–1754.
[314] D. Derkacs, S.H. Lim, P. Matheu, W. Mar, E.T. Yu, Appl. Phys. Lett. 89 (2006) 093103–093105.
[315] S. Pillai, K.R. Catchpole, T. Trupke, M.A. Green, J. Appl. Phys. 1001 (2007) 93105–93112.
[316] M. Green, Prog. Photovolt. Res. Appl. 7 (1999) 327–330.
[317] D.M. Schaadt, B. Feng, E.T. Yu, Appl. Phys. Lett. 86 (2006) 63106–63108.
[318] E.T. Yu, Nanotechnology for Photovoltaics, Taylor and Francis Group, LLC, Boca Raton (FL), 2010, pp. 391–221.
[319] K.R. Catchpole, A. Polman, Opt. Express 16 (2008) 21793–21800.
[320] D. Inns, L. Shi, A.G. Aberle, Prog. Photovolt. Res. Appl. 16 (2008) 187–194.
[321] W.G.J.H.M. van Sark, K.W.J. Barnham, L.H. Slooff, A.J. Chatten, A. Büchtemann, A. Meyer, S.J. McCormack, R. Koole, D.J. Farrell, R. Bose, E.E. Bende, A.R. Burgers, T. Budel, J. Quilitz, M. Kennedy, T. Meyer, C. De Mello Donegá, A. Meijerink, D. Vanmaekelbergh, Opt. Express 16 (2008) 21773–21792.
[322] F. Vitrone, J.A. Capobianco, Int. J. Nanotechnol. 5 (2008) 1306–1339.
[323] A.C. Pan, C. DelCanizo, E. Canovas, N.M. Santos, J.P. Leitao, A. Luque, Solar Energy Mater. Solar Cells 94 (2010) 1923–1926.
[324] R.M. Dell, D.A.J. Rand, J. Power Sour. 100 (2001) 2–17.
[325] D. Dunn-Rankin, E.M. Leal, D.C. Walther, Prog. Energy Combus. Sci. 31 (2005) 422–465.
[326] Energy Technology Perspectives: Scenarios & Strategies to 2050, International Energy Agency, 2008.
[327] E. Karden, P. Shinn, P. Bostock, J. Cunningham, E. Schoultz, D. Kok, J. Power Sour. 144 (2005) 505–512.
[328] J.W. Long, B. Dunn, D.R. Rolison, H.S. White, Chem. Rev. 104 (2004) 4463–4492.
[329] E. Karden, S. Ploumen, B. Fricke, T. Miller, K. Snyder, J. Power Sour. 168 (2007) 2–11.
[330] W.F. Pickard, A.Q. Shen, N. Hansing, Renew. Sustain. Energy Rev. 13 (2009) 1934–1945.
[331] A. Chaurey, S. Deambi, Renew. Energy 2 (1992) 227–235.
[332] H. Wenzla, I. Baring-Gouldb, R. Kaiserc, B. Yann Liawd, P. Lundsagere, J. Manwellf, A. Ruddellg, V. Svoboda, J. Power Sour. 144 (2005) 373–384.
[333] V. Neburchilov, H. Wang, J.J. Martin, W. Qu, J. Power Sour. 195 (2010) 1271–1291.
[334] F. Bensebaa, M. Khalfallah, M. Ouchene, Decentralized and direct solar hydrogen production: towards a hydrogen economy in MENA region, World Energy Congress, September 2010 Montreal. www.worldenergy.org/documents/congresspapers/107.pdf.
[335] J. Maeir, Solid State Ion. 154–155 (2002) 291–301.
[336] R. Palacin, Chem. Soc. Rev. 38 (2009) 2565–2575.
[337] A.S. Aricò, V. Baglio, V. Antonucci, In: J. Garcia-Martinez (Ed.), Nanotechnology for the Energy Challenge, Wiley-VCH, Weinheim, 2009, pp. 155–176.
[338] Y.-G. Guo, J.-S. Hu, L.-J. Wan, Adv. Mater. 20 (2008) 2878–2887.
[339] A. Patil, V. Patil, D. Wook Shin, J.-W. Choi, D.-S. Paik, S.-J. Yoon, Mater. Res. Bull. 43 (2008) 1913–1942.
[340] T. Osaka, M. Datta, Emerging Storage Systems for Electronics, Gordon and Breach Science, Amsterdam, 2000.
[341] Basic Research Needs for Electrical Energy Storage, Report of the Basic Energy Sciences Workshop on Electrical Energy Storage, US DOE, April 2–4, 2007.
[342] M.G. Kim, J. Cho, Adv. Funct. Mater. 19 (2009) 1–18.
[343] G.J. Wilson, Aust. J. Chem. 60 (2007) 964.

[344] P.G. Bruce, B. Scrosati, J.M. Tarascon, Angew. Chem. Int. Ed. Engl. 47 (2008) 2930–2946.
[345] E. Stura, C. Nicolini, Anal. Chim. Acta 568 (2006) 57–64.
[346] T. Takamura, Solid State Ion. 152–153 (2002) 19–34.
[347] V. Evan House, F. Ross, Power Management Design Line, 1 May 2007.
[348] J. Shelburne, ASES Meeting, Cleveland, 7–12 July 2007.
[349] J.P. Jan, M.S. Timothy, Method for Manufacturing High Surface Area Nano-Porous Catalyst and Catalyst support Structures, WO2007062356.
[350] http://www.batteryuniversity.com/partone-5A.htm.
[351] J. Furukawa, Y. Nehyo, S. Shiga, J. Power Sour. 133 (2004) 25–31.
[352] M.A. Fetcenko, S.R. Ovshinsky, B. Reichman, K. Young, C. Fierro, J. Koch, A. Zallen, W. Mays, T. Ouchi, J. Power Sour. 165 (2007) 544–551.
[353] M.A. Fetcenko, K. Young, S.R. Ovshinsky, B. Reichman, J. Koch, W. Mays, USPTO 6,270,719.
[354] S.R. Ovshinsky, M.A. Fetcenko, Appl. Phys. A. 72 (2001) 239–244.
[355] M. Jurczyk, L. Smardz, A. Szajek, Mater. Sci. Eng. B108 (2004) 67–75.
[356] D. Takemura, S. Aihara, K. Hamano, M. Kise, T. Nishimura, H. Urushibata, H. Yoshiyasu, J. Power Sour. 146 (2005) 779–783.
[357] J. Goodenough, Y. Kim, Chem. Mater. 22 (2010) 587–603.
[358] H. Rickert, Angew. Chem. Int. Ed. Engl. 17 (1978) 37–46.
[359] J. Maier, Nat. Mater. 4 (2005) 805–815.
[360] B. Scrosati, Chem. Rec. 5 (2005) 286–297.
[361] F. Croce, G.B. Appetecchi, L. Persi, B. Scrosati, Nature 394 (1998) 456–458.
[362] L.M. Bronstein, R.L. Karlinsey, B. Stein, Z. Yi, J. Carini, J.W. Zwanziger, Chem. Mater. 18 (2006) 708–715.
[363] J.-H. Ahn, G.X. Wang, H.K. Liu, S.X. Dou, J. Power Sour. 119–121 (2003) 422–426.
[364] M. Kuriana, M.E. Galvina, P.E. Trapa, D.R. Sadoway, A.M. Mayes, Electrochim. Acta 50 (2005) 2125–2134.
[365] F. Croce, B. Scrosati, Ann. N. Y. Acad. Sci. 984 (2003) 194–207.
[366] F.C. Fernanda, R.M. Torresi, J. Braz. Chem. Soc. 17 (2006) 627–642.
[367] P.R. Bueno, E.R. Leite, J. Phys. Chem. B 107 (2003) 8868–8877.
[368] M. Jo, Y.-S. Hong, J. Choo, J. Cho, J. Electrochem. Soc. 156 (2009) A430–A434.
[369] F.C. Bazito, R.M. Torresi, J. Braz. Chem. Soc. 17 (2006) 627–642.
[370] J. Jamnik, J. Maier, Phys. Chem. Chem. Phys. 59 (2003) 5215–5220.
[371] G. Derrien, J. Hassoun, S. Panero, B. Scrosati, Adv. Mater. 19 (2007) 2336–2340.
[372] D. Kovacheva, B. Markovsky, G. Salitra, Y. Talyosef, M. Gorova, E. Levi, M. Riboch, H.-J. Kimc, D. Aurbach, Electrochim. Acta 50 (2005) 5553–5560.
[373] V.G. Kumar, J.S. Gnanaraj, S. Ben-David, D.M. Pickup, E.R.H. van-Eck, A. Gedanken, D. Aurbach, Chem. Mater. 15 (2003) 4211–4216.
[374] R. Vacassy, H. Hofmann, N. Papageorgiou, M. Gratzel, J. Power Sour. 81–82 (1999) 621–626.
[375] J. Cabana, T. Valdes-Sol, M.R. Palacın, J. Oro-Sole, A. Fuertes, G. Marban, A.B. Fuertes, J. Power Sour. 166 (2007) 492–498.
[376] R. Vidu, P. Stroeve, Ind. Eng. Chem. Res. 43 (2004) 3314–3324.
[377] B. Markovsky, D. Kovacheva, Y. Talyosef, M. Gorova, J. Grinblat, D. Aurbach, Electrochem. Solid-State Lett. 9 (2006) A449–A453.
[378] T. Drezen, N.-H. Kwon, P. Bowen, I. Teerlinck, M. Isono, I. Exnar, J. Power Sour. 174 (2007) 949–953.
[379] T. Kawamura, M. Makidera, S. Okadaa, K. Koga, N. Miura, J.-i Yamaki, J. Power Sour. 146 (2005) 27–32.

[380] Q. Wu, W. Li, Y. Cheng, Z. Jiang, Mater. Chem. Phys. 91 (2005) 463–467.
[381] Y. Kim, J. Cho, J. Electrochem. Soc. 154 (2007) A495–A499.
[382] P. Balaya, Energy Environ. Sci. 1 (2008) 645–654.
[383] Z. Wang, W.H. Tian, X.H. Liu, Y. Lia, X.G. Li, Mater. Chem. Phys. 100 (2006) 92–97.
[384] G.G. Amatucci, F. Badway, A. Singhal, B. Beaudoin, G. Skandan, T. Bowner, I. Plitz, N. Pereira, T. Chapman, R. Jaworski, J. Electrochem. Soc. 148 (2001) 940–950.
[385] C.M. Julien, A. Mauger, A. Ait-Salah, M. Massot, F. Gendron, K. Zaghib, Ionics 13 (2007) 395–411.
[386] D. Son, E. Kim, T. Kim, M. Kim, J. Cho, B. Park, Appl. Phys. Lett. 85 (2004) 5875–5877.
[387] N. Meethong, H.-Y. Shadow Huang, W.C. Carter, Y.-M. Chiang, Electrochem. Solid-State Lett. 10 (2007) A134–A138.
[388] V. Srinivasan, J. Newman, J. Electrochem. Soc. 151 (2004) A1517–A1529.
[389] Z. Chen, J.R. Dahn, J. Electrochem. Soc. 149 (2002) A1184–A1189.
[390] N. Penazzi, S. Bodoardo, R. Bongiovanni, C. Gerbaldi, G. Meligran, G. Mulas, J. Nair, Fuel Cells 09 (2009) 273–276.
[391] H. Liu, D.T. Liu, Dingguo Tang, Solid State Ion. 179 (2008) 1897–1901.
[392] Y. Wang, Y. Wang, E. Hosono, K. Wang, H. Zhou, Angew. Chem. Int. Ed. Engl. 47 (2008) 7461–7465.
[393] Y.H. Huang, J.B. Goodenough, Chem. Mater. 20 (2008) 7237–7241.
[394] J.L. Tirado, Mater. Sci. Eng. R 40 (2003) 103–136.
[395] C.M. Park, J.H. Kim, H. Kim, H.J. Sohn, Chem. Soc. Rev. 39 (2010) 3115–3141.
[396] I.V. Barsukov, M.A. Gallego, J.E. Doninger, J. Power Sour. 153 (2006) 288–299.
[397] U. Kasavajjula, Ch. Wang, A.J. Appleby, J. Power Sour. 163 (2007) 1003–1039.
[398] Y. Wang, J.Y. Lee, T.C. Deivaraj, J. Electrochem. Soc. 151 (2004) A1804–A1809.
[399] P. Pozot, et al., J. Power Sour. 97–98 (2001) 235–239.
[400] C. Kim, M. Noh, M. Choi, J. Cho, B. Park, Chem. Mater. 17 (2005) 3297–3301.
[401] I. Grigoriants, A. Soffer, G. Salitra, D. Aurbach, J. Power Sour. 146 (2005) 185–189.
[402] H. Kim, J. Cho, J. Electrochem. Soc. 154 (2007) A462–A466.
[403] T. Morita, N. Takami, J. Electrochem. Soc. 153 (2006) A425–A430.
[404] Y.S. Jung, K.T. Lee, S.M. Oh, Electrochim. Acta 52 (2007) 7061–7067.
[405] H. Groult, B. Kaplan, F. Lantelme, S. Komaba, N. Kumagai, H. Yashiro, T. Nakajima, B. Simon, A. Barhoun, Solid State Ion. 177 (2006) 869–875.
[406] T. Doi, Y. Iriyama, T. Abe, Z. Ogumi, Chem. Mater. 17 (2005) 1580–1582.
[407] N.J. Dudney, Electrochem. Soc. Interface 17 (2008) 44–48.
[408] C.L. Li, Z.W. Fu, Electrochim. Acta 53 (2008) 4293–4301.
[409] C.H. Park, M. Park, S.I. Yoo, S.K. Joo, J. Power Sour. 158 (2006) 1442–1446.
[410] Y. Zhao, Q. Zhou, L. Liu, J. Xu, M. Yan, Z. Jiang, Electrochim. Acta 51 (2006) 2639–2645.
[411] M. Jayalakshmi, K. Balasubramanian, Int. J. Electrochem. Sci. 3 (2008) 1196–1217.
[412] B.E. Conway, Electrochemical Supercapacitors: Scientific Fundamentals and Technological Applications, Klewer Academic, New York, 1999.
[413] C. Ashtiani, R. Wright, G. Hunt, J. Power Sour. 154 (2006) 561–566.
[414] C. Tallner, S. Lannetoft, Batteries or supercapacitors as energy storage in HEVs? (2005) University of Lund, MS Thesis.
[415] E. Frackowiak, F. Beguin, Carbon 39 (2001) 937–950.
[416] N.L. Wu, Mater. Chem. Phys. 75 (2002) 6–11.
[417] F. Beguin, E. Frackowiak, In: Y. Gogotsi (Ed.), Nanomaterials Handbook, CRC Press, Boca Raton, FL, 2006, pp. 713–737.

[418] J. Chmiola, G. Yushin, Y. Gogotsi, C. Portet, P. Simon, P.L. Taberna, Science 313 (2006) 1760–1763.

[419] B.E. Conway, V. Birss, J. Wojtowicz, J. Power Sour. 66 (1997) 1–14.

[420] W. Sugimoto, K. Yokoshima, Y. Murakami, Y. Takasu, Electrochim. Acta 52 (2006) 1742–1748.

[421] K. Machida, K. Furuuchi, M. Min, K. Naoi, Electrochemistry 72 (2004) 404.

[422] K. Yokoshima, T. Shibutani, M. Hirota, W. Sugimoto, Y. Murakami, Y. Takasu, J. Power Sour. 160 (2006) 1480–1486.

[423] A. Zolfaghari, F. Ataherian, M. Ghaemi, A. Gholami, Electrochim. Acta 52 (2007) 2806–2814.

[424] S.L. Kuo, N.L. Wu, J. Power Sour. 162 (2006) 1437–1443.

[425] A. T-Raissi, D.L. Block, IEEE Power Energy 2 (2004) 40–45.

[426] M. Momirlan, T. Vezirolu, Renew. Sustain. Energy Rev. 3 (1999) 108–120.

[427] J. Turner, G. Sverdrup, M.K. Mann, P.-C. Maness, B. Kroposki, M. Ghirardi, R.J. Evans, D. Blake, Int. J. Energy Res. 32 (2008) 379–407.

[428] P. Ferreira-Aparicio, M.J. Benito, Catal. Rev. 47 (2005) 491–588.

[429] http://qsinano.com/white_papers/WaterElectrolysisApril2007.pdf (accessed 04.01.07).

[430] K. Nath, D. Das, J. Sci. Ind. Res. 66 (2007) 701–709.

[431] R.M.N. Yerga, M. Consuelo, I.G. Valle, J.A. Villoria de la Mano, J.L.G. Fierro, Chem. Sus. Chem. 2 (2009) 471–485.

[432] M.T. Lee, M. Werhahn, D.J. Hwang, N. Hotz, R. Greif, D. Poulikakos, C.P. Grigoropoulos, Int. J. Hydrogen Energy 35 (2010) 118–126.

[433] S.H. Jensen, P.H. Larsen, M. Mogensen, Int. J. Hydrogen Energy 32 (2007) 3253–3257.

[434] A. Steifeld, Solar Energy 78 (2005) 603–615.

[435] T. Kodama, N. Gokon, Chem. Rev. 107 (2007) 4048–4077.

[436] A. Steinfeld, Int. J. Hydrogen Energy 27 (2002) 611–619.

[437] R.J. Weiss, H.C. Ly, K. Wegner, S.E. Pratsinis, A. Steinfeld, AIChE J. 51 (2005) 1966–1970.

[438] H.H. Funke, H. Diaz, X. Liang, C.S. Carney, A.W. Wimer, P. Li, Int. J. Hydrogen Energy 33 (2008) 1127–1134.

[439] H.H. Funkea, H. Diaza, X. Lianga, C.S. Carneya, A.W. Weimera, P. Li, Int. J. Hydrogen Energy 33 (2008) 1127–1134.

[440] Y.H. Ikuhara, H. Mori, T. Saito, Y. Iwamoto, J. Am. Ceram. Soc. 90 (2007) 546–552.

[441] D. Barreca, P. Fornasiero, A. Gasparotto, V. Gombac, C. Maccato, T. Montini, E. Tondello, Chem. Sus. Chem. 2 (2009) 230–233.

[442] H. Kaneko, N. Gokon, N. Hasegawa, Y. Tamaura, Energy 30 (2005) 2171–2178.

[443] M. Bououdina, D. Grant, G. Walker International, J. Hydrogen Energy 31 (2006) 177–182.

[444] E.MacA Gray, Adv. Appl. Ceram. 106 (2007) 25–28.

[445] A. Sayead, D.M. Antonelli, Adv. Mater. 16 (2004) 765–777.

[446] A.J. Williamson, F.A. Reboredo, Giulia Galli, Appl. Phys. Lett. 85 (2004) 2917–2919.

[447] H.H. Funke, H. Diaz, X. Liang, C.S. Carney, A.W. Weimer, Int. J. Hydrogen Energy 33 (2008) 1127–1134.

[448] M. Conte, P.P. Prosini, S. Passerini, Mater. Sci. Eng. B108 (2004) 2–8.

[449] L. Schlapbach, A. Zuttel, Nature 414 (2001) 353–358.

[450] C. Nutzenadel, A. Zuttel, D. Chartouni, G. Schmid, L. Schlapbach, Eur. Phys. J. D 8 (2000) 245–250.

[451] E. David, J. Mater. Process. Technol. 162–163 (2005) 169–177.

[452] A. Züttel, P. Wenger, P. Sudan, P. Maurona, S.-I. Orimo, Mater. Sci. Eng. B108 (2004) 9–18.
[453] S.G. Chalk, J.F. Miller, J. Power Sour. 159 (2006) 73–80.
[454] http://www.gkss.de/Themen/W/WTP/wasserstoff/eSpeicher.html.
[455] S.S. Mao, X. Chen, Int. J. Energy Res. 31 (2007) 619–636.
[456] http://www.h2net.org.uk/Events/Miscellaneous_2004_Storage/Programme_after.html.
[457] H.-S. Kim, H. Lee, K.-S. Han, J.-H. Kim, M.-S. Song, M.-S. Park, J.-Y. Lee, J.-K. Kang, J. Phys. Chem. B 19 (2005) 8983–8986.
[458] H. Shao, H. Xu, Y. Wang, X. Li, Nanotechnology 15 (2004) 269–274.
[459] B. Bogdanovic, M. Felderhoff, S. Kaskel, A. Pommerin, K. Schlichte, F. Schüth, Adv. Mater. 15 (2003) 1012–1015.
[460] C. Sachs, A. Pundt, R. Kirchheim, M. Winter, M.T. Reetz, D. Fritsch, Phys. Rev. B 64 (2001) 075408–075417.
[461] P.D. Chalmer, The Future of Finishing, NCMS, 2008.
[462] K.O. Legg, M. Graham, P. Chang, F. Rastagar, A. Gonzales, B. Sartwell, Surf. Coat.Technol. 81 (1996) 99–105.
[463] V.S. Saji, J. Thomas, Curr. Sci. 92 (2007) 51–56.
[464] A. Hovestad, L.J.J. Janssen, J. Appl. Electrochem. 25 (1995) 519–527.
[465] A. Ammala, A.J. Hill, P. Meakin, S.J. Pas, T.W. Turney, J. Nanopart. Res. 4 (2002) 167–174.
[466] http://www.aerogel.com/features/termal.html.
[467] http://eande.lbl.gov/ecs/aerogels/satoc.htm.
[468] M. Reim, A. Beck, W. Korner, R. Petricevic, M. Glora, M. Weth, T. Schliermann, J. Fricke, C. Schmidt, F.J. Potter, Solar Energy 72 (2002) 21–29.
[469] Y. Akimov, Instrum. Experiment. Tech. 46 (2003) 287–299.
[470] J.L. Mohanan, I.U. Arachchige, S.L. Brock, Science 307 (2005) 397–400.
[471] C.G. Granqvist, Handbook of Inorganic Electrochromic Materials, Elesevier, Amsterdam, 1995 (Chapter 25).
[472] S. Svendson, J. Non-Cryst. Solids 145 (1992) 240–243.
[473] C.M. Lampert, Solar Energy Mater. 1 (1979) 319–341.
[474] W.F. Bogaerts, C.M. Lampert, J. Mater. Sci. 18 (1983) 2847–2875.
[475] O.P. Agnihotri, B.K. Gupta, Solar Selective Surfaces, Wiley, New York, 1981.
[476] M. Maaza, O. Nemraoui, C. Sella, J. Lafait, A. Gibaud, B. Baruch-Barak, A.C. Beye, Solid State Commun. 137 (2006) 166–170 and reference therein.
[477] G.B. Smith, C.A. Deller, P.D. Swift, A. Gentle, P.D. Garrett, W.K. Fisher, J. Nanopart. Res. 4 (2002) 157–165.
[478] S. Zhao, E. Wackelgard, Solar Energy Mater. Solar Cells 90 (2006) 243–261.
[479] B. Carlsson, K. MoKller, U. Frei, S. Brunold, M. Kohl, Solar Energy Mater. Solar Cells 61 (2000) 223–238.
[480] C.A. Arancibia-Bulnes, C.A. Estrada, J.C. Ruiz-Suarez, J. Phys. D Appl. Phys. 33 (2000) 2489–2496.
[481] K.S. Chou, Y.C. Lu, Thin Solid Films 515 (2007) 7217–7221.
[482] S. Papaefthimiou, E. Syrrakou, P. Yianoulis, Thin Solid Films 502 (2006) 257–264.
[483] G.M. Lampert, Mater. Today 7 (2004) 28–35.
[484] J. Livage, D. Ganguli, Solar Energy Mater. Solar Cells 68 (2001) 365–381.
[485] A. Georg, A. Georg, U.O. Krasovec, Thin Solid Films 502 (2006) 246–251.
[486] U. Felde, M. Haase, H. Weller, J. Phys. Chem. B 104 (2000) 9388–9395.

[487] J.P. Coleman, A.T. Lynch, P. Madhukar, J.H. Wagenknecht, Solar Energy Mater. Solar Cells 56 (1999) 375–394.
[488] W. Cai, M. Tan, G. Wang, L. Zhang, Appl. Phys. Lett. 69 (1996) 2981–2983.
[489] S.H. Lee, R. Deshpande, P.A. Parilla, K.M. Jones, B. To, A.H. Mahan, A.C. Dillon, Adv. Mater. 18 (2006) 763–766.
[490] T. Tatsuma, K. Suzuki, Electrochem. Commun. 9 (2007) 574–576.
[491] Basic Research Needs for Solid State Lighting, Report of the Basic Energy Sciences Workshop On Solid-State Lighting, This report is available on the web at www.sc.doe.gov/bes/reports/files/SSL_rpt.pdf.
[492] R.V. Steele, Nat. Photon. 1 (2007) 25–26.
[493] V.L. Colvin, M.C. Schlamp, A.P. Alivisatos, Nature 370 (1994) 354–357.
[494] J.M. Caruge, J.E. Halpert, V. Wood, V. Bulović, M.G. Bawendi, Nat. Photon. 2 (2008) 247–250.
[495] L.A. Kim, P.O. Anikeeva, S.A. Coe-Sullivan, J.S. Steckel, M.G. Bawendi, V. Bulovic, Nano Lett. 8 (2009) 4513–4517.
[496] Q. Sun, Y.A. Wang, L.S. Li, D. Wang, T. Zhu, J. Xu, C. Yang, Y. Li, Nat. Photon. 1 (2007) 717–722.
[497] Y. Li, A. Rizzo, R. Cingolani, G. Gigli, Microchim. Acta 159 (2007) 207–215.
[498] D.Y. Godovsky, Adv. Polym. Sci. 153 (2000) 163–205.
[499] A.H. Mueller, M.A. Petruska, M. Achermann, D.J. Werder, E.A. Akhadov, D.D. Koleske, M.A. Hoffbauer, V.I. Klimov, Nano Lett. 5 (2005) 1039–1044.
[500] M. Godlewski, E. Wolska, S. Yatsunenko, A. Opalińska, J. Fidelus, W. Łojkowski, M. Zalewska, A. Kłonkowski, D. Kuritsyn, Low Temp. Phys. 35 (2009) 48–52.
[501] S. Pillai, K.R. Catchpole, T. Trupke, G. Zhang, J. Zhao, M.A. Green, Appl. Phys. Lett. 88 (2006) 161102–161104.
[502] F. Feil, W. Furbeth, M. Schutze, Surf. Eng. 24 (2008) 198–203.
[503] C.C. Koch, Nanostruct. Mater. 2 (1993) 109–129.
[504] Z. Zeng, K. Natesan, Z. Cai, S.B. Darling, Nat. Mater. 7 (2008) 641–646.
[505] S.C. Tjong, Adv. Eng. Mater. 9 (2007) 639–652.
[506] D. Brondel, R. Ewards, A. Hayman, D. Hill, S. Mehta, T. Semerad, Oilfield Rev. (1994) 4–18.
[507] N. Voevodin, V.N. Balbyshev, M. Khobaib, M.S. Donley, Prog. Org. Coat. 47 (2003) 416–423.
[508] J. Cho, M.S. Joshi, C.T. Sun, Compos. Sci. Technol. 66 (2006) 1941–1952.
[509] B. Wetzel, P. Rosso, F. Haupert, K. Friedrich, Eng. Fract. Mech. 73 (2006) 2375–2398.
[510] J.M. Garces, D.J. Moll, J. Bicerano, R. Fibiger, D.G. McLeod, Adv. Mater. 12 (2000) 1835–1839.
[511] J. Lee, S. Mahendra, P.J.J. Alvarez, ACS Nano 4 (2010) 3580–3590.
[512] B.C. Tappan, S.A. Steiner III, E.P. Luther, Angew. Chem. Int. Ed. Engl. 49 (2010) 4544–4565.
[513] S. Ukai, T. Nishida, T. Okuda, T. Yoshitake, J. Nucl. Mater. 258 (263) (1998) 1745–1749.
[514] M. Taneike, F. Abe, K. Sawada, Nature 424 (2003) 294–296.
[515] H. Hahn, R.S. Averback, J. Am. Ceram. Soc. 111 (1991) 2918–2921.
[516] N. Malanowski, T. Heimer, W. Luther, M. Werner (Eds.), Woley-VCH, Weinheim, 2006, p. 44.
[517] V.N. Bakunin, A.Yu. Suslov, G.N. Kuzmina, O.P. Parenago, J. Nanopart. Res. 6 (2004) 273–284.
[518] L. Rapoport, N. Fleischer, R. Tenne, J. Mater. Chem. 15 (2005) 1782–1788.

[519] S.R. Taylor, K. Sieradzki, Prog. Org. Coat. 47 (2003) 169–173.
[520] V. Arul Mozhi Selvan, R.B. Anand, M. Udayakumar, ARPN J. Eng. Appl. Sci. 4 (2009) 1–6.
[521] A. Dufresne, Molecules 15 (2010) 4111–4128.
[522] R.H. Wijffels, M.J. Barbosa, Science (2010) 796–799.
[523] J.S. Valenstein, K. Kandel, F. Melcher, I.I. Slowing, V.S.-Y. Lin, B.G. Trewyn, ACS Appl. Mater. Interfaces 4 (2012) 1003–1009.
[524] S. Torkamani, S.N. Wani, Y.J. Tang, R. Sureshkumar, Appl. Phys. Lett. 97 (2010) 43703–43706.
[525] A. Meier, A. Steinfeld, Adv. Sci. Technol. 74 (2010) 303–312.
[526] M. Florescu, H. Lee, I. Puscasu, M. Pralle, L. Florescua, D.Z. Ting, J.P. Dowling, Solar Energy Mater. Solar Cells 91 (2007) 1599–1610.
[527] T.M. Tritt, H. Bottner, L. Chen, MRS Bull. 33 (2008) 366–368.
[528] C.J. Vineis, A. Shakouri, A. Majumdar, Mercouri G. Kanatzidis, Adv. Mater. 22 (2010) 3970–3980.
[529] X.B. Zhao, T.J. Zhu, X.H. Ji, in: Thermoelectrics Handbook: Macro to Nano, D.M. Rowe (Ed.), Taylor & Francis Group, LLC, 2006, 25-1.
[530] M.S. Dresselhaus, G. Chen, M.Y. Tang, R. Yang, H. Lee, D. Wang, Z. Ren, J.-P. Fleurial, P. Gogna, Adv. Mater. 19 (2007) 1043–1053.
[531] L.D. Hicks, T.C. Harman, X. Sun, M.S. Dresselhaus, Phys. Rev. B Condens. Matter Mater. Phys. 53 (1996) R10493–R10496.
[532] R. Yang, G. Chen, Mater. Integr. 18 (2005) 31–36.
[533] X. Ji, B. Zhang, T.M. Tritt, J.W. Kolis, A. Kumbhar, J. Electron. Mater. 36 (2007) 721–726.
[534] J.D. Ferguson, K.J. Buechler, A.W. Weimer, S.M. George, Powder Technol. 156 (2005) 154–163.
[535] C.E. Aumann, G.L. Skofronick, J.A. Martin, J. Vac. Sci. Technol. B 13 (1995) 1178–1183.
[536] C. Rossi, K. Zhang, D. Estève, P. Alphonse, P. Tailhades, C. Vahlas, J. Microelectrochem. Syst. 16 (2007) 919–931.
[537] H.J. Hwang, Y.C. Ahn, H.S. Shin, C.G. Lee, G.T. Kim, H.S. Park, J.J. Lee, Curr. Appl. Phys. 6 (2006) 1068–1071.
[538] C. Sus, in: Developments and Applications of Non-Newtonian Flows. ASME, FED 231/MD, vol. 66, 1995, pp. 99–103.
[539] L. Godson, R.D. Mohan Lal, S. Wongwises, Renew. Sustain. Energy Rev. 14 (2010) 629–641.
[540] X.Q. Wang, A.S. Mujumdar, Int. J. Therm. Sci. 46 (2007) 1–19.
[541] Nuclear Power International March. http://pepei.pennnet.com/display_article/355043/140/ARTCL/none/none/1/Rusty-Nanoparticles-Learn-New-Trick/, 2009 (accessed 16.03.09).
[542] L. Popa-Simil, I.L. Popa-simil, NSTI-Nanotech 4 (2007) 679–682.
[543] A. Shah, R. Fishwick, J. Wood, G. Leeke, S. Rigby, M. Greaves, Energy Environ. Sci. 3 (2010) 700–714.
[544] N.N. Nassar, M. Husein, P. Pereira-Almao, Fuel Process. Technol. 91 (2010) 169–174.
[545] R. Hashemi, N. Nassar, P. Pereira-Almao, Energy Fuel 26 (2012) 1645–1655.
[546] F. Bensebaa, L. Kotlyar, G. Pleizier, B. Sparks, Y. Deslandes, K. Chung, Surf. Interface Anal. 30 (2000) 207–211.
[547] F. Bensebaa, L. Kotlyar, G. Pleizier, B. Sparks, Y. Deslandes, K. Chung, Can. J. Chem. Eng. 78 (2000) 610–616.
[548] F. Bensebaa, A. Majid, Y. Deslandes, Spectrochim. Acta A 57 (2001) 2695–2702.

Chapter 6

Biomedical

1 INTRODUCTION

The biological and medical applications of nanoparticles (NPs) are well documented [1–5]. These applications could be grouped in three areas: (a) diagnostic and imaging; (b) drug delivery; and (c) tumor treatment. Successful precommercial applications have been reported already in biosensing [6] and drug delivery [7,8]. Hundreds of pharmaceutical products based on nanomaterials including NPs have already been approved for clinical use [9]. Liposomes, superparamagnetic, gold, polymeric, and semiconductor NPs are being used for drug delivery, imaging, therapeutics, and diagnostics [8]. Doxil, Ferumoxitol, Aurimune, Aurothol, MRX-951, and QDot-800 are a few examples of NP-based commercial products used in drug delivery [8]. Size, shape, and bulk and surface composition are critical parameters in defining the properties of these NPs and their applications.

Currently, NPs are mostly used for detecting and curing diseases. However, their characteristics and functionalities could also be used in furthering our understanding of how cells work. For example, NPs could be used as a tool for a quantitative understanding of how biological systems work at the molecular level. Typically, a functionalized NP is tagged by an antibody through a biotin group. Detection occurs through interaction with a receptor group of a cell or bioanalyte. Whether in a single particle or in a multiparticle embedded in a larger bead [10], NPs are (and could be) used in several *in vivo* (drug delivery, imaging, engineering) and *ex vivo* (diagnostic devices, biomineralization, cell activation) biomedical applications [10]. The protective layer is important to control size distribution and biocompatibility without negatively altering the physical properties of the core material. Numerous functional groups are linked to the particle via a molecule (linker). Its composition and structure will depend on the specific application. A single bead containing several NPs gives rise to a multitude of functions in parallel.

Different models are used to study the biological effects of NPs [10]. They include the whole animal and isolated organ model and cells *in vitro*. Extrapolation of these studies to humans requires *in vivo* and *in vitro* studies on

Interface Science and Technology, Vol. 19. http://dx.doi.org/10.1016/B978-0-12-369550-5.00006-9

human tissues and cells. Biotechnology, which had a significant impact on medicine and agriculture, is expected to complement nanoparticle (NP) technology. The merging of nanoscience and NP technology on one hand and biotechnology on the other has led to new flourishing and promising R&D programs and applications. This new field is commonly referred to as nanobiotechnology [11]. It will certainly have a significant impact on developing efficient health-related products.

Three NP-based systems have been widely studied for their biological applications. Gold and gadolinium NPs are used as contrast agents in histopathology tests using scanning electron and light imaging. Iron oxides are used in drug delivery and imaging, mostly because of their magnetic properties. The third group of semiconductor QDs are used in biosensing because of their multicolor emission.

A biosensor binds an analyte through a biomolecule or bioreceptor (often an enzyme or an antibody) giving rise to a physical signal. This signal is often a current or a voltage. An optical impedance system is used as the transducer mechanism modulated by the binding event. The transducer is the conversion mechanism reporting the binding event as a measurable signal. Often biosensors are classified according to the transducer system, which includes mechanical, thermal optical, and electronic [12,13]. Acoustic waves, quartz micro-balance, cantilever, and piezoelectric- are used as mechanical transducers in various sensing setups [13]. Conduction, resistance, and capacitance are used in electrical transducers [13], while voltage, current, and impedance (potentiometry, conductometry, and amperometry) are used in electrochemical transducers [13]. Frequency and intensity (optical absorption, optical emission, and fiber optic evanescent waves) are used in radiant transducers as key detection parameters [13]. Heat flow and heat content (pyroelectric, catalytic, calorimetric) are used in thermal transducers [13]. Field strength and paramagnetic properties are used in magnetic transducers [13].

Biosensing encompasses a vast application area including point-of-care (POC) diagnostic tools, immunoassay, genomics, food processing, and safety. Sensing and imaging of diseases at an early stage are important to reduce mortality. In the case of cancer, early detection can improve the survival rate by a factor of 5 or more. Widely used preventive screenings do not provide precise diagnosis. NP-based biosensing and imaging provide numerous advantages when compared to traditional techniques, because of their unique physical properties [14]. In particular, they provide an improved targeted delivery of therapeutic agents. Besides their unique electronic, optical, and magnetic properties, their tunable size in the submicron regime allows practitioners to directly target a cell site at the molecular level.

Amperometric glucose sensors based on the enzymatic catalyst and its variants dominate the biosensor market [15]. However, the low sensitivity and specificity of these sensors limit the range of their applications. Needs and challenges in biosensing have spawned a boom in the biosensing field.

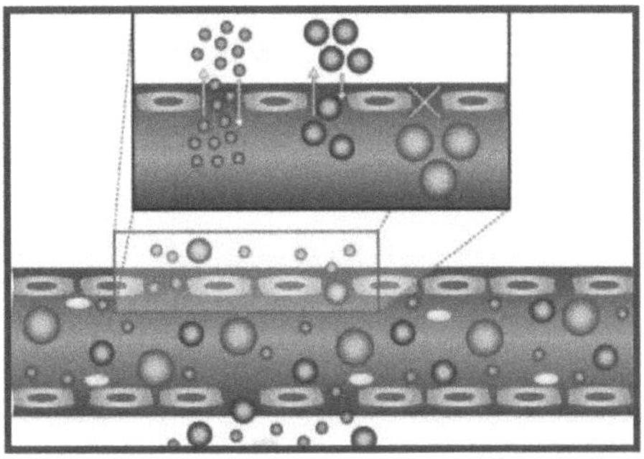

FIGURE 6.1 Delivery efficiency of nanoparticles as a function of size. Nanoparticles with different sizes are shown in different colors. *Reprinted with permission from Farokhzad and Langer [7]. Copyright 2005, American Chemical Society.* (For color version of this figure, the reader is referred to the online version of this chapter.)

Depending on the transducing mechanism, at least five categories of biosensors can be distinguished. The transducing mechanism relies on electrochemical, optical, thermal, mechanical, or electrical properties. In the optical category, three different transducing mechanisms can be distinguished. Biosensing concepts incorporating NPs have been demonstrated for each of the seven transducing mechanisms.

Size and shape have a direct impact on particle uptake (Fig. 6.1). This could be a good attribute in the case of drug delivery [7]. But it could have a negative impact in the case of toxic NPs. Indeed, the movement of NPs in and out of the vasculature is size-dependent. The shape affects the margination of the NPs to the vessel wall (Fig. 6.1).

Emulsion NPs and liposomes provide advantages in carrying hydrophilic substances and lipophilic agents, respectively. NP-based intravenous nutritional fat emulsion and liposomal products have been developed and marketed [16]. NPs may facilitate the delivery of difficult-to-formulate pharmaceutical reagents. This is particularly the case of poorly water-soluble compounds. NP-based drugs may improve bioavailability and drug exposure. For example, when NPs are used, maximum blood concentration is reached after a few hours. In the case of conventional formulation, the mean blood concentration is an order of magnitude lower even several hours after delivery [16].

Iron oxide NPs are already being used in some medical applications [17]. There are numerous preclinical and clinical examples of bioconjugated NPs. These applications take advantage of the enhanced and/or specific properties of NPs. Surface modification and bioconjugation are often used to enhance the performance of these NPs in surface plasmon resonance (SPR), luminescence, and surface-enhanced Raman spectroscopy (SERS) [1]. In most biological applications, NPs are modified with antibodies, DNA, and enzymes, using specific bioconjuguation processes [11,18].

Given the fact that the majority of the synthesis methods are undertaken in organic solvents, additional processing steps are required before the products can be used in biological media. NPs should be transferred into aqueous solution without losing their intrinsic optical properties. Transferring NPs from organic solvent (usually used during synthesis or posttreatment) into aqueous solution is important for most biological applications [4,19]. This step is particularly challenging because it may lead to agglomeration and/or surface defects. These changes will directly affect the optical properties. Several techniques have been developed to address this issue [20]. Functionalization and core–shell methodologies provide high-quality QDs dispersed in aqueous solution.

Early cancer diagnosis and treatment are critical to reducing its negative impact in terms of treatment severity and death toll. In 2001, 10.9 million new cases, 6.7 million deaths, and 24.6 million persons alive with cancer were reported [21]. NPs may improve the sensitivity and precision of cancer diagnosis. Furthermore, NPs could also provide more efficient and targeted cancer treatments. For example, Combidex, based on superparamagnetic iron oxide NPs, is used for tumor imaging [22]. Paclitaxel-albumin NP is used in lung and breast cancer treatment [22]. More rigorous testing and trial are required for health-related applications. In particular, side effects related to using these NPs should be investigated. It is worth noting that two NP (pactilaxel-albumin and gold)-based solutions are already in the market. Other products have also entered (or are in the process of entering) the market.

NPs can be used in passive and active applications. Passive applications include utilization of contrasting agents. Active applications include enzyme assays not limited to a passive role. NPs could mediate biological events, including cell function control [23]. Indeed, the activity of an enzyme elicits specific responses from a NP-based structure. This gives rise to a signal related to the enzyme activity. Current enzyme-linked immunological assays (ELISA) are limited by the requirement of producing high-quality antibodies, multiple washing steps, and low capacity for multiplexed analyte detection [24]. Homogeneous NP dispersion (e.g., DNA-structured NPs) assays allow highly sensitive analyte detection without the need for additional separation and purification [25]. These new assays are potentially cost effective, more robust, and have a wider scope for detection. When compared to conventional radioscopic techniques, NP-based assays provide more flexibility [24,25]. For example, they are accessible with routine optical methods with a single step reaction. However, more stable and robust NPs are required.

Plasma membrane is considered the portal entry into cells. It also allows segregation between the intracellular medium (cytoplasm) and the extracellular environment. Macromolecules and particles are internalized within cells through phagocytosis and pinocytosis [26]. NP internalization occurs mostly through clathrin-mediated and caveolae-mediated endocytosis. Different endocytic pathways related to the size of the endocytic vesicle, the nature of the cargo (ligands, receptors, and lipids), and the mechanism of vesicle formation have been reported [26]. Interaction of NPs with cells and lipid bilayers is mostly dictated by the surface properties of NPs [27].

In this chapter, biomedical applications of NPs, mostly in sensing and treatment, are reviewed. Specific examples are also provided in some key application areas.

2 OPTICAL BIOSENSING

The biosensor market has grown rapidly in recent years, reaching an estimated turnover of 11 billion dollars in 2007 [28]. Existing commercial biosensors have shown some limitations for some applications. The main disadvantages of the current biosensors include sensitivity, specificity, multiplexing, miniaturization, *in situ*, portability, and robustness.

Fluorophore labeling, colorimetric detection, fluorescence resonance energy transfer (FRET), SPR analysis, and scattering-based sensing are potential sensitive and specific NP-based optical sensing systems. In a typical optical biosensor, an analyte is immobilized using a biofunctional element followed by a signal detection using a transduction mechanism. Three common recognition mechanisms are used in biosensing. They include enzymatic reaction, virus-antibody, and DNA hybridization. Each recognition mechanism requires specific surface preparations to provide sensitivity and selectivity. The binding affinity in the antibody–antigen recognition mechanism depends on its surface structure. The base-pair sequence in the DNA-based sensor is critical to reach high specificity. The enzyme-catalyzed reaction is key to the transduction mechanism of the third biochemical recognition mechanism. In the case of the amperometric transducer, an enzymatic glucose oxidase catalyzes the glucose reaction in the presence of oxygen. This constitutes the basis of the majority of biosensors being marketed currently.

The substrate supporting the biofunctional elements could have a bulk or nanometric structure. In the case of NPs, the effective surface area could help increase the sensitivity of the biosensor. The optoelectronic properties of these NPs also provide a few advantages over bulk substrates in terms of multiplexing and flexibility. Numerous biosensor designs, attempting to take advantage of NPs, have been reported in the literature [29–31]. To replace currently used fluorescent proteins and organic/metallorganic dye molecules, four main groups of nanoscale materials are used in fluorescence-based biosensing [32]. This includes semiconductor, metallic, polymer/dye NPs and silica/dye hybrid nanocomposites [32].

Some bulk inorganic materials are endowed with special optical properties [33]. Their corresponding nanostructured materials provide the possibility to tune these optical properties [34]. Compared to organic dyes and fluorescent proteins (Fig. 6.3), semiconductor and metallic quantum dots are brighter over a larger multicolor emission range using a single light source, and, even more importantly, are more stable against photobleaching [2]. The main fluorescent materials have different sizes, an important factor in some applications. Organic dyes and dyed polymer particles are about 1–10 nm in size [32]. On the other hand, QD and green fluorescent protein are about 3–5 nm. QD systems covering the optical spectrum ranging from UV to IR regions have been integrated into various optical sensors [18]. The material composition of these optical sensors includes ZnS, ZnSe, CdSe, CdTe, and PbS. These materials cover the UV, blue, green, yellow/orange, and red regions of the optical spectra.

NP-based applications in the health-care sector face numerous challenges. The first challenge is the toxicity potential of these NPs. This issue is discussed in more detail in Chapter 8. The second challenge is the ability to reproducibly control the composition and properties of the NPs. For example, utilization of QD for quantitative biosensing is complicated by the difficulty in controlling surface states and properties [36]. These surface states give rise to fluorescence instabilities. These intermittent optical signals are either rationally exploited or minimized [36].

2.1 QD Fluorophore and Labeling

Submicron particles are highly fluorescent and could be used as tracer labels in clinical and biological applications [37,38]. Semiconductor [39] and silica [32] NPs have been used for fluorescence analysis. As shown in Fig. 6.2, two strategies can be used for QD-based fluorescence bioanalysis. The first consists of a functionalized QD attached to an antibody-modified sensing surface. In the second strategy, a nucleic acid-functionalized QD is attached to the sensing surface. In both cases, a fluorescence signal is obtained. These two QD fluorescence techniques provide high delivery throughput and efficiency with minimal biological effects [41].

Cell fixation and permeabilization provide high delivery throughput and efficiency. Furthermore, specific delivery and subcellular targeting are achieved. Physical (microinjection, electroporation) and chemical (lipofectamine, TAT peptide) delivery techniques have been developed [41]. These subcellular delivery and labeling methods often do not match the delivery characteristics of cell fixation. Several subcellular delivery and labeling methods based on fluorescent QD probes have been reported.

Gold NP labeling can be used as a blueprint for cellular migration pathway [42]. This is quite challenging, given the difference in dimension and scale between NP (ca. 10 nm) and cell (ca. 10 μm). Following injection into the living cells of the bloodstream, conjugated NPs bind to specific receptor structures (immunostaining). This specific binding is guided by molecular recognition.

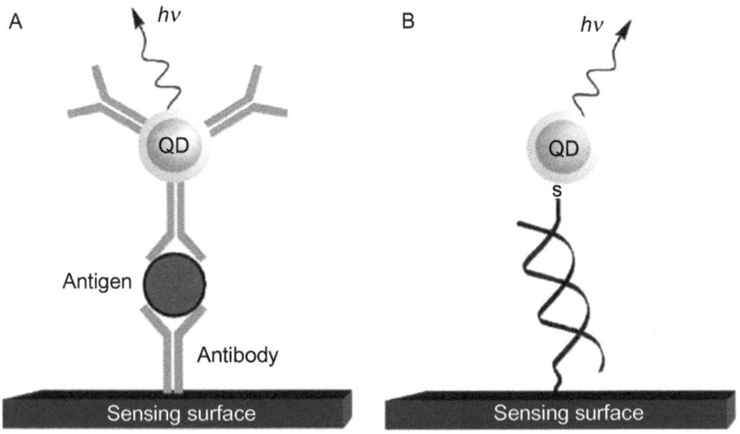

FIGURE 6.2 Schematic representation for two fluorescent analysis scenarios: (A) antibody-functionalized QDs and (B) nucleic acid-functionalized QD [40].

The surface of the nucleus containing these specific receptors becomes stained. Visualization of the animal organ locally enriched with gold particles using X-ray tomography allows tracking the cellular migration along the surface. When cells are cultured on top of the surface, they ingest the underlying NPs. Thus, the cells take all NPs along their pathway, leaving behind an area free of NPs. X-ray tomography imaging can be used to follow the migration pathway of NP cells [42].

QD-based fluorophore systems have all the attributes to overcome conventional fluorescent labeling shortcomings. The optical properties of QDs provide better sensitivity, selectivity, and multiplexing capacity, difficult to achieve with conventional molecular fluorophores. Compared to organic dyes, QDs provide high quantum yield, high molar extinction coefficients, broad and tunable absorption with narrow and symmetric photoluminescence (PL) spectra, and large effective Stokes shifts. Furthermore, high resistance to photobleaching and resistance to photochemical degradation are achieved [18,43,44].

Various capping strategies are used to address the requirements of different applications [18]. This aspect has been broadly reviewed in Chapter 1. In the case of biological applications, functionalization allowing aqueous dispersion and preservation of luminescence properties and secondary covalent modification of the surface with biomolecules are required. Thiolated or dithiolated functional monolayers, glutathione layer, cysteine-terminated peptide, thiolated siloxane, carboxylic acid-functionalizatione, a phospholipid encapsulating layer, and a diblock copolymer functionalization have been reported [40].

QDs are also used to simultaneously detect multiple species in solid tumors, which is not possible with conventional fluorophores [45]. This requires using nontoxic bioconjugated NPs. Water solubility of QD is critical in most biomedical applications [46]. Ligand exchange of TOPO stabilized

QD and encapsulation are the two main approaches to generate water solubility. Different water-soluble layer compositions have been obtained via ligand exchange. Encapsulation using bifunctional layers allows secondary covalent modification of the surface by biomolecules.

Metallic NPs with specific functionalization are also used for cancer detection. DNA-functionalized gold NPs are relatively less toxic and are currently available in the market for cancer diagnosis. QDs can be used in both *in vitro* and *in vivo* cancer diagnosis [21]. Tunable wavelength QDs provide the possibility of multiple biomarkers, screening, and detection. Several brands (Qdot®, PLxBeads®, and BioPixels®) are in the process of commercialization. However, their clinical success is impeded by cytotoxicity issues related to inorganic QDs [21].

Bright and stable near-infrared fluorescence using InAS QD containing two shell layers has been recently reported [47]. The intermediary layer in the InAs/CdSe/ZnSe structure lowers the strain between the InAs core and the top ZnSe shell. These QDs are used in *in vivo* imaging [48]. InAs–ZnSe core–shell NPs are also used for *in vivo* imaging [49]. Random QD blinking (on/off signal), particularly those prepared by wet chemistry, is hindering their application. Blending the boundary between a core of cadmium, selenium, and zinc, and a shell composed of the latter two elements [50] allowed suppression of this blinking. However, the current methodology used to obtain these modified QDs is expensive and hardly scalable.

2.2 Colorimetry

Metallic NPs are used as label-free colorimetric biosensors [51]. Mercaptoalkyloligonucleotide-modified gold NPs have been used as colorimetric probe [52]. Oligonucleotide concentration down to 10 fM has been achieved [52]. A polymeric network of NPs embedded with a red-to-pinkish/purple color change allowed signaling hybridization. Signal intensity depends on the interparticle distance in the generated aggregates. Other groups have reported similar findings [53,54]. It is possible to distinguish between single-stranded DNA and double-stranded DNA using color-induced changes from gold NPs [55]. The difference in agglomeration state leads to color change. Silica-coated gold NPs are also used as bioprobe for calorimetric DNA and protein detection [56].

2.3 Fluorescence Resonance Energy Transfer

FRET technique is based on nonradiative excitation energy transfer between two fluorophores. Luminescent NPs can be used as donor and acceptor luminophores. Tunability of QD allows optimal spectral overlap between donor PL and acceptor absorption, a critical criterion in FRET. Various fluorescence probing approaches are used [57]. QD fluorescence properties can be used directly or indirectly to build high sensitivity sensors [57]. Indeed, QD fluorescence is quenched upon binding of an analyte to the QD surface, or QD containing an organic fluorophore forms a donor–acceptor system giving rise to a FRET. The presence of an analyte displacing the fluorophore will reduce the FRET signal. QD with cell receptors can be used as a label. Yet another approach is based on cells containing QD in which PL quenching upon analyte binding is a simple but nonspecific sensing. A last approach is related to QD-based FRET (QD-FRET) design using modified QD with an organic fluorophore forming a donor–acceptor system. These QD-FRET could be used in cell labeling and probing [57–60].

PL quenching using nonorganic dyes is of interest, particularly metal NPs with a stable high extinction coefficient and a broad absorption spectrum. Oh et al. demonstrated a FRET efficiency modulation between commercial QDs and gold NPs in the presence of molecules inhibiting their interaction [61]. Avidin easily binds to biotin-modified gold NP and thus modulate the FRET signal. This technique could be used for quantitative analysis and high-throughput screening of molecules, which inhibit the specific interaction between biomolecules.

A FRET-based protease biosensor has been demonstrated [35]. First, CdSe/ZnS QDs with TOPO capping ligands are modified with a tetra peptide RGDC (R, arginine; G, glycine; R, aspartic acid; C, cysteine) molecule. Second, the peptide molecules are labeled with a rhodamine acting as acceptor. An enzymatic cleavage of the rhodamine (acceptor) suppresses the energy transfer channel [35]. Addition of increased concentration of collagenase to a rhodamine-labeled peptide-coated QDs solution increases fluorescence intensity (Fig. 6.3). The emission peak intensity of the QDs at 545 nm (donor—F_d) and the rhodamine molecules at 590 nm (acceptor—F_a) vary in opposite directions. This indicates that energy transfer between the two fluorophores may have occurred. The ratio F_d/F_a seems to increase with time for all collagenase concentrations. In the case of 5 μg/mL collagenase concentration (Fig. 6.3B), the enzymatic reaction was completed in 750 s.

2.4 Surface Plasmon Resonance Detection

SPR attributes are used to detect non-destructively the chemicals adsorbed on metallic film. Tunable SPR properties based on NPs are presently used to provide even more advantages. These NPs could be used in solution or precisely ordered and attached to a solid substrate. In the second case, these ordered metallic nanostructures give rise to localized surface plasmon resonance (LSPR). SPR and LSPR involve propagating and localized surface plasmon mode, respectively [62]. Surface plasmon resonance detection (SPRD) is used with both SPR and LSPR configurations.

Gold and other metallic nanostructures are endowed with surface plasmon properties. This property is used in affinity biosensing, where a receptor

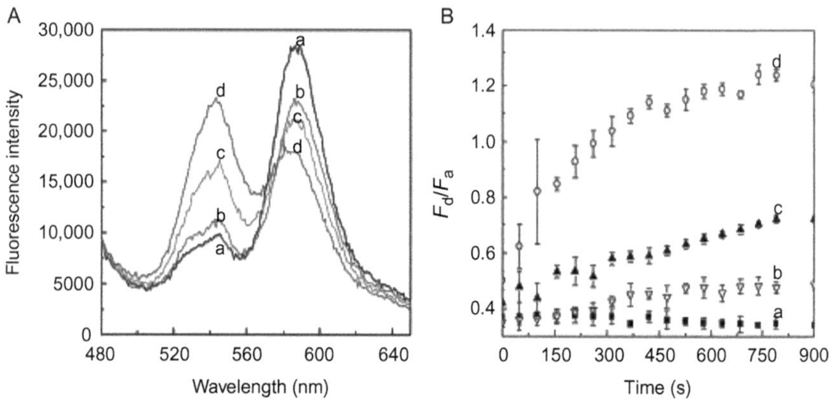

FIGURE 6.3 (A) Variation in the emission spectra of rhodamine-labeled peptide-coated QD as a function of added collagenase concentration. (B) Time dependence of the ratio F_d/F_a for different collagenase concentration. This time dependence is provided for four concentrations (bottom to top): 0 (black), 0.5 L (red), 2.5 μg/mL (green), and 5.0 μg/mL (blue). *Reprinted with permission from Shi et al. [35]. Copyright 2006, American Chemical Society.* (For interpretation of the references to color in this figure legend, the reader is referred to the online version of this chapter.)

molecule is attached to the NP with a specific affinity to a biomolecule analyte [63]. The increase in the thickness of the dielectric layer around the core NP leads to a change in the peak features (intensity and peak position) of even the extinction spectra. SPRD is very sensitive to even an average thickness change of less than 1 nm. Gold NPs functionalized with two noncomplementary oligonucleotides do not recognize each other if they are mixed. However, if they are hybridized with a complementary strand that recognizes both sequences, mixed oligonucleotide-modified 13-nm gold NPs will aggregate. Aggregation will lead to red shift in the plasmon resonance peak [64]. This process is quite reversible upon heating the solution to around 80 °C. Indeed, a reversible, thermally induced DNA dissociation has been observed.

There are numerous NP-based SPRD designs [65]. Differences in detection systems are related to the experimental layouts. Evanescent mode, front dark-field illumination, dark-field back illumination, and bright-filed/dark-field illumination designs have been used [65]. They provide high specific detection through preferential light scattering. The optimized optical detection setup allows the development of ultrasensitive and multiplex detection. Indeed, these optical techniques give rise to ultra-bright signals without photobleaching. Under ideal conditions, single-molecule detection is achievable.

One particular advantage of SPRD is the fact that it is easily amenable to integration into micrometer-sized optical fiber sensors. Fiber optic with a core diameter of less than 50 μm is used to probe a volume of 50 nL solution of protein [66]. The setup consists of gold NPs adsorbed on an endface of an optical fiber [66]. A surface coverage of about 10–30% is obtained.

A sensitivity factor of 10^{-5} RIU (refractive index unit) has been estimated using red LED. This sensitivity value is comparable to conventional SPR macro-sensors using a 50 μL volume. A multianalyte detection is feasible using the multimode capability of optic fiber [66].

Extinction spectra of gold NPs have been estimated. The peak maximum position and intensity vary gradually in the 5–100 nm size range [67]. A significant extinction peak shift is measured when replacing 50 nm NP solutions with 100 nm NP solutions [67]. Multiwavelength detection is possible using NPs with different sizes. Plasmon resonance peak characteristics depend on NP size, shape, composition, and the local environment. For example, the whole visible region could be covered. Blue (430 nm), green (540 nm), and red (670 nm) emissions are simultaneously obtained from silver NPs of different sizes, illuminated with a white light [68]. These emissions are achieved with silver NPs of diameter varying from 40 to 100 nm [68].

SPRD setup including modified gold NPs has been used in oral cancer diagnosis [69]. These gold NPs are conjugated with monoclonal anti-epidermal growth factor receptor (anti-EGFR) antibodies. Contrary to bare gold, gold NPs modified with anti-EGFR antibody bind specifically and homogeneously to the surface of the cancerous cells.

Core–shell structures with a dielectric core NP surrounded by a thin metal shell with tunable optical properties have been developed by the Halas groups [70]. NPs consisting of a dielectric core and a metallic shell could give rise to SPR modes with tunable wavelengths over a broad optical range. By varying the relative dimensions of the core and shell constituents, it is possible to design NPs that either absorb or scatter light over the visible and/or infrared regions of the electromagnetic spectrum [70]. SPR peaks positioned in the near-infrared are obtained by coating silica beads with variable gold shell thickness [67]. Simulation based on discrete dipole approximation showed that varying the gold shell thickness from 2 to 15 nm on 60 nm silica beads allowed tuning the extinction peak position from around 600 nm to more than 1000 nm [67]. Gold nanocage obtained by refluxing aqueous silver nanocubes and $HAuCl_4$ solutions allowed tuning extinction peak from 400 up to 900 nm [67]. These structures could be of interest not only for optical sensing but also for photothermal and contrast imaging.

2.5 Surface-Enhanced Raman Spectroscopy

SERS has been introduced recently for biomolecular optical detection. SERS has been widely used for material characterization over decades [71]. Particularly when combined with nanostructured substrates, SERS could provide high sensitivity and specificity. Near-field Raman microscopy could also provide high spatial resolution even below diffraction limits [72]. Another feature of Raman spectroscopy is the possibility to integrate microscopy and imaging capabilities. This provides the possibility of multiplex sensing options at high spatial resolution. Raman signal intensity following elastic scattering off the metal nanosphere with radius r_a (Fig. 6.12) is given by

$$I_{SERS} = I_{in} \cdot I_{out} = |E_{in}|^2 \cdot |E_{out}|^2,$$

where the incoming (E_{in}, I_{in}) and outgoing (E_{out}, I_{out}) electrical fields are given by $E_{in} = E_o + E_{LM}$ and $E_{out} = E_{DIP} + E_{SC}$, respectively [74]. The emitted dipolar radiation E_{DIP}, which is red-shifted in the case of Stokes scattering, is elastically scattered off the metal NPs. As shown in Fig. 6.4, E_0 is the incident electric field, E_{LM} corresponds to elastic scattering of the incident field, E_{DIP} the emitted dipolar radiation off the metal NP which is red-shifted in the case of Stokes scattering, and E_{sc} the elastic scattering. The incoming field $E_{in} = E_0 + E_{LM}$ induces a dipole moment $\mu = \alpha E_{in}$, where α is the polarizability. Both field components, E_{DIP} and E_{SC}, add up to the outgoing field E_{out}. The overall signal enhancement depends on particle size, interparticle distance, aggregation, chemical/electrochemical interactions, and excitation wavelength. Often, signal enhancement is divided into electromagnetic and electrochemical origins.

Choosing the appropriate NP is based on its optical, surface, and electrochemical properties. Composition (Ag, Au), size, and the dispersing medium of the NP have direct effect on the plasmon resonance properties. Plasmon resonance frequencies should match incident excitation laser to obtain maximum enhancement [75]. Using just Ag and/or Au materials, the maximum peak position of the plasmon resonance could be swept from less than 400 nm in the case of small silver NPs, up to 650 nm for Au/Ag core–shell structure [74]. Optical excitation of these plasmon modes could be achieved by adjusting excitation laser wavelength. This option implies high capital investment for a wavelength tunable excitation and appropriate optical bench and detection. A cost-effective alternative is to adjust size and/or composition of these NPs for a given particle.

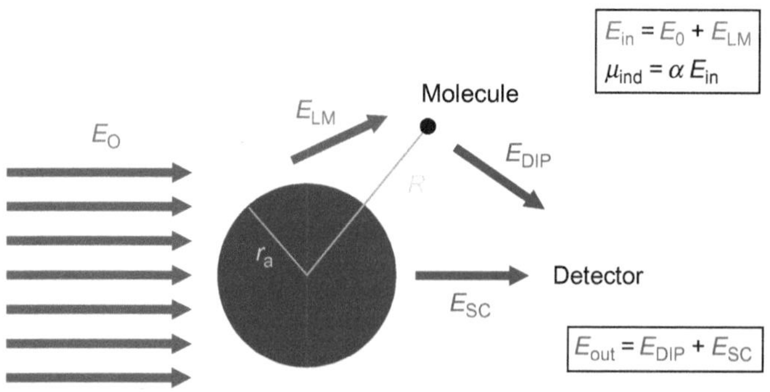

FIGURE 6.4 Schematic representation of the electromagnetic enhancement mechanism in SERS. See text for definitions of the different electric fields [74]. (For color version of this figure, the reader is referred to the online version of this chapter.)

NP-based SERS biosensing allows molecular information with and without labeling. The label-free detection of biomolecules from a solution is based on their characteristic Raman features following adsorption onto the NP surface. When using labeling, metal NPs are chemically modified with reporter molecules having specific spectral Raman features. Often, a silica-based protective layer is used to provide chemical stability and improved quantification capability [74]. Adsorption of other SERS-active molecules onto the surface has has also been reported [74]. The Raman enhancement factor will depend on the structural properties of the NP system.

For specific antigen recognition, labeled NPs are biofunctionalized using antibodies. Gold NPs modified by a nitro-derivative of mercaptobenzoic acid and an antibody show three characteristic Raman peaks [74]. A barcode representation corresponding to the Raman peak position and intensity could be used specifically for this antibody. An assay with different antibodies could be constructed and used to detect different antigens/proteins (Fig. 6.5). SERS allows differentiating the target protein from other proteins. The multiplexing capacity of SERS is at least an order of magnitude higher than that of dyes- and fluorophores-based labeling techniques [74].

With a judicious choice of NPs, SERS is a powerful diagnostic and detection tool. It can provide high sensitivity, high specificity, and high spatial resolution. Single-molecule detection has been reported by numerous groups [76,77]. Designing the proper substrate is critical to taking advantage of the SERS potential [78]. For example, the size of metallic features and the distance separating the features have significant roles in the overall Raman signal.

SERS of Au–Ag core–shell-based NPs could provide a high enhancement factor. SERS of Au–Ag NPs modified by monoclonal antibody have been used in immunoassay analysis [79]. Local SERS enhancement induced by

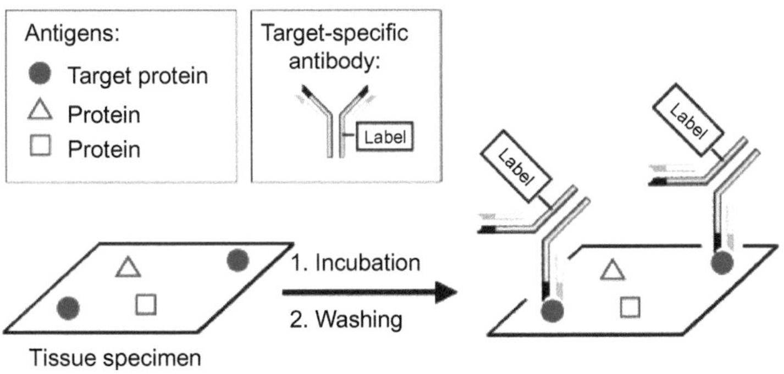

FIGURE 6.5 Schematic representation of immune-histo-chemistry. A selective antibody–antigen binding site (left) leads to a specific signal from the attached label following the incubation and washing steps [74]. (For color version of this figure, the reader is referred to the online version of this chapter.)

small NPs has been used for direct molecular imaging [80]. Far-field SERS array imager has also been reported [81].

SERS is used for noninvasive deep-tissue molecular imaging on small animals [82]. Two types of functionalized gold NPs (SERS 421 and SERS 440) are used to detect and quantify multiple tags simultaneously within the same animal [82]. SERS 421 and SERS 440 have different Raman signatures. Raman spectra of a rat deep-tissue injected with S421 SERS NPs and S440 SERS NPs are measured [82]. Furthermore, a spectrum is calculated using software to represent a mixture of the S421 and S440 SERS NPs. Besides its multiplexing capability, a sub-10 pM detection limit has been demonstrated with this modified commercial Raman microscope. Up to 20 or more multiplexing components are possible using this SERS system [82].

2.6 Other Optical Sensing Applications

An interferometric method has been developed to measure the scattering light of single-NP detection by eliminating background signal and reducing noise [83]. A two-laser interferometric technique has been recently used to detect a single NP or virus down to 10 nm [83]. After crossing the beam splitter, the laser is focused by a high numerical aperture into a microfluidic cell containing NPs. Scattered light is focalized onto a split photodetector after crossing a pinhole (diameter = 500 μm) allowing reduced background noise [83]. Polystyrene beads (15 and 50 nm) and gold NPs (7 and 20 nm) were used to detect influenza AX-31. The detection signal is based on the scattered light amplitude. Single gold (high refractive index) NP with 10 nm diameter and polystyrene (low refractive index) NP with 20 nm diameter have been shown to provide a low detection limit [83].

2.7 *In Vitro* Diagnostics

NP-based *in vitro* diagnostics could improve the sensitivity of current testing and allow the development of novel sensing platforms [84]. In particular, these new tests could facilitate the transition to POC services. As described elsewhere, NPs could be used as biomarkers for detecting pathogens.

Semiconducting, metallic, and superparamagnetic NPs are promising materials for *in vitro* diagnostic applications. Inorganic fluorescent NPs are currently considered good alternatives to fluorescent dyes. Indeed, they provide higher sensitivity, multianalyte testing, and potentially cost-effective solutions. Gold NPs have been developed for rapid pregnancy testing. Other examples of multianalyte detection are provided elsewhere in this chapter. In particular, simultaneous detection of different targets in a serum sample has been achieved using QDs of different sizes modified with different recognition moieties (Fig. 6.6). This simultaneous multianalysis is critical for fast and precise diagnosis. This platform could also be used in genomic analysis and nucleic acid detection.

3 ELECTRICAL SENSOR

Numerous detection concepts involving changes in NP electrical properties have been reported [6,85]. Array-based electrical detection of DNA using target-probe binding events has been reported by several groups [6,86]. Three

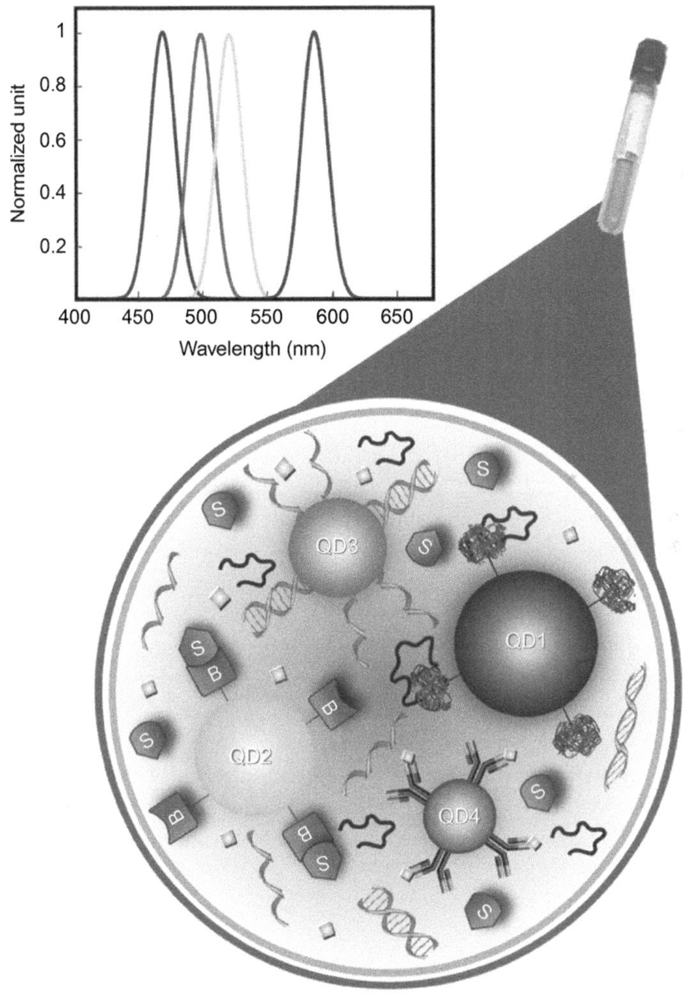

FIGURE 6.6 Schematic diagram of QDs with different sizes and surface moieties in a serum (lower). Corresponding emission spectra of different sized QDs (1–4) are shown in the upper left corner. B and S are abbreviations for biotin and streptavidin, respectively [84]. (For color version of this figure, the reader is referred to the online version of this chapter.)

process steps are required for DNA electrical sensing [6]. First, the capture probe is immobilized between two electrodes. The second is hybridization with target DNA and Au NP-labeled detection probe. The third is silver deposition to create a conductive bridge. Charge transfer between different gold NPs is thus mediated by metallic silver. A detection limit of 500 fM and a mutation selectivity ratio of about 100,000:1 have been reported. This is several orders of magnitude higher than the selectivity achieved by fluorescence-based QDs [6].

Single-nucleotide polymorphism (SNP) detection has been demonstrated using conductivity enhancement. A resistivity drop induced by structural change in gold NP-modified DNA target has been measured [87]. Following DNA probe immobilization on an insulating substrate, DNA-label is hybridized to DNA target and subsequently attached to a gold NP [87]. The gold particle film is then coated with a silver layer for signal enhancement. The hybridization of the DNA target with DNA probe is detected by contacting a movable electrode tip to the conductive silver coating surface formed around the gold NP film. A drop in electrical resistance, measured by a movable electrode tip, indicates the hybridization event.

A gold NP-modified receptor (streptavidin) inserted into the gap between two planar electrodes allows significant conductance variation [88]. By reducing the gap between the two electrodes to about 100 nm or below, two to three orders of magnitude variation in conductance has been measured. This allows detection of biotin interactions with streptavidin, even without using the silver-enhancing step [88].

Ion-sensitive field-effect transistors use the sensitivity of electron conductance between source and drain electrodes. The conductance change is induced by the gate polarization following chemical modification of the gate. Indeed, adsorption of charged species induces a change in the polarization of the underlying SiO_2/Si interface [85]. The source–drain current at a given gate–source potential (V_{gs}) and the gate–source potential required for a given source–drain voltage (V_{sd}) to retain the source–drain current (I_{sd}) are two sensing parameters for adsorbed analytes [85]. This design allows sensing of any charged molecule complexing with the adsorbed receptor. Note that even if the molecule is not electrochemically active, a change in surface polarization is observed, allowing change in electronic properties. This technique allows sensing over six orders of magnitude in the concentration range.

4 ELECTROANALYTICAL AND ELECTROCHEMICAL SENSING

Several mono- and bimetallic NP-modified electrodes have been shown to provide unique advantages over conventional macroelectrodes. Electroanalytical and electrochemical sensors based on these NPs provide high surface area, enhanced mass transport, catalytic and selective properties, and control over the electrode microenvironment [85,89,90].

In the case of macroelectrode, the peak current (I_p in amperes) and the limiting current (I_{lim}) at a hemispherical electrode for a single one-electron redox couple are given by the following equations [89]:

$$I_p = 2.69 \times 10^5 n^{3/2} D^{1/2} C_{bulk} v^{1/2}$$

$$I_{lim} = 2\pi n F C_{bulk} D r$$

where n, D, C_{bulk}, v, A, r, and F correspond to the number of electrons transferred in the rate-determining step, diffusion coefficient in cm^2/s, bulk concentration of active species in solution in mol/cm^3, scan rate in V/s, the surface area in cm^2, the radius of each hemispherical NP electrode and the Faraday constant, respectively. Interparticle distance to ensure diffusional independence for NP-based array has been estimated at a nominal value of $10r$. In this case, the total current (I_{total}) for a particular area A of the NP array is [89]:

$$I_{tot} = I_{lim} A/(10r)^2$$

The sensitivity of NP-based electrode c will be relatively high. Unfortunately, high NP reactivity often leads to electrode passivation due to oxidation. This could be prevented by surface treatment and use of appropriate support. However, these modifications could affect analytical performance. Chemically stable gold NPs are widely used to build electrochemical sensors, although they contribute to increased overall cost.

It is possible to use this technique to identify the DNA bases [91]. Gold NPs have been widely used as an electrochemical label for DNA detection [6,89]. A schematic representation of the setup is shown in Fig. 6.7. Four monobase nucleotides labeled with gold NPs were used. Different redox signals indicate the identity of mismatch [6]. Although gold NP-based electrochemical sensing is less sensitive than PCR, several recent improvements to overcome this drawback have been reported [92,93].

A stable and continuous gold film obtained by a two-step rapid deposition process has been used as electrode for the electrochemical process [94]. A gold colloid monolayer was first assembled on a poly(diallyldimethylammonium chloride)-modified glass slide. Electroless plating was then used to control the surface roughness [94]. Enhanced current density has been reported with silver NPs [95].

The photoelectrochemical properties of QD assembly on an electrode have been used for sensing. Conduction band electrons of photoexcited QDs are injected into the supporting electrodes. Alternatively, photogenerated electrons are first transferred to an electron acceptor in the electrolyte, followed by a transfer to the electrode [40]. Numerous examples of QD-biomolecule systems have been reported recently in the literature [40]. CdS NPs modified with

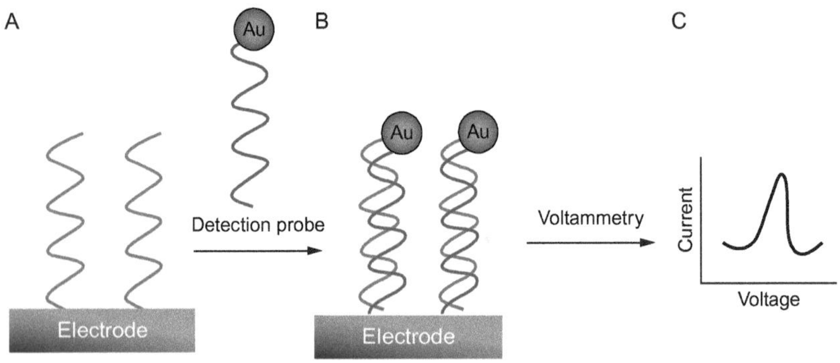

FIGURE 6.7 Pictorial representation of electrochemical detection of DNA hybridization using modified gold nanoparticle. (A) Immobilization of target DNA on the electrode surface. (B) Hybridization with modified gold nanoparticle. (C) Voltammetric (I–V) signal providing the occurrence of gold nanoparticle redox event [6]. (For color version of this figure, the reader is referred to the online version of this chapter.)

a nitrate-containing molecule are attached to a gold electrode through di-thiol molecules. Photochemical activation of lactate oxidation in the presence of LDH and reduction of nitrate by nitrate reductase involve a two-step charge transfer process [40]. Photocurrent direction (cathodic or anodic) generated at CdS–cytochrome interface could be controlled by the redox state of cytochrome [85].

5 MAGNETIC SENSING

Magnetism and magnetic devices play important roles in medicine [96]. Magnetic NPs have been mostly used as sensing probes [97–105,17,106–112]. More recent reports have demonstrated the therapeutic applications of magnetic NPs [103]. Modulation of the magnetic properties of NPs is achieved through composition, size, shape, and concentration. Three types of sensors are currently under development: magnetic relaxation switch assay-sensors, magnetic particle relaxation sensors, and magnetoresistive sensors [107]. Superparamagnetic and ferromagnetic NPs have been used. Iron oxide (magnetite or Fe_3O_4 and maghemite or γ-Fe_2O_3) and doped iron oxide (MnF_2O_4, $CoFe_2O_4$, $NiFe_2O_4$) NPs have been widely used for magnetic sensing. These magnetic NPs are believed to overcome the low sensitivity of the conventional gadolinium-based contrast agents. Furthermore, a high gadolinium dose may give rise to side effects such as nephrogenic systemic fibrosis.

5.1 Magnetic Relaxation Switches Assay Sensors

In this magnetic relaxation sensing approach, polymer-coated iron oxide superparamagnetic NPs are used as contrast agents in preclinical and targeted molecular imaging applications. Magnetic resonance imaging (MRI) provides 3D contours of living bodies at the molecular level, allowing detection of tumors at an early stage. MRI is an important noninvasive imaging technique, based on the nuclear magnetic resonance (MR) properties of the various interacting nuclei. T_1 (spin–lattice or longitudinal relaxation time), T_2 (spin–spin or transverse relaxation time), and r (spin energy) are three important factors influencing MRI signal strengths. A typical setup used for *in vivo* magnetic NP relaxation measurement is shown in Fig. 6.8.

Magnetic relaxation sensors are based on proton relaxation time (T_1 and T_2) measurements. The transverse spin–spin relaxation times (T_2) of proton water medium are affected by the structural change in NPs. The value of $1/T_1$ is only slightly affected by the aggregation and agglomeration states. In low size range, the $1/T_2$ value increases with aggregation. Target molecule adsorption on modified magnetic NPs may also play a role. Three approaches are used to enhance the detection sensitivities [107]. In one approach, the difference in particle size gives rise to a change in aggregation. Higher concentration of analyte (anti-Tag antibody) detection is achieved when NPs are used. In the second approach, a magnetic field (0.47 T) gives rise to aggregation and accelerates the interaction between the magnetic particles and analytes. The last approach uses a secondary antibody (e.g., sheep anti-mouse) to increase valency [107].

One advantage of magnetic sensors over optical sensors is related to an unperturbed penetration of radiofrequency radiations within the biological samples. Whereas most optical wavelengths are absorbed within the first few micrometers, the applied magnetic field is not affected. Furthermore, the potential miniaturization of the magnetic relaxation switches (MRS) offers great

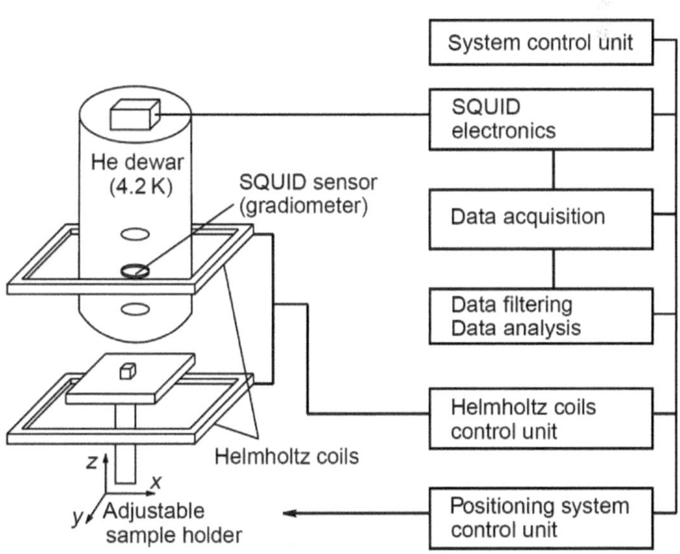

FIGURE 6.8 Schematic representation of an *in vivo* magnetic nanoparticle relaxation setup [98].

potential for POC applications [107]. Typically, a miniaturized MRS consists of a small palm-sized permanent magnet, integrated on-board RF transceiver and a microcoil array with integrated microfluidic channels [107]. A differential microcoil system consisting of magnetic coils is used for excitation; sensing coils and connection pads containing beads have been reported [100]. Magnetic relaxation data are found to depend on the particle size [100].

Noninvasive detection of apoptosis has been successfully investigated using iron oxide NPs as MRI contrast agent [110]. Inflammatory cells have also been detected [111]. Superparamagnetic iron oxide (SPIO) is the preferred magnetic choice. Dextran-coated SPIO NPs provide enhanced MR contrast, critical in molecular imaging [108].

Contrast agents (complexes of GdIII complex in liposomes or micelles) are often injected to enhance the tissue contrast. However, these systems are limited by GdIII (exogenous) ion exchange with endogenous metals (Zn, Cu). Stable and nontoxic magnetic NPs such as cross-linked iron oxide (CLIO) are better alternatives for targeted imaging. Often, paramagnetic compounds such as gadolinium chelates are used to provide a positive contrast by shortening proton T_1 (longitudinal relaxation times) or negative contrast by shortening proton T_2 (transverse relaxation time) of water MR image [100,101]. In the case of SPION, a marked shortening of T_2 and a less-marked shortening of T_1 have been observed. Usually, NPs smaller than 10 nm have larger residence time in the blood stream, whereas the particles larger than 30 nm are preferentially collected by the liver and spleen applications [10].

Conventional magnetic contrast agents are often based on iron oxide NPs. Availability, cost, and easy approval process are the main reasons for the widespread use of iron oxides. Several commercial solutions (AMI-25, CLIO, MION) have been marketed with magnetic core NP size varying from 2.8 to 6 nm and Dextran as coating materials [104]. More recently, new larger core magnetic NPs ($MnFe_2O_4$, FeCo) with improved magnetic properties have been introduced [105]. Different coating materials, including dimercaptosuccinic acid (DMSA), have also been used. These new materials give rise to an enhanced magnetic contrast effect. Size-dependent MR contrast effect has been demonstrated using both $MnFe_2O_4$ (MnMEIO) and Fe_3O_4 (MEIO) NPs [112]. Twelve-nanometer MnMEIO NPs showed the best MR contrast effect. Furthermore, the detection sensitivity of MR imaging of cancer markers using these magnetic NPs is higher than that of conventional probes.

5.2 Magnetic Particle Relaxation Sensors

Magnetic relaxation sensors are based on measuring magnetic susceptibility. This property will depend on the state of aggregation of NPs following target analyte recognition. For example, aggregated NPs will decrease relaxation constant due to a higher hydrodynamic radius that reduces Brownian movements.

The magnetic relaxation properties of NPs are a convolution of two different processes. In the first process, referred to as Brownian, the NP rotation movement is described by the following relation [107]:

$$\tau_B = 3V_H\eta/kT$$

where V_H, η, κ, and T correspond to the hydrodynamic volume, the viscosity of the medium, the Boltzmann's constant, and the absolute temperature, respectively.

In the second process, referred to as Neel, the following relation describes the effect [107]:

$$\tau_N = \tau_0 \exp(E_a/kT)$$

where E_a is the anisotropy energy of the single domain magnetic particle, and τ_0 is the preexponential factor that decreases as the anisotropy energy increases.

A detection limit of 5×10^4 NPs in a 20 mL volume assay containing 1.1×10^5 bacteria has been achieved with a superconducting quantum interference device [107]. A detection sensitivity of 0.05 μg/mL for antibodies was achieved with an AC susceptometer [107].

5.3 Magnetoresistive Sensors

Magnetic NPs adsorbed to a targeted surface give rise to perturbation of the applied magnetic field. This perturbation is sensed by changes in the surface current density. Size-tunable superparamagnetic NPs have been used in magnetoresistive biosensing [107]. A magnetoresistance-based ELISA immunoassay using antibodies linked to a target protein has been developed [107]. Streptavidin-coated magnetic particles are used for tagging the biotinylated antibodies. Following surface modification using target proteins, the analyte and the linker are incubated. A giant magnetoresistive signal is observed following specific biotin–streptavidin interaction on the streptavidin-coated magnetic NP probe [107].

Microparticle sizes up to 3 μm have been used. Recently, superparamagnetic NPs have been developed and used [107]. Stabilized NPs are not prone to aggregation under a magnetic field. Indeed, 50 nm NPs have been successfully used in cancer marker detection. Silanized high magnetic moment FeCo NPs (12.8 nm) have been recently integrated into an ELISA [107]. A detection limit of 2×10^6 molecules has been achieved for biomarker protein [107].

5.4 Therapeutics

Magnetic NP-based systems are used not only for detection but also for medical treatments, including destroying cancer cells. Magnetic NP functionalization is critical to complete both tasks [103–105,17,106]. Antibody-conjugated magnetic poly-(D,L-lactide-*co*-glycolide) (PLGA) NPs with doxorubicin (DOX) were used for detection and treatment of breast cancer [103]. More details of therapeutic applications of magnetic NPs are provided in the following sections.

6 MULTIPLEXING TECHNIQUES

Multiplexing techniques provide the ability to simultaneously measure multiple analytes in a single assay. They could provide improvements in immunology, drug screening, disease diagnosis, and defense against biological threats [113,114]. These potential improvements over conventional detection techniques are based on higher throughput, single-target systems. This will provide savings in reagents and consumables, decreased sampling errors, and easy integration of internal controls.

Two geometrical designs are used in multianalyte sensing. One is based on 2D solid substrates and the other on 3D solutions. In both cases, NPs are used to improve sensitivity and functionality. Four different solid surface–NP structures are under evaluation [115]. SERS-active film structures include dispersed NPs (island), metal-coated with shell structure, metal-coated random nanostructures, and metal NPs dispersed in a polymeric matrix [115]. Although these structures are developed for SERS-based sensing, they could also be used for other optical transducers. In all these designs, reproducibility of these nanoscale structures and the high production cost are the two biggest challenges. Specific geometry is often used to address sensing requirements and limitations in using different NPs. The first SERS structure consists of stable and functionalized NPs easily deposited to the substrate. NP-based nanocomposites may provide more flexibility for integration and utilization of SERS-based sensors.

Flat surfaces or biochip are widely used [116,117]. However, multiplexing detection directly in a homogeneous or quasi-homogeneous solution shows more potential. These solutions may include suspension array of microspheres and/or NPs [117,118]. Numerous transducers are used to encode the signal, including optical, chemical, graphical, electronic, and physical [117]. Figure 6.9 schematically summarizes the differences between a 2D microarray and a 3D suspension array. The 2D array consists of recognition molecules (antibodies, peptides, oligonucleotides) adsorbed on a well-defined position on the array surface. The 3D suspension array consists of recognition molecules adsorbed to encoded NPs. In both cases, different colors (codes) are used to discriminate the different recognition sites [117]. NPs provide a range of size, composition, and structure, allowing a wealthy library [119].

Recently, four QDs with different emission wavelengths have been used in a high-throughput, multiplexed gene expression [120]. A QD-based nanobarcoded microbead system for gene expression consists of microbeads with assigned pseudocolors. Emission spectra-based identification is achieved with

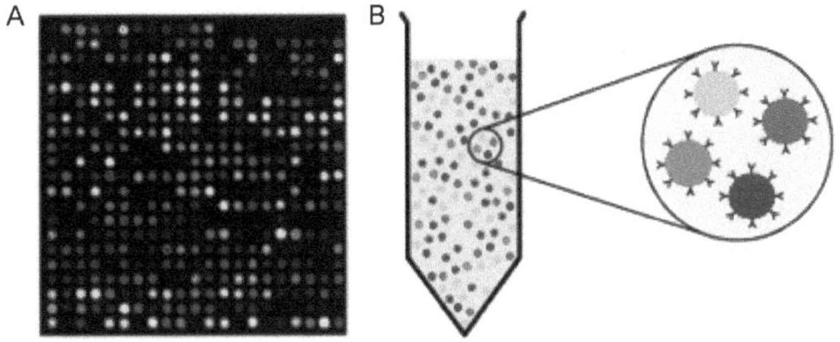

FIGURE 6.9 Schematic representation of a conventional 2D microarray (A) and a suspension 3D array (B) multiplexing [117]. (For color version of this figure, the reader is referred to the online version of this chapter.)

a nanobarcode containing spectral characteristics of four different QDs. The QDs are used to decorate the microbead system for high-throughput gene expression analysis. Each microbead consists of a distinctive ratio of four different QDs, allowing identification using characteristic spectral nanobarcode. The microbeads are bioconjugated with oligonucleotide probes endowed with a specific transcript. The gene expression monitoring and quantification assay are achieved through hybridization occurring between oligonucleotides and the cRNA. The presence of the sandwiched fluorescent streptavidin QD having different characteristic wavelengths allows quantification [120]. Using this multiplexing probe, each oligonucleotide is assigned to a specific spectral-barcoded bead. Following mixture of biotinylated cRNA and streptavidin QDs, a mosaic scanner is used to monitor the gene expression.

A sequential and simultaneous incorporation of QD and Fe_3O_4 NPs inside porous silica beads have been used as building blocks for microarray [121]. Strong interactions between hydrophobic surface ligands on the NP and the C-18 modified silica walls provide higher stability to these bifunctional silica microbeads [121].

Several other examples of multiplexing sensing using different NP compositions have been reported [122–130]. Several optical-based multiplexing techniques are suitable to provide 2D and 3D imaging. Four optical imaging techniques, including two-photon laser scanning microscopy (TPLSM), optical coherence tomography (OCT), Raman microspectroscopy (RMS), and photoacoustic tomography (PAT) have been developed [129]. These techniques are endowed with different physical characteristics, although often complementary. In particular, with penetration depths varying from 0.2 to 10 mm, a variety of tissues are accessible with these optical imaging techniques. High axial and spatial resolutions are achieved, particularly with the use of TPLSM.

Optical-based techniques should account for hemoglobin absorption below 700 nm and water absorption above 1200 nm. Thus, the region between 700 and 1200 nm is the ideal optical range for optimum sensing. In this near-infrared range, water radiation absorptions within the tissue are minimized. Near-infrared emitting fluorophore-doped calcium phosphate NPs have been developed for *in vivo* imaging of human breast cancer [130]. These NPs could also be used for therapeutics, including anticancer drugs.

As a spectroscopy technique, RMS is a suitable multiplexing transducer [122,123]. SERS of metal NPs could be a valuable tool for proteomics. A top decorative metal cluster overlayer on a resonance layer (tin nitride/oxide ranging from 0 to 500 nm) covered with a protein/receptor/ligand has been used [124]. The optical properties of the gold NP with 10–30 nm diameter allow reflected light with a distinct color. This characteristic optical property is ascribed to phase-dependent cluster–mirror–cluster resonance. Six dye-labeled Ag NP probes are used in SERS-based biochip array [122]. A flat-bed scanner image of Ag-enhanced microarrays, with corresponding characteristics of Raman spectra, is allowed. A detection limit of 20 fM has been achieved.

Robustness and automation of the different multiplexing immunoassays remain important issues. These issues need to be addressed before these immunoassays could finally prove suitable for high-throughput applications [131].

7 DRUG DELIVERY AND THERAPEUTICS

NP technology is considered an enabling platform for nanomedicine. Several concepts of NP-based drug delivery and therapeutics have been developed. For example, metallic NPs are used for gene therapy and phototherapy. In the case of cancer treatment, NPs with enhanced permeability and retention (EPR) could replace chemotherapeutics, which affects both cancer cells and normal cells.

Active and passive NP therapeutics are used. Early studies are mostly based on passive targeting. Tumors are attained and treated by the combined effect of blood circulation and diffusion mechanism. Active targeting takes advantage of functionalized NPs binding to specific antigen sites of the tumor cell.

There are numerous cellular uptake and delivery pathways of NPs. The majority of potential drug compounds and molecules require formulation to allow solubility in aqueous solutions. Major issues associated with poor water-soluble compounds include the following [16]:

- **a.** poor and variable bioavailability;
- **b.** lack of dose–response proportionality;
- **c.** suboptimal dosing;
- **d.** use of harsh excipients including extreme basic/acidic conditions and excessive use of cosolvents;
- **e.** uncontrollable precipitation after dosing;
- **f.** inconvenience of the dosage platform leading to noncompliance by the patient.

Oral drug delivery is more prone to enzymatic decomposition than subcutaneous injections. However, subcutaneous injection of drug is invasive and painful. Integration of these compounds and molecules in soluble NPs is currently being investigated [16,132–136,14,137,138]. These NPs could be modified to facilitate bioavailability and targeting specific sites [16]. Since poor bioavailability is related to dissolution-rate kinetics, size reduction of drug crystals has been proposed to improve their solubility. Indeed, reducing the crystal size from 10 μm to 100 nm increases the surface-area-to-volume ratio by two orders of magnitude. In the case of oral drug delivery, improving absorption will also enhance bioavailability.

An increased efficiency in drug delivery is achieved when both passive and active targeting are used. This is achieved by overcoming the different biological, biophysical, and biomedical barriers. Passive targeting based on EPR characteristics of solid tumors has been reported [132]. Active targeting requires modified NP to allow specific attachment to target tumor cells [132]. Depending on the application, NP-based drug carriers are formulated using different compositions and geometries. For example, modifying agents are used for improved tumor targeting.

There are four main areas in which NP-based drug delivery could have an impact [134–136,14,137,138]. First, they provide the possibility for a more efficient targeted dry delivery and overcome multiple biological barriers [139]. Second, a controlled release is possible through either an internal or an external process, thus releasing a therapeutic load in the optimal dosage range [140]. The last two areas are related to improvement of biocompatibility [141] and biodegradability of the drug systems [142].

Various polymers, ceramics, metals, and biological materials have been used to synthesize NP-based drug delivery carriers [16,132–136,14,137–152]. Natural and artificial polymeric materials have been used for the fabrication of particle-based drug carrier. This includes polysaccharides, polyacrylates, and polyesters. Various inorganic NPs have also been studied for the same purpose. Four main nanosized drug-carrier systems are currently used for mucosal vaccines: liposome, micelle/emulsion, polymeric NP, and dendrimer [153]. Nanosuspensions consisting of pure drug NPs stabilized by surfactants are not specifically included in this ranking.

Polymeric materials-based NPs and microparticles are the most widely used drug carriers [133–136,14,137–155,21,156]. They include nanogel [155], albumin and micellar structures [21], polybutylcyanoacrylate [135], and liposomes ([21], p. 417). Utilization of inorganic NPs as drug carrier has also been demonstrated [16,156].

7.1 Polymeric NPs

Polymer-based drug delivery (Fig. 6.10) includes solid polymeric NP, nanoshell, dendrimer, polymeric micelle, and polymer–drug conjugate. Liposomes, nanosuspensions, and polymeric NPs provide the highest potential for pharmaceutical applications [133]. Liposomes consisting of a lipid layer surrounding the drug are currently used. They provide enhanced drug efficiency. A few NP-based drug delivery systems are currently approved or in clinical trials.

Nanosuspensions use surfactants to stabilize drugs in submicron suspensions. Indeed, when the size of these nanosuspensions is reduced below 100 nm, the solubility of the drug is significantly increased. In the case of polymeric NPs, the drug is attached to the polymer backbone. These polymeric NPs are suitable for site-specific targeting of tumors and for the transport of drugs across biological barriers.

NP-based suspensions are potentially ideal solutions to issues related to poorly water-soluble drugs [133,143–145]. In some cases, NP-based systems could be used for detection and treatment. Fe_2O_3 and DOX embedded in PLGA NPs were used for the simultaneous targeted detection and treatment of breast cancer [103].

Besides drug delivery, NPs are also used in gene therapy. NPs combined with nucleic acids provide the best prospect for targeted therapeutics [146]. There are some differences between NP and siRNA oligonucleotide delivery systems [146]. After administration, the siRNA oligonucleotides are screened out by the multiple barriers before reaching the therapeutic site. Numerous NP-based delivery systems have been designed and considered for clinical

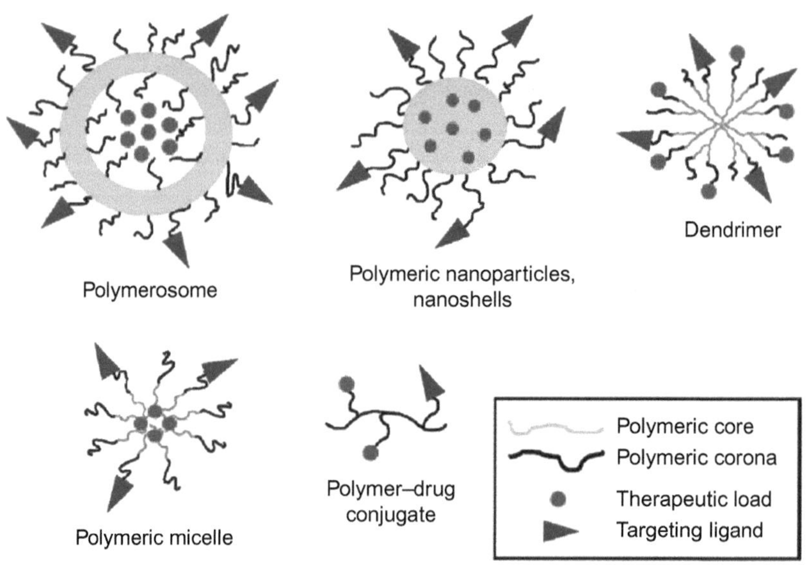

FIGURE 6.10 Schematic structure of different drug delivery systems. *Reprinted with permission from Alexis et al. [138]. Copyright 2008, American Chemical Society.* (For color version of this figure, the reader is referred to the online version of this chapter.)

testing in cancer therapeutics [146]. Although NP-based drug delivery provides some advantages, it is still quite far from reaching its maximum potential [151].

Peptide drugs can be loaded before or after NPs are prepared. The loading level should be high enough for the drug to have an impact. However, the dose should not be too high to give rise to peptide denaturation. The main role of the carrier is to protect the drug from the digestive enzyme. NP-based drug carriers overcome physiological and morphological barriers more efficiently than their microparticle counterparts [147–149]. Indeed, the screening effect of these barriers increases with size, when other parameters are kept constant. It is important to precisely report particle size parameters in the case of NPs used for drug delivery [150]. Indeed, although it is established that a smaller particle has an enhanced drug targeting ability, it is still difficult to accurately measure particle size distribution under biological conditions. For example, required size limit and other related parameters for drug delivery to a specific organ are often poorly defined.

Drug NPs and microemulsions are relatively cheaper than polymeric NPs and liposomes [151]. On the other hand, site-specific delivery is achievable only with polymeric NPs and liposomes, although local delivery is possible with other NP-based drug delivery systems [151]. When compared to other NP systems, drug NPs allow very good drug loading capability.

Given the fact that therapeutic biomolecules need to deliver drugs beyond the epithelium, interaction of NP/biomolecules with the epithelium is critical. The majority of these biomolecules navigate through the whole intestinal system without being absorbed by the epithelium. Numerous strategies have been developed to enhance interaction of these carrier biomolecules with the epithelium [154].

Different targeted and nontargeted NP-based tumor detection and treatment methods have been developed [7]. In a typical nontargeted process, NPs extravasate through the leaky vasculature, which is characteristic of solid tumors and inflamed tissue. The EPR effect promotes preferential accumulation of these NPs. In this case, the drug may be released and diffused within the target tissue for bioactivity. In the targeted process, following NP extravasation in the target tissue, ligand NPs allow active targeting of receptors present in the cell or tissue. This results in enhanced cell uptake and therapeutic efficacy of drugs. In a third case, NPs are modified with ligands to allow vascular targeting by binding to endothelial cell-surface receptors.

Several polymeric core–shell NPs obtained via the block-copolymer self-assembly process are used to carry drug and macromolecules to specific sites [21,135,156]. The hydrophobic monomer blocks form the core, whereas the hydrophilic monomer blocks form the shell. Dozens of liposome formulations are currently in clinical use or in preclinical development. Biocompatible and biodegradable liposomes allow molecule encapsulation. A liposome consists of an aqueous core and a vesicle shell, with a mean diameter of 85 nm containing DOX [157].The drug molecules are tightly packed (10,000–15,000 molecules per liposome) in a gel phase. The surface of these liposomes is modified with a polyethylene glycol (PEG), forming a protective hydrophilic layer. PEG provides stability. A hydrogenated soy phosphatidylcholine lipid bilayer membrane provides additional stability and protection from external media.

Nanogels with tunable properties and compositions are ideal drug delivery carriers for biological and biomedical applications [154]. Besides the attributes of conventional NPs, the porosity of nanogels could be used for cargo. Numerous polymers have been used to build these nanogels. A biodegradable nanogel incorporating biomolecule or inorganic NP and a disulfide-functionalized dimethacrylate has been shown as an effective tool for targeted drug delivery [155].

7.2 Magnetic NPs

Magnetic NPs could be used for drug targeting and/or tumor treatment. Under the influence of an external magnetic field, magnetic NPs are drawn to a specific target site for therapeutic applications [157]. These magnetic NPs could either contain therapeutic agents, or be used directly, taking advantage of their heat loss properties following infrared radiation.

Iron oxide magnetic NPs embedded within a hydro-gel including chemotherapeutic agents targeting ligands are used for detecting and treating cancerous tumors [137]. The iron oxide NPs have a size range of 20–40 nm, with more than 10 wt% chemotherapeutic loading. The magnetic properties and heating properties are used for targeting and treatment, respectively. Different DNA complement strands (12, 18, and 24 bp) are used during hybridization with the 30 bp parent strand. Hybridization and assembling are performed in a magnetic column. Oligonucleotide-modified magnetic NPs are used to control the delivery and release of drugs at a remote location (Fig. 6.11). Indeed, a remote electromagnetic field (EMF) is used to trigger release of magnetic NPs at a model tumor in the posterior mammary fat pad of mice. Following multiple labeling and mixing with matrigel, the 12 bp labeled, magnetic NPs are released near the tumor.

NPs are used in magnetic and/or photodynamic hyperthermia treatment of cancer and for cell sorting procedures. Magnetic fluid hyperthermia is often used [159]. The heat loss properties of magnetic NPs are useful in numerous biological applications [160]. It is possible to prepare magnetic NPs with large specific loss power to optimize the heating properties [161]. Magnetic fluid hyperthermia uses superparamagnetic NPs to treat tumors efficiently. Aminosilane-coated iron-oxide NPs are concentrated by preferential absorption by tumor cells. Using an external magnetic field, these NPs are heated above 41 °C, destroying the tumor cells.

Another critical parameter when using magnetic and magnetofluorescent NPs for biological applications is specific targeting. This is accomplished

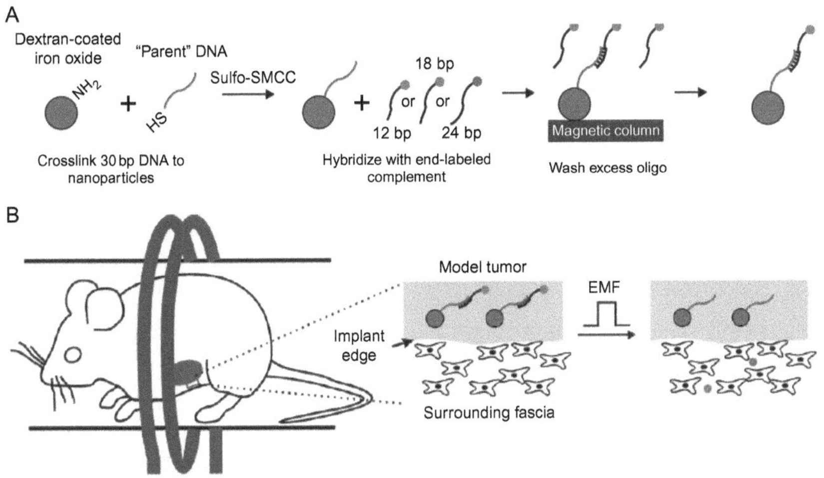

FIGURE 6.11 Magnetic nanoparticle delivery system based on remote electromagnetic field (EMF). (A) A Dextran-modified iron oxide magnetic nanoparticle is functionalized by a 30-bp parent DNA strand through thiol-amine bonding. This step is followed by assembled linking to a fluorescent complement (of 12, 18, or 24 bp); (B) Modified magnetic nanoparticles are mixed with matrigel and injected subcutaneously in the area of the model tumor (A). EMF application releases 18 bp containing nanoparticle and penetration near the targeted tissue [158]. (For color version of this figure, the reader is referred to the online version of this chapter.)

through covalent binding of antibodies and protein on the surface of NPs. Various synthetic routes have been developed to attach small molecules with different functional end groups [119]. Conjugation strategies of magnetic NPs are critical to improve their effectiveness not only as biosensors but also as therapeutics. Aminated magnetofluorescent NPs provide the possibility of conjugation with anhydride, amine, hydroxyl, carboxyl, thiol, and epoxide [119]. Upon protein binding, interaction is monitored through proton spin–spin relaxation time (T_2).

7.3 Gold NPs

Besides their inertness, functionalized gold NPs provide enhanced binding affinity and targeting selectivity. Gold NPs can be used as efficient drug carriers of poorly soluble chemotherapeutic compounds. These multifunctional gold NP scaffolds are used to increase cancer drug delivery and selective gene therapy [16]. Gold NPs modified with DNA [52] could be used in gene transfection therapy. For example, conjugated gold NPs could bind plasmid DNA and RNA and efficiently deliver them to specific sites.

Gold NPs could also be used directly for anticancer treatment. Indeed, infrared irradiation of gold NPs embedded in biological tissues will increase the overall temperature of the NP and its immediate environment. Depending on the final media temperature, targeted tumors are either killed (hot NP) or the gold NPs, being the source of controlled heat, may control their growth. Local heating using modified gold NPs with a mean diameter of 10 nm has been successfully used in tumor treatment and is referred to as hyperthermia. The temperature inside cells is raised by the radiation absorption of gold NPs, giving rise to local heating of the environment. Often, a temperature increase of only a few degrees is sufficient to kill tumor cells.

Phototherapy based cancer treatment has been developed taking advantage of conjugated gold NP absorption of infrared light. Since biological tissues are transparent to infrared radiations, localized NP heating allows obliteration of tumor cells. Local heating at controlled temperature will prevent damage to surrounding tissues and media. Nanospectra Bioscience, based in Houston (Texas, USA), is currently developing this technique for lung cancer treatment. Several studies have suggested using metal NPs for hyperthermia tumor treatment following infrared radiation. Others have suggested using these metallic NPs to enhance radiation therapy effectiveness in tumor treatment. In the second case, metal NPs induce a mild hyperthermia effect. Increased temperature increases blood perfusion throughout the tumor without damaging the cells.

Controlled bond breaking is achieved upon heat application [42]. Specific ligand binding gives rise to aggregation with red-shifted wavelengths (violet/blue appearance). Following illumination, heated gold NPs melt the receptor–ligand bonds. Therefore, the assembled components are dissolved. The average distance between the particles is increased and the particle solution appears red. When incorporated in a polymer capsule and cargo molecules, a local specific irradiation of the gold NPs raises the temperature to a point where the walls of the polyelectrolyte capsules are ruptured. The cargo molecules are thus released [42].

An *ex vivo* pulsed laser has been used to eliminate tumor cells [162]. In this technique, 30 nm gold NPs attached to a target allow the removal of toxic and clogging aggregates [163]. Irradiation with low power gigahertz EMFs provides local heat dissipation by gold NPs. Amyloid fibrils and plaque deposits have been remotely redissolved and their growth has also been affected [163].

7.4 Silver NP

The antibacterial activity of the silver NP has been well documented [164]. However, its toxicity to higher organisms should also be considered. The mechanism of silver's antibacterial properties is still not fully elucidated. Different mechanisms have been recently suggested [164]. A controlled release of silver ions and generation of reactive oxygen species (ROS) will initiate interactions with membrane proteins. Silver accumulation at the cell surface will affect membrane permeability. Once it traverses the cell membrane, ROS are generated within the cells, which could affect the DNA functions and membrane proteins [164]. Size-tunable silver NPs are obtained by saccharide reduction of $[Ag(NH_3)_2]^+$ complex cations [165]. Their antibacterial activities are found to be dependent on size over the 25–100 nm range. A very low concentration of silver (as low as 1.69 μg/mL Ag) showed antibacterial activity even on multiresistant strains.

8 OTHER MEDICAL APPLICATIONS

Other additional medical applications include gene therapy, tissue repair, and radiofrequency-controlled switching of complex biochemical processes and antibacterial agents. Numerous NP-based formulations are at different approval stages for therapeutic applications [166–173]. The vast majority of these novel medical therapies are in preclinical development. A few have moved to the clinical trial stage. For example, a NP-based magnetic fluid hyperthermia has been developed by Magforce (Germany) to treat brain cancers. Besides drug delivery systems, NPs are also used in cell therapy and in pharmaceutically active compounds.

Modified magnetic NPs are used to isolate and recover target adult stem cells in blood or bone marrow samples. Recovered cells, using an external magnetic field, are retransplanted in the patient to treat diseases such as leukemia and heart diseases [166]. Benign gold nanoshells consisting of a silica core coated with small NPs are being developed. These nanoshells are used in irreversible photothermal ablation of tumor tissue under *in vitro* and *in vivo* conditions [168]. The geometrical parameters of the nanoshells are optimized to allow high NIR-to-thermal energy conversion. NIR light is used since it is optically transparent through tissues. Furthermore, the plasmon resonance features of silica-gold nanoshells could be optimized to allow a high absorption coefficient in the NIR region. Adding gold nanoshells to human breast carcinoma cells induces morbidity following exposure to NIR light.

Numerous studies have been reported regarding the use of NPs in cancer diagnosis and treatment [170,171]. As described, polymeric NPs are often used for drug delivery [174]. Enhanced cell therapy has been achieved by grafting NPs containing adjuvant drug to the surface of therapeutic cells [170]. This therapeutic cell engineering provides a direct drug delivery to an otherwise poorly accessible site, with high therapeutic efficacy. The presence of reduced thiol group at the cell surfaces has facilitated grafting of liposomes and liposome-like synthetic sub-300 nm particles. These particles consist of drug-loaded core and phospholipid surface layer containing thiol-reactive maleimide headgroups. This generalizable approach also minimizes the systemic side effect when large doses of adjuvant drugs are used.

Table 6.1 provides examples of NP-based cancer treatments that have been approved by the FDA or are presently in clinical trials [172]. Evaluating the effectiveness of these new NP formulations and comparing them with existing solutions are challenging. Indeed, a more rigorous physicochemical characterization (uptake and clearance and cytotoxicity) of multifunctional NPs is important [172]. For example, the analytic method may interfere with the NPs. Furthermore, the catalytic properties of NPs may interfere with biological reactions.

Metallic NPs are potentially ideal candidates for immunology applications [167]. Indeed, they are endowed with numerous physicochemical attributes. The presence of gold NPs with antigens and hapten has been shown to promote immunization. For example, higher production of antibodies has been reported using gold NP as carrier. DNA vaccines have also been produced using protein (antigen)-modified gold NPs to provoke an immune response [167].

Iron oxide (Fe_3O_4) magnetic NPs embedded in poly(e-caprolactone) matrix form nanocomposite scaffolds with superior mechanical properties [173]. With just 10 wt% of NPs, the stiffness and strength of these nanocomposites are increased significantly. Furthermore, the presence of the magnetic NPs will facilitate the design of novel fixation methods and functionalization of the scaffold. These attributes have allowed advanced bone tissue engineering [173].

9 FOOD, PROTEIN, AND BIOTECHNOLOGY

The global food retail market amounts to around US$ 4 trillion [175]. A sustainable supply of safe and affordable food remains a big global challenge. Sharp increases in global food price in 2008 and 2010 [176] highlighted the major imbalance between agriculture productivity and increased demand. High energy costs, dry conditions, and degradation in agro land quality are often cited as the main causes. A plethora of advanced technologies are currently under development to address these problems. NP technology is believed to provide sustainable solutions.

Currently, nanotechnology in the food industry represents less than 0.1% of the global market [177]. Colloid science has played a key role in various areas of food science and industry [178–180]. NPs are currently used in a few applications related to food preservation, processing, and production. NPs could also be used in multilayered food packaging and improving color and flavor [177]. Latexes are used in numerous biotechnology applications [181]. The formulation of additives and supplements within NP-based systems will help their delivery and integration. Furthermore, they could provide additional benefits, including bioavailability of the nutrients and supplements.

Nanotechnology-based applications in agriculture and the food industry have received a significant coverage in the media. Recently, the potential negative features of NPs have been highlighted. However, there are very few scientific studies of NPs in this field. For example, NPs could be used as foam stabilizer in the context of food systems [179]. Biological components of living

TABLE 6.1 Examples of Nanoparticle-Based Products Developed for the Detection and Treatment of Cancer

Product	Type of nanoparticle/drug	Indication	FDA	Company
Doxil	PEGylated liposome/ doxorubicin hydrochloride	Ovarian cancer	Approved 11/ 17/1995 FDA50718	OrthoBiotech
Abraxane	Nanoparticulate albumin/paclitaxel	Various cancers	Approved 1/7/ 2005 FDA21660	American Pharmaceutical Partners
Cyclosert	Cyclodextrin nanoparticle	Solid tumors	Phase I	Insert Therapeutics
Megace ES	Nanocrystal/ megestrol acetate	Breast cancer	Approved 7/5/ 2005 FDA21778	Par Pharmaceutical Companies
INGN-401	Liposomal/FUS1	Lung cancer	Phase I	Introgen
Combidex	Iron oxide	Tumor imaging	Phase III	Advanced Magnetics
Aurimune	Colloidal gold/TNF	Solid tumors	Phase II	CytImmune Sciences
SGT-53	Liposome Tf antibody/p53 gene	Solid tumors	Phase I	SynerGene Therapeutics

Development stages (as of 2009) are provided [172].

cells, including protein, provide numerous examples of the importance of colloidal science and nanotechnology [182]. Indeed, several cell components present the geometrical features of NPs.

Small (chymotrypsin) and large (aspartate transcarbamoylase) NPs are found in typical living cells [182]. Chymotrypsin and aspartate transcarbamoylase proteins are spherical, with an average size of 4 and 7 nm, respectively. Small (ribosome) and large (virus) assemblies are obtained with a spherical structure of 20 and 100 nm, respectively. Numerous proteins qualifying as NPs play a critical role in maintaining the functioning of the cell. This includes biochemical reactions, sensing, and therapy.

Besides their use in processing, conservation, and the consumption phase of the food production chain, colloidal silvers are claimed to enhance desirable uptake [167]. Nanoemulsions consisting of oil-in-water emulsion are used in parental nutrition. Their stability at low oil droplet concentration allows the development of low fat product with the consistency of viscous cream. In the area of neutraceuticals, micelles and liposomes are efficient delivery platforms for antioxidants, vitamins, flavoring agents, and proteins. TiO_2 NPs have been licensed as food coloring agents, although the size and shape are ill defined.

Encapsulation of additives, ingredients, and supplements inside micelles and liposomes is being actively pursued. The potential applications of liposomes as an efficient vehicle to deliver nutrients and vitamins is well documented [73]. Water-soluble ingredients are incorporated inside the liposome (Fig. 6.12). Ferrous glycinate liposome NPs could be used to administrate ferrous glycinate [183]. This study showed acceptable stability under gastrointestinal juice at 37 °C for 5 h.

10 MISCELLANEOUS APPLICATIONS

There are numerous niche markets for NP applications in the biomedical sector. Using oxide particles with size ranging from 10 to 1000 nm is probably the most controversial application. These particles, used to improve homogeneity and UV protection of cosmetic products, could end up in the circulation system. The same risk could also be found in other products.

10.1 Dental Nanofilling and Adhesives

Numerous nanofilled composites with superior characteristics over universal microfilled composites have been reported [184–186]. The enhanced mechanical properties, combined with their high aesthetic properties, could be used as dental resin composites [184].

Ormocers containing functionalized silica NPs [187] have been shown to improve the mechanical properties of dental nanocomposites. Furthermore, utilization of amine or amide dimethacrylate trialkoxysilanes without diluents also improves the biocompatibility of these nanocomposites.

A dental adhesive containing 10 wt% of 5 nm diameter silica filler has been developed and marketed by 3 M [188]. Activation of this adhesive is performed under light. Hydroxyapatite NP deposition with a gradient structure has been obtained using EPD on titanium substrate. The resulting film is a stable, thick, uniform, and crack-free HA film [189].

10.2 Improvements of Mechanical Properties

Carbonated hydroxyapatite NPs [190] imbedded in a matrix structure could be used to produce materials with high toughness and stiffness [191]. It is the dream of most engineers to have biomaterials with these mechanical properties. Highly stiff and tough materials are required for dentin and bone structure.

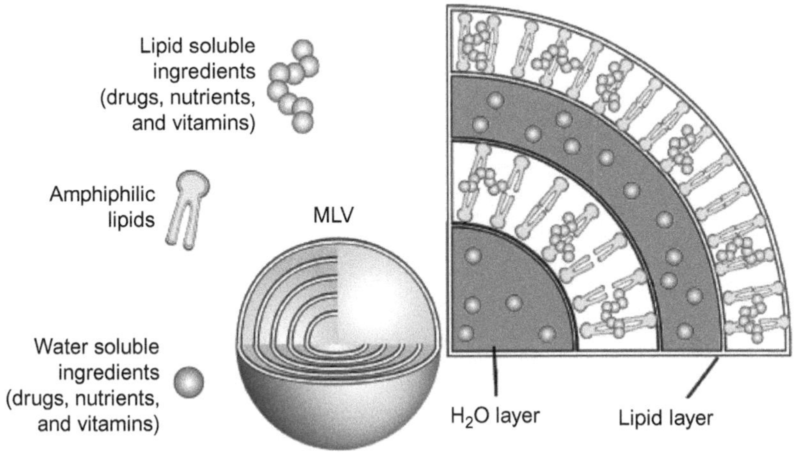

FIGURE 6.12 Schematic representation of liposome cross section [73].

10.3 Magnetic Bioseparation

Specific magnetic bioseparation has been used in immunoassay techniques [192,99]. This is usually performed by immobilization of a magnetic label (e.g., NP) to a biological target through a ligand. For example, enzymes could be immobilized on magnetic carriers to facilitate their removal from the reactor body. Magnetic bioseparation is often used to enhance the sensitivity of a biosensing technique. Indeed, utilization of the magnetic microsphere allowed the development of sensitive and high-throughput immunoassays [192]. Magnetic immunoassay labeled with superparamagnetic particles has been used. Selective enrichment of cancer cells using iron oxide NP are being developed by some companies for an early detection [192].

A kit has been developed by Chemicell [193] for DNA/RNA separation (Fig. 6.13). Five simple process steps are used to separate DNA. First, a simple washing step is used to clean metabolites and macromolecular cellular components. Functionalized silica beads containing magnetic NPs are added. The DNA is then isolated from the rest of the solution using an external magnetic field. Pure DNA is finally eluted under low ionic strength conditions with RNase-free water. The supernatant containing DNA is thus transferred to a new tube.

10.4 Antibacterial NPs

Embedded high surface area silver NPs are considered for antibacteria and antifungi applications. Released ions from silver NPs are effective in this regard. Two silver NP-based products are presently being marketed by Smith & Nephew PLC and by JR Nanotech PLC to kill bacteria. Silver NPs are believed to be more convenient, easy to apply, nonvisible, and provide both short- and

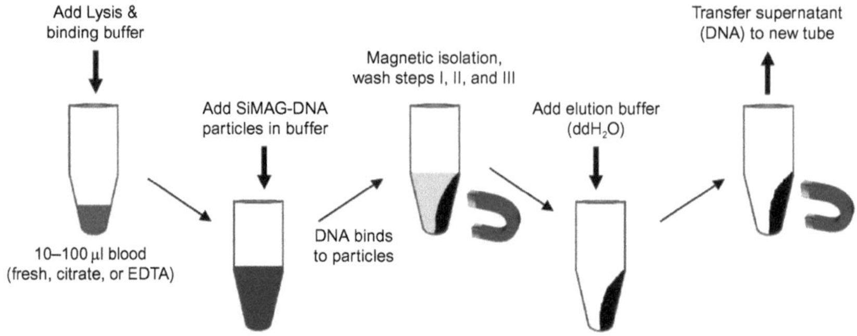

FIGURE 6.13 Different steps for isolation of DNA using functionalized magnetic silica beads [193]. (For color version of this figure, the reader is referred to the online version of this chapter.)

long-term disease prevention. The antibacterial activity of silver NPs decreases with increasing particle size [194]. The optimized antibacterial activity of silver NPs could help remediate increasing bacterial resistance to classic antibiotics [195]. Zinc oxide and TiO_2 NPs are other candidates for antimicrobial applications. Indeed, besides their UV light blocking, TiO_2 NPs generate free radicals under UV light [196]. These free radicals are excellent antibacterial agents.

10.5 Biodetoxification

Specific antidotes are often required to treat life-threatening overdoses of drugs or other toxic chemicals. A toxicity level beyond the minimum toxic level is maintained until the toxic chemicals are eliminated from the tissue [197]. This elimination is facilitated by diffusion and/or metabolism mechanisms. Reducing this toxicity level is required to eliminate death risk. NP-based carriers allowing removal of toxic agents from intoxicated tissues have been used [197]. Magnetic NPs, polymeric NPs, nanoemulsions, and liposomes have been used as nonimmune carriers for biodetoxification investigations. Emulsions obtained from natural or synthetic lipids are particularly attractive. Emulsion NP could be used as nanocarrier for biodetoxification [197]. Nanocarriers are coated with a hard polymeric shell to improve stability and also control drug uptake kinetics [197].

10.6 Bioavailability

Using NPs directly as active materials in pharmaceuticals, cosmetics, and food-related applications is difficult. Precise functionalization to meet application requirements is often required. Increasing the bioavailability and properties of pharmaceutical and cosmetic substances could be achieved via particle size and structure tuning [198]. Improved peroral bioavailability of insoluble active compounds has been achieved using synthetic carotenoids

[198]. Indeed, aqueous nanodispersions of emulsion, hydrosol, and micelles have been obtained from crystallite and synthetic carotenoids [198].

10.7 *In Vivo* Imaging

Shifting from the treatment of symptoms to prevention of outbreak of diseases has been "rediscovered" as the future paradigm in health care. *In vivo* molecular imaging of pathogens is currently being developed to improve our understanding of diseases at the nanoscopic level. Currently used imaging technologies (ultrasound, MRI, nuclear, and optical imaging) are not suitable for molecular imaging. Indeed, these techniques do not provide the required specific and chemical resolution.

Several techniques have been developed to address these issues. Some of these techniques have already been described in the previous sections. The suitability of several optical imaging techniques, including TPLSM, OCT, RMS, and PAT, for intravascular applications has been studied. Association with NPs improves the sensitivity and spatial resolution of these optical techniques. As discussed elsewhere, SERS-based microspectroscopy will improve the spatial resolution below 100 nm with potentially single molecular detection. Unfortunately, the majority of these optical diagnostic techniques are not yet clinically accepted.

There are several clinically accepted diagnostic techniques. For example, NP-based magnetic contrast agents have been used in MRI to improve sensitivity to small constituents of the pathological processes. Indeed, site-targeting is achieved by properly bioconjugated NPs.

Combining ultrasound with IR excitation, photoacoustic imaging provides *in vivo* capability [199]. Following NIR excitation of tissues containing NPs, nonradiative relaxation results in the generation of heat. Metallic NPs endowed with tunable SPR could provide enhanced contrast imaging. Ligand-targeting NPs provide the possibility of early disease detection by *in vivo* assessment.

REFERENCES

[1] Q. Huo, Colloids Surf. B Biointerfaces 59 (2007) 1–10.

[2] R.E. Bailey, A.M. Smith, S. Nie, Physica E 25 (2004) 1–12.

[3] E. Katz, I. Willner, Angew. Chem. Int. Ed. Engl. 43 (2004) 6042–6108.

[4] W.J. Parak, D. Gerion, T. Pellegrino, D. Zanchet, C. Micheel, S.C. Williams, R. Boudreau, M.A. Le Gros, C.A. Larabell, A.P. Alivisatos, Nanotechnology 14 (2003) R15–R27.

[5] M. De, P.S. Ghosh, V.M. Rotello, Adv. Mater. 20 (2008) 4225–4241.

[6] N.C. Tansil, Z. Gao, Nanotoday 1 (2006) 28–37.

[7] O.C. Farokhzad, R. Langer, ACS Nano 3 (2009) 16–20.

[8] M.A. Dobrovolskaia, S.E. McNeil, Nat. Nanotechnol. 2 (2007) 469–478.

[9] M. Powers, NanoBiotech News (2006) 1–69 4 January.

[10] G. Schmidt, Nanotechnology: Assessment and Perspectives, Springer, Berlin, 2006, pp. 67–286.
[11] S.J. Rosenthal, D.W. Wright (Eds.), Nanobiotechnology Protocols, Humana Press, Totowa, 2005.
[12] G. Marshall, PhD thesis, University of Sherbrooke, 2011.
[13] C. Drake, S. Deshpande, D. Bera, S. Seal, Int. Mater. Rev. 52 (2007) 259–317.
[14] D.F. Emerich, C.G. Thanos, Biomol. Eng. 23 (2006) 171–184.
[15] J.D. Newman, A.P. Turner, Biosens. Bioelectron. 20 (2005) 2435–2453.
[16] E.M. Merisko-Liversidge, G.G. Liversidge, Toxicol. Pathol. 36 (2008) 43–48.
[17] A.K. Gupta, M. Gupta, Biomaterials 26 (2005) 3995–4021.
[18] I.L. Medintz, H. Tetsuo Uyeda, E.R. Goldman, H. Mattoussi, Nat. Mater. 4 (2005) 435–446.
[19] T. Pellegrino, S. Kudera, T. Liedl, A.M. Javier, L. Manna, W.J. Parak, Small 1 (2005) 48–63.
[20] M. Sastry, Curr. Sci. 85 (2003) 1735–1745.
[21] K.S. Soppimath, G.V. Betageri, K.E. Consalves, G.R. Halberstadt, C.T. Laurencin, L.S. Nair (Eds.), Biomedical Nanostructures, Wiley, Hoboken, 2008, pp. 409–437.
[22] K.J. Morrow, R. Bawa, C. Wei, Med. Clin. N. Am. 91 (2007) 805–843.
[23] W. Jiang, B.S. Kim, J.T. Rutka, W.C.W. Chan, Nat. Nanotechnol. 3 (2008) 145–150.
[24] J.E. Ghadiali, M.M. Stevens, Adv. Mater. 20 (2008) 4359–4363.
[25] C.A. Mirkin, R.L. Letsinger, R.C. Mucic, J.J. Storhoff, Nature 382 (1996) 607–609.
[26] S.D. Conner, S.L. Schmid, Nature 422 (2003) 37–44.
[27] A. Verma, F. Stellacci, Small 6 (2010) 12–21.
[28] C.T. Lin, S.M. Wang, Front. Biosci. 1 (2005) 99–106.
[29] N.L. Rossi, C.A. Mirkin, Chem. Rev. 105 (105) (2005) 1547–1562.
[30] M. Seydack, Biosens. Bioelectron. 20 (2005) 2454–2469.
[31] J. Wang, Small 1 (2005) 1036–1043.
[32] A. Burns, H. Ow, U. Wiesner, Chem. Soc. Rev. 35 (2006) 1028–1042.
[33] P.B. Johnson, R.W. Christy, Phys. Rev. B 6 (1972) 4370–4379.
[34] U. Kreibig, G. Bour, A. Hilger, M. Gartz, Phys. Stat. Sol. 175 (1999) 351–366.
[35] L.F. Shi, V. De Paoli, N. Rosenzweig, Z. Rosenzweig, J. Am. Chem. Soc. 128 (2006) 10378–10379.
[36] S.F. Lee, M.A. Osborne, Chem. Phys. Chem. 10 (2009) 2174–2191.
[37] J. Yguerabide, E.E. Yguerabide, Anal. Biochem. 262 (1998) 137–156.
[38] J. Yguerabide, E.E. Yguerabide, Anal. Biochem. 262 (1998) 157–176.
[39] M. Bruchez Jr., M. Moronne, P. Gin, S. Weiss, A.P. Alivisatos, Science 281 (1998) 2013–2016.
[40] R. Gill, M. Zayats, I. Willner, Angew. Chem. Int. Ed. Engl. 47 (2008) 7602–7625.
[41] B.Y.S. Kim, W. Jiang, J. Oreopoulos, Ch.M. Yip, J.T. Rutka, W.C.W. Chan, Nano Lett. 8 (2008) 3887–3892.
[42] R.A. Sperling, P.R. Gil, F. Zhang, M. Zanella, W.J. Parak, Chem. Soc. Rev. 37 (2008) 1896–1908.
[43] A.P. Alivisatos, W. Gu, C. Larabell, Annu. Rev. Biomed. Eng. 7 (2005) 55–76.
[44] F. Wang, W.B. Tan, Y. Zhang, X. Fan, M. Wang, Nanotechnology 17 (2006) R1–R13.
[45] M. Stroh, J.P. Zimmer, D.G. Duda, T.S. Levchenko, K.S. Cohen, E.B. Brown, D.T. Scadden, V.P. Torchilin, M.G. Bawendi, D. Fukumura, R.K. Jain, Nat. Med. 11 (2005) 678–682.
[46] W.W. Yu, E. Chang, R. Drezek, V.L. Colvin, Biochem. Biophys. Res. Commun. 348 (2006) 781–786.
[47] A. Aharoni, T. Mokari, I. Popov, U. Banin, J. Phys. Chem. 128 (2006) 257–264.

[48] X. Michalet, F.F. Pinaud, L.A. Bentolila, J.M. Tsay, S. Doose, J.J. Li, G. Sundaresan, A.M. Wu, S.S. Gambhir, S. Weiss, Science 307 (2005) 538–544.
[49] J.P. Zimmer, S.-W. Kim, S. Ohnishi, E. Tanaka, J.V. Frangioni, M.G. Bawendi, J. Am. Chem. Soc. 128 (2006) 2526–2527.
[50] X. Wang, et al., Nature 459 (2009) 686–689.
[51] N. Nath, A. Chilkoti, J. Fluoresc. 14 (2004) 377–389.
[52] R. Elghanian, J.J. Storhoff, R.C. Mucic, R.L. Letsinger, C.A. Mirkin, Science 277 (1997) 1078–1082.
[53] J. Liu, Y. Lu, J. Fluoresc. 14 (2004) 343–354.
[54] C.L. Schofield, A.H. Haines, R.A. Field, D.A. Russell, Langmuir 22 (2006) 6707–6711.
[55] H. Li, L. Rothberg, PNAS 101 (2004) 14036–14039.
[56] S. Liu, M. Han, Adv. Funct. Mater. 15 (2005) 961–967.
[57] C.A.J. Lin, T. Liedl, R.A. Sperling, M.T. Fernandez-Arguelles, J.M. Costa-Fernandez, R. Pereiro, A. Sanz-Medel, W.H. Chang, W.J. Parak, J. Mater. Chem. 17 (2007) 1343–1346.
[58] A.R. Clapp, I.L. Medintz, H. Mattoussi, Chemphyschem 7 (2006) 47–57.
[59] I.L. Medintz, J.R. Deschamps, Curr. Opin. Biotechnol. 17 (2006) 17–27.
[60] M.T. Fernandez-Arguelles, A. Yakovlev, R.A. Sperling, C. Luccardini, S. Gaillard, A. Sanz Medel, J.-M. Mallet, J.-C. Brochon, A. Feltz, M. Oheim, W.J. Parak, Nano Lett. 7 (2007) 2613–2617.
[61] E. Oh, M.-Y. Hong, D. Lee, S.-H. Nam, H.C. Yoon, H.-S. Kim, J. Am. Chem. Soc. 127 (2005) 3270–3271.
[62] K.A. Willets, R.P. Van Duyne, Annu. Rev. Phys. Chem. 58 (2007) 267–297.
[63] A.J. Haes, D.A. Stuart, S. Nie, R.P. Van Duyne, J. Fluoresc. 14 (2004) 355–366.
[64] C.A. Mirkin, Inorg. Chem. 39 (2000) 2258–2272.
[65] D.A. Schultz, Curr. Opin. Biotechnol. 14 (2003) 13–23.
[66] K. Mitsui, Y. Handa, K. Kajikawa, Appl. Phys. Lett. 85 (2004) 4231–4233.
[67] M. Hu, J. Chen, Z.Y. Li, L. Au, G.V. Hartland, X. Li, M. Marquez, Y. Xia, Chem. Soc. Rev. 35 (2006) 1084–1094.
[68] S. Schultz, D.R. Smith, J.J. Mock, D.A. Schultz, Proc. Natl. Acad. Sci. U.S.A. 97 (2000) 996–1001.
[69] I.H. El-Sayed, X. Huang, M.A. El-Sayed, Nano Lett. 5 (2005) 829–834.
[70] J.L. West, N.J. Halas, Annu. Rev. Biomed. Eng. 5 (2003) 285–292.
[71] M. Moskovits, J. Raman Spectrosc. 36 (2005) 485–496.
[72] N. Anderson, A. Hartschuh, L. Novotny, Mater. Today 8 (2005) 50–54.
[73] B.C. Keller, Trends Food Sci. Technol. 12 (2001) 25–31.
[74] S. Schlücker, Chemphyschem 10 (2009) 1344–1354.
[75] A. Wei, B. Kim, B. Sadtler, S.L. Tripp, Chem. Phys. Chem. 2 (2001) 743–745.
[76] S. Nie, S.R. Emory, Science 275 (1997) 1102–1107.
[77] W.E. Doering, S. Nie, J. Phys. Chem. B 106 (2002) 311–317.
[78] R.A. Tripp, R.A. Dluhy, Y. Zhao, Nanotoday 3 (2008) 31–37.
[79] X. Ji, Sh. Xu, L. Wang, M. Liu, K. Pan, H. Yuan, L. Ma, W. Xu, J. Li, Y. Bai, T. Li, Colloids Surf. A Physicochem. Eng. Asp. 257–258 (2005) 171–175.
[80] P. Zhang, S. Smith, G. Rumbles, M.E. Himmel, Langmuir 21 (2005) 520–523.
[81] G. Laurent, N. Felidj, S. Lau Truong, J. Aubard, G. Levi, J.R. Krenn, A. Hohenau, A. Leitner, F.R. Aussenegg, Nano Lett. 5 (2005) 253–258.
[82] S. Keren, C. Zavaleta, Z. Cheng, A. de la Zerda, O. Gheysens, S.S. Gambhir, PNAS 105 (2008) 5844–5849.
[83] F.V. Ignatovich, L. Novotny, Phys. Rev. Lett. 96 (2006) 13901–13904.

[84] H.M.E. Azzazy, M.M.H. Mansour, Clin. Chim. Acta 403 (2009) 1–8.
[85] E. Katz, I. Willner, J. Wang, Electroanalysis 16 (2004) 9–44.
[86] S.J. Park, T.A. Taton, C.A. Mirkin, Science 295 (2002) 1503–1506.
[87] J. Burmeister, V. Bazilyanska, Anal. Bioanal. Chem. 379 (2004) 391–398.
[88] V. Haguet, D. Martin, L. Marcon, T. Heim, D. Stievenard, C. Olivier, O. El-Mahdi, O. Melnyk, Appl. Phys. Lett. 84 (2004) 1213–1215.
[89] C.M. Welch, R.G. Compton, Anal. Bioanal. Chem. 384 (2006) 601–619.
[90] X. Luo, A. Morrin, A.J. Killard, M.R. Smyth, Electroanalysis 18 (2006) 319–326.
[91] K. Kerman, M. Saito, Y. Morita, Y. Takamura, M. Ozsoz, E. Tamiya, Anal. Chem. 76 (2004) 1877–1884.
[92] H. Cai, Y. Wang, P. He, Y. Fang, Anal. Chim. Acta 469 (2002) 165–172.
[93] K. Kerman, Y. Morita, Y. Takamura, M. Ozsoz, E. Tamiya, Anal. Chim. Acta 510 (2004) 169–174.
[94] S. Hrapovic, Y. Liu, G. Enright, F. Bensebaa, J.H.T. Luong, Langmuir 19 (2003) 3958–3965.
[95] X. Ren, X. Meng, D. Chena, F. Tang, J. Jiao, Biosens. Bioelectron. 21 (2005) 433–437.
[96] M.A. Riley, A.D. Walmsley, J.D. Speight, I.R. Harris, Mater. Sci. Technol. 18 (2002) 1–12.
[97] L. LaConte, N. Nitin, G. Bao, NanoToday 8 (2005) 32–38.
[98] E. Romanus, M. Huckel, C. Gro, S. Prass, W. Weitschies, R. Brauerc, P. Weber, J. Magn. Magn. Mater. 252 (2002) 387–389.
[99] Q.A. Pankhurst, J. Connolly, S.K. Jones, J. Dobson, J. Phys. D Appl. Phys. 36 (2003) R167–R181.
[100] J. Mäkiranta, J.O. Lekkala, Proceedings of the 2005 IEEE, Engineering in Medicine and Biology 27th Annual Conference, Shanghai, China, September 1–4, 2005.
[101] W.J.M. Mulder, G.J. Strijkers, G.A.F. van Tilborg, A.W. Griffioen, K. Nicolay, NMR Biomed. 19 (2006) 142–164.
[102] P.J. Debouttière, S. Roux, F. Vocanson, C. Billotey, O. Beuf, A. Favre-Réguillon, Y. Lin, S. Pellet-Rostaing, R. Lamartine, P. Perriat, O. Tillement, Adv. Funct. Mater. 16 (2006) 2330–2339.
[103] M. De, P.S. Ghosh, V.M. Rotello, Adv. Mater. Adv. Mater. 20 (2008) 4225–4241.
[104] Y.-W. Jun, J.-H. Lee, J. Cheon, Angew. Chem. Int. Ed. Engl. 47 (2008) 5122–5135.
[105] C.C. Berry, A.S.G. Curtis, J. Phys. D Appl. Phys. 36 (2003) R198–R206.
[106] P. Tartaj, M. del Puerto Morales, S. Veintemillas-Verdaguer, T. Gonzalez-Carreno, C.J. Serna, J. Phys. D Appl. Phys. 36 (2003) R182–R197.
[107] I. Koh, L. Josephson, Sensors 9 (2009) 8130–8145.
[108] J.W.M. Bulte, J. Magn. Magn. Mater. 289 (2005) 423–427.
[109] E.X. Wu, H. Tang, J.H. Jensen, NMR Biomed. 17 (2004) 478–483.
[110] M. Zhao, D.A. Beauregard, L. Loizou, B. Davletov, K.M. Brindle, Nat. Med. 7 (2001) 1241–1244.
[111] M.L. Zelivyanskaya, J.A. Nelson, L. Poluektova, M. Uberti, M. Mellon, H.E. Gendelman, M.D. Boska, J. Neurosci. Res. 73 (2003) 284–295.
[112] J.-H. Lee, Y.-M. Huh, Y.-W. Jun, J.-W. Seo, J.-T. Jang, H.-T. Song, S. Kim, E.-J. Cho, H.-G. Yoon, J.-S. Suh, J. Cheon, Nat. Med. 13 (2007) 95–99.
[113] M. Nichkova, D. Dosev, S.J. Gee, B.D. Hammock, I.M. Kennedy, Anal. Biochem. 369 (2007) 34–40.
[114] H. Zeng, M. Snyder, Curr. Opin. Chem. Biol. 7 (2003) 55–63.
[115] T. Vo-Dinh, F. Yan, D.L. Stokes, T. Vo-Dinh Humana (Ed.), Protein Nanotechnology, Protocols, Instrumentation, and Applications, Press Inc., Totowa, NJ, 2005.

[116] H. Zhu, M. Snyder, Curr. Opin. Chem. Biol. 7 (2003) 55–63.
[117] R. Wilson, A.R. Cossins, D.G. Spiller, Angew. Chem. Int. Ed. Engl. 45 (2006) 6104–6117.
[118] J.P. Nolan, L.A. Sklar, Trends Biotechnol. 20 (2002) 9–12.
[119] E.Y. Sun, L. Josephson, K.A. Kelly, R. Weissleder, Bioconjug. Chem. 17 (2006) 109–113.
[120] P.S. Eastman, W.M. Ruan, M. Doctolero, R. Nuttall, G. De Feo, J.S. Park, J.S.F. Chu, P. Cooke, J.W. Gray, S. Li, F.Q.F. Chen, Nano Lett. 6 (2006) 1059–1064.
[121] T.R. Sathe, A. Agrawal, S.M. Nie, Anal. Chem. 78 (2006) 5627–5632.
[122] Y. Cao, R.C. Jin, C.A. Mirkin, Science 297 (2002) 1536–1540.
[123] M.J. Banholzer, J.E. Millstone, L. Qin, C.A. Mirkin, Chem. Soc. Rev. 37 (2008) 885–897.
[124] C. Mayer, R. Palkovits, G. Bauer, T. Schalkhammer, J. Nanopart. Res. 3 (2001) 361–371.
[125] M. Nichkova, D. Dosev, R. Perron, S.J. Gee, B.D. Hammock, I.M. Kennedy, Anal. Bioanal. Chem. 384 (2006) 631–637.
[126] C.P.Y. Chan, Y. Bruemmel, M. Seydack, K.K. Sin, L.W. Wong, E. Merisko-Liversidge, D. Trau, R. Renneberg, Anal. Chem. 76 (2004) 3638–3645.
[127] L. Wang, C.Y. Yang, W.H. Tan, Nano Lett. 5 (2005) 37–43.
[128] E.R. Goldman, A.R. Clapp, G.P. Anderson, H.T. Uyeda, J.M. Mauro, I.L. Medintz, H. Mattoussi, Anal. Chem. 76 (2004) 684–688.
[129] K. Douma, L. Prinzen, D.W. Slaaf, Ch.P.M. Reutelingsperger, E.A.L. Biessen, T.M. Hackeng, M.J. Post, M.A.M.J. van Zandvoort, Small 5 (2009) 544–557.
[130] E.I. Altinoglu, T.J. Russin, J.M. Kaiser, B.M. Barth, P.C. Eklund, M. Kester, J.H. Adair, ACS Nano 2 (2008) 2075–2084.
[131] M.F. Templin, D. Stoll, J. Bachmann, T.O. Joos, Comb. Chem. 7 (2004) 223–229.
[132] N.P. Praetorius, T.K. Mandal, Recent Pat. Drug Deliv. Formul. 1 (2007) 37–51.
[133] F.N. Bushrab, R.H. Müller, NewDrugs 5 (2003) 20–22.
[134] Y. Fukumori, H. Ishikawa, Adv. Powder Technol. 17 (2006) 1–28.
[135] J.-C. Olivier, NeuroRx 2 (2005) 108–119.
[136] A.A. Datea, V.B. Patravale, Curr. Opin. Colloid Interface Sci. 9 (2004) 222–235.
[137] C.J. Sunderland, M. Steiert, J.E. Talmadge, A.M. Derfus, S.E. Barry, Drug. Dev. Res. 67 (2006) 70–93.
[138] F. Alexis, E. Pridgen, L.K. Molnar, O.C. Farokhzad, Mol. Pharm. 5 (2008) 505–515.
[139] E. Allémann, R. Gurny, E. Doelker, Eur. J. Pharm. Biopharm. 39 (1993) 173–191.
[140] J. Williams, R. Lansdown, R. Sweitzer, M. Romanowski, R. LaBell, R. Ramaswamia, E. Unger, J. Control. Release 91 (2003) 167–172.
[141] J. Panyam, V. Labhasetwar, Adv. Drug Deliv. Rev. 255 (2003) 329–347.
[142] M.L. Hans, A.M. Lowman, Curr. Opin. Solid State Mater. Sci. 6 (2002) 319–327.
[143] J.E. Kipp, Int. J. Pharm. 284 (2004) 109–122.
[144] B.E. Rabinow, Nat. Rev. 3 (2004) 785–796.
[145] G.B. Sukhorukov, A.L. Rogach, B. Zebli, T. Liedl, A.G. Skirtach, K. Kçhler, A.A. Antipov, N. Gaponik, A.S. Susha, M. Winterhalter, W.J. Parak, Small 1 (2005) 194–200.
[146] M.C. Woodle, P.Y. Lu, Mater. Today 8 (2005) 34–41.
[147] H. Takeuch, Y. Matsui, H. Sugihara, H. Yamamoto, Y. Kawashima, Int. J. Pharm. 303 (2005) 160–170.
[148] N.N. Sanderes, S. De Smedth, J. Demeester, J. Pharm. Sci. 89 (2000) 835–849.
[149] P.G. Jenkins, K.A. Howarday, N.W. Blackball, N.W. Thomas, S.S. Davis, D.T. O'Hagan, J. Control. Release 29 (1994) 339–350.
[150] M. Gaumet, A. Vargas, R. Gurny, F. Delie, Eur. J. Pharm. Biopharm. 69 (2008) 1–9.
[151] A.A. Date, V.B. Patravale, Curr. Opin. Colloid Interface Sci. 9 (2004) 222–235.
[152] T.C. Yih, M. Al-Fandi, J. Cell. Biochem. 97 (2006) 1184–1190.

[153] A. Shahiwala, T.K. Vyas, M.M. Amiji, Recent Pat. Drug Deliv. Formul. 1 (2007) 1–9.
[154] Y.J. Yamanaka, K.W. Leong, J. Biomater, Sci. Polym. Ed. 19 (2008) 1549–1570.
[155] J.K. Oh, R. Drumright, D.J. Siegwartb, K. Matyjaszewski, Prog. Polym. Sci. 33 (2008) 448–477.
[156] S.A. Agnihotri, N.N. Mallikarjuna, T.M. Aminabhavi, J. Control. Release 100 (2004) 5–28.
[157] J. Dobson, Drug Dev. Res. 67 (2006) 55–60.
[158] A.M. Derfus, G. von Maltzahn, T.J. Harris, T. Duza, K.S. Vecchio, E. Ruoslahti, S.N. Bhatia, Adv. Mater. 19 (2007) 3932–3936.
[159] A. Jordan, R. Scholz, P. Wust, H.F. Khling, R. Felix, J. Magn. Magn. Mater. 201 (1999) 413–419.
[160] K. Hamad-Schifferli, J.J. Schwartz, A.T. Santos, S. Zhang, J.M. Jacobson, Nature 415 (2002) 152–155.
[161] R. Muller, R. Hergt, M. Zeisberger, W. Gawalek, J. Magn. Magn. Mater. 289 (2005) 13–16.
[162] D. Lapotko, E. Lukianova, M. Potapnev, O. Aleinikova, A. Oraevsky, Cancer Lett. 239 (2006) 36–45.
[163] M.J. Kogan, N.G. Bastus, R. Amigo, D. Grillo-Bosch, E. Araya, A. Turiel, A. Labarta, E. Giralt, V.F. Puntes, Nano Lett. 6 (2006) 110–115.
[164] C. Marambio-Jones, E.M.V. Hoek, J. Nanopart. Res. 12 (2010) 1531–1551.
[165] A. Panacek, L. Kvıtek, R. Prucek, M. Kola, R. Vecerova, N. Pizurova, V.K. Sharma, T. Nevecna, R. Zboril, J. Phys. Chem. B 110 (2006) 16248–16253.
[166] M.A. Maurer-Jones, K.C. Bantz, S.A. Love, B.J. Marquis, C.L. Haynes, Nanomedicine 4 (2009) 219–242.
[167] L. Dykman, N. Khlebtsov, Chem. Soc. Rev. 41 (2012) 2256–2282.
[168] L.R. Hirsch, R.J. Stafford, J.A. Bankson, S.R. Sershen, B. Rivera, R.E. Price, J.D. Hazle, N.J. Halas, J.L. West, PNAS 100 (2003) 13549–13554.
[169] E. Prodan, P. Nordlander, N.J. Halas, Nano Lett. 3 (2003) 1411–1415.
[170] M.T. Stephan, J.J. Moon, S.H. Um, A. Bershteyn, D.J. Irvine, Nat. Med. 16 (2010) 1035–1042.
[171] H. Fukumori, H. Ichikawa, Adv. Powder Technol. 17 (2006) 1–28.
[172] S.E. McNeil, WIREs Nanomed. Nanobiotechnol. 1 (2009) 264–271.
[173] R. De Santis, A. Gloria, T. Russo, U. D'Amora, S. Zeppetelli, C. Dionigi, A. Sytcheva, T. Herrmannsdorfer, V. Dediu, L. Ambrosio, J. Appl. Polym. Sci. 122 (2011) 3599–3605.
[174] M.V. Yezhelyev, X. Gao, Y. Xing, A. Al-Hajj, S. Nie, R. O'Regan, Lancet Oncol. 7 (2006) 657–667.
[175] Q. Chaudhry, R. Watkins, L. Castle, in: RSC Nanoscience & Nanotechnology No 14: Nanotechnologies in Food, Q. Chaudhry, R. Watkins (Eds.), Royal Society of Chemistry, London (2010), pp. 1–17.
[176] http://siteresources.worldbank.org/INTPOVERTY/Resources/335642-1210859591030/Food_Price_Watch_May2010.pdf.
[177] Helmut Kaiser Consultancy, Study: Nanotechnology in Food and Food Processing Industry Worldwide 2003–2006–2010–2015, 2004, available at www.hkc22.com/Nanofood.html.
[178] R. Mezzenga, P. Schurtenberger, A.D. Burbidge, M. Michel, Nat. Mater. 4 (2005) 729–740.
[179] B.S. Murray, R. Ettela, Curr. Opin. Colloid Interface Sci. 9 (2004) 314–320.
[180] P. Sanguansri, M.A. Augustin, Trends Food Sci. Technol. 17 (2006) 547–556.
[181] C. Pichot, Curr. Opin. Colloid Interface Sci. 9 (2004) 213–221.
[182] T. Vo-Dinh, T. Vo-Dinh (Ed.), Protein Nanotechnology, Protocols, Instrumentation, and Applications, Humana Press Inc., Totowa, NJ, 2005, p. 3.

[183] B. Ding, X. Zhang, K. Hayat, S. Xia, C. Jia, M. Xie, C. Liu, J. Food Eng. 102 (2011) 202–208.
[184] S. Beun, T. Glorieux, J. Devaux, J. Vreven, G. Leloup, Dent. Mater. 23 (2007) 51–59.
[185] K.S. Chan, Y.-D. Leea, D.P. Nicolella, B.R. Furman, S. Wellinghoff, R. Rawls, Eng. Fract. Mech. 74 (2007) 1857–1871.
[186] S.S. Sternstein, A.J. Zhu, Macromolecules 35 (2002) 7262–7273.
[187] N. Moszner, A. Gianasmidis, S. Klapdohr, U. Karl Fischer, V. Rheinberger, Dent. Mater. 24 (2008) 851–856.
[188] 3M EPE, AdperTM Single Bond Plus Adhesive, Technical product profile.
[189] X. Meng, T.-Y. Kwon, K.-H. Kim, Dent. Mater. J. 27 (2008) 666–671.
[190] W.Y. Zhou, M. Wang, W.L. Cheung, B.C. Guo, D.M. Jia, J. Mater. Sci. Mater. Med. 19 (2008) 103–110.
[191] P. Fratzl, R. Weinkamer, Prog. Mater. Sci. 52 (2007) 1263–1334.
[192] M. Safarikova, I. Safarik, Magn. Electric. Sep. 10 (2001) 223–252.
[193] http://www.chemicell.com/products/purification/blood/index.html.
[194] G.A. Martınez-Castanon, N. Nino-Martınez, F. Martınez-Gutierrez, J.R. Martınez-Mendoza, F. Ruiz, J. Nanopart. Res. 10 (2008) 1343–1348.
[195] P. Li, J. Li, C. Wu, Q. Wu, J. Li, Nanotechnology 16 (2005) 1912–1917.
[196] Q. Chen, L.M. Peng, Int. J. Nanotechnol. 4 (2007) 44–65.
[197] J.C. Leroux, Nat. Nanotechnol. 2 (2007) 679–684.
[198] D. Horn, J. Rieger, Angew. Chem. Int. Ed. Engl. 40 (2001) 4330–4361.
[199] X. Yang, E.W. Stein, S. Ashkenazi, L.V. Wang, WIREs Nanomed. Nanobiotechnol. 1 (2009) 360–368.

Chapter 7

Optoelectronics

1 INTRODUCTION

Technological barrier and high capital cost are two critical hurdles facing new entrants into the microelectronic industry. Furthermore, sustaining current miniaturization trends in performance/cost is quite difficult. This has raised the interest of a large number of research groups within both academia and industry to seek alternative manufacturing techniques for these devices. Furthermore, market demands for microelectronic devices with additional features such as flexibility and reduced weight are difficult to satisfy with current commercial technologies. Indeed new devices that are flexible, light, fast, durable, and have higher resolution are hardly achievable with current energy-intensive crystalline silicon-based technology.

Besides its relatively high cost, the environmental footprint of the computer desktop is quite significant. Compared to the fridge, the life-cycle carbon footprint of a computer is five times higher [1]. Current microelectronic processes also generate a significant amount of waste. This difference is mostly due to the fabrication processes used in current microelectronic Fabs. For example, electric tools account for about 65% of the total electricity, while in other microelectronic Fabs this represents only about 30% [2]. Vacuum requirements account in large part for this disparity. Thus, designing nonvacuum processes is attracting a lot of interest.

Replacing current photolithographic and silicon-based complementary metal–oxide–semiconductor (CMOS) technology has opened the door to alternative approaches. The current technology node set at around 45 nm is believed to be near fundamental physical limits. Furthermore, the price tag of current microelectronic Fabs with sub-45 nm futures is becoming quite prohibitive. Bottom-up approaches using QD are believed to soon overtake the current CMOS technology. The barrier height (E_b) and width (a) in single charge transport devices are important for QD-based electronic devices. This barrier controls electron (information carrier) position and flow. Reducing the barrier width below a fundamental limit will lead to a spontaneous transition of the electron, indicating an error state. The lower theoretical limit of the barrier has been estimated to $a_{min} = 1.45$ nm [3]. Barrier control is achievable

Interface Science and Technology, Vol. 19. http://dx.doi.org/10.1016/B978-0-12-369550-5.00007-0

using tunable semiconductor nanoparticles (NPs), although not yet at room temperature. Low-barrier transition state is obtained by applying a voltage corresponding to the energy barrier. However, it is difficult to measure the electric properties of a single nanoparticle (NP). Indeed, making electric contact on semiconducting NPs is challenging.

The relative ease with which numerous tunable QDs could be prepared in most chemistry laboratories contributed to the interest in the bottom-up manufacturing of semiconductor devices. Several review papers [4–11] have discussed potential optoelectronic applications of QD. However, the performance and durability of QD-based microelectronic devices are relatively poor. The scalability and cost effectiveness of QD-based devices are often questionable. Potential applications of NPs in different optoelectronic sectors are receiving some interest. Sensing and lasing properties have been discussed in numerous reviews. Other applications in coating and display industries are seldom reviewed.

In addition to NP technologies, advanced printing will be the key enabling technology for the introduction of flexible electronic devices [12–14]. Plastic substrates are not compatible with conventional integrated circuit processing steps. Printed electronics are expected to overtake conventional silicon-based electronics in the future (Fig. 7.1). As discussed above, silver and copper NP-based conductive inks are already being used to produce conductive coating instead of conventional physical deposition techniques. Low sintering temperature inks are currently developed for their use on flexible plastic substrates. Gold

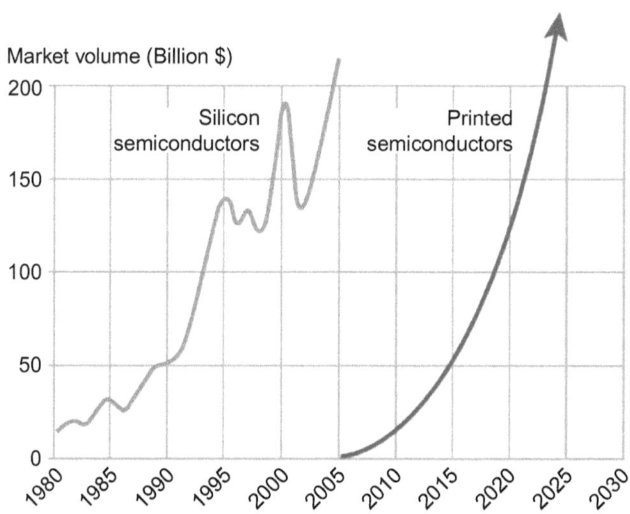

FIGURE 7.1 Comparative growth of conventional silicon electronics and printed electronics [15]. *Courtesy of Johnson Matthey & World Gold Council.* (For color version of this figure, the reader is referred to the online version of this chapter.)

NPs sintered at around 100 °C give rise to high electric conductivity. Gold NPs are currently used in touch-sensitive screens for iReaders and iPhones, as a replacement for indium-based TCO (transparent conductive oxides). Other potential printed electronic applications of NPs include data storage, pulp and paper, and textiles.

In this chapter, current and promising commercial solutions based on NPs are provided. Passive and active components using NPs are also described. A few coating films based on NPs are already in the market. Active electronic and optical devices are believed to take advantage of the special properties of NPs in the future. Futuristic applications like quantum computing using QD are not discussed here [16].

2 COATING AND COLORATION

Glass pans are omnipresent in numerous optoelectronic devices and systems. Besides their low cost, glass substrates are endowed with numerous basic physicochemical properties. Their chemical inertness, combined with their mechanical and thermal properties, makes them well sought. Their large-scale production at low cost, using the float process, is widely seen.

Metallic NPs embedded in glass have been used for centuries to control the optical properties of stained windows [17]. More recently, numerous functional devices such as switches, shutters and waveguides have been suggested for using NPs.

Glass pans are not always the best substrate, particularly when flexibility and weight are critical characteristics. Alternative substrates are often considered for some applications. Plastic substrates are preferred because of their reduced weight, flexibility, availability, and potentially reduced cost. The main drawback of plastic substrates is their low melting point. Wet coating methodologies that do not require very high deposition and annealing temperatures could address this issue.

Numerous plastic films match the transparency characteristics of glass substrates. However, their thermomechanical properties are weaker, requiring NP-based fillers to enhance these physical properties. In the case of flexible applications (electronic newspaper), this drawback is an advantage. Alternatively, NPs are used as nanofillers to form transparent nanocomposite films. The optical properties of these nanocomposites depend mostly on particle size, interparticle distances, relative refractive indices ($n_p - n_m$) of the matrix (n_m) and particles (n_p), and the optical path length (l). Relatively large light scattering is expected with composites containing micro- and macro-sized particles. The following Rayleigh law quantifies this effect [18]:

$$I/I_0 = 1/\exp\left[3\varphi_p l r^3 \left(n_p/n_m - 1\right)/4\lambda^4\right]$$

In the case of $n_p = n_m$ (index matching), there is no scattering of all particles. Thus, using well-dispersed NPs, it is possible to improve some physical properties without significant loss of transparency of the polymer film [18].

Metal-NP size, composition, and structure allow flexibility in tuning optoelectronic device properties. Using Au, Ag, and Cu NPs, the whole visible range can be covered. Orange-red and blue-violet spectra ranges are obtained using Au–Ag and Au–Cu alloyed NPs [20]. Bimetallic alloying allows filling the gap in the optical ranges not covered by single gold and silver NPs. Doping is another approach to tune the emission properties of NPs. For example, doping of silicon NP by Eu gives rise to a red shift from 394 to 613 nm [19].

2.1 Bulk Glass Coloration

The optical and thermal properties of glass are controlled through surface and bulk modification. The color of glass articles has been mastered for centuries using colloidal metal addition [17,21]. Ionic coloration by different transition metal oxides has been successfully used for centuries [22]. However, colloidal impregnation provides a better tunability in color control on a wide spectral range. The coloring based on colloidal modified flat glass is due to the surface plasmon resonance (SPR) from metal particles. For example, in the case of colloidal silver, gold, and copper, optical absorption bands around 410, 530, and 570 nm, respectively, are reported [23]. Furthermore, with today's precise colloidal synthesis approaches, it is possible to fine-tune size, shape and alloy composition to control absorption maxima from 200 up to 900 nm.

Color change of a glass containing metallic particles as a function of the viewing angle was observed more than 16 centuries ago [24]. Indeed the Lycurgus Cup appears green in daylight (reflected light), but red when light is transmitted from the inside [24]. This effect has been recently assigned to the presence of sub-100 nm metal alloy NPs [24]. Gold ruby glasses have also been reported [25]. Given the fact that light scattering is proportional to λ^{-4} (λ being the wavelength), only the blue part of the visible range (with UV) provides significant excitation. That is why highly absorbing metallic particles are required to reach the red part of the incident light. Besides, the size, color, and appearance of silver stained glass depend on NP shape and concentration. For NPs less than 20 nm in diameter, the glass tends to be transparent (no haze or light scattering) with yellow, brown, or red-brown coloring. For NPs larger than 20 nm, light scattering is significant, resulting in a more opaque appearance [26].

Float glass is a large commodity product with applications in solar, buildings, and transportation sectors. Metal-based coloration represents an increasing share of the overall market. Yellow coloration of float glass is obtained by heat treatment of sputtered silver film [27]. Larger silver NPs (less than

0.5 μm) are formed near the surface and dictate the overall visual coloration [27]. Silver layers on float glass are obtained on preformed copper colloids at the surface of float glass [28].

2.2 Surface Coating

Wet coating has several attributes as an alternative to industrial techniques such as CVD and PVD. Various wet coating techniques have been successfully used in the automotive and battery storage sectors with an overall film thickness of around 5 μm or more. For submicron film thickness, PVD and CVD techniques have mostly been used. However, there are potential coating methods that could be used even in high end optoelectronic devices. The lack of standard deposition equipment and the lack of collaboration between materials developer and equipment provider are the two most critical shortcomings preventing wider commercial use of wet-based coating solutions. Coating solutions containing NPs have a wide range of potential applications in optoelectronics. It is quite impossible to discuss all the developments in this area in this chapter. We focus on three areas: TCO, metal contact for optoelectronic devices, and antireflective coating (ARC).

The different colloidal NPs (metal, dielectric oxides, and semiconductors) obtained by sol–gel offer several technical and commercial advantages (see Chapter 2). Sol–gel solutions are easily transferred to substrates and devices using dip-coating, spin coating, screen printing, inkjet, capillary, and spray techniques. After the annealing step, they offer relatively high durability compared to standard techniques. The low sintering temperature of these NPs is a significant advantage, although its origin is not yet well understood. Lower melting of smaller NPs is often indicated, although higher surface to bulk atomic concentration could also play a significant role. As shown in Fig. 7.2, there are different atomic diffusion pathways during the sintering process. No densification is expected from lattice and surface diffusions. Grain boundary and through-diffusions are expected to contribute significantly toward the NP densification process. It is likely that the interface materials of these NPs have a much lower melting temperature than bulk materials. In the case of silver, a sintering temperature of around 150 °C has been reported, compared to a bulk melting of 961 °C [29].

The surface composition of ink NPs is critical to obtain high-density film and high conductivity. An oxide layer on Cu NPs synthesized in ambient atmosphere is controlled by adjusting the molecular weight of poly(*N*-vinylpyrrolidone) surfactant [30]. As shown in Table 7.1, higher molecular weight PVP leads to a thinner oxide layer and higher conductivity. Compared to the resistivity of bulk copper (1.7 μΩ cm), the film conductivity obtained using NPs is at least five times higher. It is also worth noting that transition temperature is lower with NPs having thinner oxide shell coating. Highly conductive

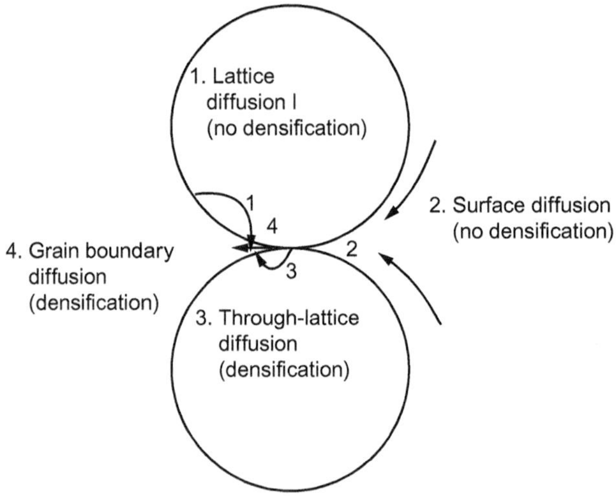

FIGURE 7.2 A schematic representation of different atomic diffusion paths contributing to interparticle densification. Only paths 3 and 4 contribute to densification [29].

TABLE 7.1 The Conductivity of Cu–PVP Nanocomposite Films Posttreated at Different Temperatures

	PVP molecular weight (g/mol)		
	10,000	*29,000*	*40,000*
Particle size (nm)	35±5	45±8	60±8
Oxide layer thickness (nm)	3.1	2.5	1.6
Resistivity at 275.8 °C (μV cm)	92	27.8	11.6
Resistivity at 325.8 °C (μV cm)	34	17.2	11.5
Exothermic peak (°C)	173	165	158

Three PVP molecular weights are used [30].

Cu patterns are obtained on a plastic substrate by inkjet-printing Cu NPs with the thinnest oxide layer [30].

Colloidal ink will certainly facilitate the transition from the current technology and capital-intensive silicon-based processes to the electronic printed era. For example, conducting patterns with high resolution is achievable using metal NP inks. These patterns are omnipresent in printed circuit boards (PCBs), light-emitting diodes (LEDs), photovoltaic cells, flat panel display (FPD) backplanes and radio-frequency identification (RFID) tags.

Furthermore, these NP inks are amenable to scalable roll-to-roll and reel-to-reel printing processes.

2.2.1 Transparent Conductive Materials

It is well known that light transmission through a glass pane can be controlled with very thin metal film. Distinct islands are formed, which transmit both visible and infrared light. This early stage of film growth corresponds to the Volmer–Weber mechanism. Upon island percolation, a relatively continuous film (20 nm) is formed. This film is transparent in the visible range and reflective in the infrared range. Once thicker layers are formed (>100 nm), the metal-coated glass reflects all incident radiation. Vacuum deposition techniques (CVD, PVD) are normally used to obtain these films.

TCO films having different compositions have been developed to address numerous industrial applications [31–33]. The most widely used TCO coatings are n-type semiconductors. Indium tin oxide or In_2O_3:Sn (ITO), fluorine doped tin dioxide or Sn_2O_3:F, antimony-doped tin dioxide, and aluminum-doped zinc oxide are readily prepared using the sol–gel route. High resistivity remains the biggest challenge of sol–gel-based TCO films because of their high porosity (low density) even after sintering at high temperature [34].

Low sheet resistance and high optical transmission TCO coating have been obtained using direct gravure printing of ITO NPs on polyethylene terephthalate and polyethylene naphthalate foils [35]. NPs with 20–30 nm diameter and a dispersing agent such as 3,6,9-trioxadecanoic acid are combined with a binder to form a homogeneous ink (Fig. 7.3). The resulting ink is deposited on PEN and PET substrates using a gravure printer. Posttreatment at temperatures between 130 and 180 °C is required to reach low sheet resistivity. Both large area and patterned coatings are obtained and could be used for OLED.

NP-based ITO films have been used in conventional LED [36,37]. High indium cost has steered interest in alternative TCO compositions. Transparent ZnO film has been obtained using the NP process [38]. Resistivity and optical characteristics of ITO coating obtained from sol and NPs are comparable to those obtained using sputtering and CVD [39]. Often, the electrical and optical performances of NP-based ITO films are less than those obtained using conventional processes [40,41]. The sol–gel process, used to prepare ITO film, has shown promising results [42].

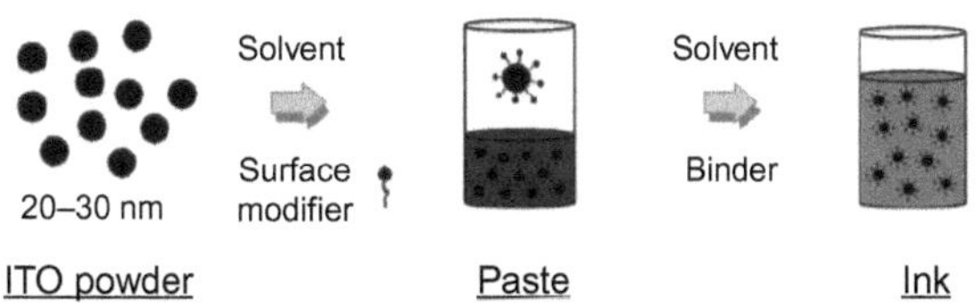

FIGURE 7.3 The two processing steps leading to ITO nanoparticle-based ink [35].

2.2.2 Nontransparent Conductive Contact

Printing processes on low cost and flexible substrates will eliminate the requirement for high-cost wafers using high vacuum and lithographic techniques. Electric wiring is important and omnipresent on electronic devices. Often, this is achieved via highly conductive layers connecting various components using complex and expensive photolithographic processes. Electrolytic plating and microparticle-based screen printing are simpler but do not address current requirements for low temperature and high-resolution patterning.

Silver NPs modified by detachable alkylamines are printed on silica to form conductive contacts following low temperature sintering [43]. These contacts are used as source/drain electrodes in organic thin film transistor (OTFT). Large and direct energy band gap gallium nitride semiconductors are used in a variety of optoelectronic applications, including LEDs, laser diodes, and photodetectors. Attempts to build p-type ohmic contacts on GaN that have high light transmittance as well as low contact resistance have not been very successful [44]. A thin interlayer of gold NPs between the GaN and ITO layer has been shown to provide the best intermediary solution [44]. The gold interlayer was obtained by spraying a solution of 5 nm diameter gold NPs on heated substrate at around 120 °C.

A patterned nontransparent film is obtained using NP deposition followed by contactless AC sintering process [45]. This process is fast since it requires only a few microseconds [45]. Pulse and square-shaped submicron conductor patterns are obtained using this method.

Controlling the oxide layer thickness on copper NP is critical. It is possible to minimize the surface oxide layer on Cu NPs synthesized in ambient atmosphere by adjusting the molecular weight of poly(*N*-vinylpyrrolidone) capping molecules [30]. Oxide layer of 1.6 nm thickness has been obtained using 40,000 g/mol PVP molecular weight. This has allowed higher Cu density and improved conductivity.

Compared to organic-based printed patterns, metal ink gives rise to much higher conductivity (Fig. 7.4). Indeed, conductivity in the range of $10^{-6}\ \Omega$ cm is achievable using metal NPs. This is not far from evaporated metal conductivities. Another advantage of metal ink is its relatively high stability compared to organic ink, and lower melting temperature compared to bulk metal. Several companies are currently marketing colloidal metal inks. This is the case of NanoMass Inc. [46].

The high-conductivity films and patterns obtained are crucial for the development of low-cost and low-energy content microelectronic devices, including interconnects, sensors, RFID tags, high-Q inductors, and capacitors. Alkanethiol-stabilized gold NPs are successfully inkjet printed and annealed on a commercial polyester-based general purpose transparency plastic [47]. The transition temperature during annealing is reduced from around 175 to

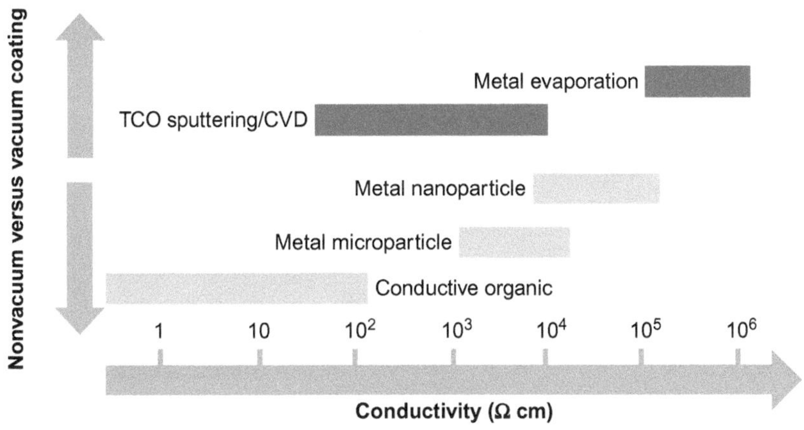

FIGURE 7.4 Conductivity ranges of films obtained using vacuum and nonvacuum deposition methods. Note that metal and organic-based inks are amenable to printing techniques. (For color version of this figure, the reader is referred to the online version of this chapter.)

120 °C when dodecanethiol (12 carbon chains) and butanethiol (4 carbon chains) are used respectively. Optimal conducting properties are obtained using 1.5 nm hexanethiol gold NPs dispersed in plastic matrix [47].

2.2.3 Antireflective Coating

ARC is used in solar cells, ophthalmic lenses, displays, and architectural windows. Presently, vacuum-based techniques (sputtering and evaporation) are the preferred methodology. In most cases, it is desirable to minimize reflectance to enhance optical device efficiency.

Recent reports have shown that it is possible to reach high efficiency ARC using multilayer coating on plastic substrate while using aqueous colloidal dispersions of metal oxide NPs [48]. They show that the refractive index for the polymer film was tuned from 1.46 to 1.54 and from 1.54 to 1.95 using silica and ceria NPs, respectively. Silica NPs self-assemble to form ARC film [49]. High-quality MgF_2 thin films have been obtained using fluoride NPs [50].

2.2.4 Ohmic Contact

Ti film embedded with gold NPs has been deposited on n- and p-type SiC substrate using aerosol techniques [51]. The observed reduction in the Schottky barrier has been assigned to the enhanced electric field at the interface provided by the presence of the metallic NP.

Gold NP has also been shown recently to lower the contact resistance between p-type GaN layers with the ITO substrate of an LED [44]. Using Zn–Ni NPs lowers contact resistance between silver metal contact and

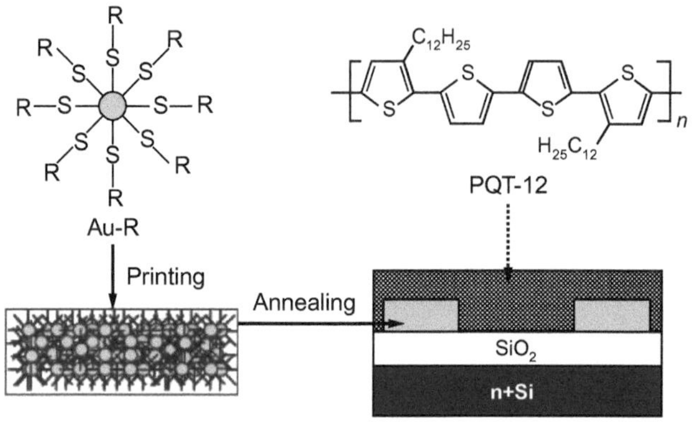

FIGURE 7.5 Schematic outline of an OTFT electrode fabrication process from a gold nanoparticle solution [53]. (For color version of this figure, the reader is referred to the online version of this chapter.)

InGaN/GaN multiquantum well of flip-chip LED. This has been achieved without losing output power [52]. Solution-printing of gold NP dispersion-based ink has been used to form highly conductive ohmic contact on OTFT. As demonstrated in Fig. 7.5, the gold NP ink could be used to form source and drain electrodes of the bottom-contact OTFT. Following ink deposition, thermal annealing gives rise to a dense and continuous film structure. Indeed, electrical resistivity deceased significantly following thermal annealing at 200 °C for 30 min [53].

Dispersion of copper NPs with an average size of 5 nm is used to form an ultrathin and conductive layer on TiN and TaN [54]. These layers are used for ULSI interconnects [54]. A spin-on-metal copper (SOM-Cu) solution including Cu NPs, an organic solvent, and a surfactant is first prepared. This solution is applied by spin coating on a Si wafer precoated by TiN or TaN barrier layer [54]. Following infrared thermal annealing at 350–400 °C, high-conductivity film with high adhesion and strength is obtained [54].

Water-based silver NP ink has been inkjet-printed on glass substrates using an Epson R210 printer [55]. Using 50 nm Ag NPs dispersed in water and diethylene glycol co-solvent system, continuous and smooth lines of 130 μm with a resistivity of 1.6×10^{-5} Ω cm were obtained. Heat treatment at 260 °C for 3 min has been used to increase density.

3 DISPLAY

Visible display is a key part of the information technology and media industries. Direct view display and projection display have been developed and used. Recently, 3D image displays are attracting a lot of interest. In this chapter, we are mostly concerned with direct display. Two display technologies

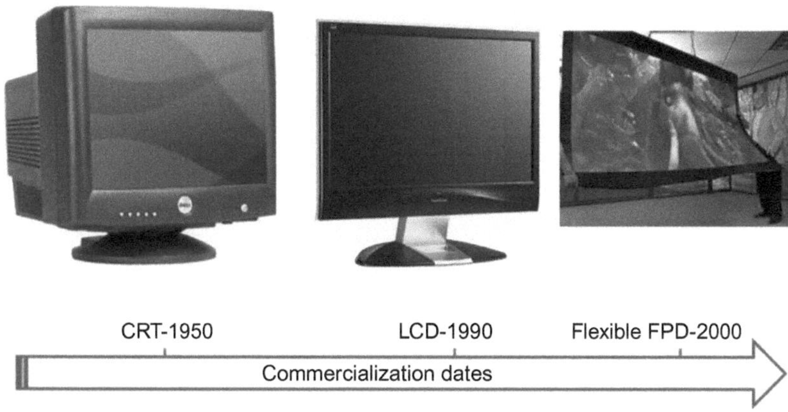

FIGURE 7.6 The three main technologies illustrating the evolution of display products over the last 100 years. (For color version of this figure, the reader is referred to the online version of this chapter.)

allowing conversion of electronic data to a visual display are used today (Fig. 7.6). The traditional display technology is currently dominated by the bulky cathode ray tube (CRT). This category is losing the market to the second category consisting of FPD (flat panel display). A third category of display technologies based on a flexible substrate has currently been developed, although it is not yet widely available. This is particularly the case of e-paper technologies. The FPDs are present in the TV, desktop, laptop, and PDAs. Although it was developed in the 1950s, only in the last two decades the commercial viability of FPD has been demonstrated. In particular, increased substrate size has allowed it to dominate the TV market completely. With its three killer applications (laptop, monitor, and TV), thin film liquid crystal display (LCD) takes around 90% of the electronic-display market. With the development of the large screen, TV application will probably continue increasing its market share. Flexible FDP is currently developed and commercialized, facilitated by recent developments in printed electronics.

Recent trends in the use of devices with multiple applications and simple user interface have added new requirements to display technology. This includes the touch screen, high ppi (pixels per inch), large size with full floor, fast response, and slim shape. All these requirements should be met using high volume and low-cost materials and processes. Several e-paper technologies are currently being developed to address these issues. Besides electrophoretic display technologies, cholesteric LCD and electrowetting techniques have been successfully demonstrated in prototypes.

FPD is presently produced using three technologies: LCD, plasma display panel (PDP), and field emission display (FED). The first two are mostly used for display up to one meter or more, whereas the last one is mostly used for screens larger than 1 m. FED-, OLED-, and PLED-based display have now

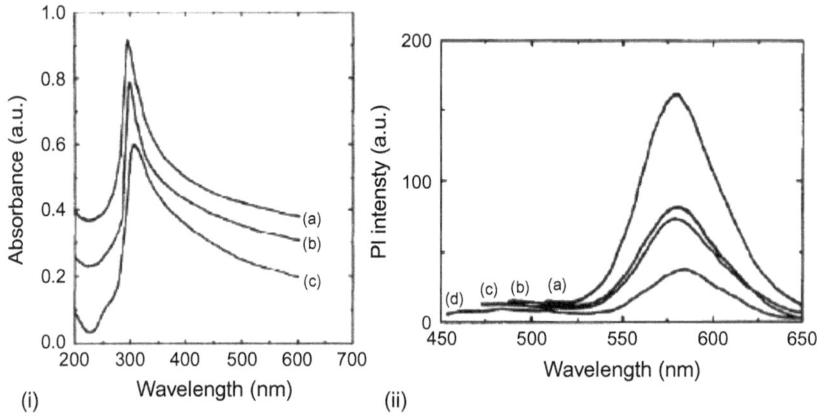

FIGURE 7.7 UV absorption (i) and PL (ii) spectra of ZnS:Mn nanoparticles of different sizes: (a) 2.3 nm, (b) 2.5 nm, (c) 2.7 nm, and (d) conventional bulk ZnS:Mn [61].

been introduced into the market, though for smaller display size. A recent report available from JISC [56] provides a nice overview for the interested reader to explore the advantages and drawbacks of each display technology. Thin film transistor (TFT)-based LCD retains the lion's share of the FDP market. Indeed, in 2010, TFT-LCD represented around 80% of the electronic-display market [57]. OLED/PLED is currently challenging this dominance, particularly in small size display.

Luminescent semiconducting NP (nanophosphor) remains the main application of NPs in display technologies. Recent developments in nanocrystalline phosphor will likely stimulate further development in the area of emission-based display [58,59]. There are two classes of nanocrystalline phosphors: semiconductor quantum dot and transition or rare-earth metal-doped phosphors. It has been well established that luminescence quantum efficiency (QE) increases dramatically with decreasing diameter [60,61]. This property is well illustrated in Fig. 7.7 in the case of Mn-doped ZnS NPs.

3.1 Cathode Ray Tube

Although the trend is toward FPD, state of the art CRT will continue to retain a relatively big market share. CRT still provides comparatively high brightness. As discussed earlier, NP-based coating allows controlled light transmission through the viewing glass. It could also enhance the quality of visible light coming out from the display set. Replacing transition metal and rare-earth compounds with alternating phosphor materials provides potential avenues for the application of NPs. Nanophosphors play a key role in enhancing the technical performance of both FED and CRT [59]. Nanophosphors are discussed below in more detail.

3.2 Plasma Display Panel

Phosphor materials fabrication and integration are critical aspects of plasma manufacturing [62]. A recent review article has summarized the physics and issues behind PDP (plasma display panel) technology [63]. PDP consists of millions of pixels addressed by metallic and TCO-based electrodes. Each pixel is made up of three fluorescent lights: a red light, a green light, and a blue light. Each pixel contains three cells (or subpixels) that are coated with phosphor material. These phosphor materials are excited with UV light emitted by rare gas (Xenon or Neon) that turn to plasma state once electric current is passed between the anode and cathode. The stimulated phosphorus will reemit visible light. Each pixel contains three color-emitting phosphors: red, green, and blue. By varying the characteristics of the current flowing through the different subpixels, one can control the individual intensity of each color-emitting subpixel. These colors are blended together to create the overall color of the pixel.

Phosphor materials used in PDP must have high QE (quantum efficiency), low reflectivity in VUV, and must be stable and highly reflective for visible light. State of the art phosphors that have achieved QE are usually above 80% [64]. The willemite (Zn_2SiO_4:Mn) is the most widely used phosphor for green emission. For blue emission, it is $BaMgAl_{10}O_{17}$: Eu_2^+ (BAM). (Y,Gd)BO_3:Eu^{3+} and Y_2O_3:Eu^{3+} are used for red emission. High power consumption is probably the main issue facing PDP. Various colloidal nanophosphors that may find use in both PDP and CRT have been developed [59].

Due to the dominant surface state passivation effect, colloids usually provide poor electroluminescent properties. Bhargava et al. [65] showed that rare-earth impurities in a QD (also referred to as doped nanocrystal or DNC) provide electroluminescent properties. ZnS and Y_2O_3 showed a sharp increase in QE for smaller particle size [66].

Compared to conventional phosphors, direct bandgap II–VI semiconductors are expected to have brighter PL with tunable emission wavelength [67–71]. However, long stability of II–VI NPs remains an issue that should be addressed. Several approaches have been developed to provide longer lifetime, using NP-based phosphors. Incorporation into core–shell silica particles [67,68] and homogeneous dispersion in SiO_2 glass matrix [69,71] seem to provide promising results.

3.3 Liquid Crystal Display

Thisayukt et al. showed that twisted nematic liquid crystal display (TN-LCD) provides good electro-optic characteristics [72]. The TN-LCD consists of nematic liquid crystal (NLC) doped with silver NPs. Using NPs coated with 5CB (Merck, K-15), which is highly compatible with NLC molecules, good frequency modulation and fast response are obtained. The electrical conductivity of the doped Ag NPs plays a significant role. NLC E7 from Merck containing silica NPs has also shown promising electro-optic properties for future use in LCD [73]. Polymer NPs filled with nematics also provide good electro-optical properties [74].

3.4 Field Emission Display

Commercially available displays including CRT, LCD, and PDP have low luminescence (lm/W) efficiency. Even fluorescent lamps are 10–100 times more efficient than these conventional display technologies. In an FED, an image is obtained by impinging electrons from an array of millions of cold emitters onto a phosphor-coated screen. Requirements for high voltage imply utilization of complex and sophisticated display components [75]. Overall, display cost and large area scalability are major issues limiting the competitivity of FED.

Various emitter materials have been developed. Diamond, composite, graphite, carbon fibers, and carbon nanotubes are materials of interest for cold cathode applications [76,77]. Nanocrystal phosphors are endowed with superior PL properties [78,79]. Nanocrystal doping has shown improved luminescence properties [80]. Although well established in CRT manufacturing, the technology of phosphor utilization has not yet matured in the case of FED systems [76]. Utilization of luminescent rare-earth-doped oxides could provide high purity, uniform composition and structure, small and monodisperse particles, and a relatively low operation voltage [81,82].

Even a small amount of contamination will affect the optical characteristics. Mn-doped ZnS nanosized phosphor has been shown to compare quite well with microsized commercial phosphor [82]. The intensities of ZnS:Tb and ZnS:Eu nanocrystals prepared by coprecipitation were found to be about three times higher than those of bulk phosphors [78]. Furthermore, degradation under electron bombardment is significantly reduced by using a core–shell structure with a silica shell coating. Surface modification of $ZnS:Mn^{2+}$ using acrylic acid monomers has been shown to enhance the PL [80].

3.5 NP-Based LED

The work of R. Friend's group at Cambridge on polymeric electroluminescence has spurred a strong industrial interest in building plastic-based LED and displays. Cambridge Display Technology and other companies have attempted to commercialize this new technology. Despite its early success, polymeric LED (PLED) has some limitations. The efficiency, ratio of light energy emitted over the electricity used, is still relatively low ($\sim$5%). The high cost and lack of stability of the semiconductor polymers are still hindering the development of PLED. Another limitation is related to the size of the deposition area. NP-based LED (nanoLED), also referred to as QD LED (or QLED), provides some improvement over organic LED (OLED) and PLED. Higher luminous efficiency and low power consumption are achievable. Quantum efficiencies as high as 50% were reported a decade ago with the use of a core–shell structure (ZnS shell on CdSe core) [83].

The size dependence of semiconductor band gap could be used to tune the fluorescence wavelength. Smaller NPs give rise to a larger optical band gap. An optical range of CdSe–CdS core–shell fluorescent NPs with tunable diameters from 1.7 (blue) to 6 nm (red) was obtained [84]. This allows, for example, for building LED in the UV and visible emission ranges.

Due to the unique luminescence properties of inorganic nanophosphors and the good processability of their dispersions in polymer solutions or monomers, inorganic/polymer nanocomposites are promising materials for different lighting and display devices. Two basic structures can be used for NP-based LEDs [18]. One consists in separating the hole transporting semiconducting polymer layer and inorganic NPs for the emission layer. In a second structure, both semiconductors are combined in a single nanocomposite layer. It is already recognized that embedding luminescent NPs in polymers offers a wide range of applications and integrations. For example, CdSe NPs embedded in a semiconducting polymer matrix give rise to an LED with tunable optical emission colors [85]. A near-infrared LED using InAs/polymer nanocomposites has been reported [86]. CdSe–ZnS core–shell NPs embedded in lauryl methacrylate covering a GaN diode were reported to produce effective diodes [87]. Under excitation of the blue LED, the particles showed photoluminescence with emission colors depending on the particle size.

Dispersed inorganic quantum dots in a polymeric matrix (nanoPLED) combine the attributes of conventional LED with the processing and mechanical advantages of PLED and QD optoelectronic properties. In particular, nanoPLED can provide the following advantages: thinner and lighter film with no need for backlighting; flexibility; lower power consumption; faster reaction time; and brightness.The addition of QD not only improves the technical characteristics but may also improve the overall durability in comparison to organic materials.

The toxicity of Cd, Se, or As could be a deterrent to the proposed nanoLED and nanoPLED. Doped ZnS NPs, including orange-yellow (manganese) and blue-green (copper), are nontoxic and could emit over a wide range. In the case of Mn-doped ZnS NPs, a stable dispersion in polymers with high efficiency is achieved [88]. These dispersions are easily amenable to integration and process using printing and other nonvacuum technologies.

Although conventional semiconductor NPs are often discussed, it is important to realize that some metallic particles behave like semiconductors when their size is reduced to below a critical value. This is certainly the case with gold, silver, and their alloys. The energy band gap of gold clusters could be tuned from 0.2 up to 1.5 eV, covering part of the visible range and most of the infrared range [89]. This is achieved when the gold cluster size is reduced from Au_{255} (0.2 eV) to Au_{13} (1.5 eV) [89].

A significant improvement in the characteristics of solution-based QD LED has been reported, by using enhanced charge properties of the transport layers [90]. The proposed LEDs consist of an ITO anode, a nickel-based hole transport layer, the active luminescent NP (QD) layer, the $ZnO:SnO_2$ electron transport layer, and the silver cathode. Using ultrathin NiO and $ZnO:SnO_2$ interlayers between the active QD-based layer and the respective electrodes leads to a 100-fold improvement over previously reported structures [90]. Indeed, this device can operate under current densities exceeding 3.5 A/cm^2 with a peak brightness of 1950 Cd/m^2 and external electroluminescence efficiency of around 0.1% [90].

Incorporation of functionalized polymeric NPs in nonemissive layers improves the efficiency of organic LEDs [91]. With the use of amino functionalized polymeric NPs, the resultant power efficiency is improved by a factor of 2. This improvement is due to modulation of the hole charge transport, which allows controlling the number of holes entering the emissive layer. The charged QDs are used as a charge blocking or trapping medium.

3.6 Electrophoretic

Paper-less information platforms are currently attracting a lot of interest because of the booming portable display market. For example, electrophoretic display technologies are currently being used in several commercial e-readers. Typically, a solution of micron-sized TiO_2 particles dispersed in organic solvent containing dark-colored dye, surfactants and surface charging agents is sandwiched between two conducting surfaces, used as front and back plates. After applying an appropriate voltage, particles migrating to the front (viewing) plate appear white because incident light is scattered back owing to the high refractive index of TiO_2. The particles migrating to the back plate appear dark because the incident light is absorbed by the dye. Usually, the front electrode consists of a TCO-coated transparent substrate, whereas the back plate may consist of a metallic layer. Dividing the back plate into pixels with specific voltages creates reflecting (white) and absorbing (dark) patterns. This technology and its application in the context of e-paper and e-readers are discussed in more details later in this chapter.

NPs may contribute to improving the electrophoretic display performance/cost ratio. In display applications, it is advantageous to use NPs having diameters substantially less than the wavelength of visible light. This will improve the electrical properties of the medium without affecting the optical properties. Nonabsorbing and conductive NP fillers of about 20 nm have been used to lower volume resistivity, increase switching rate, and image bistability [92]. These filler NPs are incorporated into the composition of the microcups supporting the display cells. A more efficient and rapid electrophoretic image display has been achieved using small-sized and monodisperse NP obtained by dispersion polymerization [93].

To match the density of the surrounding medium, while maintaining high mobility, appropriate surface medication is required. Park et al. used comonomer ethylene glycol methyl ether acrylate to encapsulate TiO_2 NPs by dispersion polymerization. This gives rise to relatively high dielectric properties and a well-dispersed solution. Electrophoretic properties are improved by adding a succinimide-type charge control agent to increase the surface charge density [94]. Sub-100 nm toluidine red (TR) pigment NPs show superior electrophoretic properties [95]. E-Ink Corporation [96] used microcapsules embedded in a liquid polymer layer on a plastic substrate. These microcapsules are filled with charged particles dispersed in colored organic solvent.

4 PHOTONICS AND OPTOELECTRONICS

Impinging electromagnetic (EM) radiation on NPs excites the electronic and vibrational states. This opens up a lot of opportunities for optical applications, including passive and active optical devices [97]. In the following secion, a few examples of potential NP applications in optoelectronics are provided.

4.1 Laser and Radiation Sources

In theory, quantum dots are ideal active materials for lasing because of their insensitivity to temperature, lower lasing threshold, and high-gain profile [98]. Eisler et al. used nanocrystalline TiO_2 to fabricate an optically pumped laser with size-tunable color output dependence [99]. Semiconductor QDs with tunable wavelength emitting in the near-infrared have been demonstrated [100].

Tunable blue semiconductor nanocrystal lasers have been achieved using core–shell CdS/ZnS NPs [101]. Composite films are first made from silica microspheres coated with core–shell CdS/ZnS NPs. Tunable room temperature-amplified spontaneous emissions are obtained at 454, 477, and 503 nm with a small full-width at half-maximum (FWHM) between 5 and 25 nm. Furthermore, a uniform resonator lasing over long periods is achieved by coupling the CdS/ZnS NPs to a spherical resonator. Colloidal CdSe/ZnS QDs incorporated into sol–gel-derived titania matrixes are also used to fabricate a robust laser [102]. The stability of this film under different chemical environments allows their integration into a microfluidic network. Er_2SiO_5 NPs have been synthesized and integrated using Si-NWs (silicon nanowires) as a template substrate [103]. Using this structure, a sharp luminescence has been obtained even at room temperature. This nanocomposite can be used for a compact high-gain silicon-based light source.

4.2 Waveguide

Integration of colloidal semiconductor NPs in photonic structures will likely enable technological breakthroughs in the area of monolithic light emission and modulation devices on chips. QDs offer potential applications in the area of telecommunications networks. This is particularly the case with II–VI materials. For example, their strong and tunable luminescent properties make them potential candidates for optical amplifier media [104]. In particular, QDs with high QE infrared-emitting materials can be used as gain media in ultra-wideband optical amplifiers. This property is important for high-capacity wavelength-division multiplexing telecommunications systems.

A multilayered film based on tunable CdTe/HgTe NPs will cover a wide PL wavelength range. Indeed, adding HgTe layers to CdTe NPs shifts the PL peak from around 600 to 1200 nm [104]. Note that this PL shift is achieved by adjusting not only the composition (alloying and multilayering) but also the size (ripening process). The absorption characteristics of silica-based transmission fibers open up possibilities of 1.3 and 1.55 μm as telecommunications wavelength windows. Based on this, other alternative materials could be considered. For example, GaAs and InP are particularly suitable in spite of their higher cost [104].

Optical gain has been reported for the first time using CdS QDs embedded in photonic crystals [105]. The photonic crystals are based on 3D synthetic opals. These QD-photonic crystal composites give a gain maximum (around 2.2 eV). This is far below the edge of the photonic stop band of the semiconductor (around 2.5 eV). This shift has been ascribed to multiple coherent Bragg scattering in the periodic 3D photonic crystals [105]. Concepts of single NP photonic switch and optical memory element have also been described [106]. These NPs could be integrated into nanostructured optical waveguides. Another option is to use these NPs as scattering centers in photonic band gaps and plasmon–polariton waveguides.

Poitras et al. showed that spontaneous emission of colloidal CdSe QDs embedded in a 1D cavity is enhanced by a factor of 2.7 [107]. They also showed that the absorption of the CdSe quantum dots is enhanced by an order of magnitude. This enhancement is ascribed to the cavity effect.

Semiconductor quantum dots have been used in polymer-based optical devices, offering the possibility of low-cost integration into various optical designs [108]. To allow good miscibility, native QD ligands are replaced by aromatic aniline capping ligands. PbSe and InAs QDs have been integrated to perfluorocyclobutane (PFCB)-based host matrix following surface modification [108]. Indeed, a TOP/TOPO-capped InAs/ZnSe QD was exchanged with 1,1,1-tris (4-trifluorovinyloxy phenyl) ethane monomer (aniline) [108].

The PFCB polymer structure formed after cyclopolymerization of the monomer is endowed with good thermal, dielectric and optical properties. The Qd–polymer nanocomposites can be used in passive and active optical waveguides. In particular, they show low optical loss in the visible and NIR ranges.

PbSe- and InAs-based nanocomposites are used to fabricate waveguide devices on a silicon substrate [108]. First, trenches are made by etching thermally grown silica glass. Then, the trenches (3 μm deep and 2 μm wide and 9 μm long) are filled with the nanocomposite by spin coating followed by an additional cover layer. Following a curing step at 120 °C, a final cladding layer having the same refractive index as the substrate is added. Since the refractive index of the intermediary nanocomposite layer is slightly higher than that of the top and bottom glass cladding, this method provides a cheap and precise process to fabricate a waveguide. Furthermore, there is no degradation of the emission features following the final curing stage (Fig. 7.8). This confirms that the optical properties of the QD are only minimally affected by the various processing steps.

A pump-probe technique has also been used to demonstrate that optical gain is achieved in this nanocomposite-based waveguide. For commercial use in lasers and amplifiers, a higher QD/polymer volume fraction should be used in the polymer hybrid system [108].

Two layers are deposited on a silicon substrate covered with its native oxide layer to build a wavelength-tunable light-emitting source [109]. The first layer is a silica cladding. The second layer is based on semiconductor NPs (CdS, CSe, CdSe/ZnS) dispersed in a ZrO_2 sol–gel film [109]. A zirconia-Glymo sol–gel film using acetylacetone as the zirconium chelating agent instead of acetic acid was used. This allowed a high degree of cross-linking between the Zr and the Glymo. The incorporation of the photoemissive CdSe/ZnS core–shell NPs within the zirconia gel waveguide will provide a means to create miniature multispectral light sources [109].

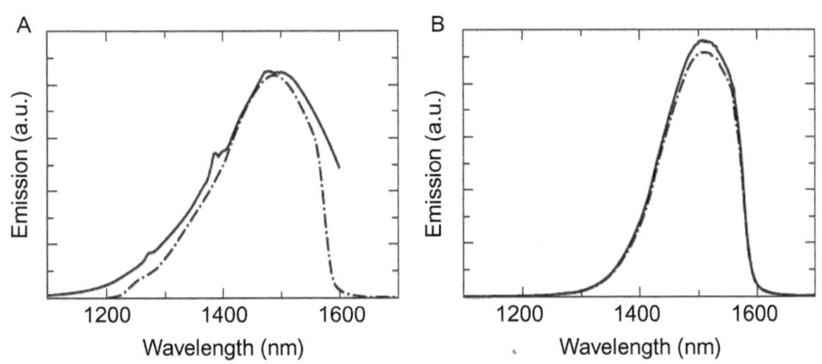

FIGURE 7.8 Emission spectra of PbSe QD (A) and InAs QD (B) before (solid lines) and after (dash-dotted lines) curing [108].

Optical switching based on PbSe NPs has been reported [110]. Optical switching on the order of 10% and a recovery time of 1.5 ps have been measured [110]. Intense photoluminescence from 1000 to over 1700 nm has been obtained with HgTe NPs [111]. HgTe NPs are a potentially cost-effective alternative to the erbium-doped fiber amplifier or ultra-broadband optical amplifier, allowing complete coverage of the telecommunications wavelengths.

4.3 Negative Refractive Index

Negative refractive materials continue to attract significant interest in academia [112]. One potential application offered by these metamaterials is the resolution of subwavelength features. This, for example, allows producing sharper images. At a given frequency (ω), refractive index is defined by the relation:

$$n^2(\omega) = \varepsilon(\omega) \times \mu(\omega)$$

where $\varepsilon(\omega)$ and $\mu(\omega)$ are the permittivity and permeability constants, respectively, of the material at frequency ω. The parameter $n(\omega)$ is a measure of its deflection power and also of the propagation speed of an EM wave in a given material. These two constants define the response of materials to the electric and magnetic field component at a given frequency and wavelength.

Engineered materials with a negative refractive index hold significant potential for various applications [112,113]. For example, they can fill gaps in radio frequencies, millimeter-waves and infrared wavelengths where the response of existing materials is limited. To obtain a negative refractive index, both $\varepsilon(\omega)$ and $\mu(\omega)$ need to be negative, a condition not observed in naturally occurring materials. There are very few engineering nanomaterials satisfying these conditions. A few artificial designs have been reported [113]. Paleari et al. have claimed that positive and negative change of refractive index up to 0.01 is obtained on SnO_2 NPs embedded in silica. These values are obtained by varying the power density of a focused infrared femtosecond laser beam [114].

4.4 Photonic Crystals

Photonic band gap (PBG) or photonic crystals [115] are receiving considerable attention because of their capability to confine and control EM waves in all three directions of space. Their potential application in building an all photonic waveguide is also considerable. Several NPs and matrix materials are available to produce opal structures. These opal structures are often achieved via self-assembling and layer-by-layer processes. Using an aqueous colloidal dispersion and a confined self-assembling process, Xia et al. obtained highly ordered and well-packed (74% density) 3D lattices [115].

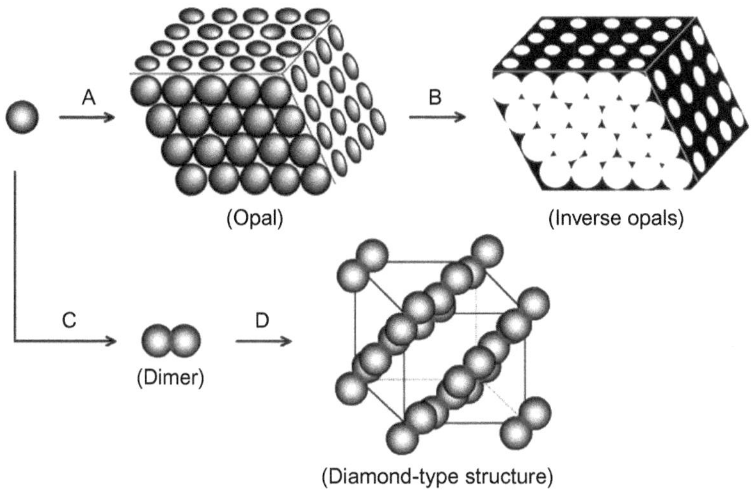

FIGURE 7.9 3D illustration of opal (A) and inverse opal (B) structures. Inverse opals are formed by selective etching of the 3D colloidal template. Formation of a dimeric nanoparticle (C) and a diamond lattice structure (D) obtained by self-assembly [115].

Preferential chemical etching of the NPs gives rise to an inverse opal (Fig. 7.9).

Achieving spatially well-defined structures is critical to optimize energy-transfer and electron-transfer processes between the particles. It is quite challenging to obtain a defect-free long-range order with submicron particles for a statistically meaningful analysis [116]. Although in most reports micron-sized colloidal particles are used, NPs in the range of 100 nm could be of interest for the UV range. A 3D PBG crystal over an area of several cm^2 has been achieved. The peak position of the primary diffraction peak is inversely proportional to the diameter of the polystyrene beads [116]. For colloidal NPs of 100 nm, the wavelength of the peak maximum is around 300 nm. It is thus possible to produce PBG crystals with tunable optical properties. This tunability can be achieved by tailoring the lattice parameters [116].

Photonic crystals have also been obtained by infiltrating gold NPs in polystyrene inverse opal [117]. This system combines the localized surface plasmon resonance (LSPR) of gold NPs and the PBG features of silica colloidal crystal templates. Building 2D and 3D QD arrays with tunable photoluminescence properties is potentially interesting [118]. Well-ordered 3D arrays of CdSe/ZnS NPs have been obtained by self-assembling (Fig. 7.10). A strong emission is obtained from single QD monolayers chemisorbed through short linkers to the silver substrate. Adding more layers leads to a monotonic but nonlinear increase in emission intensity. The peak intensity is reduced when the spacer size is increased. However, the peak position

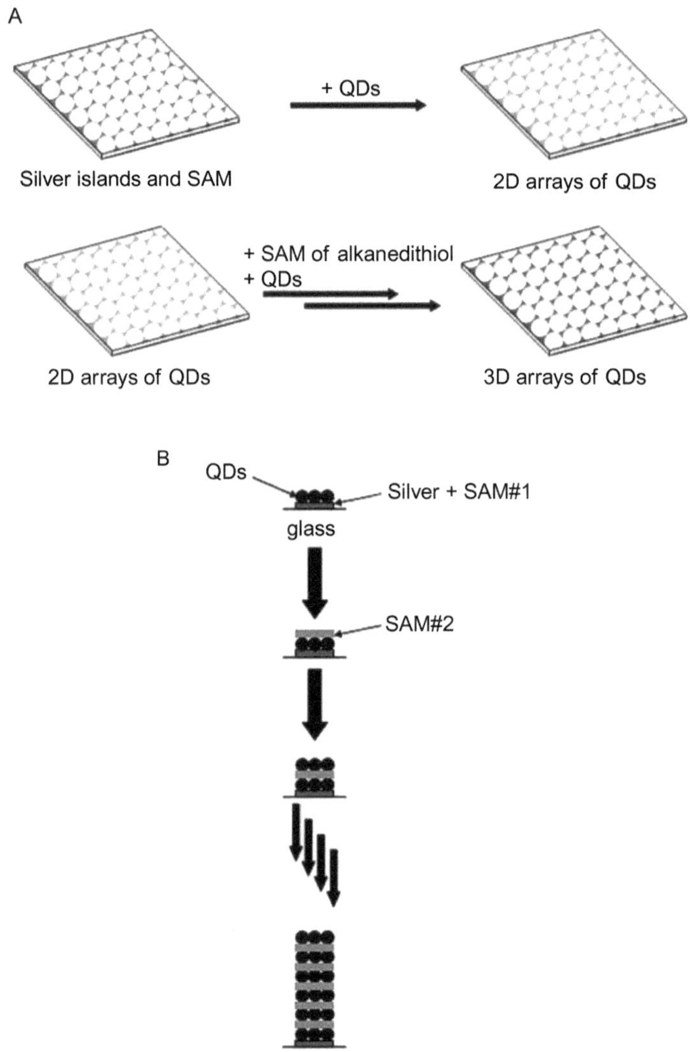

FIGURE 7.10 Schematic representation of the SAM process leading to 2D and 3D nanoparticle arrays [118]. (For color version of this figure, the reader is referred to the online version of this chapter.)

and line width are independent of the thickness of the film and the number of layers.

Most microparticle-based PBGs are obtained using materials with low refractive indices ($n < 1.6$) such as silica and polymers. Very few tunable optical 3D structures are successfully achieved using structures with high refractive indices. The functional properties of these monodiospersed microspheres can be modulated by adding materials with high refractive indices. Monodisperse titania, silica, and their composites obtained by water-in-oil emulsion droplets are used as confining geometries. In particular, composite microparticles are obtained from a mixture of TiO_2 and SiO_2 (30 nm) NPs (Fig. 7.11A). A dye molecule (TRITC) is added to the titania suspension to dope the microsphere. By varying the relative ratio of these two NPs, the refractive index of the microspheres can be modulated (Fig. 7.11B). Besides ceramic materials, this emulsion-assisted assembly can be applied to metal, polymer and semiconductor NPs.

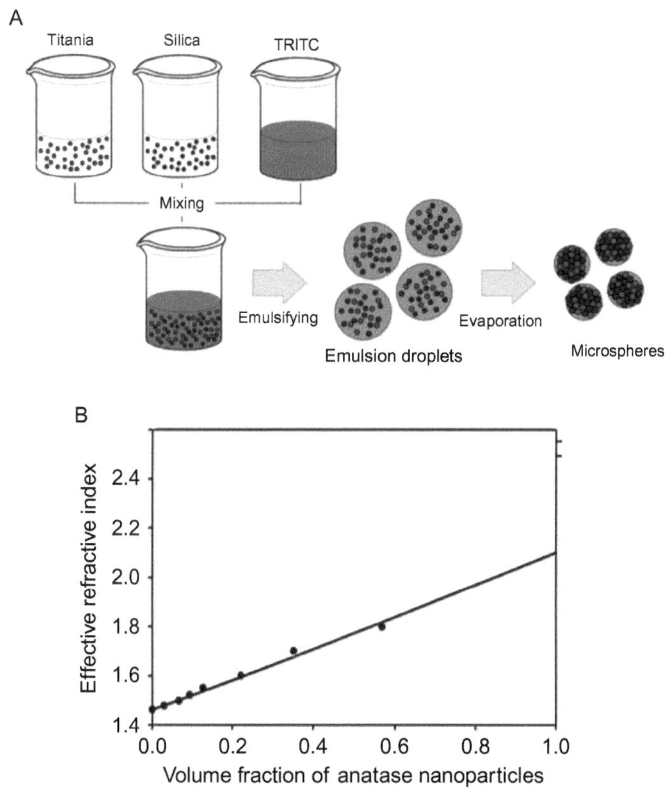

FIGURE 7.11 (A) Schematic illustration for fabricating a composite consisting of tetramethylrhodamine isothiocyanate (TRITC) microspheres containing titania particles and silica nanoparticles; (B) variation of the effective refractive index as a function of the titania volume fraction. The refractive index (2.493–2.554) of pure single-crystalline anatase phase is marked (top-right) [119]. (For color version of this figure, the reader is referred to the online version of this chapter.)

4.5 Optical Storage

High-density optical data storage can be obtained using metallic NPs [120]. The strong resonance properties of single and clustered NPs can be used to retrieve stored information with the help of spectroscopic far-field detection.

Precise PL intensity control of a thin film obtained by spin coating of CdSe/ZnS core–shell NPs has been reported [121]. Indeed, the saturated PL intensity is reversibly controlled using different radiation intensities. This property can be used to develop a rewritable storage of multi-digital data using a single excitation wavelength [121]. Following deposition of the CdSe–ZnS on a glass slide, PL and fluorescence images are obtained at different irradiations. Measurements show that the PL enhancement of CdSe/ZnS film is reversible [121].

5 MICROELECTRONIC MATERIALS

Continuing to scale down the dimensions of microelectronic devices remains the main strategy for improving the performance of very-large-scale integrated circuits and ultra-large-scale integrated circuits. Miniaturized technologies are in high demand for personal and portable devices. NPs have been shown to provide a potential solution to address this need. There are numerous electronic applications of NPs [4,122,123]. It is impossible to cover all these potential applications. Therefore, some of these applications of NPs are not discussed in this chapter.

Passive components contribute significantly to the overall volume, cost, and energy consumption of PCBs. There are different passive components that could take advantage of NP-based solutions [124,125]. NP-based composites, including high dielectric materials, soldering, interconnect, and electrically conductive adhesives, have been used as passive components in different microelectronic packaging solutions [124].

NP-based 1D, 2D, and 3D assemblies exhibiting tunable bulk properties are controlled by various procedures [126,127]. Semiconductor NPs are projected as potential building blocks for single-electron transistors (SET). Integrated circuitry using NPs and advanced soft-lithography could give rise to new bottom-up manufacturing systems for integrated circuits [128].

5.1 Single-Electron Transistor and Field-Effect Transistor

A SET consists of a quantum dot connected to an electrode through a tunneling junction and to a gate electrode through an insulating capacitor. Using SET, it is possible to increase the number of electrons in a semiconductor NP one by one. Under these conditions, the gate voltage is increased by $e/2C$. Numerous laboratory-based single-electron devices using metallic and semiconductor NPs have been reported [129]. A 10 nm NP's single-electron charging energy is 0.072 eV, corresponding to about threefold the thermal fluctuation of room temperature which is estimated at 0.025 eV [126].

A SET has been demonstrated using CdSe NPs deposited on closely spaced gold leads forming the drain and source contacts. These two leads are deposited on a silicon substrate covered by a thin SiO_2 layer [130]. Numerous SET devices are obtained following the deposition of CdSe between the two leads. This has been observed directly using SEM [130].

A SET has also been demonstrated using individual metal NPs [131]. The SET is obtained by depositing NPs (5–15 nm in diameter) into a gap created between two electrodes using electromigration (Fig. 7.12). *In situ* formation of gold NPs with an average size of 1 nm around the source–drain gap allowed the completion of a SET device [132]. A Coulomb gap has been observed on this structure up to a temperature of 200 K.

Inorganic NPs provide a perfect alternative to organic FET. They provide band gap tunability and, in some cases, long-term stability. These two properties are difficult to achieve with organic materials. To overcome

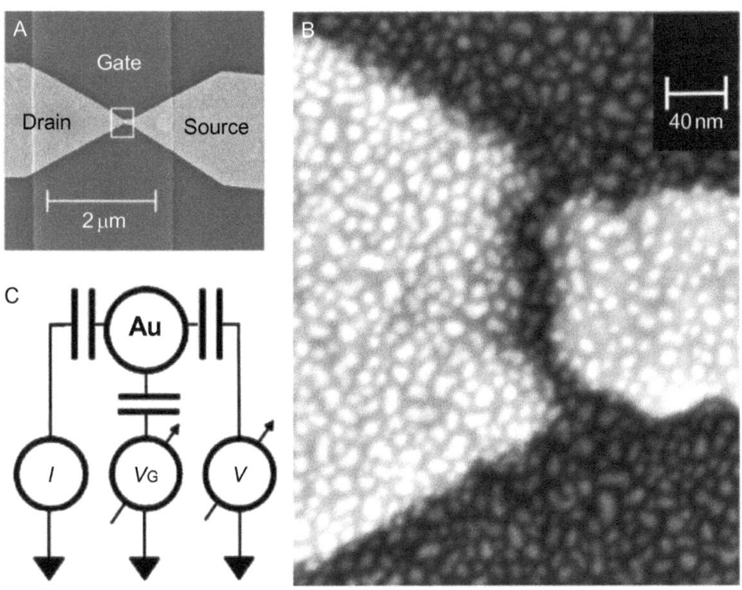

FIGURE 7.12 Schematic representation (A) and SEM image (B) of metal-nanoparticle single-electron transistors, with gold source and drain electrodes on top of an oxidized aluminum gate. Note the small gap of around 10 nm. A schematic representation of the electric circuit of the SET is also shown (C) for the Au-nanoparticle SET [131].

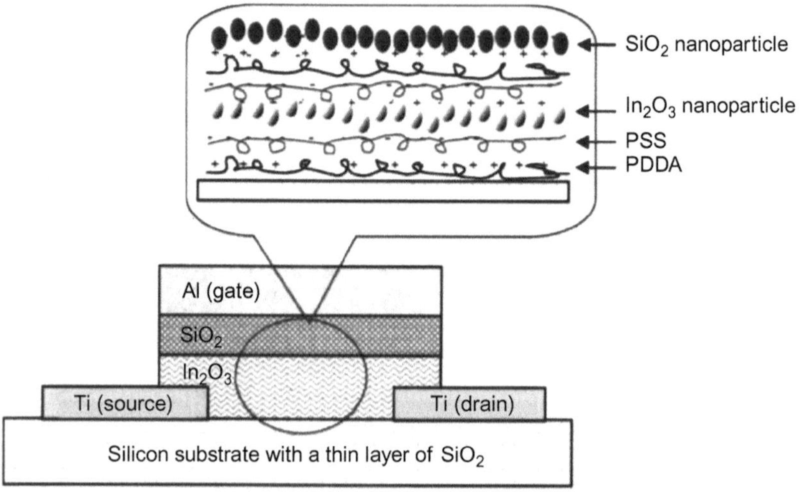

FIGURE 7.13 Schematic representation of the field-effect transistor obtained using LBL deposition of In_2O_3 and SiO_2 nanoparticle films [134].

the poor conducting properties of the native PbSe QD film, chemical treatment has been performed to achieve electron and hole mobilities of 0.9 and 0.2 $cm^2/V/s$ on n- and p-channel field-effect transistors (FETs), respectively [133]. A layer-by-layer (LBL) process has been used to build a multilayered FET device from In_2O_3 and SiO_2 NPs (Fig. 7.13). This technique is particularly suitable for inexpensive FET devices.

Fabrication of OFET using printing methodology is key to the quest for low-cost electronics. Laser printing technology can replace energy and capital-intensive lithographic processes. A FET device with highly conductive metallic contacts has been fabricated using laser sintering [135]. A sub-100 nm particles-based solution with similar physical properties could be integrated in these printing technologies. A seven-step process was used for FET fabrication from NP solution [135]:

a. inkjet printing of the NP solution;
b. selective sintering to form surface patterns and remove nonsintered NPs;
c. inkjet printing of a PVP dielectric layer;
d. inkjet printing of a second NP solution line on top of the PVP layer;
e. laser sintering to define source and drain electrodes;
f. removal of nonsintered NPs;
g. inkjet printing and thermal annealing of semiconducting polymer in the channel.

Using this process, FET devices with good spatial resolution have been obtained.

5.2 Chemical Mechanical Polishing

In chemical mechanical polishing (CMP), a wafer is placed on a carrier and pressed into contact with a slurry film on a polishing pad (Fig. 7.14). The planarization and polishing of the wafer is accomplished through a dual mechanical abrasion and chemical etching. Fumed aqueous solutions containing silica particles are often used for polishing. In the quest for a precision polishing process, finer abrasive particles are used more in the semiconductor industry. Polishing slurry consisting of NP suspension in water with some dissolved chemical additives has been developed. Silica and alumina NPs are used for polishing metal surfaces. Silica and ceria NPs are mostly used to polish oxide surfaces. Colloid composition, size, shape, and charge have a direct effect on the polishing properties [136]. Silica, alumina, and, more recently, ceria NPs with an average size down to 20 nm have been added to the slurry solution. CMP is also used to remove unwanted metal deposition during formation of interconnect in logic and memory devices. According to a 2003 report [137], NP-based polishing slurries, used in silicon chips, hard drive disks, and optical lens processes, had captured a half-billion dollar market. For example, ammonia-stabilized 40 nm colloidal silica (sold under the name OX-200) is widely used for semiconductor polishing [138]. Alumina and ceria NPs are also produced in thousands of tons for wafer polishing applications.

Besides titania and carbon black, silica and alumina NPs are also used in the microelectronic industry [138]. The effect of size on the polishing rate

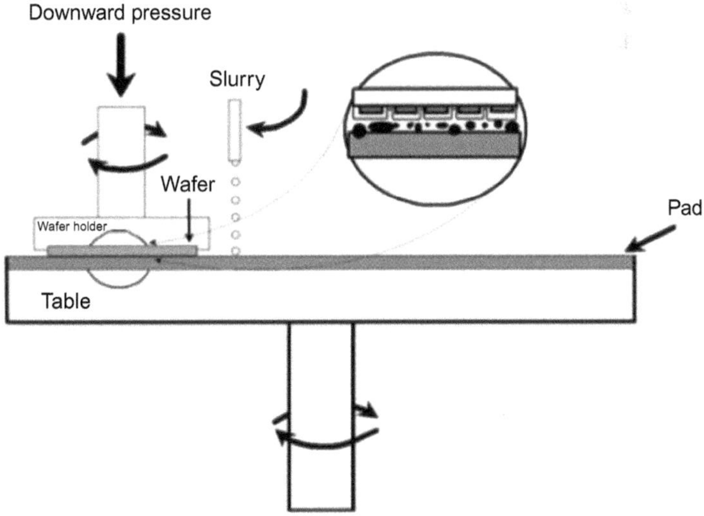

FIGURE 7.14 Simplified representation of a CMP setup [136].

in CMP remains somewhat controversial [139]. The polishing rate is gradually increased when silica particle size is reduced from around 200 to 50 nm [140]. Mixing with other abrasive slurries, surface coatings, and irregular shapes have been shown to increase the polishing rates [141]. CMP process introduces defects onto the polished surface, requiring post cleaning process. Indeed, postprocessing is often required when submicron particles are used in the CMP cleaning step [142].

5.3 Spintronics

Charge and spin of electrons confer special electronic and magnetic properties, respectively. Until recently, the spin state of electrons had been ignored in microelectronic devices. With the reduction in the size of today's electronic devices, the magnetoelectronic attributes of electron spins will certainly have a significant impact on the overall properties. The spin state adds another degree of freedom to existing electronic devices, allowing more capability and performance. Spintronics (also referred to as magnetoelectronics or spin-based electronics) is an emerging technology taking advantage of the magnetic (spin) and electronic (charge) attributes of a single electron. When compared with conventional semiconductor devices, NPs provide nonvolatility, increased data processing speed, decreased electric power consumption, and increased integration densities [143,144].

Spintronics materials include manganites, double perovskites, spinel ferrites, and diluted magnetic semiconductors [144]. Recent developments in NP technologies may help address some of the challenges (e.g., efficient charge injection) for spintronics. For example, QDs may help to control and enhance the ferromagnetic interactions.

Since the discovery of the giant magnetoresistance (GMR) effect in 2D magnetic multilayers [145] a variety of materials have been developed and evaluated for spintronics [144]. Indeed, it was observed that the resistance of a film of alternate ferromagnetic and nonmagnetic layers is lowest when the magnetic moments (spin) are aligned in parallel and highest when they are antiparallel. CrO_2 with a high Curie temperature of $T_c = 390$ K and S-moment up to $2\mu B$ is a potentially good candidate for GMR. Furthermore, it has good optical properties, which provides the possibility of building the whole device from a single material. Magnetoresistance (MR) of CrO_2 obtained using different methods has been reported [146]. Improved MR values have been reported with the use of CrO_2 nanopowder that compare quite well with the values obtained using other sources of CrO_2 [146].

Recently, Coulomb blockade effects have been observed in 2D self-assembled NPs on a thin alumina layer [147]. Indeed, single-electron tunneling through gold NPs 2.5 nm in diameter has been achieved. Spin injection and accumulation are enhanced in NP-based structures. For example, spin-relaxation time in the tunneling regime of devices containing cobalt NPs is significantly enhanced in comparison with that in the bulk material [148]. This device consists of a 15 nm Co–Al–O granular layer sandwiched between Al and Pt electrodes [148]. The nominal thickness of the Co–Al–O film corresponds to that of about three to four Co-based NPs. However, the dominant electrical current path may consist of only one NP due to the roughness of the film [148].

Carrier injection of either Co^{2+} or Mn^{2+} into a diluted magnetic semiconductor (DMS) based on ZnO nanocrystals has been recently reported [149]. Detailed characterization of DMS nanocrystals with different doping levels revealed the presence of interactions between the magnetic dopants and the free electrons of the semiconductor. Spin-polarizable excitonic photoluminescence, magnetic circular dichroism, exciton storage, and excitonic magnetic polaron formations have been reported using Mn^{2+}-doped CdSe QDs [150]. For example, a strong dopant-exciton magnetic exchange coupling with a spin-polarizable exitonic PL has been observed in Mn^{2+}-doped CdSe NPs. A giant Zeeman splitting has also been observed in colloidal Mn^{2+}-doped CdSe NPs.

5.4 Soft Magnetic Shielding

High-performance soft magnetic materials are required in modern electronic devices. Their applications span numerous sectors. NP-based materials provide a unique combination of soft magnetic properties and scalable production [151–156]. In particular, NP-based coatings may provide superior magnetic shielding properties. FINEMET materials based on the nanosized FeCuNbSiB alloy family are widely used [155].

The low sintering temperature of NP-based coatings allows integration in high-throughput printed electronics manufacturing. Stabilized magnetic Co NPs with an average size of 28 nm are mixed with organic additives to form a stable ink. Following screen printing and curing steps at 110 °C, high-quality printed patterns are obtained. High permeability properties were obtained over a large frequency range using cobalt content of around 70 wt/% [157].

5.5 Magnetic Storage

Numerous designs using NPs for memory and storage applications have been reported [158–160]. Iron NP-based materials with high coercivity have the largest commercial application in magnetic recording media [161]. Indeed, backup tapes of very high capacity are obtained using sub-100 nm particles with a relatively large aspect ratio. On the other hand, transformer core and inductors materials require very low coercivity properties. Fortunately, iron NPs can be tailored to have either very high or very low coercivity.

Magnetic powder-based film for tape recording is a relatively mature technology [162]. However, there is still room for improving the recording characteristics such as capacity and noise reduction. Geometrical and compositional characteristics of the particles provide avenues for improvements. Backup tapes have been prepared from ultrafine acicular metal particulate composites and hexagonal platelet barium ferrite particles [163,164].

Numerous methods have been developed to prepare patterned surfaces for high-density magnetic recorders [165,166]. Focused electron and ion beams have been used to directly deposit 2D arrays of single-domain NPs on magnetic layers. Using these techniques, a bit density on the order of Tbpsi (terabit per square inch) can be achieved [165]. Soft lithography techniques have also been used to imprint multilayered structures based on Co/Pt NPs [165].

5.6 Dynamic Random-Access Memory

Dynamic random-access memory (DRAM) or flash memory uses a capacitor in an integrated circuit to store data bits. Ideal DRAM requires nonvolatility, low write and erase times, high density, and low power consumption. Conventional DRAM, or flash memory, has shortcomings such as low density and high operation voltage [167]. DRAM based on semiconductors [168] and metallic [169,170] NPs has been proposed.

Discrete traps in nanostructured films are probably more immune to leakage [169]. A memory cell consisting of semiconductor NPs is grown on a thin oxide layer (1.1–1.8 nm thick) of n-channel silicon (Fig. 7.15). A 4.5-nm-thick oxide (control oxide) separates the NP from the control gate. Such structures are quasi-nonvolatile with write times of about 100 ns and refresh times in hours or more at 125 °C.

5.7 Dielectric Nanocomposites

Achieving a dielectric constant (low or high) as a function of frequency, temperature, and voltage is important but challenging [125]. Several compositions including bulk ferroelectric ($PbZrTiO_3$), ceramic nanocomposites (e.g., $BaTiO_3$ NP embedded in high-κ epoxy), metallic nanocomposites (e.g., silver NPs in a polymeric matrix), and dielectric NPs have been investigated [125].

Materials with high capacitance are achievable with the use of high-κ solution-processable materials. Inorganic NPs embedded in a polymeric matrix are used for this purpose. To improve processability, these NPs are often modified by adding a shell layer. Nanocomposite solutions amenable to printing techniques for high-capacitance flexible films are limited by the incompatibilities between the ceramic and the polymer matrix. Polymers and ceramics are the most widely used dielectrics. While polymers allow good

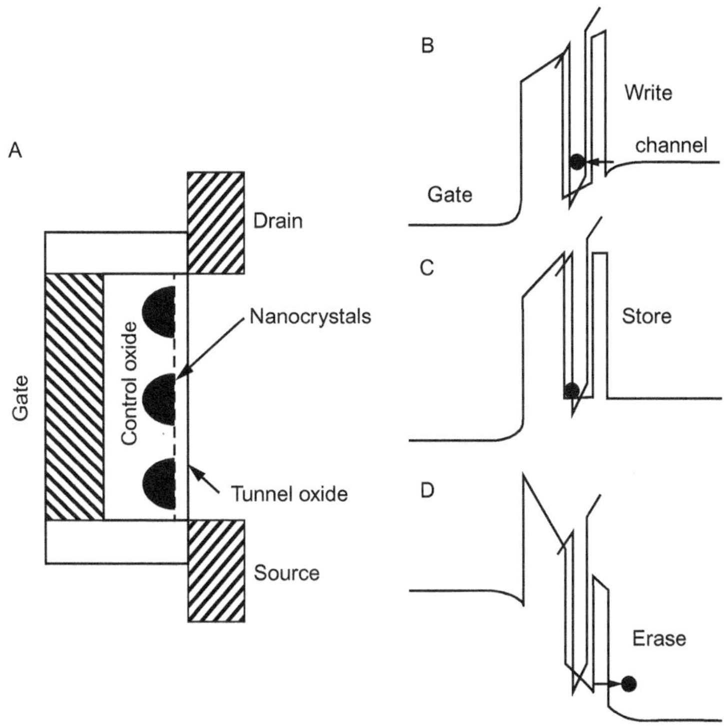

FIGURE 7.15 (A) Schematic representation of a memory cell used in DRAM, consisting of discrete charge storage traps. (B), (C), and (D) are the band diagrams corresponding to injection, storage, and removal of an electron from a nanoparticle, respectively [168].

mechanical properties but low dielectric constants, ceramics give rise to high dielectric constants, but they are brittle materials and difficult to incorporate into printing methods.

Nanocomposites consisting of ceramic NPs uniformly dispersed in a plastic matrix may provide high-κ and flexible film. NPs can also enable the development of low-temperature cofired ceramics technology. This allows fabrication of 3D ceramic modules with low dielectric loss and embedded silver electrodes [171]

Composite ceramics are important and widely used in electronics [172]. Applications include their use in multilayer capacitors, piezoelectric transducers, and packaging. High dielectric constant (high-κ) materials have been intensively investigated for various applications. For example, ferroelectric ceramics-based nanocomposites have, on the one hand, high dielectric constants and are flexible and easy to process, on the other [173]. High dielectric constants with a high breakdown field strength are achieved by dispersing 500 nm $PbMg_{1/3}Nb_{2/3}O_3$–$PbTiO_3$ powders in a solution of P(VDF/TrFE) copolymer in methyl ethyl ketone solvent. In addition, films with tunable

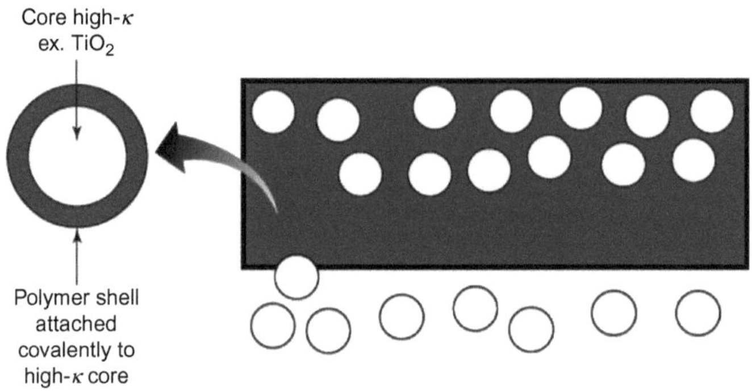

FIGURE 7.16 Schematic representation of core–shell nanoparticle architecture to produce flexible high-κ processable gate dielectric films [174]. (For color version of this figure, the reader is referred to the online version of this chapter.)

thickness are obtained by casting. Flexible, high-κ, processable gate dielectric films are obtained using core–shell TiO_2-polystyrene NP solutions. Following spin coating onto an ITO substrate, gold evaporation through a shadow mask completes the device (Fig. 7.16). TFTs have been fabricated using this new dielectric material.

To reduce leakage current observed in nanocomposite dielectrics, core–shell NPs consisting of aluminum NPs naturally coated with an oxide layer have been used [124]. A dielectric constant of around 160 has been obtained using aluminum-based nanocomposites. Modelizing NP-based nanocomposites as a three-phase system, Li et al. showed that adding NPs increases the permittivity but at the same time reduces the breakdown field [175]. Overall, the potential gain in electric energy density is relatively small. In addition, the interfacial interaction shifts the "percolation" threshold toward a lower volume fraction of NPs. To avoid these drawbacks, nanocomposite films should be carefully designed and processed. For example, it is advisable to use NPs 100 nm or smaller in diameter to obtain high electric energy density.

To respond to the requirements for larger dielectric permittivity in ever-shrinking DRAMs, a dielectric matrix containing metallic NPs has been studied. Silver NPs having diameters of 2–8 nm dispersed in lead zirconate titanate (PZT) have been shown to provide high dielectric permittivity and a low dissipation factor [176]. In this report, it was shown that smaller NPs give rise to lower dielectric permittivity but improve the dissipation factor, indicating that some compromises have to be made. It has also been recognized that particle shape and interparticle separation affect the dielectric constant [177].

A carbonaceous shell coating on silver cores is used as a filler to enhance the dielectric properties of polymers [178]. The high electric properties of the shell layer make these nanocomposites more similar to the conventional percolative materials. However, the high ceramic content within the polymer composite makes the polymer matrix less flexible. Furthermore, this three-phase nanocomposite provides tunable properties. As shown in Fig. 7.17, the relative sizes of the core and shell layers allow a wide range of dielectric, conductivity, and dielectric loss factor values.

High-κ materials as the gate insulators allow low-voltage operation [179]. In the case of OTFTs, a low deposition temperature and process are important. Using nanocomposite materials consisting of surface-modified TiO_2 NPs and cross-linked poly-4-vinylphenol, a dielectric constant higher than 11 is

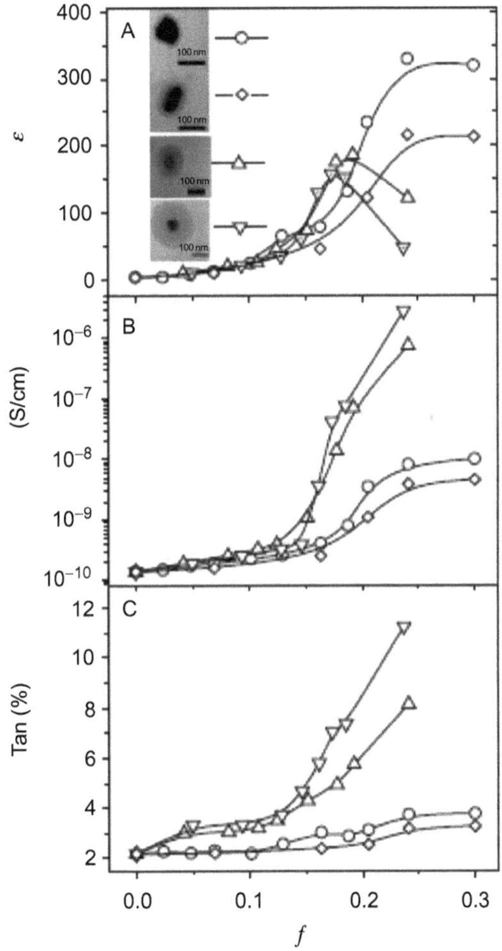

FIGURE 7.17 Variations of dielectric constant $\varepsilon(a)$, conductivity σ (S/cm) (B), and dielectric loss tan δ (%) (C) with the volume fraction f of the Ag–C nanoparticles at 1 kHz. The microscopy images in the inset (A) correspond to different core–shell nanoparticles [178].

achieved [180]. Furthermore, high leakage current is prevented by adding another poly(α-methylstyrene) layer.

Barium titanate ($BaTiO_3$) is widely used as a dielectric layer in multilayer ceramic capacitors (MLCCs). However, to reduce the thickness of the dielectric layer in MLCCs to 1 μm or less, the particulate size of the starting suspension should be reduced to a few tens of nanometers or less [181]. Furthermore, the need to develop aqueous processing to replace current state-of-the-art organic solution processing requires the development of a new process. A key process step is the reduction of the Ba^{2+} leaching by using an improved passivating agent layer. $BaTiO_3$ NPs with a size range of 70–130 nm are often successfully synthesized using a hydrothermal process [182]. $BaTiO_3$ NP passivation using different organic agents showed that Ba^{2+} leaching is reduced significantly. Furthermore, the density and dielectric constant (high-κ) of the resulting film are effectively controlled using an appropriate crystal phase with low impurities and heat treatment [181].

6 SENSORS

Nanostructured transducers have attracted increasing attention for their high sensitivity and specificity [183]. NP-based detectors, particularly oxide gas sensing, have been pursued [184,185] to increase their sensitivity and lower the integration cost. The same argument about the advantages of microsensors in comparison to traditional sensors applies to nanosensors as well. As has been the case throughout this book, so also here, the focus is on sensors using NPs as the main building blocks.

Properties of thin film incorporating NPs are often very sensitive to environmental changes. A wide variety of metallic, semiconductor, and oxide NPs have been used for sensing applications. NP-based sensors of biomolecules are discussed in Chapter 6. This chapter discusses mostly gas sensors. Gas sensors have been used to detect and monitor a large variety of gases and vapors, including toxic or explosive gases, humidity, and odors.

NP integration into various sensors (chemiresistor, optical, electrochemical, and FET) has been pursued by several groups, leading to significant developments. They are reviewed individually. Chemisensors [186] and magnetic [187] sensors have also benefited from incorporation of NPs, although to a much less extent than the rest. Magnetic and thermal sensors are not described in this chapter.

6.1 Electrical Sensor (Chemiresistor)

Electrical property (resistance, mobility) variations upon gas adsorption are widely used in gas sensing [188]. Chemiresistors are resistive gas-sensing

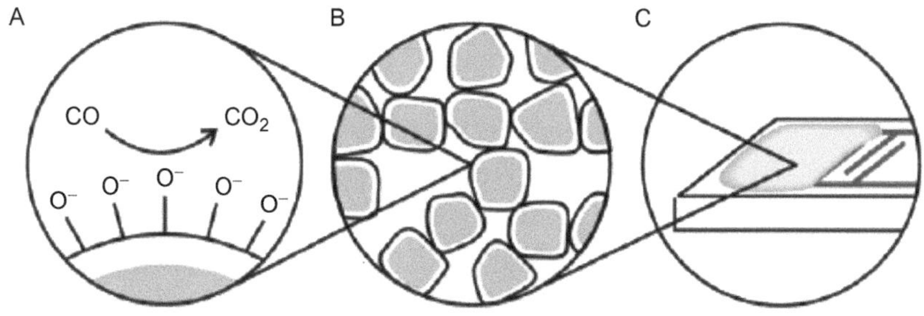

FIGURE 7.18 Schematic structure and operation of a chemiresistor gas sensor: surface chemistry allowing specific receptor function (A); microstructure of the sensing layer used for the transducing function (B); component enabling the detection of the change in output resistance of the sensing layer (C) [189].

mechanisms based on chemical reactions occurring at the surface. These reactions lead to surface oxide formation giving rise to an electric potential barrier proportional to the reactant concentration. Chemiresistors are often made from metallic or semiconducting materials. Figure 7.18 summarizes the operation principle and the structure of a chemiresistor-based CO sensor. Higher sensitivities are obtained at high temperatures. A variety of single- (ZnO, SnO_2, WO_3, TiO_2, Fe_2O_3) and multicomponent ($BiFeO_3$, $MgAl_2O_4$, $SrTiO_3$) oxides have been used in chemiresistor-based sensors. Metals, conductive polymers, and carbon black–polymer composites have also been used to produce chemiresistor films.

Hydrogen sensing has also received some attention. High sensitivity has been obtained using Pd NPs [190]. Formation of palladium hydrid gives rise to increased resistance. A three-layered structure is used in a resistance change-based transducer. The active layer providing the selectivity and sensitivity, the substrate, and the electrode layers are the main components of the sensing device. A heating arrangement is often added to improve sensitivity. Rella et al. showed that SnO_2 NPs with sizes below 10 nm provided high sensitivity to NO_2 and CO [191]. Low-level detection of ethanol and H_2S has been achieved using WO_3 NP-based sensors [192]. Volatile organic compounds and nitro-aromatic compounds have been detected using modified gold NPs [193]. A network of gold-coated NPs with mixed alkanethiolate (nonlinker ligands) and ω-carboxylate alkanethiolate (linker ligands) monolayers can be used as a sensor (Fig. 7.19). This sensor uses the change in the electronic conductivity of films consisting of a network polymer by carboxylate–Cu^{2+}–carboxylate bridges.

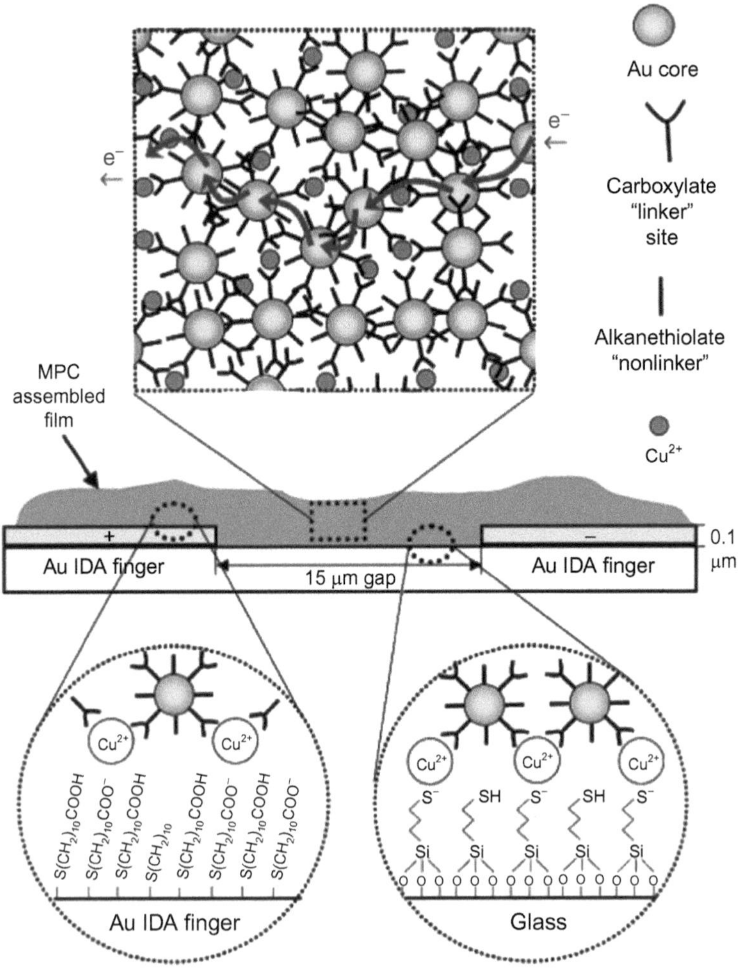

FIGURE 7.19 Macroscopic and microscopic representations of an electric sensor based on modified gold nanoparticles. *Reprinted with permission from Zamborini et al. [194]. Copyright 2002, American Chemical Society.* (For color version of this figure, the reader is referred to the online version of this chapter.)

6.2 Optical

LSPR of metal particles creates EM field enhancements with potential applications in sensing. Indeed, metallic NPs give rise to enhanced SPR, surface-enhanced Raman scattering (SERS), fluorescence, and subwavelength imaging [195]. SPR detection is often based on an ATR configuration, also known as the Kretschmann geometry. Following excitation, a plane wave inside the prism is coupled with an LSPR for a specific incidence angle and wavelength. Efficient interfaces allow reduced reflection.

The shift in the reflection minimum is the detection parameter. LSPR allows a sensitivity of about 10^{-7} refractive index units corresponding to a detection limit of about 10 pg/mL [195]. In the case of SERS, a nanostructured metal that can be obtained using NPs can provide up to 3000 times electric field enhancement, which corresponds to a 14 order-of-magnitude increase in Raman scattering. This enhancement allows sensing at the single-molecule level [195].

The unique optical properties of metal and semiconductor NPs are exploited to build sensors of high sensitivity and/or specificity [196]. SPR-based sensors are used [197]. LSPR of metallic NPs gives rise to resonance phenomena and large electric fields [198,199]. Au–Ag alloys provide a wide range of the plasmon resonance peak frequency [200]. The sensitivity of the resonance factor depends on the dielectric constant of the medium surrounding the NPs. Fiber-optic SPR with high sensitivity using gold NPs has been reported [201]. The signal-to-noise ratio depends on the NP size. The dielectric constant of the host material is also important.

The plasmon resonance frequency of various nanomaterials including NPs covers a wide range of optical spectra [202]. While silver and gold NPs are limited to the 400–550 and 600–750 nm spectral ranges, metallic nanoshells may cover optical spectra ranging from 600 up to the mid-IR [202]. For example, plasmon resonance of a 120-nm-diameter silica core coated with varying thicknesses of gold shell varies from 600 to 900 nm with 20 and 5 nm gold shell thickness, respectively [202]. Furthermore, NPs including nanoshells are of particular interest in exploiting localized and propagating surface plasmons [202].

A very high SERS enhancement factor is obtained with NP-based array structures (Fig. 7.20). Indeed, signal enhancement higher than 10^7 has been observed for SERS. This enhancement depends on particle size, excitation wavelength, and numerical aperture (NA).

SERS can be used to detect analytes directly (Fig. 7.21—top) and indirectly (Fig. 7.21—bottom). SERS NPs have all the attributes to become a key tool for the detection and analysis of dangerous chemical and biological substances [204]. Indeed, SERS can identify a wide range of known and unknown substances and provide information reproducibly in real time with a low detection limit. SERS is amenable to the analysis of substances in gas, liquid, and solid states. Despite all these advantages, SERS remains an academic tool with very few commercialized solutions. There are several reports demonstrating the attributes of SERS in the detection of chemical and biological substances of interest to homeland security [204]. Combining SERS and principal component analysis, five Bacillus spore samples have been accurately differentiated [204]. SERS induced by Au and Ag NPs has been used to detect TNT with a detection limit of 1 pg [205]. Traces of dinitrotoluene have been detected by SERS using a porous substrate decorated with metal NPs [206]. A SERS enhancement

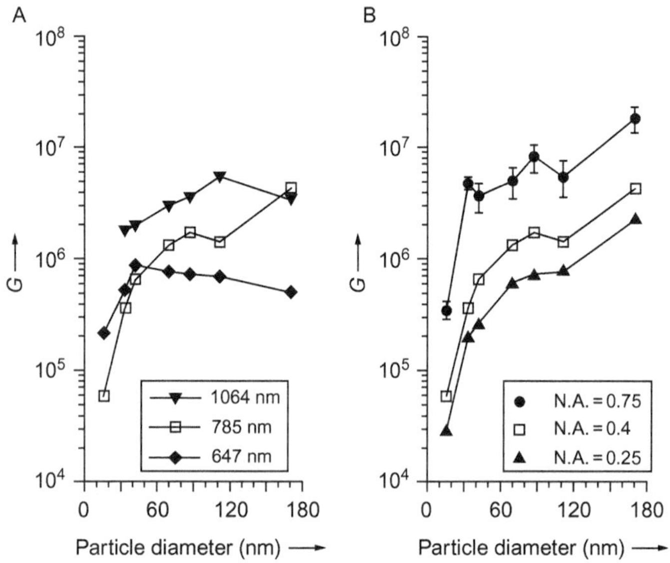

FIGURE 7.20 Variation of the SERS enhancement factor (G) as a function of the particle diameter for different excitation wavelengths (A) and numerical apertures (NA) and input power at the sample (B) [203].

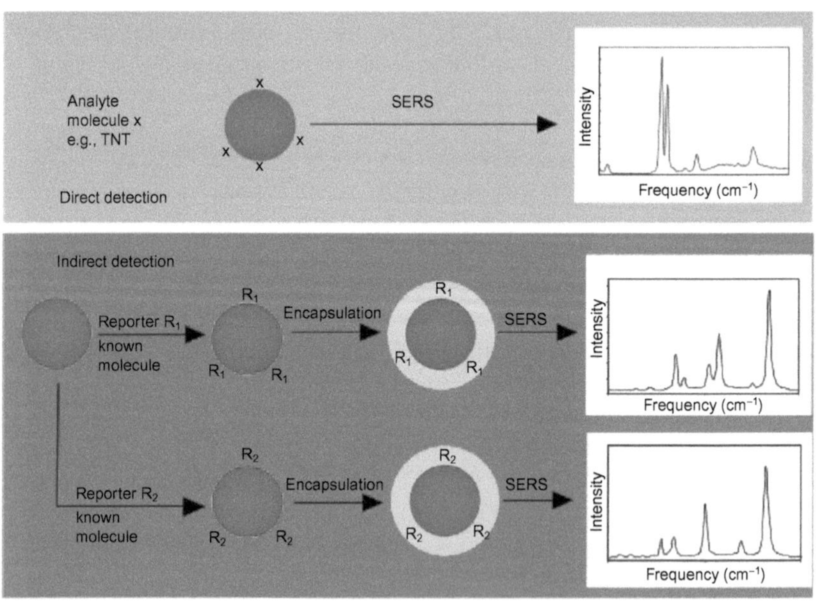

FIGURE 7.21 SERS methodologies for direct (top) and indirect (bottom) detection. Indirect detection includes label preparation. *Reprinted with permission from Golightly et al. [204]. Copyright 2009, American Chemical Society.* (For color version of this figure, the reader is referred to the online version of this chapter.)

factor of more than 10^6 has been achieved, and a detection limit of 10 fg has been reported [207].

Other optical sensing concepts have also been developed. Surface states of semiconductors have a significant effect on the photoluminescence signal [208]. This effect is expected to be more critical in the case of semiconductor quantum dots. Focal plan arrays (FPA) consisting of 2D detector arrays are commonly used in infrared cameras. These cameras are now used in state-of-the-art infrared spectrometers, night vision systems, and satellites [209].

Although thermal detectors are widely used, quantum detectors are considered mostly because of their much higher sensitivity. Compared to quantum well (QW) and bulk materials, QD provides higher responsivity and higher operation temperatures [196,210]. Indeed, QD photoexcited carriers have longer relaxation times. Furthermore, because of the 3D quantum confinement, QD detectors are predicted to have lower dark current. In spite of these theoretical advantages, in practice, QD-based IR detectors show low performance.

QD infrared photodetectors (QDIP) have been demonstrated with high-temperature (175 K) operation, large responsivity ($\sim$2.5 A/W), and high detectivity ($D^* \sim 10^{11}$ cm$\sqrt{\text{Hz}}$/W) [196]. Furthermore, QD-based infrared photodetectors have been used to demonstrate long-wavelength 640×512 pixels FPA infrared imaging [211,212]. Molecular beam epitaxy (MBE) technology has been used to build hybrid multilayer QD/QW structures. Performance of QDIP (Fig. 7.22) in dark conditions and under illumination depends on the density of QD, uniformity of the QD properties, and the doping level [213].

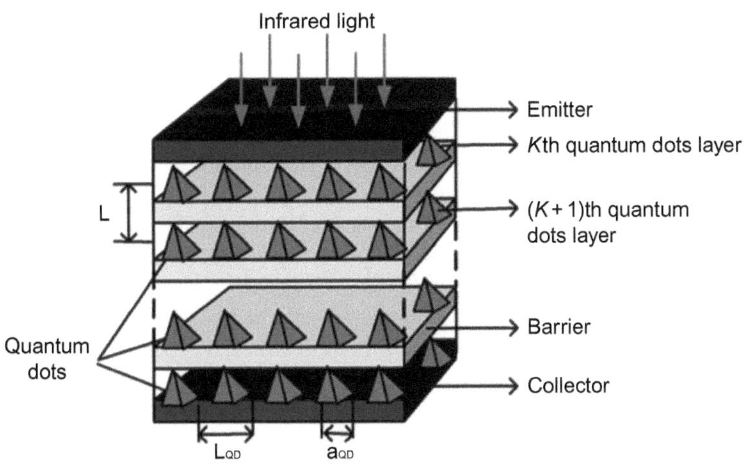

FIGURE 7.22 Schematic structure of a QD-based photodetector. The triangular pyramids in green correspond to QDs [213]. (For color version of this figure, the reader is referred to the online version of this chapter.)

LP-MOCVD has been used to build a QD FPA [214]. MBE is an expensive process for QD fabrication. Solution-processed QD could provide a cost-effective alternative in the future [215,216]. Spin-coated films of around 0.8 μm from band gap-tunable PbS nanocrystals indicate high responsivity and relatively low dark current density. It was found that early oxidation of the nanocrystals leads to low performance [216].

High photoconductivity has been obtained using spin-coated CdTe–polymer nanocomposites [217]. This high conductivity is likely due to the higher concentration of CdTe NPs following annealing. Instead of the relatively expensive MBE technique, a combination of chemical and electrochemical processes has been developed to form tunable QD directly on graphite substrates [218]. Early data showed that these photodetectors are inefficient. Their low cost and the fact that interfacial properties are not optimized could lead to a new promising platform for developing cost-effective photodetectors.

6.3 Electrochemical Sensors

Various electrochemical sensing concepts using NPs have been reported [219,220]. More details about electrochemical sensing are provided in Chapter 6 within the biosensing context. Oxide and semiconductor NPs are often used. Often, these NPs catalyze the electrochemical reaction and enhance electron transfer. Due to their high surface area, metallic NPs (gold and platinum) provide an excellent platform for catalyzing H_2O_2 sensing. Indeed, a sensitivity of 103.5 $\mu A/mM/cm^2$ has been achieved using Prussian blue NPs [220]. An electron-transfer rate of around 5000 s^{-1} has been achieved for glucose oxidase using gold NPs [220].

6.4 Electronic Sensors

FETs are currently being developed for gas sensing. FET-based sensors consist of a FET with a gate metallization exposed to the surrounding atmosphere (Fig. 7.23). The main drawback of FET sensors is the requirement for high operating temperatures. Reducing grain size enhances the sensitivity and, thus, the operating temperature [221]. Siemens developed a CO_2 sensor operating at room temperature taking advantage of the work-function change upon gas adsorption [221]. The potential drop across the double layer varies logarithmically with the CO_2 concentration when operating at $T = 50$ °C and $T = 23$ °C.

SETs [222] and ion-sensitive field-effect transistors (ISFET) [223] based on NPs also showed enhanced sensitivity. In the case of ISFET, a charged species near the gate surface (sensing interface) causes a change in polarization, which affects electron conductance between the source and drain electrodes. Adding a gold NP layer at the sensing surface of an ISFET allows detection of any charged molecules, even those that are not electrochemically active [223].

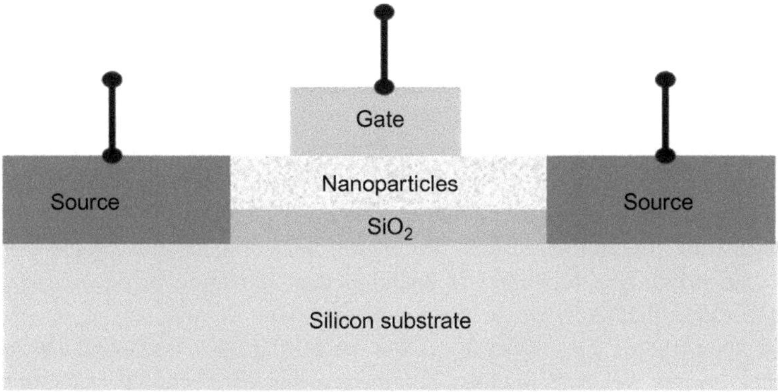

FIGURE 7.23 Schematic representation of a nanoparticle-based FET device. (For color version of this figure, the reader is referred to the online version of this chapter.)

7 PACKAGING AND INSULATORS

Packaging is equally important in microelectronics devices. Various conductive pastes are used for the formation of conductive circuits, electrodes, resistors, and dielectrics [224–226]. Pastes using relatively large particles require high annealing temperatures of over 500 °C. This limits the type of working substrates. Metallic NPs have been shown to melt at much lower temperatures than the corresponding bulk. Furthermore, NP paste and ink allow formation of much finer patterns.

Good electrical insulation has been obtained using epoxy resin nanocomposites [224]. The use of ZnO NPs with diameters of 65 nm as nanofillers provides significant improvements in breakdown resistance when compared to micro-fillers. Indeed, a breakdown time of 20,000 min is achieved with 1 wt% ZnO NP filler content [224]. When ZnO microparticles are used instead, a 1 wt% filler content allows less than 50 min before breakdown [224].

Fine circuit patterns have been obtained by screen printing of silver NP paste [225]. Posttreatment at 230 °C for 60 min is required to achieve high conductivity. Inkjet printing has also been used successfully to fabricate conductive patterns from silver NPs. Silver NPs with an average diameter of 11 nm have been successfully used to join two copper discs. With heat treatment at 300 °C, a shear strength of 25–40 MPa of the Cu–Ti–Cu joint is obtained [226].

8 PRINTED ELECTRONICS AND E-PAPER SENSORS

Printed electronics allows the production of functional electronic devices on a wide range of substrates using high volume and low-cost printing processes. Using ink as a starting material, inkjet, screen printing, gravure, and flexography offer flexibility in forming coatings, multilayers, and patterns on almost any substrate type and geometry. Conducting (metal NPs), semiconducting (II–VI and III–V semiconductors), and dielectric (inorganic oxides) inks have been developed. These printed electronics can be obtained using either batch or continuous (Roll-to-Roll) processes. Printed electronics now offers huge potential in a number of areas: flexible displays, solar cells, sensing, and lighting systems and displays. Other potential markets may include smart devices, batteries, and transistors. Synergetic development in microelectronics, NPs, and printing technologies is enabling the realization of the potential of printed electronics.

In spite of all the advantages, the performances of printed electronic devices do not yet match those obtained using conventional vacuum processes. For several decades, electronic paper, also referred to as e-paper, has been heralded as a replacement for conventional paper-based prints. This transition is facing several technico-economic challenges [227]:

a. low or no energy consumption while displaying information (bistable);
b. touching behavior like normal paper (Haptics);
c. contrast and reflection like real paper (can be read in direct sunlight);
d. limited choice of flexible substrates;
e. high cost.

Several technologies are being considered for direct view electronic paper. OLED, electrophoretic (EP), and LCD (cholesteric) technologies are currently competing for the e-paper market. All three technologies have the required attributes to address the challenges facing e-paper (Table 7.2). Furthermore, OLED and EP are light and flexible, with low power consumption. Due to its inherently slow switching speed (Table 7.2), electrophoresis is not suitable

TABLE 7.2 Comparison of the Characteristics of Different Display Technologies [227,228]

Characteristics	e-Paper	OLED	Cholesteric LC
Contrast ratio	>10:1	>10,000:1	>250:1
Response time	<500 ms	<1 ms	<2 ms
Roll to roll	Yes	Yes	Yes (future)
Colors	Thousands	Wide/millions	Thousands
Maturity	High	High	Medium
Lifetime	Good	Good	Good
Market pricing	Low, descending	High/med, descending	Low/medium

for video games, but its bistability makes it very suitable for e-books and e-paper markets. EP deposition is a simple, scalable, and cost-effective technique for the formation of the display support and process.

OLED and nanoLED technologies have already been described elsewhere in this book. They provide a platform for flexible displays that could satisfy the requirements of e-paper. LED using luminophore NPs provides high quantum yields, extremely narrow emission, spectral tunability, and high stability [229]. Large-area, full-color displays have been obtained using printed QD. Furthermore, a uniform and compact 2D QD film has been obtained using a solvent-free printing process [229]. First, SAM modified substrates are modified by QD using a spin coating technique. Then, a microcontact printing process is used to produce regular QD patterns. A flexible 4-in. full-color active matrix QD display with a resolution of 320×240 pixels has been obtained using this technique [230]. RGB-patterned flexible QD devices are thus obtained. No changes in luminous efficiency and current density have been observed with the use of this technique.

Electrophoresis was first used in different displays by Xerox in the 1970s. The electrophoresis process allows separation of particles according to their size and electric charging state under voltage. This is achieved using electrophoretic particles and suspending medium sandwiched between two electrodes (Fig. 7.24). Each microcapsule contains positively charged white NPs and negatively charged black NPs in a clear dielectric fluid. Thus, the electrophoretic displays are reflective and bistable. Typically, an e-paper may contain millions of microcapsules containing charged pigment particles in a clear fluid [231]. Displays on flexible active-matrix-array sheets with a resolution of 96 pixels/in has been achieved. A light state is achieved when the positively charged white particles are moved to the top electrode under an electric field. If the field direction is reversed, the dark state (spot) is obtained.

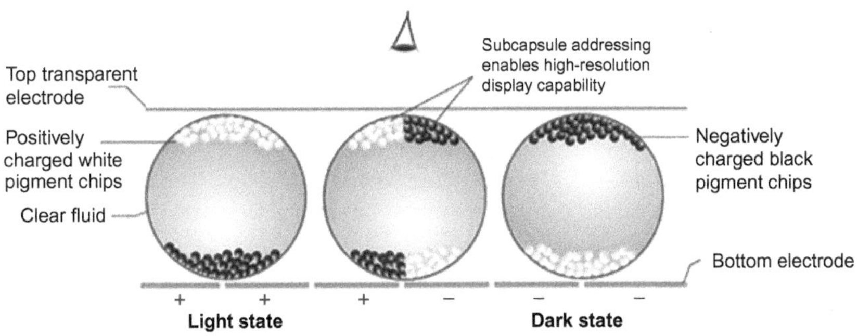

FIGURE 7.24 Schematic structure of an electrophoretic-based paper-like display including a cross section of electronic ink [230]. (For color version of this figure, the reader is referred to the online version of this chapter.)

Although early electrophoretic inks for reflective electronic displays were based on microparticles with a size range of 0.1–5 μm [232], more recent work has shown improved features using sub-100 nm particles. The size of the negatively charged ink NPs decreases with increasing surfactant concentration. Electronic ink NPs are obtained by mixing organic dyes and PVP (poly(styreneco-4VP)) in methanol/water solution containing different concentrations of ionic surfactant. These NPs are used to fabricate an electrophoretic display cell [93]. This dispersion polymerization allowed preparing NPs of small size and narrow size distribution, which provided a more efficient and rapid electrophoretic image display [93]. Furthermore, PVP particles containing organic dyes such as Acid Blue 25, Acid Red 8, and Acid Yellow 76 were also obtained using sodium dodecyl sulfate (SDS) and/or cetyltrimethylammonium bromide (CTAB) surfactants. High mobility has been achieved on dyed charged poly(methyl methacrylate-co-ethylene glycol dimethacrylate) NPs obtained by emulsifier-free emulsion copolymerization and aminolysis [233].

The recent development of high-quality flexible graphene oxide (GO) substrates provides a new platform for flexible electronics [234]. With its high surface areas, superior electrical conductivity, high charge carrier mobility, and broad electrochemical window [234], GO is an almost ideal substrate for printed electronics and e-paper. Transparent and nontransparent GO conductive patterns have been reported. Closely packed 2D assemblies of metallic NPs have been recently achieved on free standing GO [235]. Patterned grapheme has also been obtained on plastic substrates using the inkjet printing process [236]. NP-based TCOs have been described elsewhere in this chapter. Flexible TCO films are critical for display devices including e-paper.

Besides its application in producing electronics devices at a lower cost and with reduced environmental impact, printed electronics is a key platform technology for electronic paper. Readability is currently the biggest challenge for e-readers. Indeed, most people prefer printing a digital document for reading instead of reading it directly from the computer screen. Eyestrain due to poor viewing position, low contrast ration, and low resolution is often cited as the main drawback of e-readers. Furthermore, the lack of portability, even with current tablets, contributes to the quasi-status quo. Current printed electronics techniques allow a resolution of 500 ppi, which is about five times higher than the resolution of current computer displays. With a contrast ratio of 10:1, these printed electronics technologies will allow the tablet to match the conventional paper print contrast. Furthermore, a thin and lightweight e-reader may offer other advantages in comparison to conventional FPDs and conventional paper reading. Besides the ergonomics, storage capacity, and other functionality advantages, the reflective screen-based display of e-paper allows reading under any ambient light condition.

REFERENCES

[1] E. Williams, Environ. Sci. Technol. 38 (2004) 6166–6174.

[2] D.J. Lipomi, E.A. Weiss, G.M. Whitesides, In: J. Garcia-Martinez (Ed.), Nanotechnology for the Energy Challenge, Wiley-VCH, Weinheim, 2009, pp. 231–279.

[3] R.K. Cavin, V.V. Zhirnov, G.I. Bourianoff, J.A. Hutchby, D.J.C. Herr, H.H. Hosack, W.H. Joyner, T.A. Wooldridge, J. Nanopart. Res. 7 (2005) 573–586.

[4] F.E. Kruis, H. Fissan, A. Peled, J. Aerosol Sci. 29 (1998) 511–535.

[5] D.Y. Godovsky, Adv. Polym. Sci. 153 (2000) 163–205.

[6] A.P. Alibisatos, J. Phys. Chem. 100 (1996) 13226–13239.

[7] T. Trindade, P. O'Brien, N.L. Pickett, Chem. Mater. 13 (2001) 3843–3858.

[8] T.J. Bulowski, J.H. Simmons, Crit. Rev. Solid State Mater. Sci. 27 (2002) 119–142.

[9] H. Chandler, Mater. Sci. Eng. R 49 (2005) 113–155.

[10] S.P. Gubin, N.A. Kataeva, G.B. Khomutov, Russ. Chem. Bull. Int. Ed. 54 (2005) 827–852.

[11] A.L. Rogach, D.V. Talapin, H. Weller, In: F. Caruso (Ed.), Colloids and Colloid Assemblies: Synthesis, Modification, Organization and Utilization of Colloid Particles, Wiley-VCH, Weinheim, 2004, pp. 52–95.

[12] S.H. Ko, H. Pan, C.P. Grigoropoulos, C.K. Luscombe, J.M. Frechet, D. Poulikakos, Nanotechnology 18 (2007) 345202–345210.

[13] D. Huang, F. Liao, S. Molesa, D. Redinger, V. Subramanian, J. Electrochem. Soc. 150 (2003) G412–G417.

[14] S. Magdassi, A. Bassa, Y. Vinetsky, A. Kamyshny, Chem. Mater. 15 (2003) 2208–2217.

[15] Gold and Nanotechnology in the age of Innovation, World Gold Council, 2010.

[16] D. Loss, D.P. DiVincenzo, Phys. Rev. A 57 (1998) 120–126.

[17] A.J. Haes, D.A. Stuart, Sh. Nie, R.P. Van Duyne, J. Fluoresc. 14 (2004) 355–367.

[18] H. Althues, J. Henle, S. Kaskel, Chem. Soc. Rev. 36 (2007) 1454–1465.

[19] F. Iskandar, I.W. Lenggoro, B. Xia, K. Okuyama, J. Nanopart. Res. 3 (2011) 263–270.

[20] H. Ishibash, J. Vac. Soc. Jpn. 51 (2008) 737–742.

[21] N.V. Nikonorov, A.I. Sidorov, V.A. Tsekhomskii, In: D.P. Perez (Ed.), Silver Nanoparticles, InTech, Croatia, 2010, pp. 177–200.

[22] C.R. Bamford, Colour Generation and Control in Glass, Elsevier, Amsterdam, 1977.

[23] J.A. Creighton, D.G. Eadon, J. Chem. Soc. Faraday Trans. 87 (1991) 3881–3891.

[24] F.E. Wagner, S. Haslbeck, L. Stievano, S. Calogero, Q.A. Pankhurst, K.-P. Martinek, Nature 407 (2000) 691–692.

[25] J.A. Williams, G.E. Rindone, H.A. McKinstry, J. Am. Ceram. Soc. 64 (1981) 709–713.

[26] T.P. Seward, J. Non-Cryst. Solids 40 (1980) 499–513.

[27] S. Takeda, K. Yamamoto, K. Matsumoto, J. Non-Cryst. Solids 265 (2000) 133–142.

[28] F.C. Raszewski, K.A. Murphy, J.E. Shelby, J. Non-Cryst. Solids 352 (2006) 528–533.

[29] J.R. Greer, R.A. Street, Acta Mater. 55 (2007) 6345–6349.

[30] S. Jeong, K. Woo, D. Kim, S. Lim, J.S. Kim, H. Shin, Y. Xia, J. Moon, Adv. Funct. Mater. 18 (2008) 679–686.

[31] K.L. Chopra, S. Major, D.K. Pandya, Thin Solid Films 102 (1983) 1–46.

[32] H. Ohta, H. Hosono, Mater. Today 7 (2004) 42–51.

[33] B.G. Lewis, D.C. Paine, MRS Bull. 25 (2000) 22–27.

[34] C. Goebbert, R. Nonninger, M.A. Aegerter, H. Schmidt, Thin Solid Films 351 (1999) 79–84.

[35] J. Puetz, M.A. Aegerter, Thin Solid Films 516 (2008) 4495–4501.

[36] A. Cirpan, F.E. Karasz, J. Appl. Polym. Sci. 99 (2006) 3125–3129.

[37] Y.-H. Taka, K.-B. Kima, H.-G. Parka, K.-H. Leeb, J.-R. Lee, Thin Solid Films 411 (2002) 12–16.
[38] M. Berber, V. Bulto, R. Kliß, H. Hahn, Scr. Mater. 53 (2005) 547–551.
[39] N. Al-Dahoudi, M.A. Aegerter, Thin Solid Films 502 (2006) 193–197.
[40] J. Ederth, P. Johnsson, G.A. Niklasson, A. Hoel, A. Hultaker, P. Heszler, C.G. Granqvist, A.R. van Doorn, M.J. Jongerius, D. Burgard, Phys. Rev. B 68 (2003) 155410–155419.
[41] J. Ederth, P. Heszler, A. Hultaker, G.A. Niklasson, C.G. Granqvist, Thin Solid Films 445 (2003) 199–206.
[42] M. Toki, M. Aizawa, J. Sol-Gel Sci. Technol. 8 (1997) 717–720.
[43] Y. Li, Y. Yu, B.S. Ong, J. Am. Chem. Soc. 127 (2005) 3266–3267.
[44] L.C. Chen, S.-F. Lu, Phys. Stat. Sol. (a) 203 (2006) 2451–2456.
[45] M.L. Allen, M. Aronniemi, T. Mattila, A. Alastalo, K. Ojanpera, M. Suhonen, H. Seppa, Nanotechnology 19 (2008) 175201–175203.
[46] www.nanomastech.com.
[47] D. Huang, F. Liao, St. Molesa, D. Redinger, V. Subramanianb, J. Electrochem. Soc. 150 (2003) G412–G417.
[48] K.C. Krogman, T. Druffel, M.K. Sunkara, Nanotechnology 16 (2005) S338–S343.
[49] S.E. Yancey, W. Zhong, J.R. Heflin, A.L. Rittera, J. Appl. Phys. 99 (2006) 034313–034322.
[50] T. Murata, H. Ishizawa, I. Motoyama, A. Tanaka, Appl. Opt. 45 (2006) 1465–1468.
[51] S.K. Lee, C.M. Zettering, M. Ostling, I. Aberg, M.H. Magnusson, K. Deppert, L.E. Wernersson, L. Samuelson, A. Litwin, Solid State Electron. 46 (2002) 1433–1440.
[52] J.S. Kwak, J.O. Song, T.Y. Seong, B.I. Kim, J. Cho, C. Sone, Y. Park, J. Nanosci. Nanotechnol. 6 (2006) 3547–3550.
[53] Y. Wu, Y. Li, B.S. Ong, P. Liu, S. Gardner, B. Chiang, Adv. Mater. 17 (2005) 184–187.
[54] H. Iwashige, G. Kutulk, S. Hayashi, T. Suzuki, T. Yoshida, T. Abe, M. Oda, Scr. Mater. 44 (2001) 1667–1670.
[55] H.H. Lee, K.S. Chou, K.C. Huang, Nanotechnology 16 (2005) 2436–2441.
[56] P. Anderson, Advanced Display Technologies (accessed 25.02.07).
[57] P. Semenza, Nat. Photon. 1 (2007) 267–268.
[58] G. Yi, B. Sun, D. Chen, In: H.S. Nalwa (Ed.), Encyclopedia of Nanoscience and Nanotechnology, vol. 6, American Scientific Publishers, New York, 2004, pp. 465–476.
[59] H. Chander, Mater. Sci. Eng. R 49 (2005) 113–155.
[60] A.V. Dijken, J. Makkinje, A. Meijerink, J. Lumin. 92 (2001) 323–328.
[61] I. Yu, T. Isobe, M. Senna, J. Phys. Chem. Solids 57 (1996) 373–379.
[62] M. Uchidoi, IEEE Trans. Plasma Sci. 34 (2006) 287–293.
[63] J.P. Boeuf, J. Phys. D Appl. Phys. 36 (2003) R53–R79.
[64] H. Bechtel, T. Jüstel, H. Gläser, D.U. Wiechert, J. Soc. Inform. Display 10 (2002) 63–67.
[65] R.N. Bhargava, D. Gallagher, X. Hong, A. Nurmikko, Phys. Rev. Lett. 72 (1994) 416–419.
[66] R.N. Bhargava, J. Lumin. 72–74 (1997) 46–48.
[67] Y. Chan, J.P. Zimmer, M. Stroh, J.S. Steckel, R.K. Jain, M.G. Bawendi, Adv. Mater. 16 (2004) 2092–2097.
[68] Y. Yang, M. Gao, Adv. Mater. 17 (2005) 2354–2357.
[69] M. Nogami, K. Nagasaka, M. Takata, J. Non-Cryst. Solids 122 (1990) 101–106.
[70] V. Stanic, T.H. Etsell, A.C. Pierre, R.J. Mikula, Mater. Lett. 31 (1997) 35–38.
[71] D. Haranath, Nitesh Bhalla, Harish Chander, Rashmi, M. Kar, R. Kishore, J. Appl. Phys. 96 (2004) 6700–6705.
[72] J. Thisayukt, H. Shiraki, Y. Sakai, T. Masumi, S. Kundu, Y. Shiraishi, N. Toshima, S. Kobayashi, Jpn. J. Appl. Phys. 43 (2004) 5430–5434.

[73] L.O. Dolgov, O.V. Yaroshchuk, Colloid Polym. Sci. 282 (2004) 1403–1408.
[74] C.W. Marysia, H.V. Rob, D.J. Broer, H.T.A. Wilderbeek, C.W.M. Bastiaansen, Adv. Mater. 12 (2000) 753–757.
[75] A. Talin, K.A. Dean, J.E. Jaskie, Solid State Electron. 45 (2001) 963–976.
[76] A.P. Burden, Int. Mater. Rev. 46 (5) (2001) 213–5231.
[77] C.E. Hunt, A.G. Chakhovskoi, J. Vac, Sci. Technol. B 15 (1997) 516–519.
[78] M. Ihara, T. Igarashi, T. Kusunoki, K. Ohno, J. Electrochem. Soc. 149 (2002) H72–H75.
[79] S.C. Gosh, C. Thanachayanont, J. Dutta, 1st ECTI Annual Conference, 13–14 May 2004, Pattaya, Thailand, 2004, pp. 145–148.
[80] M. Konishi, T. Isobe, M. Senna, J. Lumin. 93 (2001) 1–8.
[81] J. McKittrick, L.E. Shea, C.F. Bacalski, E.J. Bosze, Displays 19 (1999) 169–172.
[82] A.D. Dinsmore, D.S. Hsu, H.F. Gray, S.B. Qadri, Y. Tian, B.R. Ratna, Appl. Phys. Lett. 75 (1999) 802–804.
[83] B.O. Dabbousi, J. Rodriguez-Viejo, F.V. Mikulec, J.R. Heine, H. Mattoussi, R. Ober, K.F. Jensen, M.G. Bawendi, J. Phys. Chem. B 101 (1997) 9463–9475.
[84] E. Roduner, Chem. Soc. Rev. 35 (2006) 583–592.
[85] V.L. Colvin, M.C. Schlamp, A.P. Alivisatos, Nature 370 (1994) 354–357.
[86] N. Tessler, V. Medvedev, M. Kazes, S. Kan, U. Banin, Science 295 (2002) 1506–1508.
[87] J. Lee, V.C. Sundar, J.R. Heine, M.G. Bawendi, K.F. Jensen, Adv. Mater. 12 (2000) 1102–1105.
[88] H. Althues, R. Palkovits, A. Rumplecker, P. Simon, W. Sigle, M. Bredol, U. Kynast, S. Kaskel, Chem. Mater. 18 (2006) 1068–1072.
[89] R.W. Murray, Chem. Rev. 108 (2008) 2688–2720.
[90] J.M. Caruge, J.E. Halpert, W. Wood, V. Bulovic, M.G. Bawendi, Nat. Photon. 2 (2008) 247–250.
[91] J.-H. Jou, W.-B. Wang, M.-F. Hsu, J.-J. Shyue, C.-H. Chiu, ACS Nano 4 (2010) 4054–4060.
[92] X. Weng, Y. Chen, S.C. Tseng, J. Hou, Modification of Electrical Properties of Display Cells for Improving Electrophoretic Display Performance, USPTO 7,504,050 B2 March 17th 2009.
[93] S.W. Oh, C.W. Kim, H.J. Cha, U. Pal, Y.S. Kang, Adv. Mater. 21 (2009) 4987–4991.
[94] J.H. Park, M.A. Lee, Y.H. Kim, B.J. Park, H.J. Choi, J. Korean Phys. Soc. 53 (2008) 50–54.
[95] W.-W. An, Y. Le, J.-F. Chen, 3rd International Nanoelectronics Conference (INEC), Hong Kong, 2010, pp. 1280–1281.
[96] http://www.eink.com.
[97] M. Dinu, R. Rapaport, G. Chen, H.R. Stuart, R. Giles, Bell Labs Tech. J. 10 (2005) 215–234.
[98] G. Yi, B. Sun, D. Chen, In: H.S. Nalwa (Ed.), Encyclopedia of Nanoscience and Nanotechnology, vol. 6, American Scientific Publishers, New York, 2004, pp. 465–476.
[99] H.J. Eisler, V.C. Sundar, M.G. Bawendi, Appl. Phys. Lett. 80 (2002) 4614–4616.
[100] A.A. Guzelian, U. Banin, A.V. Kadavanich, X. Peng, A.P. Alivisatos, Appl. Phys. Lett. 69 (1996) 1432–1434.
[101] Y. Chan, J.S. Steckel, P.T. Snee, J.M. Caruge, J.M. Hodgkiss, D.G. Nocera, M.G. Bawendi, Appl. Phys. Lett. 86 (2005) 073102–073104.
[102] Y. Chan, T. Preston, T. Snee, J. Caruge, B.K. Yen, G.P. Nair, D.G. Nocera, M.G. Bawendi, J. Am. Chem. Soc. 128 (2006) 3146–3147.
[103] K. Suh, J.H. Shin, Appl. Phys. Lett. 89 (2006) 223102–223104.
[104] S.V. Kershaw, M. Harrison, A.L. Rogach, A. Kornowski, IEEE J. Sel. Top. Quant. Electron. 6 (2000) 534–543.
[105] A. Vlasov, K. Luterova, I. Pelant, B. Honerlage, V.N. Astratov, Appl. Phys. Lett. 71 (1997) 1616–1618.

[106] N.I. Zheludev, J. Opt. A Pure Appl. Opt. 8 (2006) S1–S8.
[107] C.B. Poitras, M. Lipson, H. Du, M.A. Hahn, T.D. Krauss, Appl. Phys. Lett. 82 (2003) 4032–4034.
[108] M. Dinu, R. Rapaport, G. Chen, H.R. Stuart, R. Giles, Bell Labs Tech. J. 10 (2005) 215–234.
[109] J.L. Zhang, T.F. Kuech, J. Electron. Mater. 37 (2008) 135–144.
[110] A. Hachéa, S.-É. LeBlancb, M. LoCascio, A. Martucci, Physica E 17 (2003) 104–106.
[111] M.T. Harrison, S.V. Kershaw, M.G. Burt, A.L. Rogach, A. Kornowski, A. Eychmüller, H. Weller, Pure Appl. Chem. 72 (2000) 295–307.
[112] V.G. Veselago, E.E. Narimanov, Nat. Mater. 5 (2006) 759–762.
[113] W.J. Padilla, D.N. Basov, D.R. Smith, Mater. Today 9 (2006) 28–35.
[114] A. Paleari, E. Franchina, N. Chiodini, A. Lauria, E. Bricchi, P.G. Kazansky, Appl. Phys. Lett. 88 (2006) 131912–131914.
[115] E. Yablonovitch, J. Opt. Soc. Am. B 10 (1993) 283–295.
[116] Y. Xia, B. Gates, Z.-Y. Li, Adv. Mater. 13 (2001) 409–413.
[117] Y. Tan, W. Qian, S. Ding, Y. Wang, Chem. Mater. 18 (2006) 3385–3389.
[118] J. Pacifico, J. Jasieniak, D.E. Gomez, P. Mulvaney, Small 2 (2006) 199–203.
[119] S.H. Kim, Y.S. Cho, S.J. Jeon, T.H. Eun, G.R. Yi, S.M. Yang, Adv. Mater. 20 (2008) 3268–3273.
[120] M. Mansuripur, A.R. Zakharian, A. Lesuffleur, Sang-Hyun Oh, R.J. Jones, N.C. Lindquist, H. Im, A. Kobyakov, J.V. Moloney, Opt. Expr. 17 (2009) 14001–14014.
[121] T. Uematsu, J. Kimura, Y. Yamaguchi, Nanotechnology 15 (2004) 822–827.
[122] J.H. Fendler, Chem. Mater. 13 (2001) 3196–3210.
[123] D.J. Mowbray, M.S. Skolnick, J. Phys. D Appl. Phys. 38 (2005) 2059–2076.
[124] J.E. Moris, In: J.E. Morris (Ed.), Nanopackaging: Nanotechnologies and Electronics Packaging, Springer, New York, 2008, pp. 1–13.
[125] J. Lu, C.P. Wong, In: J.E. Morris (Ed.), Nanopackaging: Nanotechnologies and Electronics Packaging, Springer, New York, 2008, pp. 121–138.
[126] A. Zabet-Khosousi, A.-A. Dhirani, Chem. Rev. 108 (2008) 4072–4124.
[127] D. Vanmaekelbergh, P. Liljeroth, Chem. Soc. Rev. 34 (2005) 299–312.
[128] C.R. Barry, C.J. Hoon, H.O. Jacobs, Proceedings of the Foundations of Nanoscience, Utah, April 21–23, 2004.
[129] D.L. Feldheim, D. Keating, Chem. Soc. Rev. 27 (1998) 1–12.
[130] D.L. Klein, R. Roth, A.K.L. Lim, A.P. Alivisatos, P.L. McEuen, Nature 389 (1997) 699–701.
[131] K.I. Bolotin, F. Kuemmeth, A.N. Pasupathy, D.C. Ralph, Appl. Phys. Lett. 84 (2004) 3154–3356.
[132] C. Vieu, A. Pepin, J. Gierak, C. David, Y. Jin, F. Carcenac, H. Launois, J. Vac. Sci. Technol. B 16 (1998) 3789–3794.
[133] D. Talapin, C.B. Murray, Science 310 (2005) 86–89.
[134] T. Cui, Y. Liu, M. Zhu, Appl. Phys. Lett. 87 (2005) 183105–183107.
[135] S.H. Ko, H. Pan, C.P. Grigoropoulos, Ch.K. Luscombe, J.M. Frechet, D. Poulikakos, Nanotechnology 18 (2007) 345202–345209.
[136] E. Matijevic, S.V. Babu, J. Colloid Interface Sci. 320 (2008) 219–237.
[137] Nanoparticles find a smooth opportunity in chip-polishing, SmallTimes, 14 June 2003, www.smalltimes.com/articles/article_display.cfm?ARTICLE_ID=268461&p=109 (accessed 11.02.08).

[138] P.J.A. Borm, D. Robbins, S. Haubold, T. Kuhlbusch, H. Fissan, K. Donaldson, R. Schins, V. Stone, W. Kreyling, J. Lademann, J. Krutmann, D. Warheit, E. Oberdorster, Particle Fibre Toxicol. 3 (2006) 11–45.
[139] T. Zeng, T. Sun, IEEE Trans. Semicond. Manuf. 18 (2005) 655–663.
[140] Z. Lu, S.-H. Lee, S.V. Babu, E. Matijevic, J. Colloid Interface Sci. 261 (2003) 5564.
[141] Z. Lu, S. Lee, V.R.K. Gorantla, S.V. Babu, E. Matijevic, J. Mater. Res. 18 (2003) 2323.
[142] D. Ng, Nanoparticles Removal on Post-CMP Cleaning, MSc Thesis, Texas A&M University, 2005.
[143] S.A. Wolf, D.D. Awschalom, R.A. Buhrman, J.M. Daughton, S. von Molnar, M.L. Roukes, A.Y. Chtchelkanova, D.M. Treger, Science 294 (2001) 1488–1495.
[144] J. Cibert, J.-F. Bobo, U. Lüders, C. R. Phys. 6 (2005) 977–996.
[145] M.N. Baibich, J.M. Broto, A. Fert, F. Nguyen van Dau, F. Petroff, P. Etienne, G. Creuzet, A. Friedrich, J. Chazelas, Phys. Rev. Lett. 61 (1988) 2472–2475.
[146] S. Ram, S. Biswas, H.J. Fecht, In: S. Seal (Ed.), Functional Nanostructures, Springer, Berlin, 2008.
[147] A. Bernand-Mantel, P. Seneor, N. Lidgi, M. Muñoz, V. Cros, S. Fusil, K. Bouzehouane, C. Deranlot, A. Vaures, F. Petroff, A. Fert, Appl. Phys. Lett. 89 (2006) 62502–62504.
[148] K. Yakushiji, F. Ernult, H. Imamura, K. Yamane, S. Mitani, K. Takanashi, S. Takahashi, S. Maekawa, H. Fujimori, Nat. Mater. 4 (2005) 57–61.
[149] W.K. Liu, K.M. Whitaker, K.R. Kittilstved, D.R. Gamelin, J. Am. Chem. Soc. 128 (2006) 3910–3911.
[150] R. Beaulac, P.I. Archer, S.T. Ochsenbein, D.R. Gamelin, Adv. Funct. Mater. 18 (2008) 3873–3891.
[151] J. Petzold, Scr. Mater. 48 (2003) 895–901.
[152] D.L. Leslie-Pelecky, R.D. Rieke, Chem. Mater. 8 (1996) 1770–1783.
[153] U. Jeong, X. Teng, Y. Wang, H. Yang, Y. Xia, Adv. Mater. 19 (2007) 33–60.
[154] M. Pardavi-Horvath, J. Magn. Magn. Mater. 215–216 (2000) 171–183.
[155] F. Mazaleyrat, L.K. Varga, J. Magn. Magn. Mater. 215–216 (2000) 253–259.
[156] G. HerZer, In: K.H.J. Buschow (Ed.), Handbook of Magnetic Materials, vol. 10, Elsevier, Berlin, 2007, pp. 415–463.
[157] M. Nelo, A.K. Sowpati, V.K. Palukuru, J. Juuti, H. Jantunen, Prog. Electromagn. Res. 110 (2010) 253–266.
[158] I. Matsui, J. Chem. Eng. Jpn. 38 (2005) 535–546.
[159] D.A. Thompson, J.S. Best, IBM J. Res. Dev. 44 (2000) 311–322.
[160] A. Moser, K. Takano, D.T. Margulies, M. Albrecht, Y. Sonobe, Yoshihiro Ikeda, S. Sun, E.E. Fullerton, J. Phys. D Appl. Phys. 35 (2002) R157–R167.
[161] D. Huber, Small 1 (2005) 483–502.
[162] R.J. Veitch, A. Ilmer, W. Lenz, V. Richter, J. Magn. Magn. Mater. 193 (1999) 279–283.
[163] H. Nishio, H. Yamamoto, J. Appl. Phys. 101 (2007) 123904–123907.
[164] H. Nishio, Y. Ota, M. Maekawab, H. Yamamoto, J. Magn. Magn. Mater. 287 (2005) 234–238.
[165] J.C. Lodder, J. Magn. Magn. Mater. 272–276 (2004) 1692–1697.
[166] G.M. McClelland, et al., Appl. Phys. Lett. 81 (2002) 1483.
[167] B. El-Kareh, G.B. Bronner, S.E. Schuster, Solid-State Technol. 40 (1997) 89–101.
[168] S. Tiwari, F. Rana, H. Hanafi, A. Hartstein, E.F. Crabbe, K. Chan, Appl. Phys. Lett. 68 (1996) 1377–1379.
[169] Z. Liu, C. Lee, V. Narayanan, G. Pei, E.C. Kan, IEEE Trans. Electron Devices 49 (2002) 1606–1613.

[170] D. Tsoukalas, Int. J. Nanotechnol. 6 (2009) 35–45.
[171] M.T. Sebastian, H. Jantunen, Int. Mater. Rev. 53 (2008) 57–90.
[172] R.E. Newnham, Annu. Rev. Mater. Sci. 16 (1986) 47–68.
[173] R. Gregorio Jr., M. Cestar, F.E. Bernardino, J. Mater. Sci. 31 (1996) 2925–2930.
[174] E. Reichmanis, H. Katz, Ch. Kloc, A. Maliakal, Bell Labs Tech. J. 10 (2005) 87–105.
[175] J.Y. Li, L. Zhang, S. Ducharme, Appl. Phys. Lett. 90 (2007) 132901–132903.
[176] T.K. Kundu, D. Chakravorty, J. Mater. Res. 11 (1996) 200–203.
[177] G.C. Vezzoli, M.F. Chen, J. Caslavsky, Nanostruct. Mater. 4 (1994) 985–1009.
[178] Y. Shen, Y. Lin, C.W. Nan, Adv. Funct. Mater. 17 (2007) 2405–2410.
[179] C.D. Dimitrakopoulos, S. Purushothaman, J. Kymissis, A. Callegari, J.M. Shaw, Science 283 (1999) 822–824.
[180] F.-C. Chen, Ch.-S.h. Chuang, Y.-S.h. Lin, L.-J. Kung, Y.-S.h. Lin, T.-H. Chen, H.P.D. Shieh, Org. Electron. 7 (2006) 435–439.
[181] B.J. Lee, P. Badheka, D.H. Yoo, V. Magadala, M. Wang, J. Ceram. Process. Res. 5 (2004) 127–132.
[182] X. Wang, B.I. Leee, M.Z. Hu, E.A. Payzant, D.A. Bloom, J. Mater. Sci. Mater. Electron. 14 (2003) 495–500.
[183] M. Kohler, W. Fritzsche, Nanotechnology: An Introduction to Nanostructuring Techniques, Wiley-VCH, Weinheim, 2007, pp. 225–241.
[184] G. Eranna, B.C. Joshi, D.P. Runthala, R.P. Gupta, Crit. Rev. Solid State Mater. Sci. 29 (2004) 111–188.
[185] D. Kohl, J. Phys. D Appl. Phys. 34 (2001) R125–R149.
[186] Y. Su, React. Funct. Polym. 66 (2006) 967–973.
[187] I. Koh, L. Josephson, Sensors 9 (2009) 8130–8145.
[188] M.J. Toohey, Sensors Actuators B 105 (2005) 232–250.
[189] M.E. Franke, T.J. Koplin, U. Simon, Small 2 (2006) 36–50.
[190] C. Drake, S. Deshpande, D. Bera, S. Seal, Int. Mater. Rev. 52 (2007) 289–317.
[191] R. Rella, P. Sicilliano, S. Capone, M. Epifani, L. Vasanelli, A. Licciulli, Sensors Actuators B 58 (1999) 283–2888.
[192] R. Ionescu, A. Hoel, C.G. Granqvist, E. Llobet, P. Heszler, Sensors Actuators B 104 (2005) 132–139.
[193] L. Han, X. Shi, W. Wu, F.L. Kirk, J. Luo, L. Wang, D. Mott, L. Cousineau, S. Lim, S. Lu, C. Zhong, Sensors Actuators B 106 (2005) 431–441.
[194] F.P. Zamborini, M.C.L. Leopold, J.F. Hicks, P.J. Kulesza, M.A. Malik, R.W. Murray, J. Am. Chem. Soc. 124 (2002) 8958–8964.
[195] R. Gordon, IEEE Nanotechnol. Mag. 2 (2008) 12–18.
[196] P. Bhattacharya, S. Ghosh, A.D. Stiff-Roberts, Annu. Rev. Mater. Res. 34 (2004) 1–40.
[197] J. Homola, S.S. Yee, G. Gauglitz, Sensors Actuators B 54 (1999) 3–15.
[198] E. Hutter, J.H. Fendler, Adv. Mater. 16 (2004) 1685–1706.
[199] Paul Mulvaney, Langmuir 12 (1996) 788–800.
[200] S. Link, Z.L. Wang, M.A. El-Sayed, J. Phys. Chem. B 103 (1999) 3529–3533.
[201] Anuj K. Sharma, B.D. Gupta, Photon. Nanostruct. Fundam. Appl. 3 (2005) 30–37.
[202] S. Lal, S. Link, N.J. Halas, Nanophotonics 1 (2007) 641–648.
[203] A. Wei, B. Kim, B. Sadtler, S.L. Tripp, Chem. Phys. Chem. 2 (2001) 743–745.
[204] R.S. Golightly, W.E. Doering, M.J. Natan, ACS Nano 3 (2009) 2859–2869.
[205] K. Kneipp, Y. Wang, R.R. Dasari, M.S. Feld, B.D. Gilbert, J. Janni, J.I. Steinfeld, Spectrochim. Acta A 51 (1995) 2171–2715.
[206] M.K. Oo, C.F. Chang, Y. Sun, X. Fan, Analyst 136 (2011) 136–142.

[207] H. Ko, S. Singamaneni, V.V. Tsukruk, Small 4 (2008) 1576–1599.
[208] F. Seker, K. Meeker, Th.F. Kuech, A.B. Ellis, Chem. Rev. 100 (2000) 2505–2536.
[209] B.G. Vainer, J. Phys. D Appl. Phys. 41 (2008) 65102–65111.
[210] P. Boucaud, S. Sauvage, C. R. Phys. 4 (2003) 1133–1154.
[211] S.D. Gunapala, S.V. Bandara, C.J. Hill, D.Z. Ting, J.K. Liu, S.B. Rafol, E.R. Blazejewski, J.M. Mumolo, S.A. Keo, S. Krishna, Y.-C. Chang, C.A. Shott, Infrared Phys. Technol. 50 (2007) 149–155.
[212] P. Boucaud, S. Sauvage, C. R. Phys. 4 (2003) 1133–1154.
[213] H. Liu, J. Zhang, Opt. Laser Technol. 44 (2012) 1536–1542.
[214] H. Lim, S. Tsao, M. Taguchi, W. Zang, A.A. Quivy, M. Razeghi, Adv. Sci. Technol. 51 (2006) 201–208.
[215] D. Oertel, M.G. Bawendi, A.C. Arango, V. Bulovic, Appl. Phys. Lett. 87 (2005) 213505–213507.
[216] G. Konstantatos, I. Howard, A. Fischer, S. Hoogland, J. Clifford, E. Klem, L. Levina, E.H. Sargent, Nature 442 (2006) 180–183.
[217] H.-Y. Chen, F. Lo, G. Yang, H.G. Monbouquette, Y. Yang, Nat. Nanotechnol. 3 (2008) 543–547.
[218] R.M. Penner, Acc. Chem. Res. 33 (2000) 78–86.
[219] E. Katz, I. Willner, J. Wang, Electroanalysis 16 (2004) 19–44.
[220] X. Luo, A. Morrin, A.J. Killard, M.R. Smyth, Electroanalysis 18 (2006) 319–326.
[221] D. Kohl, J. Phys. D Appl. Phys. 34 (2001) R125–R149.
[222] D.L. Feldheim, C.D. Keating, Chem. Soc. Rev. 27 (1998) 1–12.
[223] A.B. Kharitonov, A.A. Shipway, E. Katz, I. Willner, Rev. Anal. Chem 18 (1999) 255–260.
[224] H.-Z. Ding, B.R. Varlow, Annual Report Conference on Electrical Insulation and Dielectric Phenomena, 2004, pp. 332–335.
[225] T. Yokoyama, In: M. Hosokawa, K. Nogi, M. Naito (Eds.), Nanoparticle Technology Handbook, Elsevier, Amsterdam, 2007, pp. 434–438.
[226] E. Ide, S. Angata, A. Hirose, K.F. Kobayashi, Acta Mater. 53 (2005) 2385–2393.
[227] S. Forge, C. Blackman, OLED and E-Paper: their disruptive potential for the European display industry, JCR Scientific and Technical Reports, 2009.
[228] T.-H. Kim, K.-S. Cho, E.K. Lee, S.J. Lee, J. Chae, J.W. Kim, D.H. Kim, J.-Y. Kwon, G. Amaratunga, S.Y. Lee, B.L. Choi, Y. Kuk, J.M. Kim, K. Kim, Nature Photonics 5 (2011) 176–181.
[229] J. Heikenfeld, P. Drzaic, J.-S. Yeo, T. Koch, J. SIF 19 (2011) 129–156.
[230] http://www.epapercentral.com/epaper-technologies-guide (accessed 24.09.11).
[231] M.-C. Choi, Y. Kim, C.-S. Ha, Prog. Polym. Sci. 33 (2008) 581–630.
[232] B. Comiskey, J.D. Albert, H. Yoshizawa, J. Jacobson, Nature 394 (1998) 253–255.
[233] T.H. Kim, Y.S. Ko, Y.K. Kwon, J. Nanosci. Nanotechnol. 6 (2006) 3450–3454.
[234] L. Grande, V. Teja Chundi, D. Wei, C. Bower, P. Andrew, T. Ryhänen, Particuology 10 (2012) 1–8.
[235] F. Xiao, J. Song, H. Gao, X. Zan, R. Xu, H. Duan, ACS Nano 6 (2012) 100–110.
[236] Lu Huang, Yi Huang, Jiajie Liang, Xiangjian Wan, Yongsheng Chen, Nano Res. 4 (7) (2011) 675–684.

Chapter 8

Environmental and Societal Applications and Implications

1 INTRODUCTION

In the last few years, the word Nano has become synonymous with miniaturization, fashion, and technological revolution. This word has been used in high-profile trademark products from Apple (iPod nano) and Samsung (Silver nano). Another, albeit misleading, example is provided by the Indian car Tata Nano. It is only meant to be the cheapest and smallest Indian car, and the name does not mean that the car contains nano-based products. Unfortunately, it raises some justified concerns, given the potential negative health and environmental effects of nanosized materials [1].

Aerosol droplets applied to glass and ceramic substrates provides oil- and water-repellent properties (Magic nano). However, this application has been the target of some controversy, particularly in Germany [2]. In fact, the German Federal Institute for Risk Assessment issued a warning regarding the potential respiratory effects of this product that led to its recall. In spite of the fact that this product does not contain small particulates besides the solvent droplets of about 10 μm, this recall has had a negative effect on the marketing of nanoparticle (NP)-based products. Therefore, careful consideration should be given to the hype created around the word nano, which, in this case, has backfired. This development has highlighted the potential limitations facing commercialization of NP-based products.

Evaluating the hazard level of nanoparticles (NPs) is critical. A hazard is defined as a risk that has the potential to cause harm. From this point of view, the hazard potential of NPs is not very different from that of any other chemical. However, what is special in the case of NPs is the difficulty in quantifying the hazard level. Indeed hazards ranging from systemic inflammation to cardiovascular diseases and brain damage have been postulated. A recent study has been aimed at categorizing and evaluating exposure of different nanomaterials in 580 consumer products [3]. Three categories of hazards have been

Interface Science and Technology, Vol. 19. http://dx.doi.org/10.1016/B978-0-12-369550-5.00008-2

considered: expected, possible, and not expected exposure. Although the majority of the analyzed products fall into the first category, low concentrations of these nanomaterials within the product have prevented a more quantitative assessment of the exposure risk.

The light weight of NPs facilitates their dispersion in air, and aerosol-like particles are formed. Their toxicity can be compared to that of natural and man-made aerosols. These aerosols operate under three modes, corresponding to different size ranges [4], which consist of nucleation mode (up to 0.1 μm), accumulation mode (0.1–1 μm), and coarse mode (above 1 μm). The first two modes (nucleation and accumulation) consist mostly of combustion, black carbon smog, and dust particles. The coarse mode consists mostly of clouds, fog, sea salts, sand, and coal dust particles. Some aerosols such as atmospheric dust may occur across the three modes. The most relevant aerosols for this chapter include combustion particles, carbon black, and smog. The toxicity of some aerosols has already been established, particularly that of submicron particles. Indeed, these nanosized aerosols are capable of penetrating deep into the respiratory tract [4]. Silver NPs are endowed with remarkably strong antimicrobial activity. However, their use in washing machines has met with some resistance owing to their potential for contaminating the water system.

To address industry and consumer needs relating to nanotechnology, extensive research on health and the societal implications of mass commercialization of NPs should be undertaken [5]. For this, a more proactive disclosure of relevant information on the potential risks of these NPs is needed. The physical state of these NPs (gas, liquid, or solid phase) will also have a direct effect on the hazard level, since the affected organ will have a different protection ability depending on the NP phase. At present, state-of-the-art equipment for characterizing the physicochemical properties of NPs, whether *in vivo* or *in vitro*, is limited.

Communication, transparency, and education will be key factors for a sustainable development of NP science and technology. Management of these three factors will reduce the level of negative perceptions related to nanotechnology. NP-based industries can learn from technological developments in other industries such as those related to genetically modified foods. Otherwise, industries using NPs in their products and processes can run the risk of facing the same opposition from the public and the media that biotechnology continues to face [6]. In this regard, the academic and industrial communities need to anticipate instead of reacting.

There should be more regulation of human- and animal-related applications of NPs. In particular, cosmetics and foods that incorporate NPs should be regulated. These regulations will not only prevent expected and unexpected toxicity effects but also preclude public overreaction. For example, TiO_2 particles having sizes above 100 nm are considered inert, although recent research has shown that they could have some toxicity implications [7–9].

Contradiction and uncertainty give rise to serious questions in the public mind. This apparent contradiction should be understood and resolved so as to allow or prevent commercial use of these particles.

Public overreaction could have unintended consequences. For example, negative perceptions can not only generate additional restrictions for all applications but also reduce support in general. NPs are vilified for their potential negative effects on health and the environment. At the same time, these two areas are the most likely to benefit from future developments in NP science and technology [10,11]. Indeed, NPs are being used and will be used to address issues related to energy and water. Energy and mining account for more than half of the water withdrawal, and unfortunately, returned water (about half) does not contribute to replenishing high-quality water sources [12]. Efficient and cost-effective water purification is needed to make affordable and safe water accessible worldwide. Water-related issues are the main cause of up to 10 million deaths [13]. Iron NPs have been shown to be effective in contaminated soil and groundwater remediation. For example, arsenic and perchlorate anions have been successfully removed from contaminated water using magnetic NPs.

The potential toxicity of NPs has been rightly singled out given the young history of this field and the hype that has been created around it [14]. Even though the existing database is insufficient, the exaggerations in some reports have led to overreaction among the general public. This situation is often fueled by interest groups who are ill-informed. Even within the scientific community, agreement on the toxicity of relatively well-studied systems such as single-wall nanotubes [15] and TiO_2 NPs is sometimes hard to find [16]. Often, this controversy stems from ill-devised experiments and/or the fact that toxicity depends on several chemical, biological, and physical parameters. Evaluating the toxicity level caused by each parameter is difficult.

Besides inhalation, there are different pathways for NP transport through membranes that lead to cellular uptake and degradation (Fig. 8.1). Skin surface with an area of around 2 m^2 is the first barrier against bacteria and viruses. Skin is not an effective barrier against lipids and lipophilic chemical

FIGURE 8.1 Pd–Fe_3O_4 magnetic nanoparticle (20–30 nm) used to separate and catalyze pollutant in sewage water [17].

penetration. Higher surface areas such as gastrointestinal (40 m^2) and epithelial cells (2000 m^2) are more effective in screening small particulates and macromolecules [18]. The gastrointestinal and respiratory tracts are the most likely routes of entry of NPs into the body. With an active surface area of 15–18 m^2 [18], the human lung also provides easy access to submicron particles and gases. The absorption process of NPs and molecules by cells is referred to as endocytosis. There are different endocytosis mechanisms of size-dependent cellular uptake [18]. The first one is phagocytosis, which uses enzymes to digest the content of larger organisms (500–1000 nm). Other processes include pinocytosis (100 nm), coated pits (150 nm), and caveoli (50 nm) [18]. Thus, a size- and chemistry-dependent endocytosis of NPs is expected.

Health risks related to NPs are mostly due to inhalation. Direct contact with surface skin (particularly nonhealthy skin) is still not well understood due to lack of controlled *in vivo* studies. Furthermore, conclusions made on the basis of animal models could hardly be translated to humans, since the skin of rats is about 11 times more permeable to tested substances than human skin [19].

Depending on NP characteristics, interaction with the blood and immune system could lead to hazardous effects [20]. NP size, bulk, and surface composition are probably the most important characteristics affecting their hazard level. Current metrological techniques quantifying NP characteristics are not yet optimal enough to quantify these effects. Depending on these characteristics, some NPs can trigger certain immune responses. Quantifying the risk of these NPs with a given characteristic should include testing for these responses. Systematic immune response studies under controlled and varied conditions are needed and they should be made public. This could help the approval process of numerous NP-based formulations for therapeutic applications [21].

Given the infancy and diversity of NP fabrication processes and their applications, the debate over their advantages and drawbacks is naturally quite complex and difficult. It is difficult to draw a general conclusion regarding a single NP composition because, as discussed in previous chapters, the properties may also depend on shape, size, crystallinity, and composition.

In the case of greenhouse gas (GHG) emissions and other toxic by-products during the product lifetime, a life-cycle analysis (LCA) should be undertaken. Lack of a database is currently a stumbling block in the development of LCA. The following process and utilization steps should be carefully analyzed [22]:

a. raw material extraction and processing;
b. NP fabrication process;
c. distribution and transport of the NP;
d. manufacturing of the NP-based product;
e. distribution and transport of the product;

f. utilization of the product;
g. disposal of the product.

Unfortunately, there are very few LCA studies on NP-based products. A few LCA studies focusing on the manufacture of NPs have been reported. For example, LCA of a novel process for producing TiO_2 NPs from an ilmenite feedstock [23] and CdSe NPs from $Cd(CH_2)/Se$ mixture have been reported [24].

2 ENVIRONMENT

NPs may have a negative impact on the environment by affecting air quality and/or composition of natural water reservoirs. A few studies discussing the interaction of NPs with the ecosystem have been reported [11,25]. However, NPs are also used to address some significant environmental issues resulting from either human actions or natural events. Indeed, NPs could be used for detection, displacement, and remediation of polluted air, water, and soil [26–28]. Furthermore, NPs have already had a positive impact by reducing the human footprint and by reducing the demand for raw materials and energy.

There are numerous environmental applications of NPs. As discussed in Chapter 5, NPs are used in several energy-related applications, including clean and renewable energy power generation, that help reduce GHG emissions. Solar cells, fuel cells, and material and fossil energy sources are the most important areas that could benefit from the attributes of NPs. NPs are used in upstream and downstream fossil fuel processing to increase efficiency and environmental impact. The following areas could benefit directly from NP utilization [10]:

a. optimized use of solvent in paints;
b. water purification and filtration;
c. treatment and remediation;
d. alternative green manufacturing.

2.1 Optimized Use of Solvent in Paints

Optimized aqueous dispersion of oxide NPs (ZnO, Al_2O_3) allows utilization of water-based paintings. Besides involving high costs, organic solvents are significant sources of pollution. A few products have been launched to take advantage of the water solubility of NP-based additives in painting processes. Rutile TiO_2 NPs have been shown to provide higher opacity allowing a more efficient UV protection in acrylic coating applications [29]. Furthermore, rutile NPs have a lower photoactivity, allowing their commercial utilization as cost-effective stabilizers with a performance that is greater than or equal to that of the conventional organic absorbers.

Supported metal NP catalysts are already in use on a commercial scale in the transportation and petrochemical sectors. A major car manufacturer has developed a precious metal dispersion concept to achieve high efficiency and a lower load of precious metal catalysts [30]. In this design, the thermal deterioration and loss of efficiency caused by the agglomeration of precious metal particles are inhibited by using even smaller catalysts (<5 nm) and catalyst supports (<100 nm). This is achieved by embedding these catalyst NPs in a ceramic base material. Furthermore, the oxygen absorption and release rates are improved, allowing enhanced GHG emission mitigation. The amount of precious metal catalysts required is reduced by up to 70–90% [30]. However, the eventual health risk related to large-scale manufacturing and use of sub-100 nm catalyst supports and sub-5 nm metal NPs has not been assessed yet.

To circumvent utilization of primer in latex-based coatings, a binder formulation in combination with an inorganic NP-based pigment has been proposed [31]. ZnO NPs with an average size of 60 nm have been reported to provide very promising results. This combination seems to yield superior tannin blocking, stain removal, corrosion resistance, and improved adhesion.

2.2 Water Purification

Water purification may impact several important socioeconomic sectors, including environment, health, food, energy, and national security. Thus, addressing problems related to water supply, source protection, conservation, and timely contamination detection and removal is of great importance. Worldwide, water equipment and chemicals represented a $76.3 billion market in 2007 [32].

NPs provide numerous opportunities for efficient and cost-effective water purification because of their high surface area and specific properties [33]. For example, zero-valent iron (ZVI) NP-based redox systems are effective in detoxification of organic and inorganic pollutants in aqueous solutions. These ZVI NPs consists of an (Fe^0) core and a relatively porous FeO(OH) shell. With higher intrinsic reactivity and a greater number of reactive density sites over a large surface area, these ZVIs are effective in degrading contaminants. TiO_2 NPs are effective and environmentally benign photocatalysts. Ag and MgO NPs are used as biocides against various bacteria. These NPs can be integrated into nanocomposites used in multicomponent packed-bed reactors [33]. Water contaminated by metal ions, organic solutes, and bacteria is purified at different stages. Each stage is designed to remove a specific contamination. Bimetallic xerogel catalysts have been used in the processing of chlorinated water [33,34].

Organic solvents (e.g., CCl_4), pesticides (e.g., lindane), and metals (e.g., Cr) are important underground contaminants. Cost-effective remediation solutions in this regard are still elusive. Iron NPs can be used on-site for remediation of a wide variety of common environmental contaminants, such as chlorinated organic solvents, organochlorine pesticides, and PCBs [35]. Dechlorination of toxic contaminants such as tetrachloroethylene proceeds via the following reduction process in aqueous solution:

$$Fe^0(s) \rightarrow Fe^{2+}(aq) + 2e^-(aq)$$
$$C_2Cl_4 + 4Fe^0 + 4H^+ \rightarrow C_2H_4 + 4Fe^{2+} + 4Cl^-$$

Injection of iron NP slurries into an aquifer is relatively easy, and they can flow with groundwater over a distance of 20 m [35]. Thus an *in situ* remediation of contaminated aquifer is feasible. These iron-based NPs have been shown to remain active in contaminated aquifer for up to 2 months [35]. Iron NPs are also inexpensive and effective in remediation of a large variety of environmental contaminants [35].

Another practical example of the use of magnetic NPs for sewage is shown in Fig. 8.1. These multicomponent NPs are endowed with two functional properties. The magnetic component (Fe_3O_4) allows the chemical harvesting. The catalyst component (Pd) chemically converts the pollutant. Indeed, organically bound halogens are selectively absorbed using functionalized magnetic NPs and then catalytically converted to biodegradable materials with absorbable organically bound halogens. As a final step, the particles containing the converted materials are magnetically separated from the solution [36].

Yavuz et al. showed that in certain cases, separation using magnetic NPs is superior to centrifugation or filtration [37]. In this work, NP size and magnetic field are shown to have a direct effect on the magnetic separation efficiency. For example, after injecting a mixture containing 4 and 12 nm of magnetite (Fe_2O_3) NPs, a field of 300 mT removed almost completely the 4 nm NPs and the majority of the 12 nm NPs remained in the solution. It is only under the influence of a weaker magnetic field (30 mT) that the 12 nm NPs are removed from the solution. It was also shown that arsenic had been effectively removed from the solutions with minimum magnetite waste [38].

Reverse osmosis for waste treatment using zeolite NP-based nanocomposites has been reported [38,39]. Photocatalysis applications in wastewater treatment are another area where NPs could play a significant role [40]. The characteristics of the main inorganic photocatalysts are summarized in Table 8.1. Energy bandgap is the most important characteristic. Particle size and, thus, surface area are also shown to have a direct effect on the efficiency. Although the bandgap of TiO_2 NPs is not optimal, it is currently the most effective photocatalyst. Lower bandgap materials (Si, WSe_2) with higher solar radiation absorption do not provide expected results. Stability is also an important attribute. Titania and most oxides are relatively stable when compared to other semiconductor materials.

Different configurations of the photochemical reactor have been used for water treatments. This includes fixed-bed continuous flow gas reactors and fluidized-bed flow gas reactors [41]. In the case of slurry reactors, the

TABLE 8.1 Characteristics of Inorganic Photocatalysts [40]

Photocatalyst	Bandgap energy (eV)	Photocatalyst	Bandgap energy (eV)
Si	1.1	ZnO	3.2
TiO_2 (rutile)	3.0	TiO_2 (anatase)	3.2
WO_3	2.7	CdS	2.4
ZnS	3.7	$SrTiO_3$	3.4
SnO_2	3.5	WSe_2	1.2
Fe_2O_3	2.2	α-Fe_2O_3	3.1

particles are collected after filtration, centrifugation, coagulation, and flocculation. Overall, this may make the process costly. In the fixed-bed reactor design with the photocatalyst embedded in a solid-supported matrix, processing of aqueous or gas-phase effluents is feasible without the need of additional postprocessing filtration. Fixed-bed reactors are limited by their low efficiency. A pilot-scale solar photocatalytic water detoxification system has been recently tested [41].

2.3 Treatment and Remediation

Velocity properties of NPs through contaminated soil could make them invaluable for bioremediation [42]. Iron NP applications for remediation have been recently reviewed [35]. They could be used for remediation of sites contaminated by polychlorinated biphenyls, chlorinated aliphatic hydrocarbon, and chlorinated pesticides. ZVI (Fe^0) NPs have been used for these purposes [43]. The remediation process involving chlorinated hydrocarbon (R-Cl) is described in the following equation [44]:

$$R-Cl+H^{+}+Fe^{0}\rightarrow RH+Fe^{2+}+Cl^{-}$$

Core–shell structure consisting of Pd on gold NPs has been used to catalyze the destructive removal of chlorinated compounds [45]. Pd coverage (Fig. 8.2) is critical to the catalysis of the remediation process. Controlled sub-monolayer surface coverage of gold NPs with Pd was obtained by adding the appropriate amount of Pd salt and a reducing agent to the gold sol. Effective hydrodechlorination of hydrodechlorinate trichloroethene in water was achieved. Pd–Au NPs were found to be considerably more active than Pd NPs. Gold NPs partially covered with Pd give rise to the highest hydrodechlorination activity [45], while lower Pd consumption certainly reduces the overall cost of the remediation.

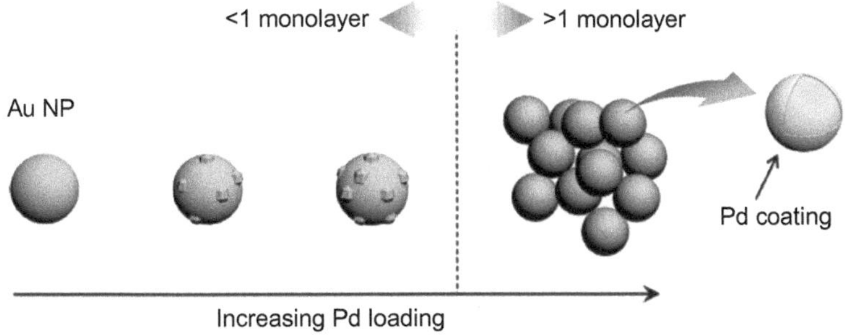

FIGURE 8.2 Schematic evolution of Pd on Au nanoparticle structure with increasing Pd coverage. *Reprinted with permission from Nutt et al. [45]. Copyright 2005, American Chemical Society.* (For color version of this figure, the reader is referred to the online version of this chapter.)

Metallic oxide NPs can be used to neutralize toxic industrial chemicals. Metallic oxide NPs have been used as destructive catalysts for fluoro-, chloro-, and bromocarbon, sulfur, and organophosphate compounds [46]. Al_2O_3 NPs have been shown to provide the ability to destroy chemical warfare agents [47].

2.4 GHG Mitigation

Increasing concentration of anthropogenic carbon dioxide is a major concern of this century. Geological CO_2 sequestration is currently being considered although long-term risks have not yet been evaluated. Photocatalysis of CO_2 in the presence of water to form hydrocarbon is probably the best avenue. Suspended semiconducting powders have been shown to reduce CO_2 to form hydrocarbon [48,49]. Several factors including size and composition could affect the overall efficiency of UV-driven photocatalysis [50]. Cu and/or Pt NPs deposited on nitrogen-doped titania nanotube arrays have been used to efficiently convert CO_2 and/or H_2O gases to fuel using sunlight to drive a photocatalysis process [51].

2.5 NO_x Decomposition

NO_x emissions from internal combustion engines and other industrial activities are significant contributors to air pollution. Smog is omnipresent in the majority of big cities of industrialized countries. It is directly related to nitrogen oxides, ozone, and hydrocarbon emissions from car exhaust. A nitrate smog component has been shown to be very detrimental to health and can increase the death rate among the elderly. Supported NP catalysts have been shown to significantly reduce NO_x emissions. A typical car exhaust catalyst contains a monolith support with a washcoat impregnated with precious metal NPs as active catalysts [52].

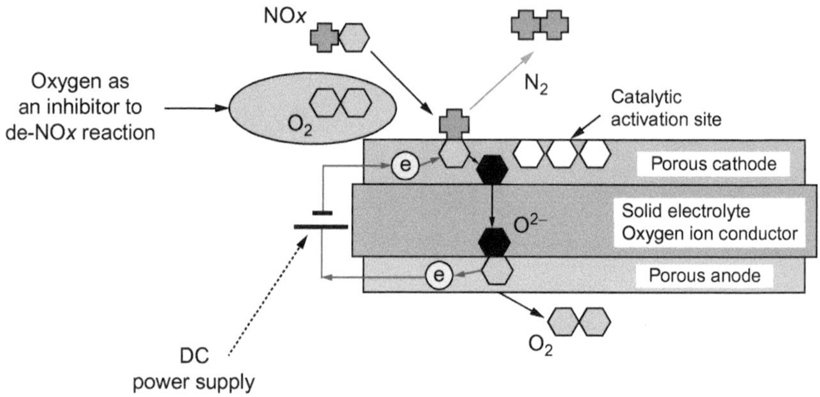

FIGURE 8.3 Solid oxide electrochemical cell and NO_x decomposition on a single-layered cathode [53].

Noble metal-based catalytic decomposition of NO_x is widely used. Solid oxide electrochemical reactors can provide some advantage in decomposing NO_x [53]. The combination of a 2D cathode structure and NiO NPs was effectively used to improve the efficiency of the NO_x-to-NO_2 conversion (Fig. 8.3). Nickel-based grain sizes are about 10–50 nm. Low energy consumption and operation under excess oxygen content are achieved using this approach. Two porous electrodes sandwiching an oxygen ion solid electrolyte were used to effectively transform NO_x to N_2 with a high yield rate (Fig. 8.3). In addition, this design is scalable.

2.6 Early Pollution Detection

Early detection of potentially harmful pollutants requires highly sensitive sensors. As described in Chapters 6 and 7 of this book, several NP-based electrochemical, magnetic, and optical sensors have been developed. These sensors could be used for both high sensitivity and multi-analyte detections. However, current sensors are often limited to known chemicals. NP-based SERS sensors provide the possibility of remote detection and molecular characterization of dangerous substances. With a detection limit down to picoliters, SERS may provide possibility for single molecule detection.

2.7 Alternative Green Manufacturing

In some cases, NP-based manufacturing is potentially energy efficient if the overall LCA is considered. Environmental impacts related to NP fabrication should be reduced in several cases, before viable commercial application can be considered. Yield production and waste remain two challenges in NP production. However, NP integration into film and other products is well established in some applications, although optimization is still required.

Silane surface treatment, thickening and strengthening by NP addition, has been shown to improve the film hardness and anti-corrosion protection of aluminum alloys [54]. If optimized, this process could be used as an alternative to the toxic chromate coating process. Surface finishing such as a hard chrome coating generates significant amounts of toxic and hazardous by-products [55]. Development of an alternative coating process and raw materials without losing current performance and cost advantages is challenging.

A lead-free soldering material and process is a hot topic within the microelectronic industry. For example, lead-free soldering was developed to replace solder alloy 95Pb5Sn (of 95 wt% Pb and 5 wt% Sn) used as solder bumps on Si chip surfaces [56]. Lead-free soldering in electronic packaging has been recently proposed based on silver NPs [57]. A paste consisting of filler silver powder (400 nm) and alkoxide-passivated silver NPs (5 nm) has been used to connect silicon diode chips to copper base. Testing following sintering at 350 °C showed that the electrical and thermal properties of this new soldering process are comparable to lead-based soldering. Low temperature sintering of silver NPs has also been reported for potential use in interconnect applications to replace lead-based metal-solder joints [58]. This lead-free process is quite significant, given the fact that just in the USA, 10^9 pounds of Pb could be found in the 600 million or so of obsolete computers [59,60].

2.8 Resource Saving

In some applications, NP-based material can provide much higher specific physicochemical properties. Aerogels have superior mechanical and thermal properties. Ultrasmall metal NPs also have high specific catalytic properties. Furthermore, these new materials provide new market opportunities. These two examples highlight the potential of NPs to provide the basis for a sustainable economic growth. NPs are presently used in anode and cathode batteries to increase the specific density and charge/dischage time in lithium-ion batteries. Furthermore, a case could be made about some NP-based materials being amenable to recycling because of the weaker bondings involved.

Conventional processes used in chemical engineering and microelectronics are resource intensive with a significant environmental footprint. For example, a 2-g DRAM chip requires 72 g of chemicals, 2000 g of fossil fuels, 20,000 g of water, and 500 g of gases [59,60]. These resource requirements are currently putting a lot of strain on several industries. Furthermore, large amounts of waste are generated from the fine chemicals used in the pharmaceutical industry. Given the increased demand for drugs, current processes for producing these fine chemicals are not sustainable.

3 TOXICITY OF NPs

NPs are considered toxic if they induce damage in cells (cytotoxicity) and/or organs (organotoxicity). The potential risks of nanomaterials including NPs have been reviewed [14,61,62]. To evaluate the toxicity of NPs, there are some key questions one needs to ask [63]:

a. Is there any contact between the NP and the human cell?
b. Is there any interaction between the NP and the protein of the cytoskeleton?
c. Could the NPs disrupt the internal organelle membranes?
d. What would be the effect of the NP on cell function?

Answering qualitatively and quantitatively to these questions will certainly help in evaluating the toxicity of NPs.

Often, NPs do not act alone and their behavior changes with agglomeration and aggregation. In certain environments such as water, their behavior is much more complex. For different reasons, clear-cut experiments showing specific effects of NPs are still rare and difficult to undertake. Toxicity of NPs must be evaluated based on the hazard level and exposure. Exposure measurement is not always straightforward in the case of NPs [64]. Indeed, the surface area (size) depends on the environment, thus requiring online surface area measurements. Unfortunately, there are no reliable online NP size measurements, particularly under *in vitro* and *in vivo* conditions.

There is a common perception that NPs are specifically more toxic than relatively larger particles with similar composition. However, as conflicting data show, it is important to make sure that when comparing small and large particles, other parameters (shape, surface composition, aggregation) are the same. Surface composition may have a significant effect on the toxicity of any particles regardless of their size. Testing for potential hazards of NPs should be based on the following criteria [65,66]:

a. potential routes (pulmonary, dermal, oral, and/or ocular) of exposures that could lead to human health effects;
b. screening for potential carcinogenic effects using established mutational and chromosomal aberration assays;
c. screening for potential toxic effects in representative aquatic organisms (rainbow trout, daphnia, and algae).

Size, charge, hydrophobicity, and targeting are four important NP properties that could affect different toxicity parameters [20]. For example, the size changes the T-helper1 (Th1):Thelper2 (Th2) activity ratio, adjuvant properties, internalization/phagocytic uptake, hapten properties, and particle clearance [20]. On the other hand, surface charge changes toxicity to immune cells, binding to plasma proteins, particle clearance, and immune cell stimulation [20]. The immediate environment of the NPs is also important. This affects the size and surface composition distributions. The state of agglomeration and aggregation of NPs depends on the characteristics of the surrounding medium including solvent and pH. The size and surface composition of the NP will dictate the area where it will deposit within the respiratory system.

Thus, the NP environment has a direct effect on its properties. In this context, *in vivo* studies are critical to assess the toxicity of NPs [67].

Each type of NP should be evaluated and tested on a case-by-case basis. For this reason, a minimum set of physicochemical characteristics of the NP should be available before drawing any significant conclusion [20,65,66]:

a. particle size, distribution, and shape;
b. degree of agglomeration and aggregation;
c. surface charge and coating;
d. surface area;
e. structure and crystallinity;
f. physicochemical state in the relevant media;
g. surface reactivity;
h. method of synthesis and posttreatment.

These studies are time consuming, requiring expertise in multiple areas and techniques that is hard to find within a single institution.

Skin, lungs, and the gastrointestinal tract are three ways of entry for natural or anthropogenic NPs into the body. Although not all NPs produce adverse health effects, it is important to understand all the factors that make NPs toxic. The relationship between size, aggregation, composition, crystallinity, and surface functionalization on one hand and the occurrence of some diseases (asthma, bronchitis, emphysema, lung cancer, and neurodegenerative diseases) on the other hand should be carefully studied [62]. Engineered and natural airborne NPs can readily enter the body through inhalation [64]. From the outset, pulmonary toxicology studies in rats showed that NPs are more toxic than larger particles of similar chemistry [68]. However, a more detailed analysis of the data established a direct correlation of NPs with surface activity, rather than with particle size and surface area [68]. For this study, cytotoxicity, histopathology, lung inflammation, and other effects are measured using four different α-quartz with different particle size (d: nm) and surface area (S: m^2/g). They comprise Min-U-Sil ($d=534$; $S=5.1$), nanoquartz I ($d=50$; $S=31.4$), nanoquartz II ($d=12$; $S=12$), and fine quartz ($d=300$; $S=4.2$).

Toxicity of NPs has now been established for some materials, although most studies are based on high level doses applied to animal models. These experiments and the findings are difficult to extrapolate to humans in real-life situations. As is the case with most chemicals, high doses are often fatal, even if the chemical is not recognized as hazardous. Toxicity of NPs can occur through direct use or via application of NP-based products. This may occur during production and packaging of NPs or during environmental release of NPs after disposal. Any study on toxicity should start with the raw materials composition used for the fabrication of the NPs. For example, the health-related effects of some QDs such as CdSe, CdTe, and PbS are challenging. The high toxicity of Cd and Pb is a significant concern as element leaching may occur during application. Therefore, extra attention should be directed toward packaging.

The health-related effects of aerosols have been studied for decades. Only recently have sub-100 nm airborne particles begun to attract special attention. Natural events are believed to constitute the main source of these aerosols. Vehicles, particularly in urban areas, are another important source. Wide size distributions of these aerosols, from a few nanometers up to several dozen micrometers, make it difficult to distinguish the effects of NPs from those of microparticles. Subsequent to exposure by inhalation or ingestion, the following body parts may be affected: skin, lung, circulatory system, brain, and gastrointestinal system [20,62]. To a lesser extent, other body parts (implant, lymphatic system) can also be affected.

Inflammatory and noninflammatory immune responses are observed following NP uptake via different pathways [20]. Once inside cells, NPs interact with subcellular structures. These interactions shape up in various forms depending on the size, structure, and composition of the NPs. These NPs can damage organelles or DNA [62]. Cell uptake of NPs occurs via mannose receptor- (MR-), complement receptor (CR-), and immunoglobulin Fcγ receptor (FcγR-). Persistence of the NPs in the body could increase the hazard even when a small dose is used. Different sampling techniques have been reported to measure exposure to NPs [69].

Photocatalytic properties of TiO_2, ZnO, and Ag-based NPs may cause damage to DNA and other components of the skin cells if they are not treated appropriately [70]. Silica and alumina coatings reduce the side effects without reducing the transparency of the photocatalytic core NPs. TiO_2 and ZnO NPs embedded in sunscreens provide UV-blocking features that can prevent cancer development. However, these NPs could be hazardous. Indeed, there is some evidence for cell uptake, oxidative cell damage or genotoxicity, and even phagocytisis of cells exposed to excessive particle concentrations [71]. Increased toxicity risk has been reported when sub-100 nm ZnO and TiO_2 NPs have been used instead of micro and macroparticles [71]. *In vitro* tests for phototoxicity, genotoxicity, and photogenotoxicity of sunscreens containing different NPs, including SiO_2 (21 nm), SiO_2/FeO (21 nm), Al_2O_3 (14–20 nm), and Al_2O_3/SiO_2 (up to 60 nm) coated and noncoated NPs showed negative results [71].

3.1 Assessment of Health Issues Related to NP Exposure

Most chemicals can be poisonous beyond a critical dose, proscribing their usage in everyday life. A few of these chemicals can be used as remedies if the right dose is applied. The dose is sometimes more critical than the hazard itself. The applied dose, derived from the exposure, depends on the exposure time, concentration, exposed surface area, and exposure route (oral, dermal, inhalation, and injection). Potential human health effects due to exposure to

NPs should be evaluated based on three key elements of the toxicity screening strategy, which include physicochemical characteristics, *in vitro* assays (cellular and noncellular), and *in vivo* assays [72]. The toxicity of some nanomaterials including NPs is not always straightforward and easy to quantify [73,74]. Interaction of NPs with biological systems is governed by their structural characteristics. NP uptake and removal are size and shape dependent [75].

It is unfair to ask the nascent NP industry to evaluate all the risks even before undertaking basic research on how to produce and control the properties of these NPs. Even in the case of the older chemical industry, quantitative assessment of the toxicity risk of all the chemicals produced, used, and commercialized is not done. In the case of NPs, it is even harder given the dependence of NP properties on size, shape, surface composition, charge, and crystallinity. This diversity could make a systematic *in vivo* study quite difficult and even impossible in some cases [72,76]. Furthermore, even existing analytical techniques and methodologies are not adequate [64,74]. Exposure pathways are considered as either direct via circulation systems or indirect via mediators released in the respiratory tract. In the case of direct pathways, NP exposure gives rise to enhanced cardiovascular responses compared to microparticle exposure. NP- and microparticle-induced cardiovascular diseases may occur via different routes (pulmonary inflammation, pulmonary reflexes, and systemic translocation to circulation and to the heart inflammatory lung response) [77]. However, the effect of particle size on the mechanism involved in particle exposure and vascular disease has not yet been established.

Toxicological experiments showed that polytetrafluoroethylene (PTEE) NPs with an average size of 20 nm are lethal upon inhaling [78]. In the case of 130 nm-sized PTFE particles, no toxic effect was observed [78]. Although there is no evidence for systemic exposure following dermal applications of oxide NP-based sunscreens, current testing methodologies are beginning to be questioned [79]. For example, studies should also be carried out on nonhealthy skin. Such studies may contradict some of the conclusions drawn thus far on particle size effects.

3.2 NPs from Natural Sources and Events

Around 90% of aerosols found in the atmosphere are related to natural events including wind erosion, earthquakes, volcanoes, dust storms, fires, sea spray, and erosion [62]. There are two types of atmospheric NPs [80]. Besides aerosols, solid particles such as ice, oxides, clays, sulpfides, and organics are found in the atmosphere, often with a core–shell structure. Liquid droplets containing water, acids, salts, and/or organics are also present in the atmosphere.

Besides being the main source of dust in the world, desert storms are a significant source of NPs and microparticles. These NPs from desert areas can be located thousands of kilometers away from their source. These dust consisting of NPs and microparticles leads to adverse health issues that could lead to asthma and emphysema [62]. These health effects occur through generation of reactive oxygen species by metal-rich dust or through dust containing viruses, bacteria, and fungi. Forest fires generate smoke NPs with an average size of about 50 nm that particularly affect patients with cardiovascular and respiratory problems [62].

3.3 Anthropogenic Sources of NPs

NPs could be also be formed in the atmosphere from natural or man-made precursors. Anthropogenic NPs result as by-products of human activity. Up to 50,000 kg/year of nanomaterials are produced from anthropogenic sources [79]. This number probably underestimates the actual level of NPs given the lack of adequate monitoring instrumentation. NPs are found mostly in the atmosphere, although traces could also be found in homes, offices, and factories. NPs or ultrafine particles less than 100 nm contributing to poor indoor air quality are frequently blamed for an illness called sick building syndrome [81]. Breathed NPs could get access to the lung, the gastrointestinal tract, and the brain. These NPs arise from different sources. Often, the source of these NPs is not properly identified. For example, natural gas- and coke-based BBQ activities are a significant source of organic-based NPs. Diesel-based power generation and automobiles also generate significant amounts of microparticles and NPs.

Since the 1950s, regulation of atmospheric particles (called primary particles) has shifted from large suspended particulate matter or PM (ca. <30 μm in diameter), through PM10 (<10 μm) to PM2.5 (<2.5 μm) [82]. Because sub-0.1 μm particles had relatively low mass, their regulation was not considered important. These ultrafine particles (UFPs) with at least one dimension less than 0.1 μm (PM0.1) correspond to the definition of NPs. They are now believed to have an adverse health risk. The major source of primary UFP is road transport (60%), followed by non-transportation combustion processes (23%).

Internal combustion engines are a significant source of NPs [83]. Diesel power generators and combustion engine cars are significant sources of particulate matter (PM). With particle sizes ranging from 20 up to around 100 nm, this PM contains highly toxic polynuclear aromatic hydrocarbons (PAHs). Following formation of PAHs from fuel molecule decomposition, crystalline and amorphous soot particles with various states of aggregation are formed [84–86]. Soot consists mostly of 100 nm particles. Given that the distribution of particle concentrations is a function of the distance from the highway, it is no surprise that the cardiopulmonary mortality is higher in the population that lives near major roads [62]. A similar conclusion can be drawn regarding indoor cooking using solid fuels.

Agglomerated carbon and mineral particles have been measured using TEM on samples collected from automotive exhaust on the highway [62]. Since both agglomerated (1 μm) and non-agglomerated (10 nm) particles are present in the air, their respective effects should be quantified separately. Based on the NP number concentrations, the majority of the particles are less than 100 nm in size [62]. The concentration of these particles decreases with receding collection distance from the highway [62]. Beyond a distance of 20 m, the concentration of these particles is significantly reduced. It is thus expected that the effect of exhaust particles will significantly increase with traffic density.

Establishing an emission standard that accounts for the specific health effects of sub-100 nm particles is quite challenging. The size of the majority of the particles emitted from engines is below 100 nm [62,85], although the mass distribution favors larger particles (Fig. 8.4). Therefore, emission standards for NP concentration should consider limiting the number of small NPs emitted. Also, sub-100 nm NPs tend to agglomerate during dilution and sampling, making it very difficult to design a standard.

It has been established that the mass of PM10 is not the critical factor leading to the adverse effects of particulate air pollution [79]. Toxicological properties of individual components such as combustion-derived NPs in PM10 are contribute to adverse effects; however, epidemiological data on this hypothesis are limited.

Besides combustion-derived NPs, nucleation contributes significantly to the formation of atmospheric NPs. These NPs contribute to cloud formation and atmosphere composition under solar radiation. Atmospheric NPs are ubiquitous, with concentrations up to 10^5 particles/cm^3 being reported near earth's atmosphere. Two nucleation processes can be distinguished [87]. Formation of atmospheric NPs may start with homogeneous nucleation following the condensation of gaseous precursors in the atmosphere. Another process, called heterogeneous nucleation, consists of gas-phase condensation on a foreign medium. In the Aitken mode, particles have a diameter of less than 50 nm [86,87]. In general, atmospheric NPs do not contribute to global warming. In some cases, as in the case of sulfates, the radiative forcing (which increases the atmospheric temperature) is even negative.

Inhaling NPs, particularly those present in polluted air, is a significant human health risk. Besides size distribution, physiological parameters related to the respiratory system could play a significant role [79]. The most acute effect of NPs from combustion is indirect. There is a direct correlation between the NPs inhaled through the nose and mouth and the NP deposition rate [79]. Indeed, whether they are inhaled through the mouth or the nose, the NPs deposited in the extrathoracic (ET) airways, the bronchial airways, and the alveolar region are size dependent. Surface interaction of combustion-derived NPs with organics and metals gives rise to free radicals. Oxidative stress leads to inflammation.

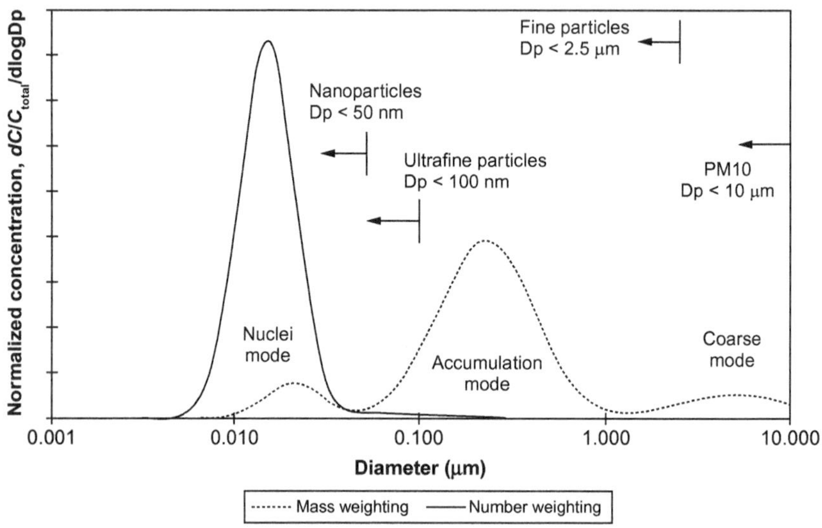

FIGURE 8.4 Number and mass distribution from an engine exhaust [83].

3.4 Engineered NPs

With over 500 nanotechnology-based consumer products [88], several questions related to potential long-term human health and environmental effects need to be answered. In particular, consumer products containing NPs that could be released on human skin or into the air and water systems may cause problems in the long term. Zinc oxide NP incorporation in cosmetic products by numerous companies has been studied, but so far, there is no report showing any adverse effects. However, NPs (TiO_2 and ZnO, functionalized fullerenes) are known for their antioxidant properties that help to maintain a youthful appearance of the skin. Furthermore, small particles with their specific optical and chemical properties are thought to smoothen wrinkles and small creases [62]. Numerous other beneficial effects have been reported and discussed in this book.

Dermal exposure to, and uptake of, NP-based cosmetics, sunscreens and detergents are still under debate. So far, no clear evidence for NP presence beyond the submucosa has been demonstrated [79]. However, the lack of evidence should not deter further investigation, particularly in the case of defective skin. This could lead to NPs entering the circulatory system, causing all the harm that one can fathom.

A more recent *in vivo* study has shown that fullerenes exhibit little pulmonary toxicity [89]. This study contradicts previous *in vitro* studies showing that aggregated underivatized C_{60} was three to four orders of magnitude more toxic than highly water-soluble C_{60} derivatives [62,90,91].

In a recent Greenpeace report, silver and semiconductor NPs have been shown to be potentially high [92]. However, these studies remain incomplete. Indeed, toxicity tests of TiO_2 NPs have been carefully studied recently [66]. NP toxicity was found to depend more on the surface composition than on the size.

4 WORKPLACE EXPOSURE

Depending on the source of the NPs, workplace exposure may have a different characteristic. Exposure could occur during NP production, NP integration, or its utilization as a commercial product. Furthermore, scientific curiosity and potential applications have led to thousands of R&D laboratories around the world attempting to produce, characterize, and use NPs. Exposure may reach hazardous levels even when only small production scales are involved. With the risk and uncertainty associated with the use of NPs, some basic precautions should be taken when handling and manipulating them. This is particularly the case when these NPs are in a dry state (dust-like state). Dustiness tests have been developed to simulate real handling scenarios [93]. These tests use a small rotating drum or the continuous single-drop principle. This test provides a dustiness index. Other tests have been developed that are appropriate at a laboratory research scale where only small amounts of NP are produced.

Manipulation must be undertaken in a fume hood and the use of an appropriate mask is recommended. In situations where it is necessary, personal protective equipment should be used. In particular, when working in a closed area with potentially high amounts of dusty NPs, wearing a full face mask with high-performance filters is recommended. Significant concentrations of sub-100 nm NPs are observed at several industrial sites using different processes (Table 8.2). In particular, granite grinding gives rise to sub-10 nm NPs. High concentrations of NPs are also observed when handling carbon, metal, and oxide material processes [93].

NPs are considered to pose the strongest emerging risk in their production and utilization stages [94]. One area that requires some attention is the training of workers involved in the production and manipulation of NPs. For this, practical handling guidelines for activities involving NPs in the workplace is needed. To provide a good assessment of workplace exposure, the following parameters should be fully characterized [93]:

1. Persistence:
   a. ready biodegradability;
   b. agglomeration;
   c. persistence of nanoscale property.
2. Primary set of safety hazards data:
   a. flammability;
   b. explosivity;

   c. incompatibility;
   d. reactivity;
   e. corrosion.
3. Secondary set of safety hazards data:
   a. stability;
   b. decomposition;
   c. polymerization;
   d. photoactivity;
   e. thermal/light stability.

Keeping an updated record of these parameters and their safety implications is critical. Unfortunately, to the best of our knowledge, there is no inventory of

TABLE 8.2 Particle Concentration, Particle Size, and Mass Concentration from Different Industrial Processes [93]

Process	Number (10^{-3} cm^{-3}) (GM)	Median diameter (nm)	Mode (nm)	Mass (mg/m^3)
Plasma cutting	40–800	40–60	—	0.3–1.3 (PM2.5,S)
Plasma cutting	40–230	140–180	120–180	4.2 (I)
Laser welding	5000–40,000	—	40–100	0.17–0.48 (I)
Aluminum welding	20–640	256–411	217–594	2.8–15 (R,P)
Soldering	28–390	43–62	36–64	<0.76 (I,P)
Cutting aluminum, minimal MWF	10–260	30–111	15–131	<0.95 (I,P)
Iron foundry	290–580	54–70	50–80	1.5–1.7 (I)
Plastic die casting	20	45	23	<0.25 (R,S)
Grinding metal	10–130	32–155	19–170	1.8 (I)
Grinding granite	20–60	<10	<10	Up to 41 (I,P)
Baking bread	5–640	55–83	32–109	0.3 (I)
Airport outdoor	26–700	<41	<45	0.07 (I,S)

I, inhalable; R, respirable; P, personal sample; S, stationary sample; MWF, metal working fluid.

such data. It is recommended that a database (life-cycle inventory) be established similar to the one developed in LCA studies in other industrial sectors.

5 TOWARD A BETTER UNDERSTANDING OF NP TOXICITY

Toxicity of NPs involves *in vivo* and *in vitro* investigations. Several physicochemical parameters may play a role. The effects of size and shape distribution along with surface composition have been studied under *in vitro* conditions. *In vitro* studies often provide information about the metabolic and membrane activities. These studies are often used as screening steps to predict *in vivo* toxicity. *In vivo* toxicity involves assessment of the immune response, reaction with the blood, toxicokinetics, and biodistribution. Little correlation has been established so far between *in vivo* and *in vitro* studies. For example, inhaled metal NPs tend to localize in the kidneys, liver, lungs, and spleen at different rates. Quantifying the health effects of this localization under *in vivo* conditions is a challenge. Well defined sub-100 nm particles are used mostly in *in vitro* toxicity studies. *In vivo* animal studies are often ill defined and are usually performed over a short term.

Establishing a relationship between the characteristics of NPs and their potential health risk is critical. Evaluation of NPs requires a multidisciplinary approach that includes materials science, biology, biocompatibility, and toxicity. Besides size, there are numerous parameters that could affect the toxicity [63]. NP characteristics depend greatly on the preparation method. For example, the surface composition and size distribution are different depending on whether wet or dry methods are used. These parameters also change depending on the application medium.

A recent review of existing NP production and integration processes concluded that exposure assessment is not adequate as of the present [95]. To better understand NP toxicity, quantification of the effects of different particle characteristics is required [96]. This issue is highlighted by recent applications of NPs in the construction industry. A variety of NPs have been added to construction materials to improve their strength, lightness, and durability. The life-cycle impacts of these NPs during the manufacturing, construction, and occupation stages are poorly understood [97].

There is no single instrument that could provide a complete analysis of NPs. The multitude of parameters requires a multi-technique approach. In the case of geometrical characterization of aerosols, more than a dozen groups of techniques have been identified [98]. Often, these different techniques require different states of materials, making interpretation even harder. Mass, surface area, and granulometric distribution are typical parameters that are readily measured, often under controlled conditions. Often, these parameters are difficult to measure *in vitro* and *in vivo*.

Besides the average size, size distribution, agglomeration state, crystal structure, shape, porosity, surface area and chemistry, surface charge, and surface and bulk compositions are possible parameters that could affect toxicity of NPs. Metrics criteria based on extrinsic characteristics and properties of NPs have been used. Agglomeration, charge, functional group, surface and bulk residual contamination, and size are the main extrinsic NP characteristics defining the health risk of different NPs [64,99]. Physicochemical characteristics and properties of different nanomaterials and NPs have been reported [99]. The NP characteristics and other criteria (bioavailability, bioaccumulation, and toxicity) can influence the potential hazards of man-made nanomaterials [99]. Agglomeration, reactivity/charge, critical function groups, and contaminant dissociation are measured as ordinal classes. For other parameters, a quantitative assessment has been used. The following ordinal classes have been used: 1 was the most favorable (least risk) value class, while 5 was the least favorable (highest risk). Based on these metrics, a classification referred to as stochastic multicriteria acceptability analysis has been developed to help decision makers assess human health and environmental risks related to different nanomaterials [99]. Although not perfect, this assessment approach is advanced enough to provide a relative ranking [99]. Potentially, CdSe and silver NPs present the highest risk, with high reactivity (4 to 5) and highest potential risk among the 5 analyzed nanomaterials [99].

Reproducing *in vivo* and *in vitro* conditions to understand and quantify NP properties and toxicity is critical. Indeed, the medium (concentration, pH, salinity, solvent type, temperature) surrounding the NPs may change their properties including the degree of dispersion. For example, in the case of TiO_2 NP dispersion in bovine serum albumin, it was found that phosphate concentration is one of the key factors governing the physicochemical properties [100]. Furthermore, it was found that fetal bovine serum was an effective dispersing agent for TiO_2 NPs. Other media also affect the size distribution of TiO_2. Increased TiO_2 NP concentration gives rise to larger average particle size in all six cell culture media [100]. The size and region of the human respiratory tract directly affect the NP deposition rate. As shown in Fig. 8.5, distribution of particles of different sizes in the ET airways varies significantly as a function of the size of the airways. Regional deposition rates are relatively high for particle size 1–10 nm in the nasal, tracheobronchial, and alveolar systems.

Most *in vivo* toxicity studies are undertaken on animals. However, there are significant differences between rodents and humans, mainly anatomical and physiological differences [77]. These differences could result in considerably lower concentrations to sensitive regions of the respiratory tract of animals compared to similar regions in humans [77].

5.1 Size and Surface Area

Size and its related properties such as surface energy and area are critical. These parameters could directly affect the deposition rate, particularly within the various sites of the respiratory system [77,96]. Although bulk gold and gold NPs of even 15 nm diameter are nontoxic, gold NPs with a diameter of 1.4 nm have been found to be toxic [101]. Toxicity studies of silver NPs (15, 30, and 55 nm) revealed size dependency. It was suggested that toxicity induced by silver NPs and the mechanism of toxicity is mediated through oxidative stress [102]. PTFE particles with sizes of 20 nm proved lethal upon inhaling, whereas 130-nm-sized PTFE particles have reduced health issues [78].

Overall surface area, directly related to particle size and to the dose, was also shown to be closely associated with lung responses, including inflammation and tissue damage, in rat lungs [64]. This study clearly shows that the bulk composition of the particles is irrelevant, but the surface area (dose and particle size) is critical. Based on data collected from different sources, it was found that the proportion of lung tumor is relatively constant with particle surface area dose up to 0.1 m^2/g lung, the critical surface area [64]. The

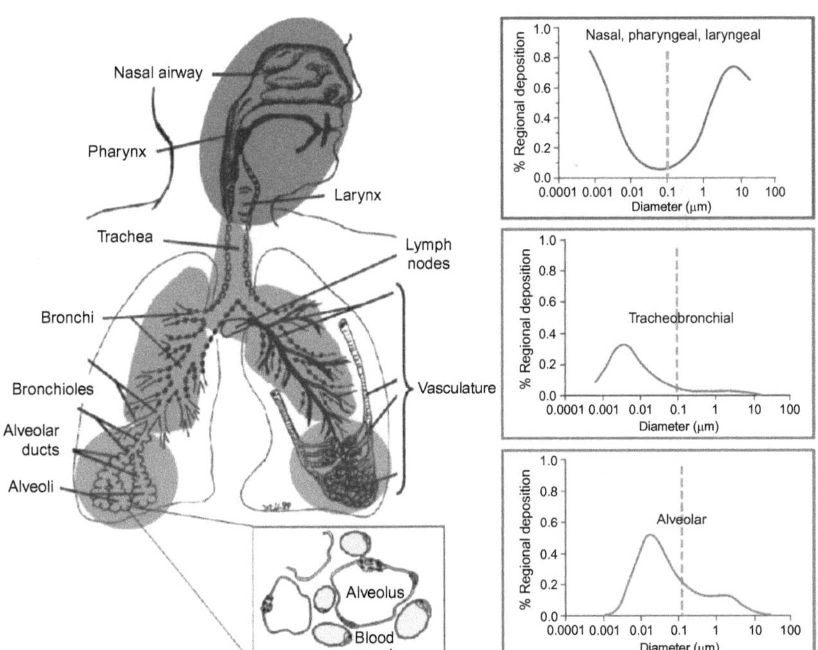

FIGURE 8.5 Nanoparticle deposition rate as a function of size in different regions of the human respiratory tract [96]. (For color version of this figure, the reader is referred to the online version of this chapter.)

lung tumor proportion increases by a factor of 6 when the surface area is increased to about 5 m^2/g lung [64].

Extra- and intracellular states of NP agglomeration are critical to the evaluation of the effect of size on toxicity [63]. Nickel ferrite particle size and surface coating effects on the cytoxicity have been studied using Neuro-2A cells [103]. No effect of particle size has been observed on noncoated materials. However, when one or two layers of oleic acid were added to the surface, large particles (150 nm diameter) showed a higher cytoxicity than smaller particles (10 nm diameter). Several factors contribute toward NP toxicity. In the case of metals, solubility and the binding energy to biological sites are already known, even in the case of bulk materials. In some cases, NP size and surface characteristics contribute to NP toxicity indirectly. For example, the toxicity of silver substances is directly related to the rate of release of Ag^+ from them. Indeed, soluble silver particles increase the rate of release with higher surface-to-bulk ratio [104].

5.2 Composition and Crystal Structure

Size dependence of different NP compositions has been reviewed [105]. In the case of MoO_3 particles sized 30 and 150 nm, moderate toxicity has been found. On the other hand, Ag NPs sized 15 and 100 nm were found to be highly toxic. Under UVA radiation, TiO_2 NPs induce strong cytotoxicity [106]. Anatase TiO_2 NPs seem to induce higher photocatalytic activity than rutile TiO_2 NPs. For larger particles (200 nm or more), the rutile phase induces hydrogen peroxide and oxidative DNA damage in the absence of light. Another study showed that both rutile and anatase phases contribute to cytoxicity, although of different types [63]. This behavior is not surprising in the case of TiO_2, given the dependency of its photocatalytic properties on the size and phase composition of the NPs [107].

The toxicity effects of transition metal NPs present in microparticle-based systems are still not clear [77]. Additional studies are required to identify any role and mode of action of UFPs. In the case of organic particles, toxicity effects were found to be more prominent in the ultrafine fraction than in their coarser counterparts. This observation is correlated with the higher contents of organic carbon and polyaromatic hydrocarbons in UFPs following combustion processes [77].

5.3 Surface Composition

Surface composition has a great impact on NP–tissue interaction [96]. For example, hydrophilic (high energy) and hydrophobic (low energy) surfaces interact with different organs [96]. The consequences of these interactions are also different. The presence of excess oxidants may activate inflammatory processes [64]. Polytetrafluoroethene (PTFE) NPs can form a highly reactive and toxic aerosol. The toxicity of aggregated PTFE NPs, with median diameters slightly greater than 100 nm and, consequently, fewer reactive surface groups, is significantly reduced [64].

The zeta potential reflects the effective charge on the particles, although it does not provide information on the surface chemical composition. Most photocorrelation spectroscopy instruments provide an option to measure the zeta potential. As it relates to the electrostatic repulsion, the zeta potential is a relevant parameter to quantify NP surface charges, which affect NP dispersions. These surface charges influence nonspecific interactions with proteins in the biological tissues. Anionically charged aqueous pore channels favor the penetration of positively charged NPs [108].

5.4 Other Factors

There are other factors that could also affect the toxicity potential of NPs [63]:

a. dispersing medium, such as serum;
b. dosimetry, including NP concentration and overall mass;
c. morphology, including spherical and fibrous structures.

Differences have been observed between phagocytosis of fine spherical and fibrous titania particles in biological responses of alveolar macrophages, although TiO_2 is considered a biologically inert material. Cytotoxicity of fibrous TiO_2 is much higher than that of spherical TiO_2 [109]. The cytotoxicity of crocidolite fibers (control) is higher than that of fibrous TiO_2 at low concentrations. However, the toxicity of fibrous TiO_2 and crocidolite are similar at higher concentrations.

All these factors affect the degree of aggregation and agglomeration of NPs. Aggregation and agglomeration before and/or after inhalation have significant effects on the physiological and toxicological properties of these NPs. Surface charge and/or ligands can be modified to control the state of aggregation.

There is a large body of work indicating a direct relationship between the solubility of NPs and their toxicity [64]. For example, insoluble but chemically active crystalline silica NPs are found to be markedly more toxic than other insoluble materials, even when normalized to surface area. Particle shape, including fiber and nanotube, may also have an effect on the toxicity. The length of the fibers is found to play a significant role.

6 RISK ASSESSMENT AND ANALYSIS

Like in any other technology, there certainly are potential risks related to NP utilization. Some technologies are inherently riskier than others. The risks should be balanced with the rewards and the existence of effective solutions to mitigate these risks. What makes a difference is the way the technology is used. Indeed, the risk level and the probability of its occurrence are very critical. It is also important to counterbalance the risk with the existence of appropriate responses to the risk and the effectiveness of the deployed responses.

There are numerous studies on the hazardous effects of NP utilization on humans and the environment. In the case of health-related risks from NP exposures, there are two main groups of scenarios that could lead to an impact on humans. These scenarios may occur via inhalation, absorption through skin, and ingestion. They can occur directly when a product containing NPs is used or after it is disposed, or indirectly when the environment is impacted, mainly through water entrainment or air release. This could occur either during product lifetime or after it is disposed.

Since risk depends on both toxicity and exposure, a quantitative evaluation of both parameters is required. There are several process steps that could provide the information required for toxicological risk assessment of NPs [98]. NP exposure data are first collected, including data on contaminant emissions, human exposure, and environmental impact. This involves emission monitoring, exposure, and modeling [98]. Second, quantification of the health effects is undertaken from pharmacokinetic modeling, structure–activity relationship and *in vitro* studies [98]. Finally, the risk assessment is based on quantification of exposure and effects.

Risk assessment study of NPs is challenging because of the numerous factors that could affect their properties. Furthermore, there is some uncertainty linked to data quality and thoroughness. The wide range of bulk and surface compositions, sizes, shapes, and mediums of NP-based systems are major hurdles to the quantification of toxicity levels. All these parameters contribute to the toxicity level. It is quite impossible to precisely quantify the toxicity of NPs. However, this should not be an excuse for not carrying out any studies on their toxicity potential.

Obviously it is difficult, if not impossible, to provide a general risk analysis for all NPs. A scientific case-by-case evaluation should be undertaken to provide a risk analysis. As discussed, this risk depends not only on the NP characteristics but also on the application and the environment. There are two important issues related to risk analysis requiring detailed assessment and discussion. First, the risks related to NP production and utilization should be compared with those of other technologies. Historically, new technologies have often been met with a lot of resistance from the general public and even from experts. These risks are either technology-inherent or technology-transcending. In the first case, the risk could be either speculative (hypothetical) or real. As discussed earlier, NPs can be less dangerous than microparticles. Whatever the potential risk that can be caused by NPs, one should ask whether it is controllable and also whether the benefits outweigh the mitigated risk.

Recently perceived risks and benefits of nanotechnology have been compared to those of 43 other technologies [110]. The results of this survey showed that nanotechnology ranks in the middle in perceived risk and benefits. However, a few bad press reports and lack of communication could change this perception. The degree of toxicity of NPs in general should be studied in relation to the chemical industry where the toxicity evaluation of thousands of chemicals has not yet been completed. Indeed, there are currently around 50 million chemicals, of which 100,000 are used daily. With some pressure from the civil society, the chemical industry is currently phasing out thousands of chemicals that are incorporated in daily consumer products. As with any chemical, the toxicity risk of NPs will depend on three parameters:

a. probability of exposure;
b. extent of exposure (time and concentration);
c. uptake route (inhalation, transdermal, ingestion).

Inconclusive data about the toxicity of NPs could be due to faulty instrumentation or our understanding limitations and should not mean that all work on NPs should be banned. Indeed, numerous examples can be cited of substances whose potential risks have generated conflicting data. Although NPs are not a defined substance, there are numerous analogies made between NPs and asbestos. Warnings about the harmful effects of asbestos were reported more than a century ago, although these warnings were challenged by others [111]. Another example is provided by the Bisphenol A (BPA)-based polymers [112]. Indeed, BPA leaching following hydrolysis leads to human exposure, although some controversy still exists over the low-dose effects of BPA being weak or otherwise. However, it is now well established that high rates of BPA leaching from food and beverage containers represents a health hazard.

Exposure control of airborne NPs should start with the large body of occupational and environmental data on aerosols. Indeed, methods to reduce aerosol exposure have already been developed, including membrane and electrophoresis filters [64]. Exposure of workers to airborne NPs could be significantly reduced by isolating and enclosing the fabrication, integration, and packaging processes. Simple rules could be easily implemented in some cases. Wet-based NP syntheses are less likely to produce airborne NPs than those based on gas phase.

Three different actions can be taken to reduce hazard risk in the workplace. First and foremost, effort should be made to prevent or contain emissions at their source. If this is not possible, hazardous emissions should be removed or proper barriers should be put in place to prevent or minimize worker exposure. NP immobilization on a stable support structure could limit the hazard of NPs while the NP is performing its function [113].

An integrated and sustainable risk assessment of NPs should be undertaken for a wide range of technologies. In the field of therapeutic drug delivery, direct administration of NPs to the patient implies drug delivery across several membranes. In addition, specific and controlled release of the drug is achievable using NPs. This will likely represent the most challenging aspect of NP toxicity, particularly in the case of nonhealthy subjects. This raises the issue of the importance of assessing the safety of these NPs. Even if the NPs are not administered, exposure assessment scenarios and their health hazards should be considered. This includes production, storage, distribution, and disposal.

To evaluate the hazard level of NPs, toxicological tests at the cellular level (multi-cell sampling) and *in vivo* testing on animals should be performed. In the case of NP- based medicine, clinical phase trials are also required before commercialization. Even if a high hazard risk is detected at one of the stages, a complete risk–benefit evaluation that includes specific preventive measures should be undertaken before consideration for commercialization. Of course, the perception of the public is an important factor requiring efforts to explain the potential benefits and mitigated risks.

A relative risk assessment has been conducted on nanomaterials and other industrial processes (Fig. 8.6). This study includes a wide variety of products. QDs rank quite high in terms of perceived risk. TiO_2-nano with commercial and precommercial products is the most developed NP technology. It is perceived as relatively risky, at the same level as lead batteries.

Quantification of the hazard potential of NPs is not always straightforward. In the case of aerosols, the hazard is assumed to be proportional to d^{α},' where d is the particle diameter, with α varying between 0 and 3. The exact value of α depends on the critical metrics affecting the hazard [115]. An average value of $\alpha = 1.5$ is obtained.

7 SOCIETAL

As in the case of other technologies, studies of the societal implications of NPs should address numerous issues related to governance, risk, ethics, law, and equity. Even if most NP technologies are in the early developmental stages, such studies should be undertaken. Societal resistance against a new technology is often due to lack of awareness about the advantages and risks.

Ethical issues related to NPs have been raised, as in the case of any emerging technology. For example, NP technology may contribute to widening the gap between rich and poor nations. Indeed, it has been postulated that the applications of these NPs will not be accessible to everyone, since advanced technologies will provide a significant edge to developed countries. However, there are numerous examples of technology and innovation helping to improve the quality of life of humankind over several centuries. This is the case of the cell phone and internet, which provide access to key services cost-effectively. Thus, a large proportion of the global population has seen the standard of life improving due to technological development.

It is thus questionable whether nanotechnology and NP technologies will contribute to widening the gap between nations. Indeed, this general statement

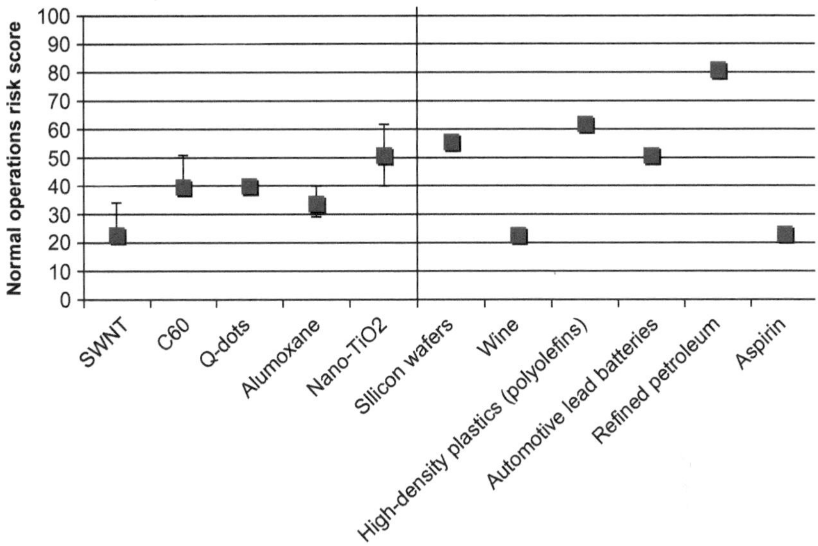

FIGURE 8.6 Relative risk score of nanomaterials and nonnanomaterials production processes. A relative risk assessment has been conducted on nanonmaterials and other industrials processes. *Reprinted with permission from Robichaud et al. [114]. Copyright 2005, American Chemical Society.*

does not apply to most NP technologies described in this book. As discussed in Chapters 2 and 3, NP synthesis is much easier to achieve with reduced capital investment. This is particularly true in the case of the technology of capital-intensive current semiconductor chip processes. NP-based technologies will provide more opportunities to the poor, given the reduction in the capital and technology gap requirements. Thus, contrary to most new technologies, NP-based technologies will provide a unique opportunity for nondeveloped countries to reduce the gap with developed countries.

However, legal issues related to commercialization of NP-based technologies may limit poor countries with weak institutions. As in the case of any other new technology, there have been, and there are going to be, several litigation issues related to freedom of use. This is particularly so in the case of NP technologies where questions remain about the composition, structure, and functions of NPs as synthesized and integrated into films and devices. Patent protection awarded to inventors may create a monopoly over all applications. This may potentially hinder their development. Legal issues related to NP IP may represent the biggest hurdle toward realizing the potential of this field. Instead of obtaining the right to "practice the invention" for disclosing the invention, patent owners may use it to "prevent others" from producing, marketing, and selling products, which may somehow infringe on the claims of the patents.

The books written by Asimov [116] and Drexler [117] have stimulated interest in nanoscopic science and technology in the general public and in academia. These science fiction reports and movies have attracted a lot of interest in this field in its early developmental stage. However, public perception has started to shift negatively with the recent reports on the risks of nanotechnology. For this reason, it is really important to step up efforts to educate the public by separating fiction from fact.

As discussed elsewhere in this book, NP technologies may have positive and negative impacts in key sectors. Indeed, NPs may improve the quantity and/or quality of food, water, energy, and environment for everyone. However, NPs may also have adverse effects on existing production and distribution chains. Calls for regulation in production and utilization of NPs are warranted, although this may slowdown the development pace. Public perception of these risks is not always based on real data. The smallness and the novelty of these NPs and their applications seem to be increasing the perception of risk among the public. Other factors may also play a role in shaping public opinion [118].

The importance and the involvement of social sciences in the development of NP science and technology are now well recognized. Decision makers are often influenced by the perceptions of the general public, particularly if they are not technically trained. Only recently, efforts have been directed toward understanding and addressing the public perception and popular misconceptions regarding nanotechnology. Potential cultural and philosophical implications of NP utilization are discussed in Ref. [119].

As recognized in the case of nanotechnology, it is important to raise the issue of societal implications with researchers and funding organizations. As in the case of any new technology, NP technologies are interacting directly with the society through enhanced productivity and quality of life. In particular, emerging technologies can also provide opportunities to the people in developing countries to improve their standard of living. Evaluation of the societal implications of NP utilization is based on the guiding principles of nanotechnology [120]:

a. risk-analysis study that balances benefit and risk;
b. public acceptance of the risk and rewards;
c. societal implications including impacts on socioeconomic sectors;
d. influence of the scientific and technological development of NPs on our understanding of naturally occurring NPs;
e. adherence to the regulations in the respective field of NP applications;
f. importance of a multidisciplinary and global perspective;
g. importance of life cycle and sustainability.

8 POLICY AND PUBLIC EDUCATION

A recent telephone survey showed that absence of relevant scientific or policy-related information did not prevent the general public from forming its opinion about the risk and benefits of nanotechnologies and other emerging technologies [121]. This study showed that opinions and attitudes are made regarding the risk and benefits of emerging technologies even in the absence of relevant scientific information. The print media seems to play a significant role in shaping the opinion of the public by emphasizing the potential benefits. However, of late, this mainstream media seems to be focusing more on the controversies, particularly those related to health and environmental risks.

The lack of understanding among the general public is related to difficulty in grasping the nanometer size scale involved in NP technologies [122]. Indeed, scaling is not a common skill requiring specific education tools. Lack of knowledge about science in general and NP- related technologies in particular is of importance. Inability of the general public to grasp the scale factor related to nano-objects contributes toward this knowledge deficit [122]. A survey has been conducted to quantify public perception of nanotechnology. Respondents have been asked to express their agreement or disagreement regarding the following two statements [123]:

a. Statement #1: Human beings will benefit greatly from nanotechnology, which works at the molecular level atom by atom to build new structures, materials, and machines;
b. Statement #2: Our most powerful twenty-first century technologies—robotics, genetic engineering, and nanotechnology—are threatening to make humans an endangered species.

Overall, this survey showed "high levels of enthusiasm for the potential benefits of nanotechnology and little concern about possible dangers" [123]. Respondents rated nanotechnology more favorably in comparison to the space program, nuclear power, and cloning research.

It is important to communicate to the public the risks and potential rewards related to NP fabrication and utilization. The media hype surrounding nanotechnologies exaggerating their commercial benefits and the controversies related to their environmental impact is not always based on facts. The potential risks of NPs were not communicated effectively at the outset. This increased the risk perceptions, giving rise to public opposition to their development. Indeed, the general public forms strong and instantaneous emotional opinion that is difficult to change even after hard empirical evidence is provided. Comparisons can be drawn with the genetically modified organisms. To promote the public's receptivity to sound scientific evidence, it is critical to provide information within a frame that neither affirms nor threatens their cultural values. It is also important to provide a context for the potential application of NPs.

The multidisciplinary nature of NP technologies makes it difficult to grasp for educating students even at graduate and undergraduate level. Perhaps the best way is to introduce NP technologies is to use in a two-step approach. In the first step, different aspects of NP technology can be introduced within the existing domains in the physical sciences, life sciences and humanities. In the second step, specialized NP- based programs can be introduced at the graduate level.

The unknown toxicity of consumer products may impose serious limitations on public acceptance of nano-enabled products. Some R&D focus is attempting to address this issue. An exhaustive study on industrial safety issues related to TiO_2 NPs has been completed by the EU Scientific Committee on Cosmetics and Non-Food Products [124]. This study has confirmed that neither NPs nor microparticles penetrate into or through the living skin. Introducing regulation in NP R&D can certainly help. Regulation with dissemination of scientific finding will build trust within the industry. However, overregulation could be harmful to this nascent industry. Overregulation could limit research and hinder commercialization.

New policies may need to be developed to deal with the expected and non-expected implications of wider NP technology use. Indeed, protection of the public and particularly the workers and scientists is very important for the long-term sustainability of this field. Unfortunately, health and safety issues related to NPs are still inconclusive because of the deficiencies and incompleteness of the methodology. The industry is willing to regulate this area if strong scientific evidence demonstrates that NPs are harmful [125]. Therefore, it is important to fund research on the health-related effects of NP exposure and make the findings available to the public.

REFERENCES

[1] M. Adeogun, Nanomaterials, SRI Strategic Business Insights, Menlo Park (CA), August 2006
[2] www.smalltimes.com/Articles/Article_Display.cfm?ARTICLE_ID=270664&p=109
[3] S.F. Hansen, A. Evan, S. Michelson, A. Kamper, P. Borling, F. Stuer-Lauridsen, A. Baun, Ecotoxicology 17 (2008) 438–447.
[4] N.-O.A. Kwamena, J.P. Reid, In: T. Cosgrove (Ed.), Colloid Science: Principles, Methods and Applications, second ed., John Wiley & Sons, Ltd., Chichester, 2010, pp. 219–244.
[5] Societal Implications of Nanoscience and Nanotechnology M.C. Roco, W.S. Bainbridge (Eds.), Kluwer Academic Publishers: Dodrecht, 2001
[6] J. Kuzma, P. VerHage, Nanotechnology in Agriculture and Food Production, Woodrow Wilson International Center for Scholars, Washington, DC, 2006.
[7] Y. Kubota, T. Shuin, C. Kawasaki, M. Hosaka, H. Kitamura, R. Cai, H. Sakai, K. Hashimoto, A. Fujishima, Br. J. Cancer 70 (1994) 1107–1111.
[8] N.A. Monteiro-Riviere, R.J. Nemanich, A.O. Inman, Y.Y. Wang, J.E. Riviere, Toxicol. Lett. 155 (2005) 377–384.
[9] J.S. Tsuji, A.D. Maynard, P.C. Howard, J.T. James, C.W. Lam, D.B. Warheit, A.B. Santamaria, Toxicol. Sci. 89 (2006) 42–50.
[10] Nanotechnologies for Environmental Applications, Technology Transfer Centre, England, October 2006.
[11] P. Biswas, C.Y. Wu, J. Air Waste Manage. Assoc. 55 (2005) 708–746.
[12] P. Torcellini, N. Long, R. Judkoff, Consumptive Water Use for U.S. Power Production, NREL/TP-550-33905, 2003.

[13] P.H. Gleick, Dirty Water: Estimated Deaths from Water-Related Diseases 200–2020, www.pacinst.org/reports/water_related_deaths/water_related_deaths_report.pdf (accessed May 2010).
[14] N. Lewinski, V. Colvin, R. Drezek, Small 4 (2008) 26–49.
[15] H.-X. Ren, X. Chen, Mater. Today 13 (2010) 6–8.
[16] S. Barillet, A. Simon-Deckers, N. Herlin-Boime, M. Mayne-L'Hermite, C. Reynaud, D. Cassio, B. Gouget, M. Carriere, J. Nanopart. Res. 12 (2010) 61–73.
[17] H. Hildebrand, K. Mackenzie, F.-D. Kopinke, Chem. Ingenieur Tech. 79 (2007) 1461–1462.
[18] G. Schmidt, Nanotechnology: Assessment and Perspectives, Springer, Berlin, 2006, pp. 67–286.
[19] B. Ravenzwaay, E. Leibold, Hum. Exp. Toxicol. 23 (2004) 421–430.
[20] M.A. Dobrovolskaia, S.E. McNeil, Nat. Nanotechnol. 2 (2007) 469–478.
[21] M.A. Maurer-Jones, K.C. Bantz, S.A. Love, B.J. Marquis, C.L. Haynes, Nanomedicine 4 (2009) 219–242.
[22] J.C. Davies, EPA and Nanotechnology: Oversight for the 21st Century, Woodrow Wilson International Center for Scholars, Washington, DC, 2007.
[23] G.F. Grubb, B.R. Bakshi, J. Ind. Ecol. 15 (2011) 81–95.
[24] H. Sengul, T.L. Theis, In: N. Savage, M. Diallo, J. Dunca, A. Street, R. Sustich (Eds.), Nanotechnology Applications for Clean Water, William Andrew Inc., Norwich, NY, 2009, pp. 561–582.
[25] D.G. Rickerbya, M. Morrison, Sci. Technol. Adv. Mater. 8 (2007) 19–24.
[26] P.V. Kamat, D. Meisel, C. R. Chim. 6 (2003) 999–1007.
[27] A. Vaseashtaa, M. Vaclavikovac, S. Vaseashtaa, G. Gallios, P. Royb, O. Pummakarnchan, Sci. Technol. Adv. Mater. 8 (2007) 47–59.
[28] P.G. Tratnyek, R.L. Johnson, NanoToday 1 (2006) 44–48.
[29] N.S. Allen, M. Edge, A. Ortega, C.M. Liauw, J. Stratton, R.B. McIntyre, Polym. Degrad. Stability 78 (2002) 467–478.
[30] M. Pagliaro, Nano-Age: How Nanotechnology Changes our Future, Wiley-VCH Verlag GmbH & Co. KGaA, Weinheim, 2010, pp. 82–83.
[31] M. Tarng, M. Minamyer, A. Ham, S. Brownell, A. Pham, A. Alexander, D. Sham, K. Nguyen, M. Pham, S. Maxey, Exterior Paint Formulation, 02.03.2006 WO 2006/023064 A1.
[32] G. Nangmenyi, J. Economy, In: N. Savage, M. Diallo, J. Dunca, A. Street, R. Sustich (Eds.), Nanotechnology Applications for Clean Water, William Andrew Inc., Norwich, NY, 2009 p. 4.
[33] N. Savage, M.S. Diallo, J. Nanopart. Res. 7 (2005) 331–342.
[34] S. Lambert, F. Ferauche, B. Heinrichs, N. Tcherkassova, J.-P. Pirard, C. Alie, J. Non-Cryst. Solids 352 (2006) 2751–2762.
[35] W. Zhang, J. Nanopart. Res. 5 (2000) 323–332.
[36] H. Goesmann, C. Feldmann, Angew. Chem. Int. Ed. Engl. 49 (2010) 1362–1395.
[37] C.T. Yavuz, J.T. Mayo, William W. Yu, A. Prakash, J.C. Falkner, S. Yean, L. Cong, H.J. Shipley, A. Kan, M. Tomson, D. Natelson, V.L. Colvin, Science 314 (2006) 964–967.
[38] B.H. Jeong, E.M.V. Hoek, Y. Yan, A. Subramani, X. Huang, G. Hurwitz, A.K. Ghosh, A. Jawor, J. Membr. Sci. 294 (2007) 1–7.
[39] N. Canter, Tribol. Lubrication Technol. 62 (2006) 30–39.
[40] D.S. Bhatkhande, V.G. Pangarkar, A.A.C.M. Beenackers, J. Chem. Technol. Biotechnol. 77 (2001) 102–116.
[41] M.R. Hoffmann, S.T. Martin, W. Choi, D.W. Bahnemannt, Chem. Rev. 95 (1995) 69–96.
[42] H.F. Lecoanet, M.R. Wiesner, Environ. Sci. Technol. 38 (2004) 4377–4382.
[43] X. Li, D.W. Elliott, W. Zhang, Crit. Rev. Solid State Mater. Sci. 31 (2006) 111–122.

[44] D.W. Elliott, H.L. Lien, W. Zhang, In: G.E. Fryxell, G. Cao (Eds.), Environmental Applications of Nanomaterials, Imperial College Press, England, 2008, pp. 25–48.
[45] M. Nutt, J.B. Hughes, M. Wong, Environ. Sci. Technol. 39 (2005) 1346–1353.
[46] S. Decker, A. Khaleel, K.J. Klabunde, Environ. Sci. Technol. 36 (2002) 762–768.
[47] G.W. Wagner, L.R. Procell, R.J. O'Connor, S. Munavalli, C.L. Carnes, P.N. Kapoor, K.J. Klabunde, J. Am. Chem. Soc. 123 (2001) 1636–1644.
[48] T. Inoue, A. Fujishima, S. Konishi, K. Honda, Nature 277 (1979) 637–638.
[49] R.L. Cook, R.C. MacDuff, A.F. Sammells, J. Electrochem. Soc. 135 (1988) 3069–3070.
[50] M. Anpo, H. Yamashita, Y. Ichihashi, S. Ehara, J. Electroanal. Chem. 396 (1995) 21–26.
[51] O.L. Varghese, M. Paulose, Th.J. LaTempa, C.A. Grimes, Nano Lett. 9 (2009) 731–737.
[52] M. Bowker, In: B. Cantor, P. Grant, C. Johnston (Eds.), Automotive Engineering Lightweight, Functional, and Novel Materials, Taylor & Francis, New York, 2008, p. 41.
[53] M. Awano, Y. Fujishiro, K. Hamamoto, Int. J. Appl. Ceram. Technol. 1 (2004) 277–286.
[54] V. Palanivel, D. Zhu, W.J. van Ooij, Prog. Org. Coatings 47 (2003) 384–392.
[55] K.O. Legg, M. Graham, P. Chang, F. Rastagar, A. Gonzales, B. Sartwell, Surf. Coatings Technol. 81 (1996) 99–105.
[56] K.N. Tu, A.M. Gusak, M. Li, J. Appl. Phys. 93 (2003) 1335–1353.
[57] M. Maruyama, R. Matsubayashi, H. Iwakuro, S. Isoda, T. Komatsu, Appl. Phys. A 93 (2008) 467–470.
[58] K.-S. Moon, H. Dong, R. Maric, S. Othukuchi, A. Hunt, Y. Li, C.P. Wong, J. Electron. Mater. 34 (2005) 168–175.
[59] E.D. Williams, R.U. Ayres, M. Heller, Environ. Sci. Technol. 36 (2002) 5504–5510.
[60] E.D. Williams, R.U. Ayres, M. Heller, Environ. Sci. Technol. 38 (2004) 1916–1917.
[61] P.J.A. Borm, D. Robbins, S. Haubold, T. Kuhlbusch, H. Fissan, K. Donaldson, R. Schins, V. Stone, W. Kreyling, J. Lademann, J. Krutmann, D. Warheit, E. Oberdorster, Particle Fibre Toxicol. 3 (2006) 11–45.
[62] C. Buzea, I.I. Blandino, K. Robbie, Biointerphases 2 (2007) MR17–MR172.
[63] S.M. Hussain, L.K. Braydich-Stolle, A.M. Schrand, R.C. Murdock, K.O. Yu, D.M. Mattie, J.J. Schlager, M. Terrones, Adv. Mater. 21 (2009) 1–11.
[64] A.D. Maynard, E.D. Kuempel, J. Nanopart. Res. 7 (2005) 587–614.
[65] www.edf.org/documents/6496_NanoRiskFramework.pdf.
[66] D.B. Warheit, R.A. Hoke, C. Finlay, E.M. Donner, K.L. Reed, C.M. Sayes, Toxicol. Lett. 171 (2007) 99–110.
[67] H.C. Fischer, Curr. Opin. Biotechnol. 18 (2007) 565–571.
[68] D.B. Warheit, T.R. Webb, V.L. Colvin, K.L. Reed, Ch.M. Sayes, Toxicol. Sci. 95 (2007) 270–280.
[69] D.H. Brower, J.H.J. Gijsberg, M.W.M. Lurvink, Ann. Occup. Hyg. 48 (2004) 439–453.
[70] L. Reijnders, J. Ind. Ecol. 12 (2008) 297–306.
[71] G.J. Nohynek, E.K. Dufour, M.S. Roberts, Skin Pharmacol. Physiol. 21 (2008) 136–149.
[72] G. Oberdörster, A. Maynard, K. Donaldson, V. Castranova, J. Fitzpatrick, K. Ausman, J. Carter, B. Karn, W. Kreyling, D. Lai, S. Olin, N. Monteiro-Riviere, D. Warheit, H. Yang, Particle Fibre Toxicol. 2 (2005) 1–35.
[73] N.A. Monteiro-Riviere, A.O. Inman, Carbon 44 (2006) 1070–1078.
[74] A.D. Maynard, Nano Today 1 (2006) 22–33.
[75] D. Chithrani, W.C.W. Chan, Nano Lett. 7 (2007) 1542–1550.
[76] G. Oberdörster, Philos. Trans. R. Soc. Lond. A 358 (2000) 2719–2740.
[77] W.G. Kreyling, M. Semmler-Behnke, W. Moller, J. Nanopart. Res. 8 (2006) 543–562.
[78] R.F. Service, Science 300 (2003) 242–243.

[79] P.J. Borm, D. Robbins, S. Haubold, T. Kuhlbusch, H. Fissan, K. Donaldson, R. Schins, V. Stone, W. Kreyling, J. Lademann, J. Krutmann, D. Warheit, E. Oberdorste, Particle Fibre Toxicol. 3 (2006) 1–35.
[80] A. Navrotskty, In: C.I. Contescu, K. Putyera (Eds.), Dekker Encyclopedia of Nanoscience and Nanotechnology, Taylor & Francis Group, LLC, New York, 2009, pp. 1262–1270.
[81] P.B. Keady, T. Halvorsen, J. Nanopart. Res. 2 (2000) 205–208.
[82] S. Hirano, H. Nitta, Y. Moriguchi, S. Kobayashi, Y. Kondo, J. Nanopart. Res. 5 (2003) 311–321.
[83] D.B. Kittelson, J. Aerosol Sci. 29 (1998) 575–588.
[84] T. Yokoyama, In: M. Hosokawa, K. Nogi, M. Naito (Eds.), Nanoparticle Technology Handbook, Elsevier, Amsterdam, 2007, pp. 477–481.
[85] D.B. Kittelson, W.F. Watts, J.P. Johnson, Atm. Environ. 38 (2004) 9–19.
[86] J.N. Smith, In: C.I. Contescu, K. Putyera (Eds.), Dekker Encyclopedia of Nanoscience and Nanotechnology, Taylor & Francis Group, LLC, New York, 2009, pp. 228–240.
[87] C. Anastasio, S.T. Martin, In: Nanoparticles and the Environment, Mineralogical Society of America, Washington, DC, 2000, pp. 293–349.
[88] http://www.nanotechproject.org/ (accessed 06.10.07).
[89] C.M. Sayes, A.A. Marchione, K.L. Reed, D.B. Warheit, Nano Lett. 7 (2007) 2399–2406.
[90] C.M. Sayes, J.D. Fortner, W. Guo, D. Lyon, A.M. Boyd, K.D. Ausman, Y.J. Tao, B. Sitharaman, L. Wilson, J. Hughes, J.L. West, V.L. Colvin, Nano Lett. 4 (2004) 1881–1887.
[91] C.M. Sayes, A.M. Gobin, K.D. Ausman, J. Mendez, J.L. West, V.L. Colvin, Biomaterials 26 (2005) 7587–7595.
[92] M. Allsop, A. Walters, D. Santillo. Nanotechnologies and Nanomaterials in Electrical and Electronic Goods: A Review of Uses and Health Concerns, Greenpeace, 2007, www.greenpeace.to/publications.asp#2007 (accessed 15.06.08).
[93] Th. Schneider, A. Jansson, K. Alstrup Jensen, V. Kristjansson, M. Luotamo, O. Nygren, K. Savolainen, V. Skaug, Y. Thomassen, T. Tuomi, H. Wallin, Evaluation and Control of Occupational Health Risks from Nanoparticles, Nordic Council of Ministers, Copenhagen, 2007.
[94] S. Kaluza, J.K. Balderhaar, B. Orthen, B. Honnert, E. Jankowska, P. Pietrowski, M.G. Rosell, C. Tanarro, J. Tejedor, A. Zugasti, Workplace exposure to NPs, European Agency for Safety and Health at Work (EU-OSHA), 2009.
[95] Nanoparticles: an occupational hygiene review, Institute of Occupational Medicine, Research Report 274, 2004.
[96] A. Elder, S. Vidyasagar, L. DeLouise, WIREs Nanomed. Nanobiotechnol. 1 (2009) 434–450.
[97] J. Lee, S. Mahendra, P.J.J. Alvarez, ACS Nano 4 (2010) 3580–3590.
[98] C. Ostiguy, B. Roberge, L. Ménard, C.-A. Endo, Best Practices Guide to Synthetic Nanoparticle Risk Management, Chemical Substances and Biological Agents, Report R-599, IRSST 2009.
[99] T. Tervonen, I. Linkov, J.R. Figueira, J. Steevens, M. Chappell, M. Merad, J. Nanopart. Res. 11 (2009) 757–766.
[100] Z. Ji, X. Jin, S. George, T. Xia, H. Meng, X. Wang, E. Suarez, H. Zhang, E.M.V. Hoek, H. Godwin, A.E. Nel, J.I. Zink, Environ. Sci. Technol. 44 (2010) 7309–7314.
[101] Y. Pan, S. Neuss, A. Leifert, M. Fischler, F. Wen, U. Simon, G. Schmid, W. Brandau, W. Jahnen-Dechent, Small 3 (2007) 1941–1949.
[102] C. Carlson, S.M. Hussain, A.M. Schrand, K. Hess, R.L. Jones, J.J. Schlager, J. Phys. Chem. B 112 (2008) 13608–13619.

[103] H. Yin, H.P. Too, G.M. Chow, Biomaterials 26 (2005) 5818–5826.
[104] S.W.P. Wijnhoven, W. Peijnenburg, C.A. Herberts, W.I. Hagens, A.G. Oomen, E.H.W. Heugens, B. Roszek, J. Bisschops, I. Gosens, D. van de Meent, S. Dekkers, W.H. deJong, M. van Zijverden, A. Sips, R.E. Geertsma, Nanotoxicology 3 (2009) 109–138.
[105] W.G. Kreyling, M. Semmler-Behnke, W. Moller, J. Aerosol Med. 19 (2006) 74–83.
[106] J.R. Gurr, A.S.S. Wang, C.H. Chen, K.Y. Jan, Toxicology 213 (2005) 66–73.
[107] H.D. Jang, S.K. Kim, S.-J. Kim, J. Nanopart. Res. 3 (2001) 141–147.
[108] J.M. Anderson, News Physiol. Sci. 16 (2001) 126–130.
[109] S. Hirano, C.D. Anuradha, S. Kanno, Am. J. Respir. Cell Mol. Biol. 23 (2000) 313–319.
[110] S.C. Currall, E.B. King, N. Lane, J. Madera, S. Turner, Nat. Nanotechnol. 1 (2006) 153–155.
[111] D. Gee, M. Greenbeer, Asbestos: From Magic to Malevolent Mineral. EEA Environmental Issue Report, No. 22, 2001, p. 61.
[112] F.S. vom Saal, C. Hughes, Environ. Health Perspect. 113 (2005) 926–934.
[113] L. Reijnders, J. Hazard. Mater. 152 (2008) 440–445.
[114] C.O. Robichaud, D. Tanzil, U. Weilenmann, M.R. Wiesner, Environ. Sci. Technol. 39 (2005) 8985–8994.
[115] A. Maynard, Ann. Occup. Hyg. 51 (2007) 1–12.
[116] Fantastic Voyage (I. Asimov), is a 1966 science fiction film produced by S. David.
[117] K.E. Drexler, Engines of Creation: The Coming Era of Nanotechnology, Anchor Books, New York, 1986.
[118] Nanotechnology Small matter, many unknowns, Swiss Re, p. 46.
[119] W. Peukert, Powder Technol. 158 (2005) 133–140.
[120] M.C. Roco, J. Nanopart. Res. 5 (2003) 181–189.
[121] D.A. Scheufele, B.V. Lewenstein, J. Nanopart. Res. 7 (2005) 659–667.
[122] C.A. Batt, A.M. Waldron, N. Broadwater, J. Nanopart. Res. 10 (2008) 1141–1148.
[123] W.S. Bainbridge, J. Nanopart. Res. 4 (2002) 561–570.
[124] Opinion of the Scientific Committee on Cosmetic Products and Non-Food Products Intended for Consumers Concerning Titanium Dioxide (SCCNFP), Brussels, European Commission, 2000, http://ec.europa.eu/health/ph_risk/committees/04_sccp/04_sccp_en.htm.
[125] A. Helland, H. Kastenholz, A. Thidell, P. Arnfalk, K. Deppert, J. Nanopart. Res. 8 (2006) 709–719.

Chapter 9

Challenges and Perspectives

1 ARE THE PROMISES REALISTIC?

For centuries, scientists and engineers have been searching for new materials with improved performance over cost ratio. Over the last few decades, numerous groups of advanced materials have been developed for different applications. They include biomaterials, ceramic composites, carbon nanotubes, sol–gel, fibers, and nanocoatings. These novel materials have been developed, to a great extent, to address the shortcomings of existing materials and also to create new market opportunities.

Depending on the application, nanoparticles (NPs) could provide several advantages. However, formulating the value proposition of nanoparticle (NP) technologies is not always straightforward. The cost advantage is difficult to quantify given the multitude of parameters and lack of techno-economic data. There are very few published studies on the economical feasibility of NP technologies. One study on zero-valent iron (ZVI) NPs for *in situ* remediation of soil and groundwater aquifer sites has been reported [1]. Typically, these ZVI NPs consist of a core/shell Fe/FeO_x structure with an average size of 5–40 nm. Pilot study data indicate high efficiency and cost effectiveness when compared to micro-sized iron particles. However, the lack of data on the long-term effects on groundwater aquifers may limit their applications.

Several other NP technologies are currently in commercial or pre-commercial stages. High-performance catalysts, lightweight materials, and high-performance coatings containing NPs are already in the market. Other promising NP technologies are under evaluation at laboratory and pilot scales. NPs provide a lot of flexibility in the design of high-performance materials. High production and integration costs are currently hindering further development of numerous potential applications. Reliability of NP-based products is also an issue in some cases.

Three broad application areas have been described in this book. They include energy and environment, biomedicine and microelectronics. Four categories of development stages are distinguished for each application (Table 9.1). The first category consists of products already in the market.

Interface Science and Technology, Vol. 19. http://dx.doi.org/10.1016/B978-0-12-369550-5.00009-4

TABLE 9.1 Examples of Nanoparticle (NP) Technologies and Their Development Status Within the Three Main Applications Areas

	Energy and environment	Optoelectronic	Health
Market	Pd NP for catalytic converter	Metal–glass nanocomposite	F_2O_3 NP contrast agent
Pre-commercial (under evaluation)	TiO_2 NP for solar cells; NP in cathode batteries	Cu/Ag NP ohmic contact	Silver NP antibacterial
Pilot stage (under development)	CIGS NP for solar cells; efficient NP photocatalysis	Electronic ink	QD sensing
Laboratory (in concept) stage	NP-based triple junction solar cells	Single electron transistor	NP-based nanorobot

The second category consists of products at the pre-commercial stage. The third category is related to products at the pilot study stage. The last category includes R&D activities at the laboratory (or concept) stage. Examples are provided in Table 9.1 for each case for illustration purposes. Given the fast pace of developments in NP technology, one can argue for moving each specific technology to a higher development stage.

As we move from the first to the fourth category of technology development, the investment risk is often significantly reduced. In spite of its challenges, the biomedical field provides numerous opportunities for significant returns of investments. However, the risks are much higher given the current scientific challenges, toxicity risks, and additional regulatory hurdles in this area. NP systems are promising solutions for early tumor detection and treatment. There are several other promising applications of NPs in biosensing and therapeutics. NP-based therapeutics are potentially more efficient when compared to conventional solutions. For example, current oral therapeutics are subject to denaturation and/or elimination by the biological environment and the metabolic system. In the case of intravenous delivery, NP drug delivery is not impeded by small capillaries. NPs are also easier to introduce in cells. Reducing the particle size of the drug below 100 nm will potentially improve the solubility and bioavailability of current drug formulations. However, toxicity risks related to the deleterious impact of newer polymers and excipients present in these NPs constitute a major hurdle.

There are several reported applications of NPs that will have difficulty matching the performance/cost of conventional solutions. This is the case when using semiconductor NPs as active photovoltaic materials in the 3D-structured solar cell. Besides the high cost of NP production, it is very difficult to produce stable 3D structures. Furthermore, power efficiency is limited by losses at the interface involving NPs. It is thus critical to develop alternative NP formulations and integration solutions for these third generation solar cells.

2 SCIENTIFIC CHALLENGES

When compared to powder materials, NPs present not only similar scientific challenges, but also new ones. Understanding and controlling the synthesis and integration processes are even more challenging for sub-100 nm particles. These challenges arise regardless of whether these NPs are under dry or wet environments. Depending on the immediate environment, several forces with similar magnitudes may play a role. Predicting the properties of these NPs is difficult when numerous forces with similar strengths are involved.

Most current laboratory synthesis methods are neither cost-effective nor sustainable. Thus novel processes of NP synthesis using environmentally friendly precursors are needed. However, development of new green processes for monodisperse NP manufacturing is a major challenge. The resulting NPs should have minimum levels of impurity and optimum size distribution. Furthermore, integration of these NPs into reliable films and devices is often challenging. This is the case of aqueous based NP solutions. Often, targeted applications dictate the manufacturing process, further limiting available choices. These novel processes should also consider safety and health effects.

Development of biodegradable NPs from renewable biomass sources should be one of the R&D focuses in academia. Understanding the limitations of existing processes should also receive some attention. For example, FDA-approved biodegradable and biocompatible polyesters (e.g., poly(epsilon-caprolactone) or PCL) are quite promising. However, NP characteristics and toxicity are not yet well understood when specific synthesis recipes and new precursors are used.

Another scientific challenge is related to understanding and controlling NP interactions with biological media under physiological conditions (blood, interstitial fluid, cell cytoplasm). For example, protein–NP interaction will modify the nature and strength of the different forces involved [2]. This change will have a significant consequence on the physiological response including signaling, kinetics, transport, accumulation, and toxicity [2]. Analyzing these properties *in situ* is quite challenging. Such analyses may require development of novel characterization methods. The physicochemical characteristics of the as-synthesized NPs will be affected when contacting with the biological media. These changes will directly affect the physiological response. Therefore, *in vitro* modeling of biological barriers should be undertaken to understand absorption at the cellular level. This could help bridge the barrier between animal testing and human clinical trials and reduce the overall development cost. It will also help accelerate the drug approval process. NP fabrication and integration should be coupled with *in vivo* and *in vitro* studies before pre-commercialization (clinical trials for health applications) is undertaken. This may require development of scientific models, allowing for a more detailed understanding, and reconciling the various *in vivo* and *in vitro* data.

The diversity of forces involved in NP integration provides opportunities to build a variety of structures and morphologies. However, this makes prediction of the assembling structure quite challenging. Understanding self-assembling processes occurring in nature could help improve accuracy of man-made NP assembly. Understanding the self-assembling process will help improve the quality and cost of NP films and materials. Indeed, spontaneous self-assembly is potentially cost-effective once optimized. Novel integration techniques should also be developed for some specific applications. Indeed, existing integration techniques do not address all the criteria of some applications.

A unified concept of the different self-assembling processes should be developed. NP assembling could benefit from the biomimetic field. In addition to the conventional colloidal, symmetric, and asymmetric packing integrations, biomimetic assembling is characterized by hierarchical, lock-and-key, directional, and strong bonding. Practical models allowing for precise quantification of the different forces involved are needed. For example, they could provide new routes for optimized colloidal 2D and 3D packing geometries. Rod-like (1D) geometry is achieved by favoring directional capillary forces over the remaining interactions. Controlling packing direction in asymmetric self-assembling is challenging.

3 ENGINEERING CHALLENGES

With a multitude of concepts, inventions, and advanced scientific understanding, NP technologies provide numerous potential applications. In most cases, these technologies require optimization and scale-up before reaching the market. Engineering challenges need to be addressed at the production, formulation, integration, and application stages.

Engineering challenges of NP fabrication and integration show some similarities to those of conventional powder technologies. In the case of powder technology, several integration processes are available. In both cases, materials formulation is critical to obtain quality products. Following green body formation, cost-effective postprocessing treatments are required to improve the film and material properties of NP and microparticle powders [3,4]. In addition, NP technologies have thrown up some specific challenges. Several issues related to NP functionalization, characterization, metrology, optimization, scale-up, and quality control have been discussed in this book. These aspects are critical to the successful manufacture and commercialization of NP-based products.

3.1 Production

Product reliability and performance reproducibility are two major obstacles that are often linked with the characteristics of NPs. High sensitivity of NP properties to environmental changes is an attribute as well as a drawback. The choice of production and formulation methods is often dictated by long-term stability issues during manufacturing, integration, and/or utilization.

Two groups of NPs have been described in this book. Dry techniques are currently used at pilot and commercial scales. They are potentially cost-effective for some applications. Often, cost effectiveness is achievable at the expense of reliability, in such cases. Wet methods offer more flexibility in terms of functionalization and integration than dry methods. However, scalability of colloidal NPs is often challenging. Several engineering solutions have been discussed in this book to address the challenges and shortcomings of both approaches.

In general, it is difficult to rank the various wet techniques on the basis of cost and quality. A ranking has been proposed for the oxide particulate synthesis [5]. Other studies have reported the advantages and disadvantages of different NP fabrication methods [6]. Mechanical attrition methods are cost- and energy-efficient, although often at the expense of poor control over composition (contamination) and size distribution (large and wide distribution) quality. At the other end of the spectrum, Tri-n-octylphosphine oxide (TOPO)-like techniques provide high-quality monodispersed NPs, but often at high financial and environmental costs. Hydrothermal and microwave techniques may provide a cost-effective solution with good quality NPs. In the case of the hydrothermal technique, some advantages include scalability, low energy consumption, reduced process steps, utilization of water as solvent, and low precursor cost. However, the requirement for safe operation, the high capital cost, and the high aggregation rate are currently hindering wider use of this technique. Sol–gel formulations also provide good quality materials ready to integrate into large-scale film and base materials. Furthermore, sol–gel formulations provide homogeneous products at lower operating temperatures. However, the air sensitivity and relatively high precursor cost of sol–gel are currently limiting sol–gel-based techniques.

Depending on the specific application, these engineering challenges could be addressed. For example, using low temperature and pressure while controlling other hydrothermal process parameters may reduce cost and size distribution. Combining low temperature process and possibility for directly integrated NPs, sol-gel could be cost effective. Cheap silica precursors have been used in the past to provide sol–gel-based solutions for industrial scale optical films.

Combining attributes of wet and dry synthesis techniques could provide opportunities for near-ideal manufacturing processes. Andres et al. have reported a continuous process starting with gold evaporation followed by gold nanocluster formation triggered by He stream condensation [7]. These nanoclusters are scrubbed from the gas phase using a solvent mist in the presence of surfactant.

The presence of surfactants prevents NP growth and agglomeration. Using this hybrid and continuous fabrication process, high-quality monodispersed gold NPs are obtained. This technique can also be applied to other chemical compositions. Another hybrid NP synthesis technique is based on supercritical fluid (SCF). SCF has been successfully employed in the food and pharmaceutical industries. Large-scale SCF production of high-quality NPs has been reported.

3.2 Formulation and Integration

New formulation solutions for NPs are needed. For example, conversion of colloidal NPs into solid state without any drop-off in the intrinsic functional properties of NPs is a significant challenge. This challenge must be addressed first to achieve commercial success. Cost-effective ink-based NPs for quality patterns are currently under evaluation. Production of NPs directly integrated into films or other systems may circumvent formulation requirements. Sol–gel and templating techniques provide direct production and integration.

3.3 Integration

Advanced concepts describing the self-assembling process of NPs into devices are needed. This involves dividing the overall manufacturing process into different operation units. A three-element motion, resulting from different sets of balanced intermolecular and colloidal forces, is used to describe the different unit operations [8]. The push (repulsive forces), pull (attractive and functional forces), and guide (repulsive, attractive, and functional forces) motions provide control over the corresponding unit operations. Self-assembling for microelectronic (and nanoelectronic) devices is a promising parallel fabrication technology. It is potentially cost-effective. However, it faces several challenges related to reliability and precise control of the properties at the nano- and/or macroscale.

Although developed in the context of nanostructure self-assembling, this three-element motion may also apply to NP-based devices. From the device operation standpoint, up to twelve different unit operations (operation mode) are distinguished [8]. They include gating and switching, directional rotation and oscillation, shafting, shuttling and elevatoring, contraction and extension, walking, tweezering or fingering, tolling and bearing, pistoning, sliding, or conveyoring, self-directional movement, capture and release, sensoring, and directional flow.

In spite of low capital and energy costs, self-assembling has still not been commercialized. It needs to overcome a few engineering challenges of which product quality and reliability are two major ones. More conventional integration techniques already being used with powder/slurry materials provide cost-effective solutions for some applications. With appropriate NP formulations and other optimizations, printing, dip-coating, and screen printing have already been used.

3.4 Applications

There are a wide variety of NP-based technologies. Some provide just an incremental improvement in existing technologies. Other applications involve new products and markets. As a new General Purpose Technology [9], NPs will not only generate new growth but will also give rise to significant structural changes. On average, technology revolution cycles in the past have lasted around 60 years, as in the case of the ICT, oil and automobile, steel, electricity, steam and railways, and industrial revolutions. It is expected that NP technology development will require up to 30 years for its development and another 5–30 years for its deployment. Thus, if we consider 1990 as the nominal birth date of NP technology [10], 2020 and 2050 will be key milestone years for their development and deployment respectively. However, whether any NP technology is considered an evolution or a revolution in technology is still a matter of debate.

Catalysis remains the leading application area for NPs. However, except for a few successful commercial applications in catalytic converters, battery, and petrochemistry, the figure of merit of new NP-based catalysts has not yet been properly assessed. In the case of colloidal NP catalysis, criteria based on efficiency, stability, sustainability, and recyclability have been suggested [11]. Often, NP size reduction provides only incremental improvements over micro-sized NPs. In gold catalysis, activity is reported only in the case of small NPs. However, no commercial application of gold NPs has yet been reported. NP technology offers possibilities for new products and markets. For example, novel nanomedicine concepts based on NPs have been reported. However this application raises issues related to possible public acceptance.

4 HEALTH AND SAFETY CHALLENGES

Using NPs as the building blocks in emerging nanotechnologies throws up several challenges related to safe production, storage, integration, and utilization. Airborne NPs (engineered or natural) are hazardous materials. *In vitro* data quantifying the health risk to humans and animals have been reported. Direct and indirect utilizations of NPs may be harmful depending on the dose. It has been established that some NPs introduced in the respiratory system may cause more damage than microparticles. For example, NPs give rise to greater inflammatory responses than microparticles in the pulmonary interstitium [12]. Often, the surface chemistry of these particles is more critical than their size in the expression of toxicity. Detailed investigation of the adverse effects of natural and engineered NPs on the airway and respiratory epithelia are required.

Often, small metal microparticles (>100 nm) are caught by the reticulo-endothelial system (RES) while sub-10 nm NPs are removed by the kidneys. Intermediary size distribution metal NP (5–100 nm) may escape the RES and kidney uptake mechanism. This situation could be beneficial and detrimental at the same time. It is beneficial when using NP-based targeted therapeutics. But it is also detrimental because these NPs may have adverse effects on tissues when biological barriers have been overcome. The potential toxicity and high persistency of these metallic NPs must be evaluated. It is important, therefore, to understand and quantify the health-related effects of size, load, and composition of these NPs. However, these studies are quite challenging for the following reasons [13]:

a. It is difficult to obtain detailed data on the physicochemical characteristics (size, coating, surface and bulk composition, structure, and purity) of NPs under dry and physiological media;
b. Usually, reported metric doses (g/L) of engineered NPs are not appropriate. Instead, surface area reactivity is a better descriptor. Few data reports allow for extracting of this information;
c. More than a dozen factors may affect NP toxicity, making a standardized comparison quite challenging;
d. The choice of the model and exposure conditions for comparing between studies conducted by different groups is difficult.

Although size is receiving a lot of attention, several other physicochemical characteristics of NPs, including surface composition, crystallinity, morphology, solubility, aggregation tendency, homogeneity of dispersions, and turbidity, may impact their overall toxicity. It is quite obvious that aerosol-based NPs are a greater risk than NPs integrated into solid state matrices [14]. NPs present in health-related products that are either orally ingested (supplements, toothpaste) or applied on the skin (crème) may carry a bigger risk. Few *in vivo* studies on NPs persistency have been reported.

Health and safety issues associated with NPs should not be limited to anthropogenic NPs. There are several naturally occurring NPs that are already ubiquitous in the environment. For example, a large number of NPs are produced at the San Andreas fault zone via mechanical grinding [15]. This amount surpasses the current NP production capacity of man.

5 SUSTAINABILITY: LCA AND RECYCLING

Assessment of the overall environmental impact of NPs requires a life cycle analysis (LCA) that includes the product's end-of-life status. Indeed, disposal, recycling, ingestion, air release, water source contamination, and landfill could contribute to the overall energy balance and emissions. There are very few detailed studies on NP-based products' end-of-life status [16,17]. An LCA study of polymer nanocomposites used in automotive body panels has been completed [18]. This study shows that replacing steel with these nanocomposites reduces the energy requirement by 10%.

Numerous NP fabrication processes of anatase-TiO_2 NPs and their applications have been reported. Anatase-TiO_2 may constitute an ideal case study for an LCA. Indeed, the LCA of anatase-TiO_2 NPs production from ilmenite hydrochloric has been recently reported [17]. These authors have considered the detailed process developed by Altairnano Inc. When compared to steel, aluminum, and carbon nanofiber, the energy requirements of TiO_2 NPs are lower.

Disposal of NPs and NP-based products after use presents significant challenges. Some metals, such as iron, are easier to biodegrade because of their high corrosion rate. Furthermore, these metal NPs could be easier to recycle given the higher surface-to-bulk ratio. Other metals such as Pb are toxic, requiring additional precaution during recycling. There are some additional complications associated with alloyed metals and composite materials.

6 IMPORTANCE OF PUBLIC AND INDUSTRIAL RELATIONSHIP IN ADDRESSING CURRENT CHALLENGES

It is important to engage the general public by providing a platform for the nonexpert contribution to policy shaping. Past experience has shown that it is always risky to force a policy decision on a general public because they may resent it. However, this is easier said than done. Indeed, there is no single approach to engaging the general public. The currently dominant Triple Helix (academia–industry–government) approach to science and technology development should be updated. A fourth dimension that will better represent the general public needs to be added. Fairness of the process that include different interest groups and public education are key elements in shaping an engagement platform. Different groups have a different understanding of what is public engagement. First, the general public is not as homogenous as is often represented. Ideally, the civil society without any political or financial influence should be the face of the public. Thus, the civil society is probably the most suitable description of this fourth axis. One has then to ask the following question: who has the right to represent the general public?

Academia, government, and industry should get the civil society involved when shaping policies related to nanoscale science and technology that can have wide-reaching societal consequences. If they fail to do this, consumer acceptance of NP products will be a challenge. Indeed, public perception could be shaped based on irrational information and apocalyptical predictions. Often, the general public is not really interested in knowing the details to make up its mind.

Balancing between the benefits and risks of new technologies is often challenging. The final decision should be made based on risk analysis. Although most experts agree on this, it remains a challenge to explain to and convince the general public. As discussed in Chapter 8, evaluating the toxicity risks of NPs is not always straightforward. Unfortunately, public debate on the risks involved opens doors to fear mongering campaigns by interest groups. Furthermore, it is always difficult to formulate government policy following public engagement exercises.

Besides issues related to public relations, NP technology proponents should pool their efforts to tackle the different technical challenges. Although some leading industrial companies are conducting in-house nanotoxicology testing, this is often not sufficient. More toxicology studies should be performed on NPs and NP-based products from different companies and not-for-profit organizations. Third party testing is often required; otherwise, there is the risk of the public remaining skeptical. Detailed risk analysis should be undertaken by the NP manufacturing industry [19] with third party verification and/or collaboration. A cleaner process and better handling procedures should be implemented [20]. Controlling the emission of NPs from industrial processes may involve installation of appropriate equipment. Electrostatic precipitators, scrubbers (venturi scrubbers), and fabric filters (baghouses) are often effective solutions [21]. The choice of solutions is often dictated by the physicochemical properties of the emitted particles.

7 NEED FOR AN OPEN DATABASE

Making an insightful decision requires having access to the right data and/or information. An open and well-organized repository/database of scientific publications and other related reports on NPs is important. An open database could have the following beneficial consequences:

a. better comparison of NP properties obtained using different synthesis techniques and/or operation conditions;
b. better comparison of NP-based films and products obtained using different integration methods;
c. reduction in the level of uncertainty and unpredictability regarding the effects of NPs on health and environment.

An open database could be used as effective instrument for shaping public opinion and gaining trust in the development of NP science and technology. It may appear odd that with the abundance of available data, more data is required. Actually, we need quality data obtained in a coordinated manner.

Today, we are entering the big data age requiring new tools to organize, qualify, and retrieve information. It is important to understand the data—the information–knowledge–expertise–insight relationship (Fig. 9.1). Information is extracted from data, which allows for building knowledge and expertise in this field. Once the insight is gained, the organization makes a business decision, builds a plan, and acts accordingly. In the case of NP technology, a well-organized database will help research organizations, the general public, and industrial communities in making proper decisions. Besides the technico-economic factors, legal and ethical issues are also important. Indeed, new technologies must be compatible with the current legislation and law, and abide with the ethical principles of the society.

This database is often considered when building a life cycle inventory (LCI), a key component of LCA. Lack of LCI is hindering LCA development. The academic and industrial communities have an important role and they should collaborate very closely to address this issue. The diversity of NPs in terms of size, bulk composition, and surface chemistry makes the LCA of any NP-based product quite challenging. There are already a large number of peer reviews and nonpeer reviewed publications. In this regard, one cannot help noticing the significant amount of duplication in some areas. At the same time, there are other areas that have not been investigated. A well-organized database will certainly highlight areas lacking data. It will also highlight shortcomings in current methodologies of conducting and reporting data. Often, decisions can be incomplete if the existing data are not sufficient, requiring additional scientific investigation. Given the time and resources required to have a complete data set, an open database will help.

Systematic study and reporting of the different biological effects (cellular uptake, accumulation, composition and size distribution, excretion, NP–biomolecule interaction) of different NPs are critical. This will help establish a relationship between the characteristics, physiological properties, and targeted effects of NPs. Better understanding of *in vitro* and *in vivo* behavior will reduce delays in the development of therapeutics in the future. Currently, there is a significant knowledge gap that makes the health and environment risk quite challenging. Silver NPs have been widely studied.

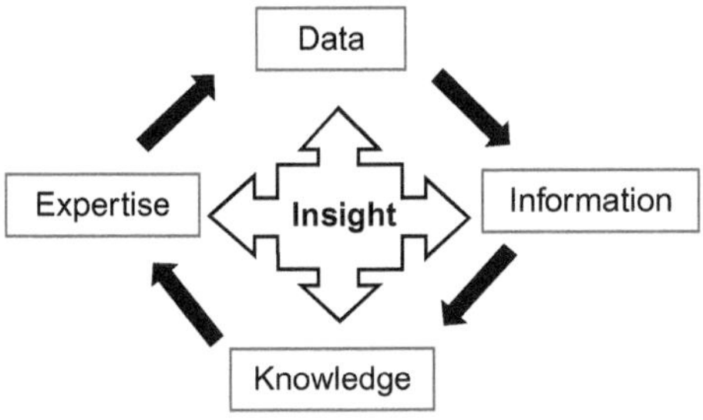

FIGURE 9.1 Schematic representation of data–information–knowledge–expertise–insight relationship.

However, there is a significant information gap related to silver NP-based products. The following information gaps have been identified [22]:

a. Little information is available on silver NP occurrence in the environment, particularly in soil;
b. Data on silver NP concentrations in various food, consumer, and medical products are not, or hardly, available;
c. Quality data on the hazard levels of silver NPs are lacking;
d. Not enough information is available on the material characteristics of silver found in the blood, organs, and tissues of organisms;
e. The factors influencing silver absorption are elusive;
f. The relation between free silver ion and silver NP concentration and activity is not clear;
g. There is a lack of consistency in *in vivo* and *in vitro* data for a given NP's characteristics.

Given the lack of systematic approaches in assessing the toxicology of NPs, the publication of a number of papers in a wide variety of journals has not helped to provide conclusive recommendations when dealing with *in vitro* studies. Indeed, dozens of parameters may impact NP toxicity. In the case of silver NPs, several experimental parameters may affect NP properties [23]:

a. silver salt precursors: $AgNO_3$, $AgC_2H_3O_2$, $AgClO_4$, Ag_2SO_4, $AgCF_3SO_3$, $AgSbF_6$, Ag_2O;
b. solvents: water, ethanol, DMF, EG, toluene, chloroform, amines, acetic acid, methanol, benzene, cyclohexane, paraffin, THF;
c. reducing agents/processes: $NaBH_4$, citrate, irradiation, organisms, amines, DMF, EG, ascorbic acid, hydrazine, aldehydes, plant extracts, hydrogen gas, sugars;
d. stabilizing agents: citrate, PVP, amines, amide, CTAB, sugars, SDS, fatty acids, PVA.

A wide range of particle size distributions has been reported using these different synthesis parameters. Although a significant number of particles have sizes less than 10 nm, particle distributions around 20, 40, and 80 nm or more have been reported. Size distribution and surface composition play a major role in the various applications and have an effect on the toxicity.

Finally, numerous integration modes have been evaluated at the laboratory scale. Most of them do not provide adequate answers to the current challenges. A more systematic structure–functionality correlation is highly desired. This will help to design improved models and products. Accumulation of well-structured data and information may be required, in particular, grading of different integration techniques based on different criteria. Easy access to formulation of materials, postprocessing requirements, film adhesion and thickness, ordering, stability, and cost are a few technico-economic criteria important to establish proper ranking.

8 CLOSING THOUGHTS

NP technology as a revolutionary concept shares some similarities with biotechnology, microtechnolgy, and information technology. However, in several applications, NP technology yields incremental improvements over existing microparticle-based products. In a few cases, it may provide solutions with significantly improved performance/cost.

Lack of long-term dialogue between engineering and materials scientists on one hand and life science practitioners on the other has so far hindered proper experiment design and reporting. In the case of toxicity of airborne NPs, *in vitro* experiments can provide reliable, high-throughput screening if designed properly [24]. Thus, research on the potentially harmful effects of the fabrication and utilization of NPs should be funded and carried out globally in a more organized fashion through a more coordinated effort.

Of the three application areas, clean energy presents a lot of potential for short-term applications. Most of the challenges in this field are related to engineering integration and optimization. High value-added optoelectronics applications face significant scientific and engineering challenges. The last application area will need to address several toxic and regulatory challenges.

It is postulated that nanotechnology and NP technology will contribute to widening the gap between the poor and the rich. However, this general statement does not apply to most NP technologies described in this book. In most cases, production and integration processes of NPs are not capital and technology intensives.

REFERENCES

[1] S.M. Cook, Assessing the Use and Application of Zero-Valent Iron Nanoparticle Technology for Remediation at Contaminated Sites, U.S. Environmental Protection Agency, 2009

[2] C.D. Walkey, W.C.W. Chan, Chem. Soc. Rev. 41 (2012) 2780–2799.

[3] M.N. Rahaman, Ceramic Processing and Sintering, second ed., New York, Marcel Dekker, 2003.

[4] G.S. Upadhyaya, Powder Metallurgy Technology, Cambridge International Science Publishing, England, 2002.

[5] M. Yoshimura, J. Mater. Sci. 41 (2006) 1299–1306.

[6] T. Athar, W. Ahmed, M.J. Jackson (Eds.), Emerging Nanotechnologies for Manufacturing, Elsevier, UK, 2009, pp. 325–385.

[7] R.P. Andres, J.D. Bielefeld, J.I. Henderson, D.B. Janes, V.R. Kolagunta, C.P. Kubiak, W.J. Mahoney, R.G. Osifchin, Science 273 (1996) 1690–1693.

[8] Y.S. Lee, Self-Assembly and Nanotechnology: A Force Balance Approach, Wiley, New Jersey, 2008 pp. 311–334.

[9] J. Youtie, M. Iacopetta, S. Graham, J. Technol, Transfer 33 (2008) 315–329.

[10] C.B. Murray, D.B. Norris, M.G. Bawendi, J. Am. Chem. Soc. 115 (1993) 8706–8715.

[11] N. Yan, C. Xiao, Y. Kou, Coord. Chem. Rev. 254 (2010) 1179–1218.

[12] G. Oberdörster, Int. Arch. Occup. Environ. Health 74 (2000) 1–8.

[13] Y.F. Li, C. Chen, Small 7 (2011) 2965–2980.
[14] A.D. Maynard, E.D. Kuempel, J. Nanopart. Res. 7 (2005) 587–614.
[15] O. Dor, B.Z. Yehuda, T.K. Rockwell, J. Brune, Earth Planet. Sci. Lett. 245 (2006) 642–654.
[16] G.F. Grubb, B.R. Bakshi, J. Ind. Ecol. 15 (2011) 81–95.
[17] E. Asmatulu, J. Twomey, M. Overcash, J. Nanopart. Res. 14 (2012) 720–727.
[18] V. Khanna, B.R. Bakshi Environmental, Sci. Technol. 43 (2009) 2078–2084.
[19] C. Robichaud, D. Tanzil, U. Weilenmann, M. Wiesner, Environ. Sci. Technol. 39 (2005) 8985–8994.
[20] L. Reijnders, J. Clean. Prod. 14 (2006) 124–133.
[21] L. Theodore, R.G. Kunz, Nanotechnology: Environmental Implications and solutions, Wiley, New Jersey, 2015 (Chapter 3).
[22] S.W.P. Wijnhovenen, W.J.G.M. Peijnenburg, C.A. Herberts, W.I. Hagens, A.G. Oomen, W.H.W. Heugens, B. Roszek, J. Bisschops, I. Gosens, D. Van De Meent, S. Dekkers, W.H. De Jong, M. Van Zijverden, A.J.A.M. Sips, R.E. Geertsma, Nanotoxicology 3 (2009) 109–138.
[23] T.M. Tolaymat, A.M. El Badawy, A. Genaidy, K.G. Scheckel, T.P. Luxton, M. Suidan, Sci. Total Environ. 408 (2010) 999–1006.
[24] H.R. Paur, et al., J. Aerosol Sci. 42 (2011) 668–692.

Index

Note: Page numbers followed by "*f*" indicate figures, and "*t*" indicate tables.

R

S

T

U

V

W

Z

FSC
www.fsc.org
MIX
Papier | Fördert
gute Waldnutzung
FSC® C083411

Kolibri 360 GmbH
Ferdinand-Jühlke-Straße 7
99095 Erfurt, Deutschland
produktsicherheit@kolibri360.de